Preparing for the
ACT
Mathematics &
Science Reasoning

SECOND EDITION

Dr. Robert D. Postman

AMSCO

AMSCO SCHOOL PUBLICATIONS, INC.

315 Hudson Street, New York, N.Y. 10013

Robert Postman is a college professor who is an expert in test preparation and subject-matter study. Dr. Postman holds a doctorate from Columbia University, where he received a full fellowship to pursue his graduate study. He is the author of over 30 books, which are found in schools and in bookstores throughout the United States. Dr. Postman has served as a consultant for many school districts and is widely recognized in numerous bibliographic publications, including *Who's Who in the World, Who's Who in America,* and *American Men and Women of Science.* He is also recognized for his work as a dean and department chair and for his faculty affiliation with Teachers College, Columbia University. An active participant in the community, he served on various boards, including special education boards, and as an elected member of the Board of Education.

Reviewers

Karen M. Brunner
Okemos High School
Okemos, MI

Mona Busch
Professional Development Institute
Decatur, IL

Carol A. Goehring, NBCT
Wekiva High School
Orlando, FL

Dr. Yolanda Mendoza
Miami Dade County Public Schools
Miami, FL

Janie Mueller
Cheyenne Mountain High School
Colorado Springs, CO

Jason D. Reissig
Larkin High School
Elgin, IL

Clayton P. Smith
Larkin High School
Elgin, IL

Dr. Maria J. Vlahos
Barrington Community Unit School District 220
Barrington, IL

TI calculator images reprinted by permission of the copyright owner, Texas Instruments

Please visit our Web site at: *www.amscopub.com*
When ordering this book, please specify:
either **R 649 W** *or* PREPARING FOR THE ACT: MATHEMATICS & SCIENCE REASONING, SECOND EDITION

ISBN 978-1-56765-717-3/*NYC Item 56765-712-2*

1 2 3 4 5 6 7 8 9 10   15 14 13 12 11 10

Preface

Preparing for the ACT: Mathematics & Science Reasoning, Second Edition will help you get your highest possible score on these sections of the ACT. The result of a three-year effort, the book includes a thorough review of the subject matter, extensive practice exercises and problems, and effective strategies for taking the ACT. This book will improve your chances of being admitted to the school of your choice, and help you get the most out of college. This is an important opportunity, and I wish you well as you prepare to continue your education.

My special thanks go to my wife Betty Ann, who has been a constant source of support. I could not have completed this project without her. My children—Chad, Blaire, Ryan, and my grandson Quinn—have been an inspiration as I worked on this and other books over the years.

I would like to thank my editors, Pat Wilson and Uriel Avalos, at Amsco for their extraordinary efforts and commitment. I am also grateful to the teachers around the country who reviewed the manuscript and offered helpful suggestions. In addition, I am grateful to my son Ryan Postman, a mathematics teacher and an ACT tutor, and my wife Betty Ann, a mathematics teacher, who contributed significantly to the development of the book.

Special thanks go to the staff at ACT, who were very helpful as I worked on the manuscript. It was wonderful to speak with people who are truly interested in the students taking their test.

Robert D. Postman

Contents

Section III ▪ Model Mathematics Tests 403

Section IV ▪ Science Reasoning 475

Section I ■

Overview and Introduction

Chapter 1 ■

Preparing for the ACT

■ The ACT Assessment

The ACT Assessment is a college admissions test. Colleges use ACT scores to help determine which students will be admitted as freshmen or as transfer students. The ACT consists of four separate multiple-choice tests: English, Reading, Mathematics, and Science Reasoning.

Each test has a different number of items. The composite score is an average of the four reported scores. ACT score reports show the composite score, the score for each test, and subscores for groups of items that show achievement in particular areas.

You'll find more detailed information about these tests, scores, test preparation and test-taking strategies, subject reviews, and practice tests starting in the next chapter. This book will lead you through the preparation you need to get your absolute best ACT score.

■ An ACT Website for Students

Check out the website ACT has just for students, at www.actstudents.org. This is our favorite first stop for students who want to take the ACT.

You can complete your registration and receive your scores online through the site, and you can hear from other students and their ACT experiences. This site also contains useful information about financial aid and college planning.

We like that ACT is also on twitter.com, which will update you regularly about ACT information. You can access the ACT twitter site from the ACT student web page.

■ Comparison of the ACT and the SAT

There are two national college admissions tests, the ACT from the American College Testing Program and the SAT from the College Board and the Educational Testing Service. Let me explain why you should take the ACT whether or not you take the SAT. The ACT focuses more on achievement and is related to the high school curriculum. ACT test makers are very clear about the material covered on the test and about the number of test items devoted to each area. If those at the ACT say there will be six grammar and usage items, that's exactly how many there will be. Since items on the ACT are related to the curriculum, you *can* effectively prepare for this test.

All the items on an ACT Assessment count toward your final score. On the SAT, one of the sections is experimental and does not count.

The ACT Writing Test is optional. You must take the SAT Writing Test.

ACT score reporting and other policies are people-friendly. The ACT reports your scores quickly, which gives you plenty of time to decide about retaking the test. You can even decide which ACT scores will be reported to colleges, even after you have seen the scores.

The SAT penalizes you for incorrect answers. There is no incorrect answer penalty on the ACT, so you should guess whenever you can't determine the correct answer.

■ Registering for the ACT

You should register in advance for the ACT. You can register on the web at www.actstudent.org. ACT registration packets should be available in your high school. If you can't find a registration packet, ask your guidance counselor, advisor, or teacher. You can also order a free registration packet on the ACT website or contact ACT directly:

> ACT Registration
> 301 ACT Drive
> P.O. Box 414
> Iowa City, IA 52243-0414
>
> (319) 337-1270 (Monday–Friday, 8:00–8:00 Central Time)
> TDD (319) 337-1701 (for hearing-impaired persons
>   calling from a TDD)

You can also call the ACT number to check on a late or delayed admission ticket, or to change your test date or test center. I called the ACT offices dozens of times while I worked on this book. Everyone I talked to was extremely helpful and pleasant. They want to help you, and you should feel very comfortable about calling.

The ACT has a website for students (www.actstudent.org). This website has complete information about the ACT, including registration information, test dates, and test sites. If you sign up for an ACT student web account, you have access to various services, for example, making changes to your registration. This site will be updated regularly. If you're online, drop in to see what additional features or services have been added.

Regular ACT administrations occur on a Saturday in September, October, December, February, April, and June. Check the registration packets for test dates and registration deadlines. Registration ends about a month before the test date. Late registration for an additional fee ends about 15 days before the test date.

When, Where, and What ACT to Take

You have to make four important registration decisions: (1) where to take the test, (2) in which school year to take the test, (3) when during the school year to take the test, and (4) what version to take: the regular ACT or the ACT Plus Writing.

You should take the ACT as close to home as possible. The test may even be given in your high school. The ACT is not given at every site on every test date. Check the registration packet to be sure the test is given at one of your preferred sites on the date you will take the test.

You should first take the ACT in your junior year. You can always take the test again in your senior year. Besides, application deadlines for many colleges and scholarship programs require you to take the ACT as a junior. Take the test toward the end of your junior year. I recommend the April test date. Since the ACT is closely tied to course content, junior-year classes will probably help. If you are taking the test in your senior year, take it early so the test scores are available to colleges.

Not all colleges require or recommend taking the writing test. Check with your college or guidance counselor first before signing up.

Special Scoring Dates. You can receive a copy of the test items, your scored answer sheet, and the correct answers (Test Information Release) if you take the test in December, April, or June at a national test center. This scoring information can be a valuable diagnostic tool. You can request this service on the registration form or when you receive your test scores in the mail. The test items, answers, and your answer sheet will be mailed to you eight to twelve weeks after the test date.

Forms of Identification

You must bring an acceptable form of identification to the test center, or you probably won't be able to take the test. Acceptable forms of identification include an up-to-date official photo ID or a picture from a school yearbook showing your first and last name. Unacceptable forms of identification include unofficial photo ID, learner's permit or driver's license without a photograph, a birth certificate, or a social security card. If you are not sure whether or not you have an acceptable ID, check the ACT student website or call the ACT ID Requirements Office at (319) 337-1510.

Standby Registration

You may be able to register standby at an ACT test center. Needless to say, you should do everything you can to avoid standby registration. There is a good chance that there will be no room for you.

Show up at a center on test day with a valid ID, a checkbook, a credit card, and some hope. All those registered at that center are seated first. If there's room, those registered at other centers are seated next. If there's still room, you can fill out an application on the spot and take the test. There is an extra fee for this service.

Alternate Testing Arrangements

If you live more than 50 miles from a test center, are confined at home or a hospital because of an illness, are in a correctional facility, or live in a country where there is no testing facility, you may be eligible for arranged testing. Check out the ACT student website for more information. Note that arranged testing is *not* available if you miss the test because of a schedule conflict.

If your religious beliefs prevent you from taking a Saturday test, you may take the ACT on a Sunday or a Monday. A limited number of sites in each state offer non-Saturday testing. If you live within 50 miles of those sites, you must take the test there *on the date it is offered*. Go to the ACT student website for more information.

If you have a diagnosed disability, you may qualify for special accommodations (such as permission to eat snacks if you are a diabetic), extra testing time, and/or special test dates. Please visit the ACT student website for details.

■ Scoring

The maximum reported score for each test is 36, although each test has a different number of items. The composite score is an average of the four reported scores. The maximum composite score is 36. ACT score reports show the composite score, the score for each test, and subscores for groups of items that show achievement in particular areas. Many colleges use these subscores for placement. If you take the optional writing test, you will receive two additional scores: the combined English/writing score and the writing subscore. Your score on the writing test does not affect the composite score.

Score Reporting. Those at the ACT treat your scores as though they were your property. That means you decide who sees your scores, which scores they see, and when they see them. I discuss this score-reporting policy as a part of the overall testing strategy later in this section.

You can even take the ACT several times until you get a score you like and have only that score sent to colleges. Do not list the colleges for which you think a particular score is needed or required. Wait for the ACT to report the scores to you. Then decide whether and where to send the scores. I prefer this option, particularly if you know that a college may require a minimum score for admission. Consult with your guidance counselor to see if this is a good strategy for your case.

■ ACT Realities

You take the ACT because it is required for college admissions or because it will help you get admitted to a college of your choice.

Tests can be unfair. A lucky guesser may occasionally do better on a multiple-choice test than someone who knows the material. Someone who knows the answers may get a lower score because he or she mismarks the answer sheet. Students who are sick the day of the test may do more poorly than they would have otherwise.

Some students may get a higher score than they have any right to expect. Other students may get a lower score than they need and deserve to receive. Students who know strategies for taking multiple-choice tests will often do better than students who don't know these strategies. You've got to make the best of it and get your highest score. This book will show you how.

It's Just People.   The ACT is designed and written by people who have their own personal strengths and weaknesses. They are not perfect and neither is the test they create.

Consultants throughout the country make recommendations to the ACT test designers about the content that should be included on the test. These recommendations are based on the consultants' knowledge of the subject matter and on the topics currently taught in American high schools.

The final list of topics is sent to test writers who actually prepare the test items. The test writers may be full-time employees of the American College Testing Program or they may be freelance writers from all over the country. The writers submit the items to the ACT, where the items are reviewed and edited. Then each item is reviewed further and tried out. The items that pass this review process are used on an ACT. Each item is used only once in its original form, but some items are revised and used in other ACTs.

Chapter 2 ■

The ACT

The ACT consists of four separate tests: English, Mathematics, Reading, and Science Reasoning. The tests are always given in that order and they must be taken together. You have 2 hours and 55 minutes to answer the items on these tests. On the typical test day you will check in at about 7:30 A.M., and leave around 12:15 P.M.

This book provides you with a complete subject and strategy review, along with four ACT Mathematics practice tests and four ACT Science Reasoning practice tests. An overview of the Mathematics and Science Reasoning tests is given below. A brief description of the English and Reading tests follows.

■ Mathematics Test Overview

The Mathematics Test consists of 60 multiple-choice items. Each item has five answer choices. You have 60 minutes to complete the test. The test measures mathematics skills in the six broad areas shown below.

ACT Mathematics: 60 minutes — 60 items

AREA	NUMBER OF ITEMS
Pre-Algebra	14
Elementary Algebra	10
Intermediate Algebra	9
Coordinate Geometry	9
Plane Geometry	14
Trigonometry	4

Scores Reported: Pre-Algebra and Elementary Algebra
Intermediate Algebra and Coordinate Geometry
Plane Geometry and Trigonometry
Total Number Correct

The mathematics topics in each area are shown on the following pages. This book contains a review of all of these topics. Most ACT items test a combination of these topics. For example, an elementary algebra question may require you to understand and use pre-algebra skills or concepts.

Pre-Algebra (14 Items)

Adding, subtracting, multiplying, and dividing whole numbers and integers
Adding, subtracting, multiplying, and dividing fractions and decimals
Whole number and decimal place value
Ordering numbers
Absolute value
Factors, divisibility, and primes
Square roots, exponents, and scientific notation
Order of operations
Percent
Ratio and proportion
Mean, median, and mode
Data collection, representation, and interpretation
Probability
Counting techniques
Writing linear expressions and equations
Solving linear equations

An ACT pre-algebra item might look like this:

1. In a certain family, children receive an allowance beginning at age 8. Each child receives an allowance equal to $.50 times his or her age. How much more allowance will a 15-year-old receive than a 9-year-old?

   A. $ 2.00
   B. $ 3.00
   C. $ 4.50
   D. $ 7.50
   E. $15.00

Elementary Algebra (10 Items)

Evaluating algebraic expressions
Properties of exponents and square roots
Algebraic operations
Factoring polynomials
Solving quadratic equations by factoring

An ACT elementary algebra item might look like this:

2. What is the solution set to the quadratic equation $x^2 - 64 = 0$?

   F. $x = \{2, -2\}$
   G. $x = \{4, -4\}$
   H. $x = \{8, -8\}$
   J. $x = \{16, -16\}$
   K. $x = \{32, -32\}$

Intermediate Algebra (9 Items)

Solving inequalities
Equations and inequalities with absolute value
Systems of equations
Rational and radical expressions
Quadratic formula
Quadratic inequalities
Complex numbers
Sequences and patterns
Matrices

An ACT intermediate algebra item might look like this:

3. Solve for t: $|3t - 4| = 8$

   A. $4\frac{1}{2}$

   B. $-\frac{1}{4}$

   C. $\frac{3}{4}, \frac{4}{5}$

   D. $4, -\frac{4}{3}$

   E. $-\frac{2}{3}$

Coordinate Geometry (9 Items)

Graphing inequalities on a number line
Graphs of points and lines
Slope
Graphing equations and systems of equations and inequalities
Distance and midpoint formulas
Graphs of circles, ellipses, parabolas, and hyperbolas

An ACT coordinate geometry item might look like this:

4. What is the distance between the points $(6, -5)$ and $(2, 3)$ in a normal (x, y) coordinate plane?

   F. 7

   G. $4\sqrt{3}$

   H. $4\sqrt{5}$

   J. 8

   K. 10

Plane Geometry (14 Items)

Angles
Perpendicular and parallel lines
Quadrilaterals
Triangles
Proof and proof techniques
Circles
Transformations
Geometric formulas
Three-dimensional geometry

An ACT plane geometry item might look like this:

5. Triangle *ABC* and triangle *ADE* are congruent right triangles. Both point *B* and point *D* lie on line *XY*. The measure of $\angle ACB = 70°$ and the measure of $\angle BAD = 80°$. What is the measure of $\angle CBX$?

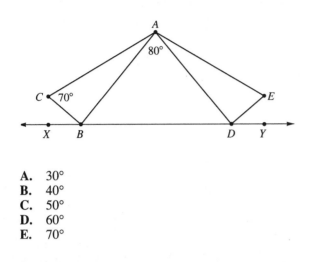

A. 30°
B. 40°
C. 50°
D. 60°
E. 70°

Trigonometry (4 Items)

Trigonometric relationships in right triangles
Values and properties of trigonometric functions
Using trigonometric identities
Trigonometry of the unit circle
Graphing trigonometric functions

An ACT trigonometric item might look like this:

6. Given the diagram below, what is the cot θ?

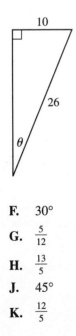

F. 30°

G. $\frac{5}{12}$

H. $\frac{13}{5}$

J. 45°

K. $\frac{12}{5}$

■ Science Reasoning Test Overview

The Science Reasoning Test consists of 40 multiple-choice items. Each item has four answer choices. You have 35 minutes to complete the test. The test measures science reasoning skills in three broad areas: data representation, research summaries, and conflicting viewpoints. Questions in these areas focus on biology, physical sciences, chemistry, and physics. The number of test items in each of these broad areas is shown below.

ACT Science Reasoning: 35 minutes — 40 items

AREA	NUMBER OF ITEMS
Data Representation	15
Research Summaries	18
Conflicting Viewpoints	7
Score reported: Total Number Correct	

Data Representation (15 Items)

An ACT data representation item might look like this:

Atmospheric Concentration of Gases in 1995

Gas	Concentration
Carbon dioxide	345 ppmv
Methane	1.7 ppmv
Nitrous oxide	0.304 ppmv
CFC-11	0.22 ppbv
CFC-12	0.38 ppbv

ppmv: parts per million by volume
ppbv: parts per billion by volume

1. If China eliminated its use of fossil fuels in the next century, one might reasonably expect:
   A. a reduction in global CFC-11 emissions.
   B. the sun to heat up.
   C. a reduction in carbon dioxide emissions.
   D. the sun to cool down.

Research Summaries (18 Items)

An ACT research summary item might look like this:

PASSAGE 1

Scientists conducted an experiment to determine how fast PCBs are degraded by microbes. PCB-degrading microbes were isolated from soils and mixed with river sediment previously contaminated with PCBs. The percent degradation of each sample was measured after 1 day, 5 days, and 10 days. Scientists classified the PCBs according to the chlorine content of the PCB isomer.

Experiment 1   The degradation of a sample that contained both dichloro-isomers and trichloro-isomers was measured.

Table 1

Sample name	Number of chlorines	Percent degradation after		
		One day	*Five days*	*Ten days*
E	2 + 3	55	81	74

2. Which of these conclusions is supported by Experiment 1?
   F. When a sample contains isomers with two different chlorine values the degradation pattern most closely resembles the isomer with the higher value.
   G. When a sample contains isomers with two different chlorine values the degradation pattern most closely resembles the isomer with the sum of the values.
   H. When a sample degrades between 70% and 80% after 10 days that sample contains isomers with two different chlorine values.
   J. When a sample degrades more than 70% after 10 days the number of chlorines in all the isomers in the sample is 1 or 2.

Conflicting Viewpoints (7 Items)

An ACT conflicting viewpoints item might look like this:

In the following paragraphs, two archaeologists discuss their theories about the age of the Sphinx.

Archaeologist 1

The Sphinx is about 4,500 years old. From 2700 BC to 2150 BC the Old Kingdom of Egypt flourished in the Nile River Valley. Pharaohs ruled over a civilization with writing and a calendar. During this period Egyptians built massive pyramids and the Sphinx in the desert region called the Valley of the Kings. Blocks cut away to form the Sphinx were used to create temples at the front of the Sphinx.

The face of the Sphinx bears a striking similarity to a pharaoh of this time and it is believed that the face of the Sphinx was created in his image. We know from artifacts found in the area that the Sphinx was carved during the time of a civilization much like the one present during the Old Kingdom.

Radiocarbon dating of pieces from the hindquarters of the Sphinx indicate that it dates from about 4,000 to 5,000 years ago. Other dating techniques confirm these dates.

Archaeologist 2

The Sphinx is about 10,000 years old. About 10,000 years ago the head of the sphinx was a yardank (an outcropping of rock). It protruded from what was then a lush, green valley in Egypt. During that time, an advanced civilization carved the yardank into some human or animal form. Over time this carving was covered over and then uncovered during the time of the Old Kingdom when that region was a desert. Workers during that time added the paws and a body to the Sphinx.

The face of the Sphinx shows very significant water erosion caused by long periods of persistent rain. We know that the Sphinx has been in a desert climate for the past 4,500 years. It was about 4,000 years prior to that when that part of Egypt would have had significant amounts of rainfall.

Radar readings also indicate that the ground in front of the Sphinx is weathered to a depth twice as great as the back of the Sphinx. These data suggest that the front is twice as old as the back.

3. Assume that Archaeologist 2 is correct. What could account for the radiocarbon dating mentioned by Archaeologist 1?

   A. The part dated was constructed before the head was carved.
   B. The face did not actually look like the pharaoh.
   C. The part dated was constructed after the head was carved.
   D. A similar civilization had existed about 5,000 years earlier.

◾ English and Reading Tests Overview

The following gives a brief overview of the ACT English and Reading tests. Get a copy of the Amsco book *Preparing for the ACT: English & Reading* for a thorough review and sample tests.

English Test

The English Test consists of 75 multiple-choice items. Each item has four answer choices. You have 45 minutes to complete the test. The test measures English skills in two broad areas, shown below along with the number of test items in each of these areas.

ACT English: 45 minutes — 75 items

AREA	NUMBER OF ITEMS
Usage/Mechanics	**40**
Punctuation	10
Grammar and Usage	12
Sentence Structure	18
Rhetorical Skills	**35**
Strategy	12
Organization	11
Style	12

Scores Reported: Usage/Mechanics
                         Rhetorical Skills
                         Total Number Correct

Writing Test

The optional ACT Writing Test gives you 30 minutes to write a persuasive essay in response to a prompt. The test gives the topic for your essay and asks you to convince someone or some group of your position on the topic. For example, you may write an essay about whether or not a school should have a dress code.

Two readers evaluate your essay holistically, and assign a score from 1 to 6. Holistic scoring means a reader's evaluation is based on his or her informed impression of your writing. The readers do not go into detailed analysis. If the readers' scores differ by more than 1 point, a third reader evaluates the essay.

Here is an example of a writing prompt.

PROMPT

Some parents asked the Town Council to impose a curfew requiring students under the age of 18 to be off the streets by 10:00 P.M. to reduce disciplinary problems and to help ensure children's safety. Other parents do not agree with a curfew. They believe that imposing a curfew will not necessarily ensure students' safety and it should be up to parents to decide what time their children should be off the streets. In your opinion, should the Town Council impose a curfew for students under the age of 18?

Take a position on the issue outlined in the prompt. Choose one of the two points of view given in the prompt, or you may present your own point of view on this issue. Be sure to support your position with specific reasons and details.

Reading Test

The Reading Test consists of 40 multiple-choice items. Each item has four answer choices. You have 35 minutes to complete the test. The test measures reading skills in four broad areas shown below, along with the number of test items in each area.

ACT Reading: 35 minutes — 40 items

AREA	NUMBER OF ITEMS
Prose Fiction	10
Humanities	10
Social Studies	10
Natural Sciences	10

Scores Reported: Arts Literature (Prose and Humanities)
Social Studies (Social Studies and Natural Sciences)
Total Number Correct

Chapter 3 ▪

Test Strategies

▪ Test-Preparation Strategies

Use these strategies and the ACT Review below as you prepare to take the ACT. They take you right up to test day.

- **Start early.**

    If you are going to take the test in April or June, start preparing in September. Do some work each week rather than cramming just before the test.

- **Eliminate stress.**

    Stress reduces your effectiveness. Moderate exercise is the best way to reduce stress. Try to find some time each day to walk, run, jog, swim, or play a team sport. Remember to exercise within your limits and stay hydrated.

- **Be realistic.**

    You don't need to answer every item correctly to get a high score on the ACT. The composite score is the total score for the entire test, and the highest ACT composite score is 36. The national average ACT composite score has been just below 21.

    About 56 percent correct on the entire test will likely earn you an above-average score. The percent correct on each test shown below would earn a composite ACT score of about 21.

English	60 percent correct
Mathematics	45 percent correct
Reading	55 percent correct
Science Reasoning	60 percent correct

    Other combinations of test scores could also earn an above-average composite score. You can always take the ACT over again.

■ ACT Review Checklist

Complete the following steps in the order shown to take the test in April of your junior year. Follow these steps but adjust the time line to take the test on other dates. Take all tests under simulated conditions. That means strict adherence to timing, taking the test in a quite area—no television or music. Do not look at the answers. Use a digital watch.

September

☐ Review this chapter.

☐ Complete the Mathematics Topic Inventory on pages 25–32.

☐ Start work on the Mathematics section (page 44).

October

☐ Continue work on the Mathematics section.

November

☐ Complete work on the Mathematics section.

☐ Take the Diagnostic Mathematics ACT on page 311 under simulated test conditions.

☐ Use the Diagnostic Study Chart and review indicated problem areas.

December

☐ Start work on the Science Reasoning section.

January

☐ Continue work on the Science Reasoning section.

February

☐ Complete work on the Science Reasoning section.

☐ Take the Diagnostic Science Reasoning ACT on page 511 under simulated test conditions.

☐ Review problem areas noted on the Diagnostic Science Reasoning ACT.

March

☐ Review the problem areas noted on the Diagnostic Mathematics and Science Reasoning Checklists.

☐ Review the test-taking strategies on pages 21– 22.

☐ Register for the April ACT. List only the colleges you want the scores sent to immediately after your test is scored. I recommend not listing colleges if you know that a particular minimum score is required.

The Saturday seven weeks before the test

☐ Take Model Mathematics ACT II and Model Science Reasoning ACT II under simulated test conditions.

☐ Score the tests. Review the answer explanations.

Six weeks to go

☐ Review the problem areas noted on the Model ACTs.

The Saturday five weeks before the test

☐ Take ACT Model Test III under simulated test conditions.

☐ Score the test.

Four weeks to go

☐ Review the problem areas noted on the Model ACT.

☐ Review the test-taking strategies on pages 21–22.

April

Saturday two weeks before the April test date (This Saturday may come in March.)

☐ Take Model ACT IV under simulated test conditions.

☐ Score the test.

Two weeks to go

☐ Review the problem areas from the Model ACTs. Review the answer explanations and refer back to the review sections. Get up at the same time every day that you will the morning of the test. Work for a half hour each morning on items from one of the Model ACTs.

Test Week

You may continue your review through Thursday, if you want.

☐ Monday
   Make sure you have your admissions ticket.
   Make sure you know where the test is given.
   Make sure you have valid forms of identification. If you are not
   sure whether or not you have a valid form of ID, check the ACT
   student website or call the ID Requirements Office at (319) 337-1510.
   They will help you.

☐ Tuesday
   Visit the test site, if you haven't done it already.

☐ Wednesday
   Set aside some sharpened No. 2 pencils, a digital watch or clock, a good
   eraser, and the calculator you will use for the mathematics test.

☐ Thursday
   Complete any forms you have to bring to the test.

☐ Friday
   Relax. Your review is over.
   Get together any snacks or food for test breaks.
   Get a good night's sleep.

☐ Saturday — TEST DAY
   Dress in comfortable clothes.
   Eat the same kind of breakfast you've eaten every morning.
   Don't overeat!
   Avoid food that is hard to digest.

Get together things to bring to the test, including the registration ticket, identification forms, pencils, eraser, calculator, and snacks or food.

Get to the test check-in site about 7:30 A.M.

You're there and you're ready.

Follow the test-taking strategies, pages 21–22.

After the Test

April/May

☐ Your scores should be ready online in about two and a half weeks after the test. You will receive your scores in the mail 3 to 8 weeks after the test. Discuss the scores with your guidance counselor, advisor, or teacher. You need YES or NO answers to these two questions:

1. Should ACT send these scores to colleges I did not list on my registration form?

   NO — wait until next time.

   YES — arrange to have Additional Score Reports (ASRs) sent to those colleges.

2. Should I take the test again?

   Lots of people take the ACT several times. If you have a bad day test day or you are sick, you might not do your best. You may just feel that you can improve your score through further review. You should consider taking the ACT again if you believe you could improve your score enough to make a difference in college admissions.

   NO — you're finished with this book.

   YES — decide if you want to take the test again in June or the following September. A June test date gives you limited opportunity for further review, but the test scores can reach colleges by September. A September test date gives you time to get your scored answer sheet along with the test questions and correct answers (Test Information Release), but test scores won't reach colleges before the following October.

June Test Date. You have about a month to prepare. Be sure to register for the test. Go back to "Four weeks to go" on the ACT Review Checklist (page 18) and follow the checklist from there.

September Test Date. Order the test questions and answers and your answer sheet from ACT. You will receive a copy of the test items, your scored answer sheet, and the correct answers. A Test Information Release can be ordered from the web.

Late May, June, or early July

The scoring information will arrive.

☐ Compare your answer sheet to the correct answers to make sure the sheet was marked correctly. Look also for any patterns that indicate that you may have mismarked your answer sheet.

☐ Check the answers and note the types of problems that you had difficulty with.

August

☐ Register for the September ACT.

☐ Show the questions, correct answers, and your answers to teachers or others who can explain the types of errors you made. Use the review sections of this book, and help from teachers, to review problem areas on the test.

Four weeks to go

☐ Take the actual ACT that was returned to you again under simulated test conditions.

☐ Mark the test and review any remaining problem areas.

☐ Review the test-taking strategies.

Two weeks to go

☐ Go back to "Two weeks to go" on the ACT Review Checklist (page 18) and follow the checklist from there.

■ Test-Taking Strategies

There is nothing better than knowing the subject matter for the ACT. But these test-taking strategies can help you get a better score.

- **Relax.**

    Get a comfortable seat. Don't sit near anyone or anything that will distract you. If you don't like where you are sitting, move or ask for another seat. You have a right to favorable test conditions.

- **You're going to make mistakes.**

    You are going to make mistakes on this test. The people who wrote the test expect you to make them. Remember, the average score for the ACT is about 55 percent correct.

- **Eliminate and guess.**

    If you can't figure out the correct answer, eliminate the answers you're sure are incorrect. Cross them off in the test booklet. **Guess** the answer from those remaining choices.

    NEVER leave any item blank. There is no penalty for guessing on the ACT.

- **All that matters is which circle you fill in.**

    A machine will score your test. The machine detects whether or not the correct place on the answer sheet is filled in. Concentrate on filling in the correct circle. The machine can't tell what you are thinking.

- **You can know the answer to an item but be marked wrong.**

    If you get the right answer but fill in the wrong circle, the machine will mark it wrong.

- **Plug in the answers.**

    Just try out the answers to find the correct choice.

- **Save the hard items for last.**

    You're not supposed to get all the items correct, and some of them will be too difficult for you. Work through the items and answer the easy ones. Pass the other ones by. Do these items the second time through. If an item seems really hard, draw a circle around the item number in the test booklet. Save these items to the very end.

- **They try to trick you.**

    Test writers often include distracters. Distracters are traps—incorrect answers that look like correct answers. It might be an answer you're likely to get if you're doing something wrong. It might be a correct answer to a different item. It might just be an answer that catches your eye. Watch out for these trick answers.

- **Watch out for _except_ or _not_.**

    ACT items can contain these words. The answer to these items is the choice that does not fit in with the others.

- **Do your work in the test booklet.**

    The test booklet is not scored. You can write anything in it you want. Use it for scrap paper and to mark up diagrams and tables in the booklet. You may want to do calculations, underline important words, or draw a figure. Do your work for an item near that item in the test booklet. You can also do work on the cover or wherever else suits you.

- **Write the letter for the answer choice in your test booklet.**

    Going back and forth from the test booklet to the answer sheet is difficult and can result in a mismarked answer sheet. Use this approach to help avoid mismarking the answer sheet:

    Write the letter for the answer choice big next to the item number in the test booklet. See the example below. When you have written the answer choice letters for each two-page spread, transfer the answer choices to the answer sheet.

EXAMPLES

1. Solve for t: $|3t - 4| = 8$

    A. $4\frac{1}{2}$

    B. $-\frac{1}{4}$

    C. $\frac{3}{4}, \frac{4}{5}$

    D. $4, -\frac{4}{3}$

    E. $-\frac{2}{3}$

2. What is the distance between the points $(6, -5)$ and $(2, 3)$ in a normal (x, y) coordinate plane?

    F. 7

    G. $4\sqrt{3}$

    H. $4\sqrt{5}$

    J. 8

    K. 10

Section 2 ■

Mathematics

Chapter 4 ∎

Mathematics Topic Inventory

∎ Introduction

This Mathematics Topic Inventory will help you decide which mathematics skills you need to review as you prepare for the ACT. You may complete the entire test at one time or complete the test section by section. Write the answers in the space provided. Don't guess. If you don't know the answer to an item, circle the item number and move on. You may use a calculator on this Topic Inventory. Chapter 5 contains suggestions for the best ways to use a calculator on the ACT.

This is the only place in this book that we don't use the ACT multiple-choice format. That's because you need to know which concepts you should study further.

∎ Topic Inventory

Pre-Algebra

Compare these numbers using the symbols for greater than, less than, or equal to.

1. 23 _____ 18
2. 98,776 _____ 137,165
3. 73 _____ 67 _____ 59
4. Give the total value of the digit 8 in the numeral 47$\underline{8}$,906,325. _____

Compute.

5. $2,709 + 18,365 + 907 = $ _____
6. $50,208 - 41,235 = $ _____
7. $309 \times 163 = $ _____
8. $6,901 \div 67 = $ _____
9. Compare using $<$, $>$, or $=$. 72.3989 _____ 72.41
10. Round 176.4382 to the nearest tenth. _____

Compute.

11. $20.168 + 243.4 + 10.058 = $ _____
12. $27.3084 - 0.9392 = $ _____
13. $3.04 \times 0.0092 = $ _____
14. $266.825 \div 8.21 = $ _____

15. Give all the factors of 18. _____

16. List all the single-digit numbers that evenly divide 2,048. _____

17. Find $0.81^{\left(\frac{1}{2}\right)}$. _____

18. Express 3,162 using scientific notation. _____

19. Write $\dfrac{39}{126}$ in simplest form. _____

20. Write these fractions in order from least to greatest: $\dfrac{7}{9}, \dfrac{11}{14}, \dfrac{5}{6}$. _____

21. $1\dfrac{1}{3} + \dfrac{7}{8} + \dfrac{1}{5} =$ _____

22. $4\dfrac{1}{3} - 1\dfrac{3}{8} =$ _____

23. $2\dfrac{1}{5} \times 3\dfrac{5}{8} =$ _____

24. $\dfrac{3}{7} \div 2\dfrac{5}{8} =$ _____

25. $|7 - 9| =$ _____

26. Compare using $<$, $>$, or $=$. -16 ____ 11

27. $-16 + (-94) =$ _____

28. $-53 - (-71) =$ _____

29. $-34 \times (-81) =$ _____

30. $-2{,}158 \div (-83) =$ _____

31. Evaluate: $4 \times 3 - (15 \div 3 + 8 \times 2 - 3^3) - 7$. _____

32. Express $\dfrac{2}{3}$ as a decimal and as a percent. _____

33. Express 0.065% as a fraction and as a decimal. _____

Solve.

34. Tony's 10-acre plot of land is 20% the size of Pat's plot of land. How big is Pat's plot of land? _____

35. After a 30% discount, an item sells for $10.50. What was the price before the discount? _____

36. Solve the proportion for x: $\dfrac{3x}{6} = \dfrac{8}{4}$. _____

37. Use a proportion to solve this problem.

At exactly the same time and place a 6-foot-high stick casts a 9-foot shadow while a pole casts a 12-foot shadow. How tall is the pole? _____

38. Represent these data on a stem-and-leaf diagram: 10, 10, 31, 40, 46, 57

Find the mean, median, and mode of this set of data.

321, 32, 86, 51, 167, 241, 51, 82, 119, 72

39. Mean _____

40. Median _____

41. Mode _____

42. What is the probability of flipping a dime and a penny and getting two heads? _____

43. How many different 5-person basketball teams can a coach choose from among 7 players? _____

44. Write an equation to represent the statement, "A boy is two years younger than twice his sister's age." _____

45. Given the equation $3x - 2y = 4$, list the ordered-pair solutions that are found on the x-axis and the y-axis. _____

Elementary Algebra

46. In the equation below, find the value of x if $t = 7$. _____

$x = 5t - 38$

47. Evaluate the formula $I = PRT$ for $P = \$2,000$, $R = 8.5\%$, and $T = 3$ years. _____

48. $28^{19} \div 28^8 =$ _____

49. Simplify: $\sqrt{567}$. _____

50. Divide. $\dfrac{\sqrt{21}}{\sqrt{14}}$ _____

51. Combine similar terms. $-3x - 2x^2y - 2xy^2 + 3x^2y - 2y + 4x$ _____

52. Subtract. $(6x^2y - 4xy^2 + 7xy - 3) - (-3xy^2 + 2xy - x)$ _____

53. Multiply. $(3x + 4)(7x - 9)$ _____

54. Factor. $12x^7y^8 + 4x^2y^5 - 28x^8y^3 + 20x^5y^9$ _____

55. Factor. $x^3 + y^3$ _____

56. Factor and solve the equation $8x^2 + x = 7$. _____

Intermediate Algebra

Solve the inequality, equation, or system of equations.

57. $3x + 5 < 23$ _____

58. $-2x - 8 \geq -17$ _____

59. $|x| = 6$ _____

60. $|x + 5| < 7$ _____

61. $x - 2y = -5$

$3x + 2y = 21$ _____

62. $3x - 3y - 1 = 0$

$4x + 6y = -2$ _____

63. Simplify. $\sqrt[3]{x^5} \div \dfrac{1}{x^{\left(-\frac{3}{4}\right)}}$ _____

64. Use the quadratic formula to solve the equation for x: $5x^2 = 6x + 3$ _____

65. Solve. $x^2 - 16 > 0$ _____

66. Add. $(2 + 3i) + (4 + 5i)$ _____

67. Multiply. $(2 + 3i)(4 + 5i)$ _____

68. Write the next two terms in this sequence. 18, 6, 2, ... _____

69. Add. $\begin{bmatrix} 2 & -3 & 5 \\ 6 & 8 & -1 \\ 4 & 0 & 3 \end{bmatrix} + \begin{bmatrix} -8 & 5 & 6 \\ 2 & -6 & 4 \\ -8 & 1 & 0 \end{bmatrix} = \begin{bmatrix} & & \\ & & \\ & & \end{bmatrix}$

70. Find the scalar product. $4\begin{bmatrix} 0 & -3 & 8 \\ 6 & 0 & 2 \\ 14 & 6 & -5 \end{bmatrix} = \begin{bmatrix} & & \\ & & \\ & & \end{bmatrix}$

Coordinate Geometry

Graph each inequality on the number line.

71. $x > -2$

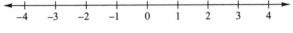

72. $4 \geq x > -3$

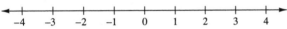

73. Graph the equation $y = 3x - 2$.

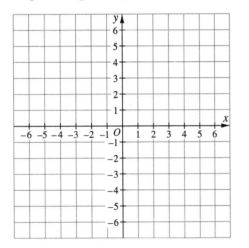

74. Find the distance between points $(-4,-3)$ and $(-8,6)$ on the coordinate plane. _____

75. Find the midpoint of the line segment joining $(-4,-3)$ and $(-8,6)$. _____

76. Graph the inequality $y < 2x + 1$.

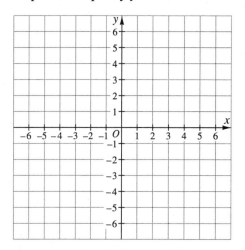

77. Shade the portion of the plane that shows the solutions to both inequalities
$y < 2x + 1$ and $y \leq 3x - 2$.

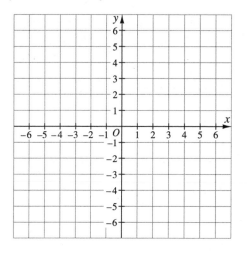

78. Graph the equation $x^2 + 6x + 10 + y^2 - 2y = 25$.

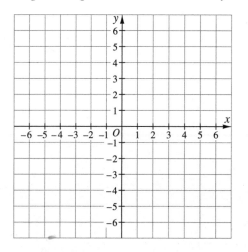

79. Graph the equation $y - 2 = 2(x^2 - 8x + 16)$.

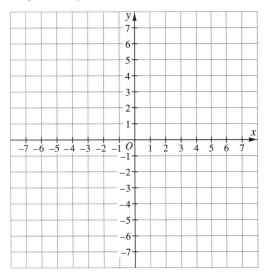

Plane Geometry

80. Use symbols to show that line *AB* and line *CD* are parallel. _____

81. Use symbols to show that ray *PQ* is perpendicular to ray *RS*. _____

Use the figure below to answer questions 82–85.

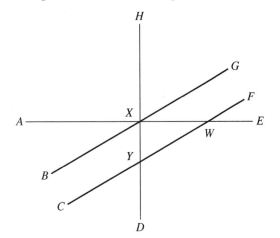

$\overline{AE}$ and $\overline{DH}$ meet at right angles.

$\overline{BG}$ and $\overline{CF}$ are parallel.

82. What is the measure of $\angle AXH$? _____

83. If the measure of $\angle BXD$ is 75°, what is the measure of $\angle XYC$? _____

84. If the measure of $\angle BXD$ is 68°, what is the measure of $\angle CYD$? _____

85. What is the sum of the measures of $\angle FYH$ and $\angle AWC$? _____

86. A four-sided figure whose opposite sides are parallel and congruent is called a(n) _____ .

87. The measure of one angle in a triangle is 36°. What is the sum of the measures of the other two angles? _____

88.

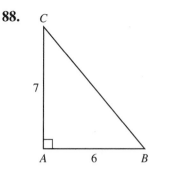

What is the length of $\overline{BC}$? _____

89. $\triangle ABC$ is similar to $\triangle DEF$. The pairs $\overline{AB}$ and $\overline{DE}$, and $\overline{BC}$ and $\overline{EF}$ are corresponding sides. $\overline{EF}$ is 1.5 times as long as $\overline{BC}$. What do you know about the relationship of the lengths of $\overline{AB}$ and $\overline{DE}$? _____

90. Given $\triangle PQR$ and $\triangle STU$, with $\angle P \cong \angle S$, $\angle Q \cong \angle T$, and $\angle R \cong \angle S$, prove that these triangles are congruent or explain why they are not. _____

_____

_____

What are the area and circumference of a circle with a diameter equal to 6 cm?

91. Area _____

92. Circumference _____

93. Point A is at $(-3,2)$. If you flip the point over the y-axis and flip the resulting point over the x-axis, what are the coordinates of the final point? _____

94. A circle is inscribed in a square whose area is 1 cm². What is the area of the circle? _____

95. What is the surface area of a cylinder if the height of the cylinder and the diameter of the base are both 1 cm? _____

Trigonometry

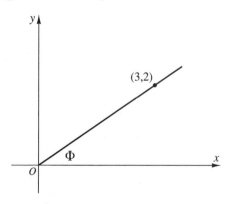

In the figure above, what are sin Φ and cos Φ?

96. sin Φ = _____

97. cos Φ = _____

98. The sine of $\theta = \sqrt{\dfrac{1}{8}}$. What is the cosine? _____

99. In which quadrant of the coordinate plane are tangent and cotangent positive, while all other trigonometric ratios are negative? _____

100. The graph of which trigonometric function is shown below? _____

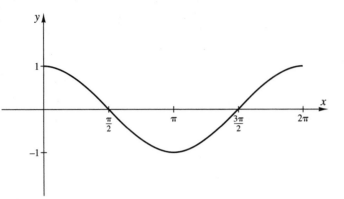

(Answers to the Topic Inventory appear on the following pages.)

Topic Inventory Answers

Check your answers to the Topic Inventory items against the correct answers below. If your answer is incorrect, circle the number for that question. After you check your answers, turn to the Study Chart on page 36. The chart shows the important study topics for each test item. If you have missed any of the items for a study topic, check the box and then carefully study that topic.

1. $23 > 18$

2. $98{,}776 < 137{,}165$

3. $73 > 67 > 59$

4. 8 million

5. 21,981

6. 8,973

7. 50,367

8. 103

9. $72.3989 < 72.41$

10. 176.4

11. 273.626

12. 26.3692

13. 0.027968

14. 32.5

15. 1, 2, 3, 6, 9, 18

16. 1, 2, 4, 8

17. 0.9

18. 3.162×10^3

19. $\dfrac{13}{42}$

20. $\dfrac{7}{9}, \dfrac{11}{14}, \dfrac{5}{6}$

21. $2\dfrac{49}{120}$

22. $2\dfrac{23}{24}$

23. $7\dfrac{39}{40}$

24. $\dfrac{24}{147} = \dfrac{8}{49}$

25. 2

26. $-16 < 11$

27. -110

28. 18

29. 2,754

30. 26

31. 11

32. $0.\overline{6},\ 66\dfrac{2}{3}\%$

33. $\dfrac{13}{20{,}000},\ 0.00065$

34. 50 acres

35. $15

36. $x = 4$

37. $\dfrac{6}{9} = \dfrac{p}{12}$

   $9p = 72$

   $p = 8$ feet

38.
1	0, 0
2	
3	1
4	0, 6
5	7

39. 122.2

40. 84

41. 51

42. $\dfrac{1}{4}$

43. 21

44. $b = 2s - 2$

45. $(0, -2)\ (1\dfrac{1}{3}, 0)$

46. $x = -3$

47. $I = \$510$

48. 28^{11}

49. $9\sqrt{7}$

50. $\dfrac{\sqrt{6}}{2}$

51. $x + x^2y - 2xy^2 - 2y$

52. $6x^2y - xy^2 + 5xy + x - 3$

53. $21x^2 + x - 36$

54. $4x^2y^3(3x^5y^5 + y^2 - 7x^6 + 5x^3y^6)$

55. $(x + y)(x^2 - xy + y^2)$

56. $(8x - 7)(x + 1) = 0$

$x = \dfrac{7}{8}$ or $x = -1$

57. $x < 6$

58. $x \le 4.5$

59. $x = \pm 6$

60. $-12 < x < 2$

61. $x = 4, y = 4.5$

62. $x = 0, y = -\dfrac{1}{3}$

63. $x^{\left(\frac{11}{12}\right)}$

64. $x = \dfrac{-b \pm \sqrt{b^2 - 4ac}}{2a}$

$x = \dfrac{3}{5} + \dfrac{2}{5}\sqrt{6} \approx 1.58$ and

$x = \dfrac{3}{5} - \dfrac{2}{5}\sqrt{6} \approx -0.38$

65. $x < -4$ or $x > 4$

66. $6 + 8i$

67. $-7 + 22i$

68. $\dfrac{2}{3}, \dfrac{2}{9}$

69. $\begin{bmatrix} -6 & 2 & 11 \\ 8 & 2 & 3 \\ -4 & 1 & 3 \end{bmatrix}$

70. $\begin{bmatrix} 0 & -12 & 32 \\ 24 & 0 & 8 \\ 56 & 24 & -20 \end{bmatrix}$

71.

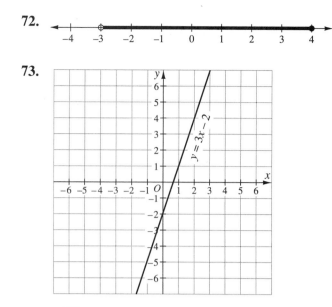

72.

73.

74. $\sqrt{97} \approx 9.85$

75. $(-6, 1.5)$

76.

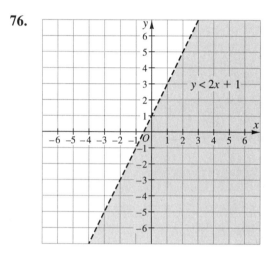

77.

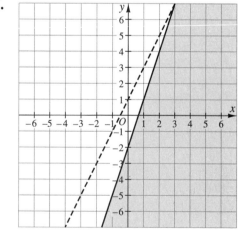

78.

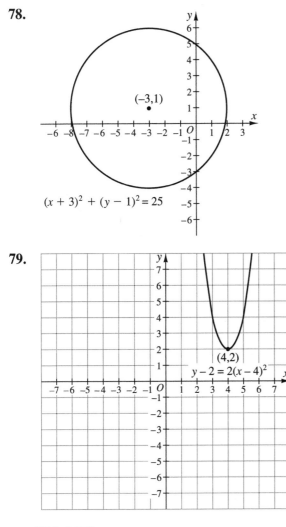

$(x + 3)^2 + (y - 1)^2 = 25$

79.

$y - 2 = 2(x - 4)^2$

80. $\overleftrightarrow{AB} \parallel \overleftrightarrow{CD}$

81. $\overrightarrow{PQ} \perp \overrightarrow{RS}$

82. 90°

83. 105°

84. 68°

85. 90°

86. parallelogram

87. 144°

88. $\sqrt{85} \approx 9.2$

89. $\overline{DE}$ is 1.5 times the length of $\overline{AB}$.

90. Since corresponding angles are congruent, the triangles are similar. No information is given about the sides, so the triangles may or may not be congruent.

91. 9π cm^2

92. 6π cm

93. $(3, -2)$

94. $\dfrac{\pi}{4}$ cm^2

95. $\dfrac{\pi}{2} + \pi$ cm^2

96. $\dfrac{2}{13}\sqrt{13}$

97. $\dfrac{3}{13}\sqrt{13}$

98. $\pm\sqrt{\dfrac{7}{8}} = \pm\dfrac{\sqrt{14}}{4}$

99. quadrant III

100. The basic cosine function or the sine function after a horizontal shift.

■ Study Chart

STUDY TOPIC	QUESTION NUMBERS	PAGES TO STUDY
Coordinate Geometry (Chapter 10)		
☐ Graphing Inequalities on a Number Line	71, 72	188–190
☐ Graphing Equations on the Coordinate Plane	73	191–199
☐ Distance and Midpoint Formulas	74, 75	199–202
☐ Graphing Systems of Inequalities on the Coordinate Plane	76, 77	202–208
☐ Graphing Conic Sections	78, 79	208–215
Plane Geometry (Chapter 11)		
☐ Basic Elements of Plane Geometry	80, 81	224–227
☐ Angles	82, 83, 84, 85	227–231
☐ Quadrilaterals	86	232–235
☐ General Properties of Triangles	87	235–238
☐ Right Triangles	88	238–243
☐ Similar Triangles	89	244–248
☐ Concept of Proof and Proof Techniques	90	249–254
☐ Circles	91, 92	255–260
☐ Transformations in the Plane	93	260–267
☐ Geometric Formulas	94	267–271
☐ Geometry in Three Dimensions	95	271–277
Trigonometry (Chapter 12)		
☐ Right Triangle Trigonometry	96, 97	283–287
☐ Trigonometric Identities	98	288–291
☐ Unit Circle Trigonometry	99	292–296
☐ Graphs of Trigonometric Functions	100	297–301

Chapter 5 ▪

Calculators

▪ Calculators and the ACT

ACT test makers like to say that you don't NEED a calculator to complete the math test. That's technically correct, but everyone knows it makes sense to bring a calculator to the test. Just remember that no matter what kind of calculator you use, it cannot tell you how to solve a problem. As a general rule, you should use a calculator you're most familiar with. But there are some exceptions, and we give some specific guidelines below.

You may use a calculator only on the ACT Mathematics Test. You can't use the calculator on any other ACT test, including the ACT Science Reasoning Test. You can't use a calculator that is on the list of prohibited calculators. You'll be dismissed from the test if you use a prohibited calculator, or if you try to use the calculator on any but the math test. It is your responsibility to know if your calculator is allowed on the math test. The ACT says that using the prohibited TI-89 calculator is one of the main reasons that students are dismissed from the test. Don't bring the TI-89. By the way, you can't use the calculator on your phone.

CALCULATOR TIP

Look for these "Calculator Tips" in the math section. They will give you hints and advice on how to use a calculator to help you on the ACT.

The people administering the ACT will not provide calculators or batteries. You may bring extra batteries or a backup calculator, just in case yours fails. Leave it with the test proctor. Forget the games on your calculator, you're not allowed to use them during the test. It makes sense to follow all the rules. You don't want to have to do this all over again.

▪ Permitted Calculators

The ACT says you may use any four-function, scientific, or graphing calculator, unless it has features described in the prohibited list. Some calculators are allowed after modification.

We give specific recommendations for the best calculators in the following section. But we can tell you this right now—bring a graphing calculator only if you really know how to use it. Otherwise, we suggest calculators similar to the TI-34 Multiview or the Casio FX-115ES. These calculators have multi-line displays and most of the capability you will use on the ACT.

■ Prohibited Calculators

You can bring most calculators to the ACT, but some calculators are prohibited. The best way to be sure is to visit the ACT student website. At press time, a list of prohibited calculators is listed at the following link:

http://www.actstudent.org/faq/answers/calculator.html

Below we list some prohibited calculators that, based on teacher reports, are most likely to get you into trouble.

TI-89 or TI-89 Titanium

You may not use the TI-89, even if it is used in your mathematics class. We mention this prohibited calculator again because teachers tell us that bringing the TI-89 to the ACT is the biggest calculator-selection mistake that students make. Teachers report that students have been dismissed from the test for trying to use the TI-89. Don't let that happen to you.

TI-Nspire CAS Version

There are two versions of the TI-Nspire: the CAS and the non-CAS. The TI-Nspire CAS is the more advanced (and expensive) version. You can bring the non-CAS version of the TI-Nspire but not the CAS version.

It goes without saying that you can't bring any type of computer or PDA. You are not allowed to have any means of communicating with someone during the test, so you are not allowed to use phones that have calculator functions.

We know that the calculators we list in the next section can be used on the ACT, but it's best not to guess.

CALCULATOR TIP

Visit the ACT student website at http://www.actstudent.org/faq/answers/calculator.html for a specific list of prohibited calculators.

Talk with your mathematics teacher if a calculator you prefer is on the prohibited list.

■ Calculator Choices

Bring a graphing calculator only if you really know how to use it. You should bring a calculator that has a display of at least two lines. That lets you see what you entered along with the answer. Many calculator errors are key-entry errors, so it is important to see the key entries. New calculators emerge every few years. Consult with your mathematics teacher to learn about the current preferred calculators.

Do Not Bring a Calculator Like This

The TI-108

This small basic calculator has a significant flaw. It does not automatically follow the correct order of operations. Enter this expression to find out if your calculator has this flaw: 3 [+] 4 [×] 5 [=]. If the answer is 35, your calculator has this problem. Don't use it. Using this type of basic calculator actually causes difficulties. Even if you use this calculator every day, we definitely recommend that you get another calculator.

Try Not to Bring a Calculator Like This

The TI-36X

This type of calculator, even though it has many built-in functions, has only one line of display. The calculator display does not show characters as they look in your textbook. There are two-line versions of this calculator, which are fine to bring, but they are not as good as the next calculators on our list.

Bring a Calculator Like These, Unless You Are an Experienced Graphing Calculator User

TI-34 Multiview or Casio FX-115ES

The best-selling TI-34 and Casio FX-115ES are pictured at the top of the next page. These calculators are very friendly, particularly for a scientific calculator. You can use them like a basic calculator, and turn to other functions when you want to. That means you have the best of both worlds. The screen display looks just like the print in your textbook, with raised exponents and all.

These calculators let you use fractions and mixed numbers, and let you easily convert among fractions, decimals, and percents. These calculators include a percent key and some built-in functions, including the formulas for permutations and combinations. It's easy to forget these formulas and it's nice to have a calculator do the work for you. There are trig functions too, but you won't need them on the ACT.

The TI-34 can show answers in pi form, while the Casio can show answers in pi form and in radical form.

Bring a Graphing Calculator Like These, If You Are an Experienced Graphing Calculator User

TI-83 or TI-84 Family of Calculators

You'll probably bring a TI-83, TI-83 Plus, TI-84 Plus, or a TI-84 Plus Silver Edition. These are the most popular graphing calculators you can bring to the ACT test. Any of them will be more than enough, so bring the one you know best. You will not use all the capabilities of any of these graphing calculators.

Think Carefully Before You Bring This Calculator

TI-Nspire (non-CAS)

The TI-Nspire is the next generation calculator from Texas Instruments. Our survey has shown that you should bring this calculator only if this is the calculator used in your school. Consult with your mathematics teacher before bringing this calculator to the test.

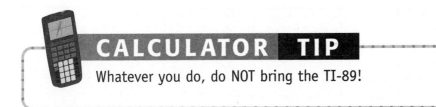

CALCULATOR TIP

Whatever you do, do NOT bring the TI-89!

Graphing calculators make an excellent classroom resource. However, a graphing calculator, by itself, will not help you get a better score than if you use a scientific calculator. Bring a graphing calculator only if you really know how to use it. Otherwise, don't even think of bringing one. It will be a disaster.

Estimate First

Many calculator errors are caused by key-entry mistakes, even on a multi-line calculator. You think you put in one number, but you really put in another.

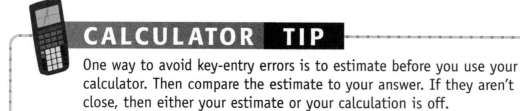

CALCULATOR TIP

One way to avoid key-entry errors is to estimate before you use your calculator. Then compare the estimate to your answer. If they aren't close, then either your estimate or your calculation is off.

Use Your Calculator to Check Your Work

CALCULATOR TIP

A calculator can be helpful after you have worked out a problem using paper and pencil. Use your calculator to check your work.

■ Recognizing When Your Calculator Will Be Helpful

ACT mathematics items fall into three categories: the calculator doesn't help at all, the calculator is somewhat helpful (usually by performing arithmetic operations), or the calculator is very helpful. Use your calculator only when it will help you.

A review of past ACT exams shows that the calculator was not helpful for about 50 to 60 percent of the test items and that the calculator became less useful as the questions became more difficult. I never found an ACT question for which a trig function on the calculator was helpful. The calculator could be helpful for about 30 to 35 percent of the test items. The calculator could give you an advantage with only about 5 to 15 percent of the test items. Given the time constraints of the test, it might be better to use the basic features of the calculator along with mental math and pencil-and-paper techniques with some items on the test.

When a Calculator Doesn't Help

Here are the kinds of ACT questions for which the calculator will be no help at all.

1. Which of the expressions below is equivalent to $(2x^2y)(4xy^2)^2$?

   A. $8x^4y^4$

   B. $32x^4y^4$

   C. $8x^4y^5$

   D. $32x^4y^5$

   E. $32x^5y^5$

   A scientific or a graphing calculator will not combine terms. You must use the rules for multiplying exponents to find that the correct answer is D.

2. Throw two six-sided dice numbered 1 through 6 and you get a 7 when the numbers represented on the dice are 6 and 1, 5 and 2, or 4 and 3. What is the probability of throwing a 7 if you roll two six-sided dice numbered 1 through 6?

   F. $\frac{1}{2}$

   G. $\frac{7}{12}$

   H. $\frac{1}{6}$

   J. $\frac{7}{36}$

   K. $\frac{1}{4}$

   You must count to find that six combinations of numbers from two dice will equal 7 and that the correct answer is H.

When a Calculator Is Somewhat Helpful

Here are the types of questions for which the calculator could be of some help.

3. What is the area in square feet of a triangular pennant with a base of 1 foot and a height of 5 feet?

   A. $\sqrt{5}$

   B. 2

   C. $2\frac{1}{2}$

   D. 5

   E. 10

   You could use your calculator to multiply $0.5 \times 1 \times 5$ to find the area of the pennant and to find that the correct answer is C. However, you could just as easily think that half of 5 is $2\frac{1}{2}$. So the calculator could be somewhat helpful if you were not sure about the calculations.

4. What is the area in square units of the rectangle shown below?

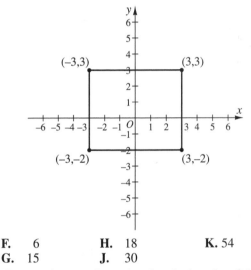

   F. 6   H. 18   K. 54
   G. 15   J. 30

   First you have to determine that the lengths of the sides of the rectangle are 5 and 6. Then multiply 5×6 to find that the answer is J. You could use the calculator to do this multiplication, but you could also multiply without the calculator.

When a Calculator Can Be Very Helpful

Here is a question for which the calculator could be a big help.

5. The compound interest formula $A = P(1 + r)^t$ represents the value of a bank account after t years where the interest r is compounded annually and P is the initial deposit. What is the approximate value of a savings account after 10 years if $2,000 is deposited at 3% compounded annually?

   A. $2,100   D. $3,000
   B. $2,600   E. $3,700
   C. $2,700

   Use the calculator to enter the values into the formula given in the problem and you will see that C is the correct choice.

Chapter 6 ▪

Pre-Algebra I

- Fourteen ACT questions have to do with pre-algebra.
- Easier pre-algebra questions may be about a single skill or concept.
- More difficult questions will often test a combination of skills or concepts.
- The pre-algebra review in Chapters 6 and 7 covers all the material you need to answer ACT questions.
- Use a calculator for the ACT-Type Problems. Do not use a calculator for the Practice exercises.

CALCULATOR TIP

Many ACT questions are easiest to answer without a calculator. This section shows how to estimate and do paper and pencil computations, along with tips for using a calculator.

▪ Whole Numbers

The **whole numbers** are 0, 1, 2, 3, 4, ... (The dots mean the numbers go on forever.)

Comparing Whole Numbers

Think of the whole numbers on a number line.

Greater numbers are to the right on the number line.
Lesser number are to the left on the number line.
Use these symbols to compare whole numbers.

$<$ less than	$4 < 9$	4 is less than 9.
$>$ greater than	$4 > 2$	4 is greater than 2.
$=$ equal to	$4 = 4$	4 is equal to 4.

Place Value

Use place value to write whole numbers.

Look at the place-value chart shown below.

hundred-millions	ten-millions	millions	hundred-thousands	ten-thousands	thousands	hundreds	tens	ones
2	6	1	5	3	8	7	4	9

The number in the chart is read: two hundred sixty-one million, five hundred thirty-eight thousand, seven hundred forty-nine.

The **6** is in the ten-millions place. The value of the digit **6** is **sixty million**.

The **4** is in the tens place. The value of the digit **4** is **forty**.

Comparing Larger Numbers

To compare larger whole numbers, line up the place values. Start at the greatest place value and compare the digits until you find the place where the digits differ.

EXAMPLES

1. Compare 341,062 and 329,989.

Line up the place values.

341,062
329,989

Start at the greatest place value.   3 = 3

Move one place to the right.   4 > 2

Since 4 > 2, then 341,062 > 329,989.

2. Compare 101,345 and 98,879.

Line up the place values.

101,345
98,879

Start at the greatest place value.

101,345 has a 1 in the hundred-thousands place.

98,879 has no digit in the hundred-thousands place.

So 101,345 > 98,879.

CALCULATOR TIP

Subtract to compare numbers on a calculator.
If the answer is negative, the number being subtracted is greater.
If the answer is positive, the number being subtracted is less.
If the answer is 0, the two numbers are equal.

Between

Use the greater than or less than symbols to show that numbers are between other numbers.

7 > 5 > 1 means that 5 is between 7 and 1.

23,457 < 24,523 < 24,532 means that 24,523 is between 23,457 and 24,532.

Rounding Whole Numbers

Use these steps for rounding whole numbers.

- Locate the place you are rounding to.
- Look at the digit to the right of that place.
- If the digit is less than 5, round down. If the digit is 5 or greater, round up.

E X A M P L E

Round 347,294 to the ten-thousands place.

Locate the ten-thousands place. 347,294

Look to the right. The digit 7 is greater than 5, so you round up.

347,294 rounded to the ten-thousands place is 350,000.

347,294 rounded to the hundreds place is 347,300.

347,294 rounded to the thousands place is 347,000.

347,294 rounded to the hundred-thousands place is 300,000.

MODEL ACT PROBLEM

Which of the following statements is true?

I. $346{,}783 < 46{,}783$

II. $100{,}000 > 99{,}999$

III. $52{,}893 > 52{,}983$

A. I only
B. II only
C. I and II only
D. I and III only
E. II and III only

SOLUTION

Line up the pairs of numbers by place value and compare the digits starting at the left.

I.   346,783
      46,783

Notice that 346,783 has a 3 in the hundred-thousands place and 46,783 has no digit in the hundred-thousands place. That means $346{,}783 > 46{,}783$ and statement I is false. You can eliminate choices A, C, and D.

II.   100,000
      99,999

Notice that 100,000 has a 1 in the hundred-thousands place and 99,999 has no digit in the hundred-thousands place. That means $100{,}000 > 99{,}999$ and statement II is true. The answer can still be choice B or E.

III. 52,893
      52,983

At the greatest place value, $5 = 5$. Moving one place to the right, $2 = 2$.

Moving another place to the right, $8 < 9$. That means $52{,}893 < 52{,}983$ and statement III is false.

Statement II is the only true statement.

The correct answer is B.

Practice

Write < or > in the blank.

1. 16 ____ 24

2. 31 ____ 91

3. 1,031 ____ 976

4. 1,891 ____ 2,001

5. 18 ____ 24 ____ 36

6. 58 ____ 40 ____ 38

7. Write the total value of each digit in the numeral 6,803,795,142.

3 _____

0 _____

2 _____

4 _____

1 _____

9 _____

6 _____

8 _____

7 _____

5 _____

Write < or > in the blank.

8. 10,037 ____ 9,869

9. 23,568 ____ 23,602

10. 728,315 ____ 728,298

11. 101,834,561 ____ 98,898,789

12. Use < or > to compare the numbers 234, 83, and 196. _____

13. Use < or > to compare the numbers 108; 1,961; 1,147; and 562.

_____

Round 3,057,814 to the

14. millions place _____

15. thousands place _____

16. tens place _____

17. hundreds place _____

18. hundred-thousands place _____

Round 17,648 to the

19. hundreds place _____

20. tens place _____

(Answers on page 334)

ACT-TYPE PROBLEMS

1. After she graduated from college, Brianna began teaching at a salary of $28,750 a year. Which of the following choices shows Brianna's salary rounded to the thousands place?

A. $28,000
B. $28,800
C. $28,900
D. $29,000
E. $30,000

2. In which of the following choices would the symbol > NOT make the statement true?

F. 34 ____ 30
G. 30 ____ 31
H. 29 ____ 20
J. 25 ____ 23
K. 25 ____ 21

3. What is the sum of the digits in the hundred-thousands place, the ones place, and the ten-thousands place in the numeral 6,825,074?

    A. 14
    B. 15
    C. 16
    D. 17
    E. 18

4. What is the total value of the digit 3 in 473,961?

    F. three hundred thousand
    G. thirty thousand
    H. three thousand
    J. one thousand
    K. three hundred

(Answers on page 334)

5. Which of the following choices has the largest digit in the ten-thousands place?

    A. 45,823
    B. 369,082
    C. 3,782,956
    D. 172,297,269
    E. 1,926,052,736

▪ Whole-Number Computation

Follow these steps to add, subtract, multiply, or divide whole numbers.

- Estimate first.
- Compute.
- Check your answer against the estimate to be sure your answer is reasonable.

Estimating will help you avoid computational errors. Sometimes an estimate may be enough to answer the question.

CALCULATOR TIP

Use your calculator to compute or to check whole-number computation. Estimate first so you can check for key-entry errors.

Addition

EXAMPLE

123,184 + 17,672

Estimate first.

123,184 rounded to the nearest ten-thousand is 120,000.

17,672 rounded to the nearest ten-thousand is 20,000.

120,000 + 20,000 = 140,000

The answer should be about 140,000.

Enter: 123184 [+] 17672 [=]

Display: 140856

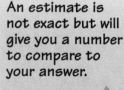

An estimate is not exact but will give you a number to compare to your answer.

Subtraction

207,536 − 152,345

Estimate first.

207,536 rounded to the nearest ten-thousand is 210,000.

152,345 rounded to the nearest ten-thousand is 150,000.

210,000 − 150,000 = 60,000

The answer should be about 60,000.

Enter: 207536 $\boxed{-}$ 152345 $\boxed{=}$

Display: $\boxed{55191}$

Keywords can help you decide what operation to use.

Addition–sum, and, more

Subtraction–less, difference

Multiplication–of, product, times

Division–per, quotient

Multiplication

6,542 × 191

Estimate first.

6,542 rounded to the nearest hundred is 6,500.

191 rounded to the nearest hundred is 200.

6,500 × 200 = 1,300,000

The answer should be about 1,300,000.

Enter: 6542 $\boxed{\times}$ 191 $\boxed{=}$

Display: $\boxed{1249522}$

Estimation can help when multiplying. Multiplication errors often result in answers that are off by a multiple of 10.

Division

23,717 ÷ 78

Estimate first.

78 rounded to the nearest ten is 80.

Round 23,717 to a number that can be divided by 80 easily. 23,717 → 24,000

24,000 ÷ 80 = 300

The answer should be about 300.

Enter: 23717 $\boxed{\div}$ 78 $\boxed{=}$

Display: $\boxed{304.0641026}$

Estimation can help when dividing. Division errors often result when the digits are not properly aligned.

Use the integer division key to show the remainder as a whole number. If your calculator does not have an integer division key, multiply the decimal part of the quotient by the divisor to find the whole number remainder. For example, entering 78 $\boxed{\times}$ 0.0641026 will give you 5. 5 is the whole number remainder of $23{,}717 \div 78$.

EXAMPLE

$99 \times 39 - 21 \div 5$

Estimate first.

99 rounded to the nearest hundred is 100.

39 rounded to the nearest ten is 40.

Round 21 to a number that can be divided by 5 easily. $21 \to 20$

$100 \times 40 - 20 \div 5 = 4{,}000 - 4 = 3{,}996$

The answer should be about 3,996.

Enter: 99 $\boxed{\times}$ 39 $\boxed{-}$ 21 $\boxed{\div}$ 5 $\boxed{=}$

Display: $\boxed{3856.8}$

MODEL ACT PROBLEM

At Emily's Auto Repair shop the mechanics work on both foreign and domestic cars. During one year, 1,974 foreign cars and 2,245 domestic cars were repaired. How many more domestic cars than foreign cars were repaired at Emily's shop?

A. 271
B. 275
C. 281
D. 285
E. 291

SOLUTION

Subtract: domestic − foreign
$2{,}245 - 1{,}974 = 271$

The correct answer is A.

Practice

Estimate first. Then compute.

1. $256 + 178$
2. $20{,}987 + 289$
3. $189{,}036 + 40{,}965$
4. $702{,}008 + 282{,}896$
5. $806 - 427$
6. $70{,}937 - 1{,}948$
7. $200{,}183 - 94{,}086$
8. $803{,}002 - 216{,}135$
9. 23×41
10. 401×38
11. 507×101
12. $3{,}017 \times 207$
13. $168 \div 6$
14. $7{,}536 \div 24$
15. $8{,}346 \div 107$
16. $29{,}596 \div 98$
17. $67{,}942 + 3{,}260$
18. $7{,}942 - 3{,}815$
19. 35×456
20. $40{,}365 \div 65$

(Answers on page 334)

ACT-TYPE PROBLEMS

1. There are 11 players on each of 12 teams in a tournament. What is the total number of players on all 12 teams?

   A. 23
   B. 122
   C. 131
   D. 132
   E. 133

2. A group of 36 friends takes a trip. If they travel with 4 people in each car, how many cars do they need for the trip?

   F. 8
   G. 9
   H. 10
   J. 11
   K. 12

3. Andy is 27 years old. Drew is 5 years younger than Andy. What is the sum of their ages?

   A. 22
   B. 27
   C. 32
   D. 39
   E. 49

4. If Danny earns $8 per hour, how much money will he make working 24 hours?

   F. $190
   G. $191
   H. $192
   J. $193
   K. $194

5. Matt averages 26 points a game. If he scores 19 points in one game, how many points from his average is he?

   A. 9 points
   B. 8 points
   C. 7 points
   D. 6 points
   E. 5 points

(Answers on page 334)

Decimals

Use place value to write decimals. Look at the place-value chart below.

thousands	hundreds	tens	ones	tenths	hundredths	thousandths	ten-thousandths	hundred-thousandths
6	5	1	2	3	8	7	4	9

The decimal in the chart is read: six thousand, five hundred twelve and thirty-eight thousand, seven hundred forty-nine hundred-thousandths.

The **6** is in the thousands place. The value of the digit **6** is **six thousand**.

The **7** is in the thousandths place. The value of the digit **7** is **seven thousandths**.

Comparing Decimals

Compare decimals the same way you compare whole numbers. Write the numbers and line up the decimal points and place values. Start at the greatest place value and compare the digits until you find a place where the digits differ.

E X A M P L E S

1. Compare 0.34203 and 0.34198.

    Line up the decimal points and place values.

    $$0.34203$$
    $$0.34198$$

Start at the greatest place value.	$0 = 0$
Move one place to the right.	$3 = 3$
Move one more place to the right.	$4 = 4$
Move one more place to the right.	$2 > 1$

    Since $2 > 1$, then $0.34203 > 0.34198$.

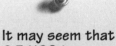

It may seem that 0.34198 is greater when you look only at the digits 98.

2. Compare 384.1806 and 384.21.

Line up the decimal points and place values.	384.1806
Write extra zeros if needed.	384.2100
Start at the greatest place value.	$3 = 3$
Move one place to the right.	$8 = 8$
Move one more place to the right.	$4 = 4$
Move one more place to the right.	$1 < 2$

    Since $1 < 2$, then $384.1806 < 384.21$.

It may seem that 384.1806 is greater because it has more digits.

Between

Use the greater than and less than symbols to show that decimals are between other decimals.

0.2894 > 0.2860 > 0.2843 means that 0.2860 is between 0.2894 and 0.2843.

23.4572 < 24.5238 < 24.532 means that 24.5238 is between 23.4572 and 24.532.

Rounding Decimals

Use the same steps to round decimals as you use to round whole numbers.

- Locate the place you are rounding to.
- Look at the digit to the right of that place.
- If the digit is less than 5, round down. If the digit is 5 or greater, round up.

EXAMPLE

Round 19.0463 to the thousandths place.

Locate the thousandths place. 19.04**6**3

Look to the right. The digit 3 is less than 5, so you round down.

19.0463 rounded to the thousandths place is 19.046.

19.0463 rounded to the hundredths place is 19.05.

19.0463 rounded to the tenths place is 19.0.

19.0463 rounded to the ones place is 19.

19.0463 rounded to the tens place is 20.

Drop all the digits to the right of the thousandths place.

MODEL ACT PROBLEM

The following shows a list of people and their times for a 100-meter dash.

Name	Time
Nathan	11.63 seconds
Andy	11.15 seconds
Julia	11.16 seconds
Thomis	11.21 seconds
Ann	11.28 seconds

Who had the fastest time?

A. Nathan
B. Andy
C. Julia
D. Thomis
E. Ann

SOLUTION

The fastest time is the lowest number. Line up all the times by place value and see which is the smallest.

> 11.63 seconds, Nathan
> 11.15 seconds, Andy
> 11.16 seconds, Julia
> 11.21 seconds, Thomis
> 11.28 seconds, Ann

Compare each place starting at the left. The first place we see a difference is in the tenths place. Julia's and Andy's times each have a 1 in the tenths place. The digits in the tenths place for the times of the other runners are all greater than 1. In the hundredths place, Andy's time has a 5 and Julia's time has a 6. Andy's time is the smallest number, so his time is the fastest.

The correct answer is B.

Practice

Write < or > in the blank.

1. 0.32 _____ 0.289

2. 1.91 _____ 9.1

3. 20.347 _____ 20.351

4. 387.9036 _____ 38.79039

5. 408.246 _____ 389.978

6. 0.73846 _____ 0.73921

7. 16.4 _____ 16.43 _____ 16.432

8. 0.56893 _____ 0.56981 _____ 0.5699

9. Write the value of each digit in the numeral 16,439.02857.

1 _____   6 _____

2 _____   7 _____

3 _____   8 _____

4 _____   9 _____

5 _____   0 _____

10. Use < or > to compare the numbers 18.61, 18.097, 18.598. _____

11. Use < or > to compare the numbers 402.5163, 403.8917, 402.5191, and 403.8799.

_____

Round 3,245.60537 to the

12. thousandths place _____

13. thousands place _____

14. tens place _____

15. hundredths place _____

16. tenths place _____

17. ones place _____

18. ten-thousandths place _____

19. hundreds place _____

20. What is the place value of 4 in the number 3,256.3745? _____

(Answers on page 335)

> Allow yourself one minute for each ACT math problem to reflect real ACT test conditions

ACT-TYPE PROBLEMS

1. What is the sum of the digits found in the tenths place and in the ten-thousandths place of 10,324.96176?

    A.   5
    B.   7
    C. 10
    D. 15
    E. 16

2. The symbol < can be used to fill in the blank to make which choice true?

    F.   345.982 _____ 345.992
    G. 356.792 _____ 356.782
    H. 272.81 _____ 272.18
    J.   3.9182 _____ 3.9082
    K. 576.91 _____ 567.91

3. In the number 10,475.9362, which digit is in the thousandths place?

    A. 1
    B. 6
    C. 4
    D. 3
    E. 2

5. Which of the following choices has the largest digit in the tenths place?

    A. 7,256.176
    B. 3,589.629
    C. 534.529
    D. 14.973
    E. 1.3

4. In which of the following choices is 3,482.9647 rounded to the hundredths place?

    F. 3,500
    G. 3,482.965
    H. 3,482.96
    J. 3,482
    K. 3,480

(Answers on page 335)

■ Decimal Computation

Follow these steps to add, subtract, multiply, or divide decimals.

- Estimate first.
- Compute.
- Check your answer against the estimate to be sure your answer is reasonable.

Estimating will help you avoid computational errors. Errors are often caused by improper placement of the decimal point. Sometimes an estimate may be enough to answer the question.

CALCULATOR TIP

Use your calculator to complete or to check decimal computations. It is very easy to make key-entry errors when inputting decimals. Be sure to estimate before you calculate so that you can check whether the answer is reasonable.

Addition

EXAMPLE

321.09 + 43.564

Estimate first.

321.09 rounded to the nearest ten is 320.

43.564 rounded to the nearest ten is 40.

320 + 40 = 360

The answer should be about 360.

Enter: 321.09 [+] 43.564 [=]

Display: | 364.654 |

Subtraction

$93.08 - 42.195$

Estimate first.

93.08 rounded to the nearest ten is 90.

42.195 rounded to the nearest ten is 40.

$90 - 40 = 50$

The answer should be about 50.

Enter: 93.08 $\boxed{-}$ 42.195 $\boxed{=}$

Display: $\boxed{50.885}$

Multiplication

20.108×1.14

Estimate first.

20.108 rounded to the nearest one is 20.

1.14 rounded to the nearest one is 1.

$20 \times 1 = 20$

The answer should be about 20.

Enter: 20.108 $\boxed{\times}$ 1.14 $\boxed{=}$

Display: $\boxed{22.92312}$

Division

$53.04 \div 7.8$

Estimate first.

53.04 rounded to the nearest one is 53.

7.8 rounded to the nearest one is 8.

$53 \div 8$ is between 6 and 7.

The answer should be between 6 and 7.

Enter: 53.04 $\boxed{\div}$ 7.8 $\boxed{=}$

Display: $\boxed{6.8}$

Sometimes the answer choice for a division problem may be rounded.

$(10.54 + 2.52) \div (8.1 \times 19.86)$

Estimate first.

10.54 rounded to the nearest one is 11.

2.52 rounded to the nearest one is 3.

8.1 rounded to the nearest one is 8.

19.86 rounded to the nearest one is 20.

$(11 + 3) \div (8 \times 20) = 14 \div 160$

$14 \div 160$ is between 0 and 1.

The answer should be between 0 and 1.

Enter: (10.54 + 2.52) ÷ (8.1 × 19.86) =

Display: 0.081185583

MODEL ACT PROBLEM

Mike makes $500 in 80 hours. On average, how much does Mike earn per hour?

A. $6.50
B. $6.35
C. $6.30
D. $6.25
E. $6.15

SOLUTION

Divide. $500 \div 80 = \$6.25$

The correct answer is D.

Practice

Estimate first. Then compute.

Add.

1. $0.06 + 19.803$

2. $19.803 + 541$

3. $11.882 + 6.49 + 0.083$

4. $3.4289 + 5.005 + 31 + 8.57$

Subtract.

5. $46.71 - 22.52$

6. $0.039 - 0.0271$

7. $388.6 - 97.86$

8. $229 - 5.04876$

Multiply.

9. 99.7×10.06

10. 0.08×58.6

11. 6.0810×148.3

12. 8.46×2.5

Divide.

13. $1006.83 \div 1.13$ **14.** $486 \div 0.391$ **15.** $0.909 \div 1.35$ **16.** $5.005 \div 0.095$

Compute.

17. $(56.4 \times 7) + (8 \div 2)$ **18.** $(34.9 - 24.5) \times 2.6$ **19.** $(19.4 + 5.4 - 8.3) \times 2$ **20.** $53.4 \div 17.8 - 3$

(Answers on page 335)

ACT-TYPE PROBLEMS

1. One day the high temperature is 92.6°; the low is 77.8°. What is the difference between the high and low temperatures on this day?

 A. 16.6°
 B. 14.8°
 C. 14.4°
 D. 13.8°
 E. 13.6°

2. On a mathematics test, Jordan scored 87.7 and Corey scored 79.4. Find the average of Jordan's and Corey's test scores.

 F. 85
 G. 84.5
 H. 83.55
 J. 83
 K. 82.25

3. Alex is a carpenter and he has fifty 1.5-inch nails and fifty 0.75-inch nails. What would be the length of all these nails if they were lined up end to end?

 A. 112.5 inches
 B. 115 inches
 C. 117.5 inches
 D. 119 inches
 E. 120 inches

(Answers on page 335)

4. Gene weighed 170 pounds. Then he worked out for 5 days, followed a new diet, and lost 0.8 pound a day. How much did Gene weigh after the 5 days?

 F. 163 pounds
 G. 164 pounds
 H. 165 pounds
 J. 166 pounds
 K. 167 pounds

5. The odometer on Eileen's car reads 2,004.7 miles. What will the odometer read at the end of the day on Friday after she drives the distances shown below?

Day	Miles
Monday	5.6 miles
Tuesday	10.4 miles
Wednesday	7.8 miles
Thursday	11.2 miles
Friday	22.7 miles

 A. 2,006.2 miles
 B. 2,062.4 miles
 C. 2,074.7 miles
 D. 2,045.6 miles
 E. 2,004.5 miles

◼ Factors, Divisibility, and Primes

Factors

A **factor** of any number divides the number exactly, with no remainder. Since $45 \div 9 = 5$ and $45 \div 5 = 9$, 5 and 9 are factors of 45. Thinking of it another way, $5 \times 9 = 45$, so 5 and 9 are factors of 45.

Every number has 1 and itself as factors.

 Factors of 1 are: 1

 Factors of 2 are: 1 and 2

 Factors of 4 are: 1, 2, and 4

 Factors of 15 are: 1, 3, 5, and 15

When discussing factors and primes, we just consider the whole numbers beginning with 1.

Divisibility Rules

Divisibility rules can be used to determine whether a number is divisible (can be divided exactly so there is no remainder) by another number. You can also use your calculator.

Use these rules to determine whether a number is divisible by 2, 3, 4, 5, 6, 8, 9, or 10.

1   Every number is divisible by 1.

2   Even numbers, which end in 0, 2, 4, 6, or 8, are divisible by 2.

3   If the sum of the digits is divisible by 3, then the number is divisible by 3.

Examples

2,481
sum of the digits:
2 + 4 + 8 + 1 = 15
15 is divisible by 3, so
2,481 is divisible by 3.

1,033
sum of the digits:
1 + 0 + 3 + 3 = 7
7 is not divisible by 3, so
1,033 is not divisible by 3.

4   If the last two digits are divisible by 4, then the number is divisible by 4.

Examples

26,347,4<u>64</u>
64 is divisible by 4, so
26,347,464 is divisible by 4.

45,344,6<u>94</u>
94 is not divisible by 4, so 45,344,694
is not divisible by 4.

5   If the last digit is 0 or 5, then the number is divisible by 5.

6   If a number is divisible by 2 and 3, then the number is divisible by 6.

7   Divisibility rule is more complex than just dividing by 7.

8   If the last three digits are divisible by 8, then the number is divisible by 8.

Examples

91,384,<u>656</u>
656 is divisible by 8, so
91,384,656 is divisible by 8.

56,400,<u>686</u>
686 is not divisible by 8, so
56,400,686 is not divisible by 8.

9   If the sum of the digits is divisible by 9, then the number is divisible by 9.

Examples

86,715
sum of the digits:
8 + 6 + 7 + 1 + 5 = 27
27 is divisible by 9, so
86,715 is divisible by 9.

93,163
sum of the digits:
9 + 3 + 1 + 6 + 3 = 22
22 is not divisible by 9, so
93,163 is not divisible by 9.

10   If a number ends in 0, then the number is divisible by 10.

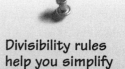

Divisibility rules help you simplify fractions and equations.

Prime and Composite Numbers

Factors of a number exactly divide that number.

A **prime number** has exactly two factors, 1 and itself.

A **composite number** has more than two factors.

Look at some numbers, starting with 1, to see if they are prime or composite.

1 has only one factor, itself. 1 is neither prime nor composite.

2 has exactly two factors, 1 and 2. 2 is a prime number, the only even prime.

3 has exactly two factors, 1 and 3. 3 is a prime number.

4 has more than two factors: 1, 2, and 4. 4 is a composite number.

5 has exactly two factors, 1 and 5. 5 is a prime number.

6 has more than two factors: 1, 2, 3, and 6. 6 is a composite number.

7 has exactly two factors, 1 and 7. 7 is a prime number.

8 has more than two factors: 1, 2, 4, and 8. 8 is a composite number.

9 has more than two factors: 1, 3, and 9. 9 is a composite number.

The prime numbers less than 30 are: 2, 3, 5, 7, 11, 13, 17, 19, 23, and 29.

MODEL ACT PROBLEM

Which of the following numbers is not a prime number?

A. 37
B. 43
C. 51
D. 67
E. 89

SOLUTION

A prime number has only itself and 1 as factors. Use divisibility rules to find that 51 is divisible by 3—3 is a factor of 51. Since 51 has more than two factors, it is not a prime number.

The correct answer is C.

Practice

Determine whether 1, 2, 3, 4, 5, 6, 7, 8, 9, and 10 are factors of each number.

1. 105   **2.** 111   **3.** 96   **4.** 176

List all the factors of each number.

5. 54   **6.** 67   **7.** 87   **8.** 120

Is the number prime or composite? Explain your answer.

9. 1   **10.** 205   **11.** 41   **12.** 111,111,111

13. 127   **14.** 123,123   **15.** 119   **16.** 675,201

(Answers on page 335)

ACT-TYPE PROBLEMS

1. What is the sum of all the factors of 20?

    A. 12
    B. 21
    C. 31
    D. 33
    E. 42

2. Which of the following is the product of the numbers from 1 to 9 that evenly divide 473,124?

    F. 144
    G. 192
    H. 1,152
    J. 8,064
    K. 10,368

3. All of the following choices are composite numbers EXCEPT:

- A. 17
- B. 18
- C. 81
- D. 116
- E. 117

5. Which of the following choices is a factor of 235,783,784?

- A. 3
- B. 4
- C. 5
- D. 6
- E. 7

4. What is the sum of the prime numbers between 30 and 40?

- F. 0
- G. 64
- H. 68
- J. 76
- K. 107

(Answers on page 336)

▪ Square Roots, Exponents, and Scientific Notation

Square Roots

The square root of a number multiplied by itself equals the number.

This symbol means the square root of 64: $\sqrt{64}$

The square root of 64 is 8 because $8 \times 8 = 64$.

A square root may be a whole number. The numbers with whole number square roots are called **perfect squares**.

$$\sqrt{1} = 1 \qquad \sqrt{4} = 2 \qquad \sqrt{9} = 3 \qquad \sqrt{16} = 4 \qquad \sqrt{25} = 5 \qquad \sqrt{36} = 6$$

$$\sqrt{49} = 7 \qquad \sqrt{64} = 8 \qquad \sqrt{81} = 9 \qquad \sqrt{100} = 10 \qquad \sqrt{121} = 11 \qquad \sqrt{144} = 12$$

You can also find the square root of a decimal.

$$\sqrt{1.44} = 1.2 \text{ because } 1.2 \times 1.2 = 1.44 \qquad \sqrt{1.69} = 1.3 \text{ because } 1.3 \times 1.3 = 1.69$$

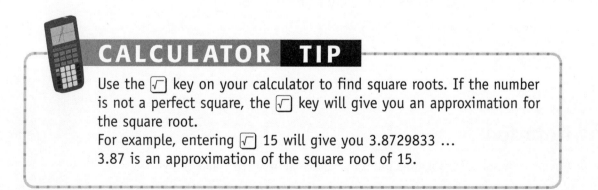

CALCULATOR TIP

Use the $\sqrt{\ }$ key on your calculator to find square roots. If the number is not a perfect square, the $\sqrt{\ }$ key will give you an approximation for the square root.

For example, entering $\sqrt{\ }$ 15 will give you 3.8729833 ...

3.87 is an approximation of the square root of 15.

Exponents

An exponent shows repeated factors.

$$\text{base} \rightarrow 3^{\overset{\displaystyle \text{exponent}}{\downarrow}4}$$

The **base** shows the factor. The **exponent** shows how many times the factor is repeated.

$$3^4 = 3 \times 3 \times 3 \times 3 = 81$$

The fractional exponent $x^{\frac{1}{2}}$ is another way of writing square root.

$$81^{\frac{1}{2}} = \sqrt{81} = 9$$

$$3.61^{\frac{1}{2}} = \sqrt{3.61} = 1.9$$

Any non-zero number raised to the zero power is equal to 1.

$$15^0 = 1$$

$$237^0 = 1$$

Negative exponents show fractions and decimals.

$$x^{-n} = \frac{1}{x^n} \ (x \neq 0)$$

$$11^{-1} = \frac{1}{11^1} = \frac{1}{11}$$

$$4^{-2} = \frac{1}{4^2} = \frac{1}{16}$$

$$10^{-1} = \frac{1}{10^1} = \frac{1}{10} = 0.1$$

$$10^{-2} = \frac{1}{10^2} = \frac{1}{100} = 0.01$$

$$10^{-3} = \frac{1}{10^3} = \frac{1}{1,000} = 0.001$$

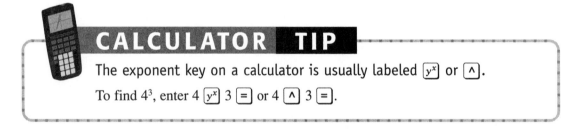

CALCULATOR TIP

The exponent key on a calculator is usually labeled $\boxed{y^x}$ or $\boxed{\wedge}$.

To find 4^3, enter 4 $\boxed{y^x}$ 3 $\boxed{=}$ or 4 $\boxed{\wedge}$ 3 $\boxed{=}$.

Scientific Notation

To write a positive number in scientific notation, multiply a decimal between 1 and 10 by a power of 10.

$6{,}728 = 6.728 \times 10^3$

Move the decimal point **three** places to the **left** and use 10^3.

$0.056 = 5.6 \times 10^{-2}$

Move the decimal point **two** places to the **right** and use 10^{-2}.

MODEL ACT PROBLEM

Light travels about 186,756 miles in a second. Which of the following shows about how far light travels in 100 seconds?

A. 1.8675600×10^{-8} miles
B. 1.86756×10^{-7} miles
C. 1.86756×10^7 miles
D. 1.86756×10^8 miles
E. 1.8675600×10^9 miles

SOLUTION

Calculate how far light travels in 100 seconds.

$186{,}756 \times 100 = 18{,}675{,}600$ miles

Write 18,675,600 in scientific notation.

Move the decimal point seven places left and use 10^7.

1.86756×10^7

The correct answer is C.

Practice

Write the value.

1. $\sqrt{400}$ 2. $\sqrt{0.25}$ 3. $\sqrt{324}$ 4. $\sqrt{2.89}$

5. 10^3 6. 4^3 7. 12^0 8. $49^{\frac{1}{2}}$

9. 2^{-1} 10. 6^{-2} 11. 10^{-3} 12. $16^{\frac{1}{2}}$

Write each number in scientific notation.

13. 2,325,000,000 14. 0.0175 15. 2.75 16. 5,659

17. 0.062 18. 50,790 19. 0.00257 20. 6,894,590

Write each as a whole number or decimal.

21. 4.2×10^5 22. 5.01×10^3 23. 2.35×10^{-2} 24. 6.6×10^2

25. 9.09×10^{-3} 26. 6.7×10^4 27. 4.05×10^{-4} 28. 1.9×10^4

(Answers on page 336)

1. The area of a square is 5.76 square inches. What is the length of each side of the square?

    A. 1.44
    B. 2.3
    C. 2.4
    D. 2.5
    E. 2.73

2. Which of the following numbers, multiplied by itself 4 times, equals 2,401?

    F. 6
    G. 7
    H. 8
    J. 9
    K. 10

3. The width of a virus is 0.0000001 meter. Which of the following correctly represents that measurement?

    A. 10^{-1} meters
    B. 10^{-5} meters
    C. 10^{-6} meters
    D. 10^{-7} meters
    E. 10^{-8} meters

(Answers on page 336)

4. Which of the following is the closest approximation of $15^{\frac{1}{2}}$?

    F. 3.6
    G. 3.7
    H. 3.8
    J. 3.9
    K. 4.0

5. It is about six billion kilometers from Earth to Pluto. Which of the following choices correctly represents that distance?

    A. 10^9 kilometers
    B. 6×10^9 kilometers
    C. 10^{10} kilometers
    D. 6×10^{10} kilometers
    E. 6^{10} kilometers

■ Fractions

Fractions are numerals that can name part of a whole. The **denominator** of a fraction shows how many equal parts or objects in all. The **numerator** shows how many parts are being discussed.

$$\frac{\text{numerator}}{\text{denominator}} \to \frac{7}{8}$$

The fraction $\frac{7}{8}$ can mean 7 out of 8 objects, 7 of 8 equal parts of a whole, or seven-eighths of the way from 0 to 1 on a number line. A stock price of $\frac{7}{8}$ means seven-eighths of a dollar.

Equivalent Fractions

The fraction $\frac{7}{8}$ is a name for a number. Other fractions that name the same number are called **equivalent fractions**. To find an equivalent fraction, multiply or divide the numerator and denominator by the same number.

A fraction can't have a denominator of 0. If a fraction turns up with a denominator of 0, we say it is undefined. For example, $\frac{7}{0}$ is undefined.

$$\frac{7 \times 5}{8 \times 5} = \frac{35}{40} \qquad \frac{7}{8} \text{ is equivalent to } \frac{35}{40}.$$

$$\frac{8 \div 2}{12 \div 2} = \frac{4}{6} \qquad \frac{8}{12} \text{ is equivalent to } \frac{4}{6}.$$

Simplest Form

A fraction is in **simplest form** when the numerator and denominator have no common factors greater than 1.

$\frac{3}{8}$ is in simplest form. No number greater than 1 divides both 3 and 8 exactly.

$\frac{14}{35}$ is not in simplest form. The numerator and the denominator are both divisible by 7.

$\frac{14 \div 7}{35 \div 7} = \frac{2}{5}. \frac{2}{5}$ is in simplest form.

CALCULATOR TIP

Use a calculator that can represent fractions and mixed numbers. The calculator can also convert fractions to simplest form and convert improper fractions to mixed numbers.

Comparing Fractions

Use the terms equivalent to ($a = b$), less than ($a < b$), greater than ($a > b$), and between ($a < b < c$) to compare fractions. If two fractions have the same denominator, the fraction with the greater numerator is greater.

$$\frac{4}{8} = \frac{4}{8} \qquad \frac{1}{8} < \frac{3}{8} \qquad \frac{5}{8} > \frac{4}{8} \qquad \frac{3}{8} < \frac{6}{8} < \frac{7}{8}$$

EXAMPLE

Compare $\frac{3}{8}$ and $\frac{5}{12}$.

You can always cross multiply to compare fractions.
The greater cross product appears next to the greater fraction.

$$^{36}\frac{3}{8} \diagdown\mathllap{\diagup} \frac{5}{12}^{\,40}$$

$$\frac{3}{8} < \frac{5}{12}$$

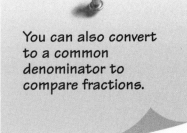

You can also convert to a common denominator to compare fractions.

Convert fractions to decimals for easy comparison. A calculator that can represent fractions will convert quickly between fractions and decimals. If you do not have this kind of calculator, divide the numerator of the fraction by the denominator to change it to a decimal.

MODEL ACT PROBLEM

In lowest terms, what is the difference between the largest and smallest fraction listed below?

$$\frac{1}{2}, \frac{3}{5}, \frac{4}{7}, \frac{2}{3}$$

A. $\frac{2}{21}$

B. $\frac{1}{8}$

C. $\frac{1}{6}$

D. $\frac{9}{14}$

E. $\frac{41}{42}$

SOLUTION

Use a calculator for this problem.

Enter:

1 ÷ 2 =	0.5
3 ÷ 5 =	0.6
4 ÷ 7 =	0.5714285714
2 ÷ 3 =	0.6666666667

The largest fraction is $\frac{2}{3} = 0.\overline{6}$ and the smallest fraction is $\frac{1}{2} = 0.5$. Subtract the smallest from the largest. Entering 2 ÷ 3 − 1 ÷ 2 = gives $0.1\overline{6}$ or $\frac{1}{6}$. The answer is C.

Practice

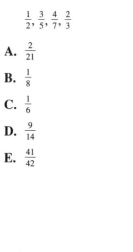

Write True if the fractions are equivalent and False if they are not equivalent.

1. $\frac{4}{7}$ is equivalent to $\frac{12}{21}$

2. $\frac{3}{5}$ is equivalent to $\frac{6}{15}$

3. $\frac{5}{9}$ is equivalent to $\frac{20}{36}$

4. $\frac{2}{3}$ is equivalent to $\frac{12}{18}$

5. $\frac{1}{5}$ is equivalent to $\frac{4}{25}$

6. $\frac{4}{9}$ is equivalent to $\frac{20}{45}$

7. $\frac{3}{4}$ is equivalent to $\frac{12}{18}$

8. $\frac{5}{7}$ is equivalent to $\frac{35}{49}$

Write each fraction in simplest form.

9. $\frac{27}{81}$

10. $\frac{8}{12}$

Fill in the blank with $<$, $>$, or $=$.

11. $\dfrac{3}{5}$ —— $\dfrac{4}{7}$

12. $\dfrac{2}{9}$ —— $\dfrac{5}{12}$

13. $\dfrac{2}{3}$ —— $\dfrac{1}{2}$ —— $\dfrac{3}{8}$

14. $\dfrac{3}{4}$ —— $\dfrac{6}{7}$ —— $\dfrac{7}{8}$

Write in order from least to greatest.

15. $\dfrac{7}{8}, \dfrac{5}{6}$

16. $\dfrac{3}{5}, \dfrac{61}{100}$

17. $\dfrac{2}{7}, \dfrac{4}{9}$

18. $\dfrac{3}{5}, \dfrac{11}{20}, \dfrac{4}{7}$

19. $\dfrac{3}{4}, \dfrac{5}{8}, \dfrac{7}{10}$

20. $\dfrac{4}{5}, \dfrac{1}{2}, \dfrac{7}{16}, \dfrac{5}{9}$

(Answers on page 336)

ACT-TYPE PROBLEMS

1. All of the following fractions are equivalent to $\dfrac{4}{7}$ EXCEPT:

   A. $\dfrac{8}{14}$

   B. $\dfrac{12}{35}$

   C. $\dfrac{16}{28}$

   D. $\dfrac{36}{63}$

   E. $\dfrac{24}{42}$

2. Which of the following choices correctly lists the fractions $\{\dfrac{3}{5}, \dfrac{5}{6}, \dfrac{3}{10}, \dfrac{1}{3}, \dfrac{8}{15}\}$ from least to greatest?

   F. $\dfrac{5}{6}, \dfrac{3}{10}, \dfrac{1}{3}, \dfrac{8}{15}, \dfrac{3}{5}$

   G. $\dfrac{5}{6}, \dfrac{3}{5}, \dfrac{3}{10}, \dfrac{8}{15}, \dfrac{1}{3}$

   H. $\dfrac{3}{10}, \dfrac{1}{3}, \dfrac{8}{15}, \dfrac{3}{5}, \dfrac{5}{6}$

   J. $\dfrac{3}{5}, \dfrac{5}{6}, \dfrac{3}{10}, \dfrac{1}{3}, \dfrac{8}{15}$

   K. $\dfrac{3}{5}, \dfrac{3}{10}, \dfrac{8}{15}, \dfrac{1}{3}, \dfrac{5}{6}$

3. For which of the following choices would the symbol $=$ in the blank make the statement true?

   A. $\dfrac{2}{5}$ —— $\dfrac{4}{15}$

   B. $\dfrac{7}{13}$ —— $\dfrac{21}{39}$

   C. $\dfrac{5}{9}$ —— $\dfrac{20}{27}$

   D. $\dfrac{3}{7}$ —— $\dfrac{12}{21}$

   E. $\dfrac{5}{12}$ —— $\dfrac{50}{1,200}$

4. Joel has 5 different kinds of salad to choose from. He has \$2 to spend. Which of the 5 salads listed below should Joel choose to get the most salad for his money?

Salad	Amount for \$2
Mediterranean salad	$\dfrac{1}{2}$ pound
Caesar salad	$\dfrac{2}{3}$ pound
Garden salad	$\dfrac{5}{7}$ pound
Oriental salad	$\dfrac{4}{9}$ pound
Potato salad	$\dfrac{3}{8}$ pound

   F. Mediterranean salad
   G. Caesar salad
   H. Garden salad
   J. Oriental salad
   K. Potato salad

5. Five college students, Brad, Alex, Scott, Jim, and Tony, live in the same dorm. The following is a list of the distances each walks to class. Which student walks the farthest?

Name	Distance
Brad	$\dfrac{7}{15}$ of a mile
Alex	$\dfrac{8}{13}$ of a mile
Scott	$\dfrac{9}{17}$ of a mile
Jim	$\dfrac{4}{9}$ of a mile
Tony	$\dfrac{5}{12}$ of a mile

   A. Brad
   B. Alex
   C. Scott
   D. Jim
   E. Tony

(Answers on page 337)

Addition and Subtraction of Fractions and Mixed Numbers

Follow these steps to add or subtract fractions and mixed numbers.

- Estimate first.
- Compute.
- Remember to write the answer in simplest form.
- Check your answer against the estimate to be sure your answer is reasonable.

Sometimes an estimate may be enough to answer the question.

A **mixed number** has a whole number part and a fraction part.

$2\frac{3}{4}$ is a mixed number.

CALCULATOR TIP

On a calculator that can represent mixed numbers, estimate first, then complete the keystrokes to show the answer in simplest form.

If you do not have this kind of calculator, express the mixed fraction as the sum of the integer part and the fractional part. For example, $7\frac{1}{5} = 7 + \frac{1}{5}$.

Addition

EXAMPLES

1. $\frac{2}{3} + \frac{3}{5}$

   Estimate first.

   Both fractions are closer to 1 than to 0.

   The answer should be between 1 and 2.

Use common denominators.	Add the numerators.	Write in simplest form. Is the answer reasonable?
$\frac{2 \times 5}{3 \times 5} = \frac{10}{15}$	$\frac{10}{15}$	
$\frac{3 \times 3}{5 \times 3} = \frac{9}{15}$	$+\frac{9}{15}$	$\frac{19}{15} = 1\frac{4}{15}$
	$\frac{19}{15}$	$1\frac{4}{15}$ is between 1 and 2.
		The answer is reasonable.

2. $1\frac{3}{4} + 7\frac{1}{3}$

Estimate first.

One fraction is less than a half and the other is more than a half.

The sum of the whole numbers is 8.

The answer should be close to 9.

Use common denominators.	Add the fractions and the whole numbers.	Write in simplest form. Is the answer reasonable?
$1\frac{3}{4} = 1\frac{9}{12}$	$1\frac{9}{12}$	
$7\frac{1}{3} = 7\frac{4}{12}$	$+ 7\frac{4}{12}$	
	$8\frac{13}{12}$	$8\frac{13}{12} = 8 + 1\frac{1}{12} = 9\frac{1}{12}$

The answer is reasonable.
It is close to 9.

Subtraction

1. $14\frac{2}{3} - 3\frac{1}{7}$

Estimate first.

The fractional part of the difference is a little less than $\frac{2}{3}$.

The difference between the whole numbers is 11.

The answer should be between 11 and $11\frac{2}{3}$.

Use common denominators.	Subtract the numerators and the whole numbers.	Write in simplest form. Is the answer reasonable?
$14\frac{2}{3} = 14\frac{14}{21}$	$14\frac{14}{21}$	
$3\frac{1}{7} = 3\frac{3}{21}$	$- 3\frac{3}{21}$	
	$11\frac{11}{21}$	The answer is in simplest form. The answer is reasonable.

It is between 11 and $11\frac{2}{3}$.

2. $5\frac{1}{5} - 2\frac{5}{6}$

Estimate first.

$\frac{5}{6}$ is larger than $\frac{1}{5}$ so you will have to borrow from 5 to subtract.

The difference between the fractions will be less than 1.

Borrowing 1 from 5, you are left with 4. The difference between 4 and 2 is 2.

The answer should be between 2 and 3.

| Write fractions with common denominators, then rename again. | Subtract. | Write in simplest form. Is the answer reasonable? |

$5\frac{1}{5} = 5\frac{6}{30} = 4\frac{36}{30}$

$2\frac{5}{6} = 2\frac{25}{30} = 2\frac{25}{30}$

$$\begin{array}{r} 4\frac{36}{30} \\ -2\frac{25}{30} \\ \hline 2\frac{11}{30} \end{array}$$

The answer is in simplest form.
The answer is reasonable.
It is between 2 and 3.

MODEL ACT PROBLEM

The following table shows how far Frank rode his bike this week.

Day	Mileage
Monday	$2\frac{3}{4}$ miles
Wednesday	$\frac{7}{8}$ mile
Thursday	$\frac{3}{5}$ mile
Friday	$3\frac{1}{2}$ miles

What is the total number of miles that Frank rode on Monday, Wednesday, and Friday?

A. $6\frac{3}{8}$

B. $6\frac{1}{2}$

C. $7\frac{1}{8}$

D. $7\frac{1}{2}$

E. 8

SOLUTION

Use only the distances for Monday $\left(2\frac{3}{4}\right)$, Wednesday $\left(\frac{7}{8}\right)$, and Friday $\left(3\frac{1}{2}\right)$.

Use your calculator to add the fractions. Or write the fractions with common denominators and add.

$2\frac{3}{4} = 2\frac{6}{8}, \qquad \frac{7}{8}, \qquad 3\frac{1}{2} = 3\frac{4}{8}$

Add the fraction parts and then add the whole number parts.

$$\begin{array}{r} 2\frac{6}{8} \\ \frac{7}{8} \\ +3\frac{4}{8} \\ \hline 5\frac{17}{8} = 5 + 2\frac{1}{8} = 7\frac{1}{8} \end{array}$$

The correct answer is C.

Practice

Estimate first. Then compute.

Add.

1. $\dfrac{3}{7} + \dfrac{2}{3}$

2. $\dfrac{4}{9} + \dfrac{5}{18}$

3. $\dfrac{6}{13} + \dfrac{1}{3}$

4. $\dfrac{11}{28} + \dfrac{6}{7}$

5. $\dfrac{5}{8} + \dfrac{2}{3}$

6. $\dfrac{1}{2} + \dfrac{1}{6} + \dfrac{5}{8}$

7. $\dfrac{3}{4} + \dfrac{1}{5} + \dfrac{7}{10}$

8. $\dfrac{2}{15} + \dfrac{2}{3} + \dfrac{1}{5}$

9. $8\dfrac{1}{3} + 6\dfrac{3}{7}$

10. $6\dfrac{2}{3} + 9\dfrac{5}{7}$

11. $14\dfrac{7}{8} + 6\dfrac{3}{4}$

12. $27\dfrac{5}{6} + 14\dfrac{3}{4}$

Subtract.

13. $\dfrac{9}{17} - \dfrac{1}{2}$

14. $\dfrac{12}{19} - \dfrac{7}{18}$

15. $\dfrac{4}{7} - \dfrac{3}{10}$

16. $\dfrac{8}{15} - \dfrac{2}{5}$

17. $\dfrac{2}{3} - \dfrac{1}{10}$

18. $4\dfrac{7}{8} - 2\dfrac{5}{6}$

19. $25 - 8\dfrac{4}{5}$

20. $6\dfrac{5}{8} - 2\dfrac{2}{3}$

21. $11\dfrac{3}{5} - 6\dfrac{3}{4}$

22. $10\dfrac{4}{5} - 3\dfrac{1}{3}$

23. $46 - 31\dfrac{3}{4}$

24. $21\dfrac{3}{8} - 16\dfrac{5}{6}$

(Answers on page 337)

ACT-TYPE PROBLEMS

1. Fast and Thrifty Shipping Company charges \$6 a pound for packages. A company has a shipment of three packages weighing $\dfrac{3}{5}$ pound, $1\dfrac{4}{15}$ pounds, and $\dfrac{23}{30}$ pound. How much will it cost to ship these packages by Fast and Thrifty?

   A. \$15.00
   B. \$15.80
   C. \$16.00
   D. \$16.20
   E. \$16.60

2. Jill bought some fruit at the grocery store. She bought $2\dfrac{1}{2}$ pounds of apples, $\dfrac{3}{4}$ pound of bananas, $1\dfrac{2}{5}$ pounds of peaches, and $5\dfrac{7}{10}$ pounds of watermelon. What is the total number of pounds of fruit that Jill bought at the store?

   F. 9
   G. $9\dfrac{1}{2}$
   H. $10\dfrac{5}{8}$
   J. $10\dfrac{7}{20}$
   K. $11\dfrac{13}{30}$

3. Alice needs a rope at least 8 feet long. She has four smaller ropes of the following lengths:

   I. $4\dfrac{2}{3}$ feet

   II. $4\dfrac{2}{5}$ feet

   III. $3\dfrac{3}{4}$ feet

   IV. $3\dfrac{2}{3}$ feet

   Alice can tie the ropes together. However, when she does so she loses a total of $\dfrac{1}{4}$ of a foot. Which of the following combinations of rope will NOT give Alice a piece of rope at least 8 feet long?

   A. I and II
   B. I and III
   C. II and III
   D. I and IV
   E. I, II, and III

4. All of the following equal 1 EXCEPT:

  F. $\frac{1}{2} + \frac{3}{12} + \frac{9}{36}$

  G. $\frac{2}{3} + \frac{2}{6}$

  H. $\frac{3}{8} + \frac{20}{24} - \frac{1}{6}$

  J. $\frac{13}{7} - \frac{12}{14}$

  K. $\frac{3}{9} + \frac{1}{3} + \frac{4}{12}$

5. Rich, Brian, Jack, and Chad ordered a pizza that is cut into eight slices of the same size. Rich ate two slices, Brian ate one slice, and Jack ate three slices. What fraction of the pie was left for Chad to eat?

  A. $\frac{3}{8}$

  B. $\frac{1}{2}$

  C. $\frac{1}{8}$

  D. $\frac{1}{4}$

  E. $\frac{3}{4}$

(Answers on page 337)

Multiplication and Division of Fractions and Mixed Numbers

Follow these steps to multiply or divide fractions and mixed numbers.

- Estimate first.
- Compute.
- Remember to write the answer in simplest form.
- Check your answer against the estimate to be sure your answer is reasonable.

Sometimes an estimate may be enough to answer the question.

You use multiplication to solve both multiplication and division problems.

Multiplication

EXAMPLES

1. $\frac{2}{3} \times \frac{3}{8}$

Estimate first.

Both fractions are less than 1.

The answer will be less than either of the fractions.

Multiply numerators.
Multiply denominators.

$\frac{2}{3} \times \frac{3}{8} = \frac{6}{24}$

Write the answer in simplest form.

$\frac{6}{24} = \frac{1}{4}$

Check to be sure your answer is reasonable.

$\frac{1}{4}$ is less than and $\frac{2}{3}$ and $\frac{3}{8}$, so the answer is reasonable.

2. $1\frac{1}{5} \times 2\frac{7}{8}$

Estimate first.

You are multiplying a little more than 1 by almost 3.

The answer will be between 3 and 4.

Write mixed numbers as fractions.	Multiply numerators. Multiply denominators.	Write in simplest form. Is the answer reasonable?
$1\frac{1}{5} = \frac{6}{5}$	$\frac{6}{5} \times \frac{23}{8} = \frac{138}{40}$	$\frac{138}{40} = 3\frac{18}{40} = 3\frac{9}{20}$
$2\frac{7}{8} = \frac{23}{8}$		$3\frac{9}{20}$ is between 3 and 4.
		The answer is reasonable.

Division

To divide fractions, invert the divisor and multiply.

E X A M P L E S

1. $\frac{3}{4} \div \frac{1}{3}$

Estimate first.

Think: Invert $\frac{1}{3}$ to get 3. You are multiplying a little less than 1 by 3. The answer should be between 2 and 3.

Invert the divisor.	Multiply.	Write in simplest form. Is the answer reasonable?
$\frac{3}{4} \div \frac{1}{3} = \frac{3}{4} \times \frac{3}{1}$	$\frac{3}{4} \times \frac{3}{1} = \frac{9}{4}$	$\frac{9}{4} = 2\frac{1}{4}$
		$2\frac{1}{4}$ is between 2 and 3, so the answer is reasonable.

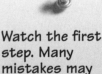

Watch the first step. Many mistakes may occur when you invert the divisor.

2. $6\frac{1}{2} \div 1\frac{3}{4}$

Estimate first.

$6\frac{1}{2}$ is near 6 and $1\frac{3}{4}$ is near 2.

$6 \div 2 = 3$. The answer should be near 3.

Write mixed numbers as fractions.	Invert the divisor.	Multiply.	Write in simplest form. Is the answer reasonable?
$6\frac{1}{2} = \frac{13}{2}$	$\frac{13}{2} \div \frac{7}{4} = \frac{13}{2} \times \frac{4}{7}$	$\frac{13}{2} \times \frac{4}{7} = \frac{52}{14}$	$\frac{52}{14} = 3\frac{10}{14} = 3\frac{5}{7}$
$1\frac{3}{4} = \frac{7}{4}$			$3\frac{5}{7}$ is close to 3, so the answer is reasonable.

Which of the following statements is *false* about (a) $\frac{1}{2}$ and (b) $\frac{1}{2}$?

A. The product of (a) and (b) is less than (a).

B. The sum of (a) and (b) is greater than the product of (a) and (b).

C. (a) divided by (b) is equal to the sum of (a) and (b).

D. (a) minus (b) is less than 0.

E. The product of (a) and (b) is less than (a) divided by (b).

SOLUTION

Work out each choice.

A. $\frac{1}{2} \times \frac{1}{2} = \frac{1}{4}$ True

B. $\frac{1}{2} + \frac{1}{2} = 1$ and 1 is greater than $\frac{1}{4}$.   True

C. $\frac{1}{2} \div \frac{1}{2} = 1$ and $\frac{1}{2} + \frac{1}{2} = 1$. 1 equals 1.   True

D. $\frac{1}{2} - \frac{1}{2} = 0$ and 0 is NOT less than 0.   False

E. $\frac{1}{2} \times \frac{1}{2} = \frac{1}{4}$ and $\frac{1}{2} \div \frac{1}{2} = 1$. $\frac{1}{4} < 1$.   True

The correct answer is D.

Practice

Multiply.

1. $\frac{2}{3} \times \frac{1}{5}$

2. $\frac{7}{8} \times \frac{4}{9}$

3. $15 \times \frac{4}{5}$

4. $3\frac{3}{4} \times \frac{5}{8}$

5. $5\frac{1}{8} \times \frac{24}{25} \times 10\frac{1}{2}$

6. $4\frac{2}{7} \times 5\frac{1}{6}$

7. $3\frac{1}{8} \times 4\frac{1}{5}$

8. $2\frac{1}{4} \times 6\frac{1}{2} \times \frac{12}{39}$

Divide.

9. $\frac{2}{3} \div \frac{3}{5}$

10. $\frac{3}{4} \div \frac{3}{8}$

11. $\frac{5}{6} \div 10$

12. $\frac{8}{\frac{1}{8}}$

13. $\frac{\frac{2}{3}}{\frac{3}{8}}$

14. $5\frac{1}{3} \div 2\frac{2}{3}$

15. $7\frac{3}{5} \div 6\frac{1}{2}$

16. $5\frac{1}{3} \div 4\frac{1}{6}$

17. $\frac{2}{3} \times \frac{1}{2} \div 3$

18. $\frac{4}{5} \div \frac{3}{10} \times \frac{1}{2}$

19. $\frac{11}{15} \times 10 \div 2 \times \frac{1}{3}$

20. $\frac{4}{7} \times \frac{28}{3} \div 4$

(Answers on page 338)

1. Bret runs $\frac{2}{3}$ of a mile. Tony runs half as far as Bret. How many miles does Tony run?

   A. $1\frac{1}{2}$

   B. $1\frac{1}{3}$

   C. $1\frac{1}{6}$

   D. $\frac{1}{3}$

   E. $\frac{1}{6}$

2. Jim bench-presses 225 pounds and does arm curls with a weight $\frac{1}{3}$ the amount he bench-presses. What weight does he arm-curl?

   F. 65 pounds
   G. 70 pounds
   H. 75 pounds
   J. 80 pounds
   K. 85 pounds

3. $\frac{2}{5} + \left(\frac{25}{9} \div \frac{1}{3}\right) = ?$

   A. $\frac{3}{10}$

   B. $\frac{13}{25}$

   C. $3\frac{1}{3}$

   D. $8\frac{11}{15}$

   E. $10\frac{1}{5}$

4. A college math professor has to grade 160 exams. He gives $\frac{1}{2}$ of the exams to his assistant to grade. Then the assistant gives $\frac{1}{4}$ of her exams to a student worker to grade. How many exams does the student worker get to grade?

   F. 20 exams
   G. 40 exams
   H. 60 exams
   J. 70 exams
   K. 80 exams

5. $\frac{5}{7} \div \frac{15}{3} \times \frac{11}{13} = ?$

   A. $\frac{12}{51}$

   B. $\frac{11}{91}$

   C. $\frac{5}{71}$

   D. $\frac{23}{57}$

   E. $\frac{17}{73}$

(Answers on page 338)

▪ Positive and Negative Numbers

You can visualize positive and negative numbers on a number line. A plus sign (+) before a number means that the number is to the right of zero. A negative sign (−) before a number means that the number is to the left of zero.

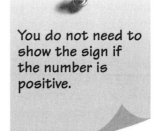

You do not need to show the sign if the number is positive.

Write positive fifteen.   15   Write negative eleven.   −11

Absolute Value

The absolute value of a number is positive and is the distance from zero to the number. The symbol for the absolute value of a is $|a|$.

EXAMPLES

$|5| = 5$ $|-5| = 5$ $|0| = 0$ $|19 - 37| = |-18| = 18$

Comparing Positive and Negative Numbers

Integers, decimals, and fractions can be positive or negative. Use these rules to compare positive and negative numbers.

- If the signs of the numbers are different, the positive number is greater.

$$\frac{1}{16} > -28 \qquad -53 < 7.8 \qquad 10.3 > -807$$

- If both signs are positive, compare the numbers.

$$203.3 > 203.198 \qquad \frac{1}{2} < \frac{3}{4} \qquad 71 > 65\frac{1}{2}$$

- If both signs are negative, the larger numeral represents the smaller number.

$$-908 < -8 \qquad -\frac{1}{7} > -\frac{7}{8} \qquad -108.53 < 6.01$$

MODEL ACT PROBLEM

Which of the following statements is false?

A. $-123.45 > -90.7$
B. $3 < 6.7$
C. $-34 < 2$
D. $17 > -600$
E. $17.89 < 17.9$

SOLUTION

In choice A, both signs are negative. Therefore, the larger numeral represents the smaller number.
$-123.45 < -90.7$

The correct answer is A.

Practice

Find the absolute value.

1. $|-8|$
2. $|58 - 41|$
3. $|13 + 12|$
4. $|17|$

5. $|41 - 7|$
6. $|-54|$
7. $|13.5 - 13|$
8. $|39,000|$

Write $<$ or $>$ in the blank.

9. -8 ____ 3
10. 54.6 ____ 39.897
11. $-\frac{1}{6}$ ____ $-\frac{5}{6}$

12. $7\frac{1}{4}$ ____ $5\frac{8}{9}$
13. -20.876 ____ -20.8759
14. $-\frac{7}{16}$ ____ $-\frac{37}{79}$

15. -15 ____ -17
16. -0.05 ____ 0.05
17. $|-45|$ ____ 46

18. -20 ____ $|-19|$
19. $|74 - 32|$ ____ 33
20. 65 ____ $|-64|$

(Answers on page 338)

1. The symbol $>$ could be used to make all of the following true EXCEPT:

   A. 45 ____ -56
   B. $|-35|$ ____ -36
   C. -34 ____ 2
   D. $|-15|$ ____ 4
   E. 5 ____ -6

2. Given the following:

   I. $|-3| < 2$
   II. $-6 > -1$
   III. $14 > -16$

   Which of the following choices contain all true statements?

   F. I
   G. III
   H. I and III
   J. II and III
   K. I, II, and III

3. $5 + |-3| - 8 + 3 = ?$

   A. -3
   B. -1
   C. 0
   D. 1
   E. 3

(Answers on page 338)

4. Which of the following choices lists the numbers $\{-3, |-9|, 4, -5, |-10|\}$ in order from least to greatest?

   F. $-3, 4, -5, |-9|, |-10|$
   G. $4, -3, -5, |-10|, |-9|$
   H. $|-10|, |-9|, -5, -3, 4$
   J. $-5, -3, 4, |-9|, |-10|$
   K. $-5, |-9|, |-10|, -3, 4$

5. Which of the following is the largest number?

   A. -110
   B. $|-120|$
   C. 105
   D. -200
   E. $|105|$

■ Computation With Positive and Negative Numbers

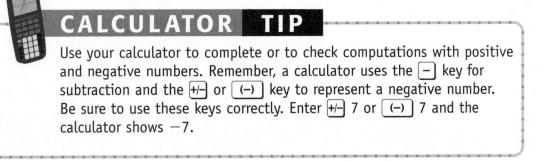

CALCULATOR TIP

Use your calculator to complete or to check computations with positive and negative numbers. Remember, a calculator uses the $\boxed{-}$ key for subtraction and the $\boxed{+/-}$ or $\boxed{(-)}$ key to represent a negative number. Be sure to use these keys correctly. Enter $\boxed{+/-}$ 7 or $\boxed{(-)}$ 7 and the calculator shows -7.

Follow these rules to work with positive and negative numbers.

Addition

- If the signs are the same (both positive or both negative), add and keep the same sign in the sum.
- If the signs are different (positive and negative), disregard the signs and subtract. Use the sign of the number with the greater absolute value in the sum.

EXAMPLES

Same Sign

1.
$$\begin{array}{r} 23.867 \\ +173.4 \\ \hline 197.267 \end{array}$$
Both numbers are positive so the answer is positive.

Add the numbers.

2.
$$\begin{array}{r} -4\frac{1}{4} \\ +(-2\frac{3}{8}) \\ \hline -6\frac{5}{8} \end{array}$$
Both signs are negative so the sum is negative.

Add the numbers.

Different Signs

3.
$$\begin{array}{r} 2,380 \\ +(-967) \\ \hline 1,413 \end{array}$$
The signs are different.
Subtract: $2,380 - 967 = 1,413$. 2,380 has the greater absolute value.
The sum is positive.

4.
$$\begin{array}{r} 108.7 \\ +(-218.9) \\ \hline -110.2 \end{array}$$
The signs are different.
Subtract: $218.9 - 108.7 = 110.2$. 218.9 has the greater absolute value.
The sum is negative.

Subtraction

- Change the sign of the number being subtracted, then add.

EXAMPLES

1. $1,652 - 1,824$

Write the problem. Both numbers are positive.	Change the sign of the number being subtracted and add.	Use the rules for adding numbers with different signs.
$\begin{array}{r} 1,652 \\ -1,824 \\ \hline \end{array}$	$\begin{array}{r} 1,652 \\ +(-1,824) \\ \hline \end{array}$	$1,824 - 1,652 = 172$ The answer is negative. $1,652 - 1,824 = -172$

2. $-6\frac{1}{3} - \left(-2\frac{5}{6}\right)$

Write the problem. Both numbers are negative.

$$-6\frac{1}{3}$$

$$-\left(-2\frac{5}{6}\right)$$

Change the sign of the number being subtracted and add.

$$-6\frac{1}{3}$$

$$+2\frac{5}{6}$$

Use the rules for adding numbers with different signs.

$$6\frac{1}{3} - 2\frac{5}{6} = 3\frac{3}{6} = 3\frac{1}{2}$$

The answer is negative.

$$-6\frac{1}{3} - \left(-2\frac{5}{6}\right) = -3\frac{1}{2}$$

3. $1,408 - (-2,318)$

Write the problem. The signs are the different.

$$1,408$$
$$-(-2,318)$$

Change the sign of the number being subtracted and add.

$$1,408$$
$$+2,318$$

Use the rules for adding numbers with the same sign.

$1,408 + 2,318 = 3,726$
The answer is positive.
$1,408 - (-2,318) = 3,726$

Multiplication

- If the signs are the same (both positive or both negative), multiply and the product is positive.
- If the signs are different (one positive and one negative), multiply and the product is negative.

EXAMPLES

Same Sign

1. $\frac{1}{3} \times \frac{4}{5} = \frac{4}{15}$ Both numbers are positive so the answer is positive.

2. $-4 \times (-16) = 64$ Both numbers are negative so the answer is positive.

Different Signs

3. $1.83 \times (-2.06) = -3.7698$ The numbers have different signs so the answer is negative.

4. $-34 \times 904 = -30,736$ The numbers have different signs so the answer is negative.

Division

- If the signs are the same (both positive or both negative), divide and the quotient is positive.
- If the signs are different (one positive and one negative), divide and the quotient is negative.

Same Sign

1. $18 \div 4 = 4.5$ Both numbers are positive so the answer is positive

2. $-\dfrac{3}{4} \div \left(-\dfrac{1}{8}\right) = 6$ Both numbers are negative so the answer is positive.

Different Signs

3. $1.8 \div (-30) = -0.06$ The numbers have different signs so the answer is negative.

4. $-24{,}745 \div 707 = -35$ The numbers have different signs so the answer is negative.

MODEL ACT PROBLEMS

1. Alice borrowed $85 from her roommate. When she got her paycheck for $225 she paid back the loan and paid her $70 telephone bill. How much money does Alice have left from her paycheck?

    A. $50
    B. $60
    C. $70
    D. $80
    E. $90

SOLUTION

Subtract the amounts Alice owed from the amount of her paycheck.

    $225 - $85 - $70 = $70

The correct answer is C.

2. $-4 \times 8 \div (-2) \times 6 \div (-3) = ?$

    F. 32
    G. -32
    H. 64
    J. -64
    K. -78

SOLUTION

$$-4 \times 8 \div (-2) \times 6 \div (-3)$$
$$= -32 \div (-2) \times 6 \div (-3)$$
$$= 16 \times 6 \div (-3) = 96 \div (-3) = -32$$

The correct answer is G.

Practice

Compute. You can use your calculator.

1. $-52 + (-74)$ **2.** $24 + (-10)$ **3.** $-12 - (-9)$

4. $-17\dfrac{1}{5} - 6\dfrac{2}{5}$ **5.** -61×23 **6.** 4.2×-4.5

7. -71×-8 **8.** $-4\dfrac{1}{4} \times -2\dfrac{3}{8}$ **9.** $72 \div -4$

10. $-108 \div -6$ **11.** $-30 \div \dfrac{1}{6}$ **12.** $-5.85 \div 0.9$

13. $43.4 - (-136)$ **14.** $-2.25 + 2.25$ **15.** $-1.4(3.4)$

16. $-63.7 \div -0.007$ **17.** $4 \times (-2) + 3$ **18.** $(-12 \div 4) - (3 \times 2)$

19. $5 - 2 + (7 \times 5)$ **20.** $(13 \times 2) - [6 \div (-2)]$

(Answers on page 338)

1. $1,364 - 1,200 - 64 + 100 = ?$

   A. 100
   B. 150
   C. 200
   D. 250
   E. 300

2. All of the choices below equal -5 EXCEPT:

   F. $-5 + (8 \times 2) - (2 \times 8)$
   G. $3 + [2 \times (-4)]$
   H. $3 + [5 \times (-2)] + 8$
   J. $11 - 8 + [16 \div (-2)]$
   K. $[17 \times (-2)] + (14 \times 2) + 1$

3. Lou worked for 16 hours earning \$7 an hour, and worked 10 hours earning \$9 an hour. How much did Lou earn?

   A. \$189
   B. \$202
   C. \$208
   D. \$214
   E. \$222

4. $(-5 \times 2) + [6 \div (-2)] + 5 = ?$

   F. 10
   G. 9
   H. 8
   J. -8
   K. -9

5. $(2.5 \times 14) - 30 + [2 \times 5 \div (-2)] = ?$

   A. 2
   B. 1
   C. 0
   D. -1
   E. -2

(Answers on page 339)

▪ Order of Operations

When doing computation, you must do arithmetic operations in a particular order.

Follow this order of operations:

P • Parentheses. All operations inside parentheses are done first using the order of operations shown below.

E • Exponents. Raise numbers to any given powers before going on.

MD • Multiply or divide. Complete all multiplication and division calculations as you come to them going from left to right.

AS • Add or subtract. Complete all addition and subtraction calculations as you come to them going from left to right.

The sentence "Please Excuse My Dear Aunt Sally" stands for Parentheses, Exponents, Multiply & Divide, Add & Subtract.

CALCULATOR TIP

You should use a calculator that incorporates the correct order of operations; check to see whether or not yours does. Be sure that you enter any parentheses that appear in the expression and wait to press ENTER or = until the entire expression has been input.

1. Compute. $3 - 4 \times 6 + 8 \div 4$

$3 - 4 \times 6 + 8 \div 4$	Multiply and divide from left to right.
$= 3 - 24 + 2$	Add and subtract from left to right.
$= -19$	The answer is -19.

2. Compute. $17 - 4 \times 6^2$

$17 - 4 \times 6^2$	Evaluate exponents first.
$= 17 - 4 \times 36$	Multiply.
$= 17 - 144$	Subtract.
$= -127$	The answer is -127.

3. Compute. $(3 + 4) \times 6$

$(3 + 4) \times 6$	Work inside the parentheses first.
$= 7 \times 6$	Multiply.
$= 42$	The answer is 42.

4. Compute. $3 + 5 \times 4$

$3 + 5 \times 4$	Multiply first.
$= 3 + 20$	Then add.
$= 23$	The answer is 23.

5. Compute. $2^3(5 + 7 \times 4)$

$2^3(5 + 7 \times 4)$	Parentheses first. Use the order of operations. Multiply.
$= 2^3(5 + 28)$	Add.
$= 2^3 \times 33$	Evaluate the exponent.
$= 8 \times 33$	Multiply.
$= 264$	The answer is 264.

MODEL ACT PROBLEM

Compute. $3 - (16 \div 2^3 + 6 \times 3 - 8) + 3 \times 4$

A. -48
B. -40
C. -24
D. 3
E. 48

SOLUTION

$3 - (16 \div 2^3 + 6 \times 3 - 8) + 3 \times 4$	Parentheses first. Use the order of operations.
$= 3 - (16 \div 8 + 6 \times 3 - 8) + 3 \times 4$	Multiply and divide inside the parentheses.
$= 3 - (2 + 18 - 8) + 3 \times 4$	Add and subtract inside the parentheses.
$= 3 - 12 + 3 \times 4$	Multiply.
$= 3 - 12 + 12$	Add and subtract.
$= 3$	The answer is 3.

The correct answer is D.

Practice

Compute.

1. $35 - 14 \div 2$
2. $54 \div 6 + 18 \times 2$
3. $44 + 17 - 5 \times 2$
4. $5 \times 7 - 2 \times 3$
5. $4^2 - 16 \div 2$
6. $3^2 - 2^2 \div 8$
7. $5 - 4^3 \div 8$
8. $2^3 + 3 \times 5$
9. $-18 \div (2 \times 3) + 6$
10. $-18 \div 2 \times 3 + 6$
11. $8 - 2^2 \times 3$
12. $12 - 2^3 \times 3 \div 2$
13. $15 + 3^2 \times 2 \div 6$
14. $4^2 - 18 \div 3^2 \times 8$
15. $(15 + 3^2) \times 2 \div 6$
16. $(12 - 2^3) \times 3 \div 2$
17. $4 - 1 + (5 - 7) \times 3$
18. $8 \times 2 \div 4 + 3$
19. $3 \div 2 - \dfrac{1}{2} - 1$
20. $7 \times 3 - (22 + 4)$

(Answers on page 339)

ACT-TYPE PROBLEMS

1. $(3 - 2) \times 15 - 4 \div 3 = ?$
   - A. 10
   - B. $12\frac{1}{3}$
   - C. 13
   - D. $13\frac{2}{3}$
   - E. 14

2. $3 \times 4 + (7 - 4) + 3 = ?$
   - F. 19
   - G. 18
   - H. 17
   - J. 16
   - K. 15

3. $(16 - 4) \div 3 + 7 - 2 \times 5 = ?$
   - A. 1
   - B. 2
   - C. 3
   - D. 4
   - E. 5

4. $7 \times (5 + 7) - 8 \times 2 = ?$
   - F. 28
   - G. 32
   - H. 54
   - J. 62
   - K. 68

5. $(21 - 7) \times (25 - 17) - 4 \times 3 = ?$
   - A. 50
   - B. 75
   - C. 100
   - D. 125
   - E. 150

(Answers on page 339)

Complete this Cumulative ACT Practice in 15 minutes to reflect real ACT test conditions. This Cumulative ACT Practice gives you an additional opportunity to practice pre-algebra concepts in an ACT format. If you don't know an answer, eliminate and guess. Circle the number of any guessed answer. Then check your answers on page 339. You will also find explanations for the answers and suggestions for further study.

1. Which of the following statements are true?

   I.   534,916 < 534,619
   II. Both 534,916 and 534,619 rounded to the thousands place are 535,000
   III. The value of the 4 in both 534,996 and 534,969 is 4,000

   A. I only
   B. I and II
   C. II only
   D. II and III
   E. I and III

2. Which of the following choices has the smallest digit in the thousands place?

   F.   54,713
   G. 3,861,932
   H.   36,734
   J. 2,769,236
   K.   8,787

3. At a party, 12 people can be seated at each table. To seat everyone, 13 tables are needed. How many people attended the party if all the tables are full except one at which 7 people are seated?

   A. 145
   B. 151
   C. 156
   D. 154
   E. 163

4. Jim's car can travel 300 miles on a full tank of gas. If Jim starts a trip with his car's gas tank full and drives 178 miles, how much farther can he travel without filling up the gas tank again?

   F. 187 miles
   G. 178 miles
   H. 152 miles
   J. 132 miles
   K. 122 miles

5. What is the sum of the digits in the tenths place and the hundreds place of the number 2,593.7365?

   A. 10
   B. 11
   C. 12
   D. 13
   E. 14

6. Which of the following choices shows 78,616.874825 rounded to the thousandths place?

   F. 78,616.8749
   G. 79,000.000
   H. 78,616.875
   J. 78,600.000
   K. 78,616.88

7. $\frac{3}{5} \div \frac{6}{5} \cdot \frac{1}{3} = ?$

   A. $\frac{1}{3}$

   B. $\frac{1}{2}$

   C. $\frac{2}{3}$

   D. $\frac{1}{6}$

   E. $\frac{5}{6}$

8. Anita places one dollar in coins in her jacket pocket on Monday. Each day of the week Anita takes some of the coins and donates them to a charity box in her school's lunchroom. Listed below are the amounts Anita donates each day.

Day	Amount Donated
Monday	$0.10
Tuesday	$0.09
Wednesday	$0.15
Thursday	$0.07
Friday	$0.20

   How much money remains in Anita's jacket pocket at the end of the week?

   F. $0.39
   G. $0.41
   H. $0.46
   J. $0.51
   K. $0.55

9. {1, 2, 3, 4, 5, 6, 10, 12, 15, 20, 30, 60} is the complete list of factors for which of the following numbers?

   A. 40
   B. 60
   C. 80
   D. 120
   E. 180

10. Which of the following fractions is the largest?

    F. $\frac{3}{6}$

    G. $\frac{2}{3}$

    H. $\frac{7}{10}$

    J. $\frac{1}{2}$

    K. $\frac{4}{5}$

11. Which of the following has the largest result?

    A. $\frac{1}{2} + \frac{3}{2} + 1$

    B. $\frac{2}{3} + \frac{5}{6}$

    C. $\frac{10}{8} + \frac{12}{24} + \frac{3}{12}$

    D. $\frac{11}{2} - \frac{7}{14}$

    E. $\frac{10}{9} + \frac{1}{3} + \frac{6}{4}$

12. When Frank went to the store on Monday he bought $\frac{4}{5}$ of a pound of grapes. How many pounds of grapes did Frank eat on Monday if he has $\frac{2}{3}$ of a pound of grapes left on Tuesday?

    F. $\frac{3}{13}$

    G. $\frac{2}{15}$

    H. $\frac{4}{11}$

    J. $\frac{6}{17}$

    K. $\frac{7}{12}$

13. Jim and Joni needed money for their weekend trip. At the bank, Jim withdrew $60 for himself, while Joni withdrew only $\frac{5}{6}$ of that amount. How much money did they withdraw in total?

    A. $ 30
    B. $ 50
    C. $ 60
    D. $ 90
    E. $110

14. Every month Kate gets a $1,600 paycheck. She uses $\frac{6}{16}$ of the paycheck for rent; $\frac{1}{8}$ of the paycheck for utilities; $\frac{1}{4}$ of the paycheck for food, and $\frac{1}{8}$ of the paycheck for other bills. How much money does Kate have left to put into her savings account after these expenses?

    F. $100
    G. $150
    H. $200
    J. $225
    K. $250

15. Janette got a paycheck for $450. She loaned her brother $175 and bought $150 worth of groceries. If she already had $50 in her bank account, how much does she need back from her brother in order to pay her rent, which is $200?

    A. $100
    B. $ 75
    C. $ 50
    D. $ 25
    E. $ 0

Chapter 7 ▪

Pre-Algebra II

- Fourteen ACT questions have to do with pre-algebra.
- Easier pre-algebra questions may be about a single skill or concept.
- More difficult questions will often test a combination of skills or concepts.
- The pre-algebra review in Chapters 6 and 7 covers all the material you need to answer ACT questions.
- Use a calculator for the ACT-Type Problems. Do not use a calculator for the Practice exercises.

▪ Percent

Percent means per one hundred or out of one hundred.

15% means 15 out of 100.

1.5% means 1.5 out of 100.

Decimals, percents, and fractions can all be used to name the same number.

- To write a decimal as a percent, move the decimal point two places to the right and write the percent sign.

    $0.78 = 78\%$ $0.09 = 9\%$ $0.0524 = 5.24\%$ $28.634 = 2,863.4\%$

- To write a percent as a decimal, move the decimal point two places to the left. Write zeros if necessary.

    $36\% = 0.36$ $7\% = 0.07$ $0.034\% = 0.00034$ $386.29\% = 3.8629$

- To write a fraction as a decimal, divide the numerator by the denominator. To change this decimal to a percent, follow the rule above.

Write $\dfrac{3}{8}$ as a percent.

$$
\begin{array}{r}
0.375 = 37.5\% \\
8\overline{)3.000} \\
\underline{2\,4} \\
60 \\
\underline{56} \\
40 \\
\underline{40} \\
0
\end{array}
$$

Write $\dfrac{2}{3}$ as a percent.

$$
\begin{array}{r}
0.6666\ldots = 66.\overline{6}\% \\
3\overline{)2.00}
\end{array}
$$

> You can write $66.\overline{6}\%$ as $66\dfrac{2}{3}\%$.

The decimal for $\dfrac{2}{3}$ is a repeating decimal. The digit 6 repeats.

- To write a percent as a fraction, write the percent as a fraction with 100 in the denominator. Then simplify the fraction if possible.

  Write 56% as a fraction.

  $$56\% = \frac{56}{100} = \frac{14}{25}$$

  Write 0.84% as a fraction.

  $$0.84\% = \frac{0.84}{100} = \frac{84}{10,000} = \frac{21}{2,500}$$

  It will help you to know the fraction and percent equivalents in the following table.

Fractions and Percents

$\frac{1}{4} = 25\%$	$\frac{1}{2} = 50\%$	$\frac{3}{4} = 75\%$	
$\frac{1}{5} = 20\%$	$\frac{2}{5} = 40\%$	$\frac{3}{5} = 60\%$	$\frac{4}{5} = 80\%$
$\frac{1}{6} = 16\frac{2}{3}\%$	$\frac{1}{3} = 33\frac{1}{3}\%$	$\frac{2}{3} = 66\frac{2}{3}\%$	$\frac{5}{6} = 83\frac{1}{3}\%$
$\frac{1}{8} = 12\frac{1}{2}\%$	$\frac{3}{8} = 37\frac{1}{2}\%$	$\frac{5}{8} = 62\frac{1}{2}\%$	$\frac{7}{8} = 87\frac{1}{2}\%$

CALCULATOR TIP

Use a calculator to convert among percents, decimals, and fractions. Many calculators have special keys for this.

MODEL ACT PROBLEM

Lucy made 9 out of 15 basketball free-throw shots. Ann made 18 of 25 basketball free-throw shots. What is the difference between Lucy's free-throw percentage and Ann's free-throw percentage?

A. 9%
B. 12%
C. 18%
D. 25%
E. 60%

SOLUTION

Use a calculator. Find each free-throw percentage by changing the fraction to a decimal and writing the decimal as a percent.

$$\frac{9}{15} = 0.60 = 60\% \qquad \frac{18}{25} = 0.72 = 72\%$$

Subtract. 72% − 60% = 12%

The correct answer is B.

Practice

Complete the table. The three numbers across in each row should be equal.

	Fraction	Decimal	Percent
1.		2.	18%
3.		0.036	4.
	$\frac{5}{8}$	5.	6.
7.		8.	0.84%
9.		0.0002	10.
	$\frac{5}{6}$	11.	12.

13. Jennifer ran $\frac{3}{8}$ of a mile. What percent of a mile did she run?

14. Bob is 75% of Chad's height. What fraction of Chad's height is Bob?

15. Jim's gas tank was $\frac{1}{4}$ full. What percent of his gas tank was full?

16. Frank wins 30% of the time. What fraction of the time does Frank win?

17. Dave rides his bike to work $\frac{2}{3}$ of the time. What percent of the time does Dave ride his bike to work?

18. Eric works out 5 days a week and he jogs on 4 of those days. On what percent of the workout days does he jog?

19. Rachel has 6 dresses and 50% of them are red. What fraction of the dresses are red?

20. Matt makes 80% of the money that Andy makes. What fraction of Andy's money does Matt make?

(Answers on page 340)

ACT-TYPE PROBLEMS

1. Leticia works in the personnel department at a company. Two out of every five people she interviews are female. What percent of the people that Leticia interviews are female?

   A. 20%
   B. 30%
   C. 40%
   D. 50%
   E. 60%

2. Of the people who went to see a movie, 65% were over 18. What fraction of the people who went to see the movie were under 18?

   F. $\frac{7}{10}$
   G. $\frac{13}{20}$
   H. $\frac{6}{13}$
   J. $\frac{7}{20}$
   K. $\frac{3}{10}$

3. Francine cooks vegetable stir-fry 15 days in 4 weeks and chicken stir-fry 6 days in 4 weeks. What percent of the days does Francine cook stir-fry?

    A. 25%
    B. 40%
    C. 50%
    D. 70%
    E. 75%

4. On one math test Beth got 84% of the questions correct. On another test she got 6 of 15 answers correct. What is the difference between the fraction of questions Beth answered correctly on the two tests?

    F. $\dfrac{2}{5}$

    G. $\dfrac{11}{25}$

    H. $\dfrac{13}{25}$

    J. $\dfrac{3}{5}$

    K. $\dfrac{32}{75}$

(Answers on page 340)

5. There are 8 chairs around one dining room table. Three of the chairs have arm-rests. There are 8 chairs around another table, and 7 of them have arm-rests. What percent of the 16 chairs have arm-rests?

    A. 47.5%
    B. 50%
    C. 55%
    D. 62.5%
    E. 70%

■ Percent Problems

The basic relationship for percent problems is

    Percent $\times$ base = percentage

    25% of 80 = 20

In a percent problem you need to find one of the three quantities.

Finding a Percent of a Number

Write an equation and solve to find the percentage.

EXAMPLES

1. Bob drove 30% of his 235-mile trip. How far is that?

    Decide which quantity is missing. $0.3 \times 235 = ?$

    To find the percent, multiply. $0.3 \times 235 = 70.5$

    Bob drove 70.5 miles.

2. Liz saved $33\dfrac{1}{3}$% off the list price of $96. How much did she save?

    Recall that $33\dfrac{1}{3}\% = \dfrac{1}{3}$ $\dfrac{1}{3} \times \$96 = ?$ $\dfrac{1}{3} \times 96 = 32$

    Liz saved $32.

Finding the Percent One Number Is of Another

Write an equation and solve to find the percent.

1. Ben completed 6 miles of a 16-mile trip. What percent of the trip is completed?

   $? \times 16 = 6$ $\qquad$ percent $= \dfrac{?}{100}$

   $\dfrac{n}{100} \times 16 = 6$

   $6 \div 16 = 0.375$. Write 37.5 as a percent.

   Ben completed 37.5% of the trip.

2. The wholesale price of a ring was $28. A jeweler sold it for $42. What percent of the wholesale price was the selling price?

   $? \times 28 = 42$

   $42 \div 28 = 1.5 = 150\%$

   The selling price was 150% of the wholesale price.

The base is not always the larger number. The percent may be greater than 100%.

Finding a Number When a Percent of It Is Known

Write an equation and solve to find the base.

When it is 30% full a container holds 6 gallons. How much does the container hold?

30% of ? = 6 $\qquad$ 30% $\times$? = 6

$6 \div 30\% = 6 \div 0.3 = 20$

The container holds 20 gallons.

Sales Tax and Discount

Everyday transactions often involve percents. A sales tax adds a certain percentage of the cost of an item to the price you have to pay. A discount reduces the price of an item by a percentage of its original cost.

Sales Tax

Aaron bought a new television for $1,199. What is the total cost of the television including 6% sales tax?

6% $\times$ $1,199 = 0.06 \times$ $1,199

0.06 $\times$ $1,199 = $71.94

$71.94 + $1,199 = $1,270.94

The total cost of the television is $1,270.94.

You can use either a decimal or a fraction for a percent.

Discount

Kerine saved 25% off the list price of $96. How much did she pay?

25% of $\$96 = ?$ $\qquad$ $25\% = \dfrac{1}{4}$

$\dfrac{1}{4} \times 96 = 24$

$\$96 - \$24 = \$72$

Kerine paid $72.

Percent of Increase or Decrease

Use these formulas to find the percent of increase and the percent of decrease.

$$\text{Percent of increase} = \frac{\text{New amount} - \text{Original amount}}{\text{Original amount}}$$

$$\text{Percent of decrease} = \frac{\text{Original amount} - \text{New amount}}{\text{Original amount}}$$

Write the result as a percent.

CALCULATOR TIP

Most calculators have a fully functional percent key. This key calculates the amount of decrease or increase (discount or tax) and the final price including tax and/or discount.

Percent of Increase

The price of an item increases from $30 to $34.80. Find the percent of increase.

1. Subtract to find the amount of increase. $\qquad$ $\$34.80 - \$30 = \$4.80$
    The amount of increase is $4.80.

2. Divide the amount of increase by the
    original amount. $\qquad$ $\dfrac{\text{Amount of increase}}{\text{Original amount}} \begin{array}{l} \rightarrow \$4.80 \\ \rightarrow \$30 \end{array}$

3. Write as a percent. $\qquad$ $30\overline{)4.8} \quad \begin{array}{l} 0.16 = 16\% \end{array}$

    The percent of increase is 16%.

Percent of Decrease

The price of an item decreases from $25 to $21.75. Find the percent of decrease.

1. Subtract to find the amount of decrease. $25 − $21.75 = $3.25
   The amount of decrease is $3.25.

2. Divide the amount of decrease
   by the original amount.

$$\frac{\text{Amount of decrease}}{\text{Original amount}} \to \frac{\$3.25}{\$25}$$

3. Write as a percent.

$$25\overline{)3.25} \quad 0.13 = 13\%$$

The percent of decrease is 13%.

MODEL ACT PROBLEMS

1. Steve stopped for a drink of water when he had completed 60% of his jog. He had traveled 3 miles. What is the total distance Steve jogged?

   A. 2 miles
   B. 3 miles
   C. 4 miles
   D. 5 miles
   E. 6 miles

   SOLUTION

   Use the equation: percent × base = percentage
   $$0.6 \times \text{base} = 3 \text{ miles}$$
   $$\text{base} = \frac{3}{0.6} = 5 \text{ miles}$$

   Steve jogged 5 miles.

   The correct answer is D.

2. Cassandra paid $35 for a shirt that originally cost $50. What percent of the original price did the shirt cost?

   F. 40%
   G. 50%
   H. 65%
   J. 70%
   K. 85%

   SOLUTION

   The percentage is the price after the sale ($35). The base is the original price of the shirt ($50).

   percent × base = percentage

   percent × 50 = 35

   $$\text{percent} = \frac{35}{50} = 0.70 = 70\%$$

   Cassandra paid 70% of the original price for the shirt.

   The correct answer is J.

3. Jeff bought a car originally priced at $10,700. However, there was a sale so he got the car for $9,630. What was the percent of decrease?

   A. 10%
   B. 20%
   C. 30%
   D. 40%
   E. 50%

   SOLUTION

   Use the equation:

   $$\text{percent of decrease} = \frac{\text{Original amount} - \text{New amount}}{\text{Original amount}}$$

   $$\text{percent of decrease} = \frac{\$10,700 - \$9,630}{\$10,700} = \frac{\$1,070}{\$10,700}$$
   $$= 10\%$$

   The percent of decrease is 10%.

   The correct answer is A.

4. Tessa bought a car that cost $15,800. What was the total cost of the car including an 8% sales tax?

   F. $14,536
   G. $15,800
   H. $17,064
   J. $17,500
   K. $18,000

   SOLUTION

   Find the 8% sales tax.

   Use the equation:

   percent × base = percentage

   $$0.08 \times \$15,800 = \$1,264$$

   Add the sales tax to the price of the car.

   $$\$15,800 + \$1,264 = \$17,064$$

   The total cost of the car was $17,064.

   The correct answer is H.

Practice

1. What is the cost of a $99 item selling for 75% of that price?

2. A tank is at $62\frac{1}{2}\%$ of its 640-gallon capacity. How many gallons are in the tank?

3. An acorn grew 0.5% from its weight of 2.5 grams. How much did it grow?

4. After two weeks on a diet Raymond weighed 180 pounds, which was 90% of his original weight. What was Raymond's original weight?

5. What percent of a 120-gallon tank is full if it contains 90 gallons?

6. A 39-foot pole casts a 26-foot shadow. What percent of the pole's height is the shadow's length?

7. Nine out of 60 cans of dog food have been eaten. What percent of the dog food has been eaten?

8. Chris is 20 years old and Susan is 50 years old. What percent of Susan's age is Chris?

9. When 19% full, a tank contains 136.8 gallons. How much does the full tank hold?

10. When a cup is 75% filled it contains 75 ounces. How much can the cup hold?

11. A certain baseball stadium is 60% filled when it has 15,000 people in attendance. How many people can the stadium hold?

12. A price increased from $15 to $18.30. What is the percent of increase?

13. An $82 item is on sale for $50.84. What is the percent of decrease?

14. The number of students in a school decreased from 4,850 to 3,104. What is the percent of decrease?

15. A balloon went from 13,500 feet to 19,008 feet. What is the percent of increase?

16. What is the total cost of a $28.50 garden rake including an 8% sales tax?

During the End-of-Summer Sale, an air conditioner that originally sold for $510 was discounted 25%.

17. How much would you save if you waited for this sale?

18. What is the sale price of the air conditioner?

Phil bought a computer for $2,000.

19. What is the final cost of Phil's computer including a 6.5% sales tax?

20. How much money would Phil save if he did not have to pay an 8% shipping charge? (The shipping charge is applied before the tax.)

(Answers on page 341)

ACT-TYPE PROBLEMS

1. 21 is 40% of which of the following numbers?

   A. 40
   B. 46.5
   C. 50.2
   D. 52.5
   E. 55

2. Mylin's Magic Club collects $4,000 from the sale of tickets when there is a full house. If all the tickets cost the same amount, how much money will Mylin collect from the purchase of tickets if the club is 75% full?

   F. $2,000   J. $3,250
   G. $2,500   K. $3,500
   H. $3,000

3. Joe has to pay $600 rent. He has only 80% of the money. If Joe takes home $6 per hour, how many hours must he work to collect the rest of the rent money?

   A. 10 hours
   B. 15 hours
   C. 20 hours
   D. 25 hours
   E. 30 hours

4. Ellen walks 0.75 mile to school each day. She stops to get breakfast after completing 60% of her walk. How far has Ellen walked when she stops for breakfast?

   F. 0.35 mile
   G. 0.45 mile
   H. 0.50 mile
   J. 0.55 mile
   K. 0.65 mile

5. Dan plays basketball on his high school team. During the last game Dan scored 18 points, while for the season he averages 20 points per game. What percent of Dan's average did he score during this game?

   A. 60%
   B. 65%
   C. 75%
   D. 80%
   E. 90%

6. During a soccer game Chad scored 2 of his team's 6 goals. What percent of his team's goals did Chad score?

   F. 33%
   G. $33\frac{1}{3}$%
   H. 66%
   J. $66\frac{2}{3}$%
   K. 77%

(Answers on page 341)

7. Shannon bought a house for $150,000 and then she had to pay a 9% tax. What was the total cost of the house including the tax?

   A. $160,000
   B. $162,500
   C. $163,500
   D. $165,000
   E. $165,500

8. Emma found a barbecue grill she wanted that originally cost $120. She discovered it was on sale for 10% off the original price. Emma bought the grill on sale and paid a 6% sales tax. What was the final cost of the grill?

   F. $112.54
   G. $113.50
   H. $114.48
   J. $115.86
   K. $116.35

9. The original price of a television was $225. During a sale the price went down to $180. What was the percent of decrease?

   A. 10%
   B. 15%
   C. 20%
   D. 25%
   E. 30%

10. In one month Dennis' puppy grew from 3 pounds to 5 pounds. What was the percent of increase?

   F. 22%
   G. $22\frac{1}{2}$%
   H. 33%
   J. 66%
   K. $66\frac{2}{3}$%

▪ Ratio and Proportion

A ratio can be expressed in three ways.

$$5 \text{ to } 6 \qquad \frac{5}{6} \qquad 5:6$$

Proportions

A **proportion** shows that two ratios are equal. If you write the ratios as fractions, the cross products are equal.

$$\frac{23}{37} \bowtie \frac{299}{481}$$
$$^{11,063} \qquad ^{11,063}$$

Since the cross products are equal, these two fractions form a proportion.

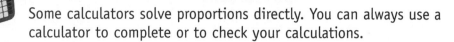

CALCULATOR TIP

Some calculators solve proportions directly. You can always use a calculator to complete or to check your calculations.

Solving a Proportion

If one of the values in the proportion is unknown, cross multiply to solve the proportion.

EXAMPLES

1. Two fractions are equivalent. One of the fractions is $\frac{13}{15}$ and the numerator of the second fraction is 156. Solve a proportion to find the denominator of the second fraction.

    First, write a proportion.

    first fraction $\rightarrow \dfrac{13}{15} = \dfrac{156}{x} \leftarrow$ second fraction

    Solve the proportion.

    Cross multiply.   Cross products are equal.   Divide to find x.

    $\overset{13x}{\dfrac{13}{15}} \bowtie \overset{2{,}340}{\dfrac{156}{x}}$ $13x = 2{,}340$ $x = 2{,}340 \div 13$

    $x = 180$

    The denominator of the second fraction is 180.

2. A builder estimates that she will use 20 bricks to cover 3 square feet of wall. How many square feet can be covered with 2,090 bricks?

    Write a proportion. $\dfrac{20 \text{ bricks}}{3 \text{ square feet}} = \dfrac{2{,}090 \text{ bricks}}{x \text{ square feet}}$

    Solve the proportion.

    Cross multiply.   Cross products are equal.   Divide to find x.

    $\overset{20x}{\dfrac{20}{3}} \bowtie \overset{6{,}270}{\dfrac{2{,}090}{x}}$ $20x = 6{,}270$ $x = 6{,}270 \div 20$

    $x = 313.5$

    The builder can cover 313.5 square feet with 2,090 bricks.

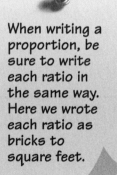

When writing a proportion, be sure to write each ratio in the same way. Here we wrote each ratio as bricks to square feet.

3. A 6-foot-tall person casts an 8.4-foot shadow. A telephone pole right next to the person casts a 20.3-foot shadow. How tall is the telephone pole?

Write a proportion.

person → $\dfrac{\text{6-foot-tall person}}{\text{8.4-foot shadow}} = \dfrac{x\text{-foot-tall pole}}{\text{20.3-foot shadow}}$ ← telephone pole

Solve the proportion.

Cross multiply.   Cross products are equal.   Divide to find x.

$\dfrac{6}{8.4} \bowtie \dfrac{x}{20.3}$ $^{121.8}$ $^{8.4x}$

$8.4x = 121.8$

$x = 121.8 \div 8.4$

$x = 14.5$

> The proportion could also be written $\dfrac{8.4}{6} = \dfrac{20.3}{x}$. The result will be the same.

The telephone pole is 14.5 feet high.

MODEL ACT PROBLEM

1. Ed is a baseball player who gets a hit 6 times out of every 20 times at bat. How many hits will Ed expect to get if he is at bat 200 times?

   A. 50 hits
   B. 55 hits
   C. 60 hits
   D. 65 hits
   E. 70 hits

SOLUTION

Set up a proportion.

$\dfrac{6 \text{ hits}}{20 \text{ times at bat}} = \dfrac{x \text{ hits}}{200 \text{ times at bat}}$

Cross multiply and solve for x.

$6 \times 200 = 20x$

$1{,}200 = 20x$

$60 = x$

Ed would expect to have 60 hits after 200 times at bat.

The correct answer is C.

2. In two weeks the Postman family drinks 3 gallons of milk. How many gallons of milk will the Postmans drink in 9 weeks?

   F. 3 gallons
   G. 6 gallons
   H. 13.5 gallons
   J. 18 gallons
   K. 27 gallons

SOLUTION

Set up a proportion.

$\dfrac{3 \text{ gallons}}{2 \text{ weeks}} = \dfrac{x \text{ gallons}}{9 \text{ weeks}}$

Cross multiply and solve for x.

$3 \times 9 = 2x$

$27 = 2x$

$x = 13.5$

The Postmans will drink 13.5 gallons of milk in 9 weeks.

The correct answer is H.

Practice

Rewrite each ratio in two different ways.

1. $\dfrac{5}{7}$ **2.** 3:13   **3.** 4 to 9   **4.** $\dfrac{2}{5}$

Solve the proportion.

5. $\dfrac{2}{3} = \dfrac{x}{9}$ **6.** $\dfrac{x}{5} = \dfrac{3}{25}$ **7.** $\dfrac{4}{x} = \dfrac{10}{13}$ **8.** $\dfrac{3}{7} = \dfrac{x}{49}$

9. $\dfrac{4}{8} = \dfrac{7}{x}$ **10.** $\dfrac{1}{5} = \dfrac{x}{100}$ **11.** $\dfrac{5}{11} = \dfrac{x}{121}$ **12.** $\dfrac{12}{x} = \dfrac{66}{27.5}$

13. Solve the proportion $\dfrac{5}{13} = \dfrac{20}{x}$.

14. Show $\dfrac{4}{9}$ written as a ratio in two other ways.

15. It costs David $25 to fill up his tank of gas twice. How much will it cost David to fill up his tank of gas 20 times?

16. Samantha has two cats, Cal and Hob. For every scoop of food that Hob eats, Cal eats 3 scoops. If Hob eats ten scoops of food, how many scoops of food will Cal eat?

17. You get 12 eggs for every carton you buy. How many eggs will you get if you buy 7 cartons?

18. A roofer estimates that 15 tiles are needed to cover 4 square feet. How many tiles will be needed to cover 120 square feet of the same roof?

19. A flagpole that stands 25 feet high casts a shadow of 40 feet. Someone standing right next to the flagpole is 6 feet tall. How long is the person's shadow?

20. Jack went to the grocery store and bought 8 apples for $2. How many apples could Jack have bought if he had $5?

(Answers on page 342)

ACT-TYPE PROBLEMS

1. John is making tacos for himself and 17 friends. For every 3 people John uses 5 tomatoes. How many tomatoes will John use in all?

   A. $10\dfrac{1}{5}$ tomatoes

   B. $10\dfrac{4}{5}$ tomatoes

   C. $28\dfrac{1}{3}$ tomatoes

   D. 30 tomatoes
   E. 85 tomatoes

2. Some say that a person should drink 8 glasses of water a day. If each of 45 people follows this advice, how much water will they drink?

   F. 300 glasses of water
   G. 340 glasses of water
   H. 350 glasses of water
   J. 360 glasses of water
   K. 390 glasses of water

3. Given the ratio $\dfrac{5}{6}$.

   I. 5 to 6
   II. 6:5
   III. 5:6

   Which of the following choices is a complete list of the ways to properly rewrite the ratio?

   A. I
   B. II
   C. II and III
   D. I and III
   E. I, II, and III

4. Jeff went to a book sale where 6 books cost $13. How many books could Jeff buy at that rate for $32.50?

   F. 10 books
   G. 15 books
   H. 17 books
   J. 19 books
   K. 23 books

5. A trainer uses 0.75 roll of medical tape on each soccer player. If there are 15 players on the team and 80% get taped up, how many rolls of tape does the trainer need?

   A. 8 rolls of tape
   B. 9 rolls of tape
   C. 10 rolls of tape
   D. 11 rolls of tape
   E. 12 rolls of tape

(Answers on page 342)

■ Statistics—Mean, Median, and Mode

The mean, median, and mode help describe sets of data.

The **mean** is the sum of the items divided by the number of items.

The **median** is the middle item, or, if there are an even number of items, the average of the 2 middle items, in a set of items arranged in order.

The **mode** is the item that occurs most frequently.

Mean (Arithmetic Average)

Follow these steps to find the mean:

- Add the items.
- Divide the sum by the number of items.

EXAMPLE

Find the mean of these scores: 34, 78, 62, 19, 71, 53, 86, 23, 78.

Add the items.

Divide by the number of scores.

$$34 + 78 + 62 + 19 + 71 + 53 + 86 + 23 + 78 = 504$$

$$504 \div 9 = 56$$

The mean is 56.

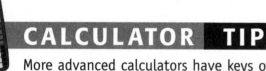

CALCULATOR TIP

More advanced calculators have keys or menus for directly calculating the mean, median, and mode. Check your calculator for this feature.

Median

Follow these steps to find the median:

- Arrange the items in order from least to greatest.
- Odd number of items: find the middle score.
- Even number of items: find the mean of the two middle scores.

EXAMPLES

1. Find the median of these scores: 34, 78, 62, 19, 71, 53, 86, 23, 78.

   Arrange the scores in order.   Odd number of scores.

   19 23 34 53 <u>62</u> 71 78 78 86   62 is the middle score.

   The median is 62.

CAREFUL. Be sure to arrange the scores in order first.

2. Find the median of these scores: 23, 98, 45, 16, 72, 20, 106, 72.

Arrange the scores in order.   Even number of scores.

16 20 23 45 72 72 98 106   45 and 72 are the two middle scores.

$$45 + 72 = 117 \div 2 = 58.5$$

The median is 58.5.

Mode

The mode is the item that occurs most often.

EXAMPLE

Find the mode of these numbers: 23, 98, 45, 16, 72, 20, 106, 72, 81, 45, 72.

72 occurs most often. 72 is the mode.

Some sets of items may have no mode or several modes. A set of items with two modes called "bi-modal."

It may help to arrange the scores in order: 16, 20, 23, 45, 45, 72, 72, 72, 81, 98, 106.

MODEL ACT PROBLEMS

1. Below is a list of times it took Simon to run 1 mile on 7 different days.

**Simon's times
for a 1-mile run**

7 minutes
6.5 minutes
7.5 minutes
7.25 minutes
6.75 minutes
7.5 minutes
6.5 minutes

What was Simon's average time for the 1-mile run on these days?

A. 6.25 minutes
B. 6.5 minutes
C. 6.75 minutes
D. 7 minutes
E. 7.25 minutes

SOLUTION

Add up all the times and divide the sum by the number of times.

$$\frac{7 + 6.5 + 7.5 + 7.25 + 6.75 + 7.5 + 6.5}{7}$$

$$= \frac{49}{7} = 7 \text{ minutes}$$

Simon's average time was 7 minutes.

The correct answer is D.

2. In which of the following lists of numbers are the mean, median, and mode all equal?

F. 5, 3, 6, 4, 7, 5, 2, 8, 5
G. 7, 2, 5, 9, 2, 4, 8, 2
H. 6, 9, 4, 7, 6, 3, 5
J. 1, 6, 3, 7, 3, 8
K. 1, 9, 3, 7, 2, 8, 7

SOLUTION

Find the mean of the first list.

$$\frac{5 + 3 + 6 + 4 + 7 + 5 + 2 + 8 + 5}{9} = \frac{45}{9} = 5$$

Find the median and the mode.

2, 3, 4, 5, 5, 5, 6, 7, 8

The median, 5, is the number in the middle after you arrange the numbers in order.

The mode is 5, the number that appears most often.

mean = median = mode

The correct answer is F.

Practice

Find the mean, median, and mode for each set of numbers.

2, 5, 8, 9, 4, 3, 6, 11, 6   80, 94, 64, 69, 79, 94   458, 361, 467, 297, 467, 178, 621

1. Mean _____   4. Mean _____   7. Mean _____

2. Median _____   5. Median _____   8. Median _____

3. Mode _____   6. Mode _____   9. Mode _____

Find the mean, median, and mode for each set of data. Round to the nearest tenth.

10. 5, 9, 9, 5, 4, 3, 2, 4, 10   11. 4.0, 2.4, 6.1, 5.0, 2.5, 6.0, 4.9

12. 589.7, 400, 475, 65.4, 475, 523.9

13. 18, 37, 56, 29.8, 17.4, 24.2, 1.1, 41.6, 32.8, 37, 37.1, 29.2

14. The mean of a list is 25.6 and there are 12 items. What is the sum of the items?

15. The sum of the items in a set is 761.8 and the mean is 58.6. How many items are there?

16. In one county 6 farms have an average crop of 47 bushels of corn per acre. Another 9 farms have an average crop of 49 bushels of corn per acre. What is the average number of bushels of corn per acre for these farms?

17. In a group of 15 students 6 students are 61 inches tall, 5 students are 60 inches tall, and the remaining students are 59 inches tall. What is the average height of these students?

18. On a recent quiz 3 students received 100, 4 students received 90, 6 students received 80, 5 students received 70, and 2 students received 60. What was the average of all the grades in the class?

19. Tim drove an average of 27.8 miles a day for 7 days. How many miles did he drive in that week?

20. From Monday to Friday Brenda spends an average of 1.5 hours a day on the phone with her customers. On Saturday and Sunday Brenda spends an average of 1 hour a day on the phone. What is the average amount of time that Brenda spends on the phone each day for the whole week, rounded to the nearest hundredth?

(Answers on page 342)

ACT-TYPE PROBLEMS

1. What is the sum of the mean, median, and mode of the numbers 4, 9, 4, 7, 1, 3, 10, 2?

   A. 11
   B. 12
   C. 13
   D. 14
   E. 15

2. Below is a list of 5 people and the number of glasses of milk they drink each day:

Person	Number of glasses of milk
Beth	3 glasses
Chris	5 glasses
Mike	2 glasses
Emily	6 glasses
Julia	8 glasses

   What is the average number of glasses that these people drink each day, rounded to the ones place?

   F. 3
   G. 4
   H. 4.8
   J. 5
   K. 5.8

3. Evan scored a total of 93 points, with an average of 18.6 points per game. How many games did Evan play?

   A. 4
   B. 5
   C. 6
   D. 7
   E. 8

4. The average weight of 4 men is 172 pounds and the average weight of 6 women is 134 pounds. What is the average weight of all 10 people?

   F. 145
   G. 148.6
   H. 149.2
   J. 151
   K. 153.4

5. Ed was able to do an average of 56.5 push-ups during 16 different attempts. What is the total number of push-ups that Ed completed?

   A. 875
   B. 880
   C. 895
   D. 900
   E. 904

(Answers on page 343)

■ Data Collection, Representation, and Interpretation

ACT test items ask you to read and interpret tables and graphs, not to create them. Graphs and charts help us read and interpret data. This section reviews bar graphs, line graphs, circle graphs, stem-and-leaf diagrams, and box-and-whisker plots.

The table below shows the number of new cars a dealer sold over six months.

New Cars Sold in Six Months

April	30
May	45
June	37
July	34
August	29
September	35

Bar Graph

A bar graph is used to compare data. The length of each bar represents a number. The bar graph below displays the data about new car sales. You can use a bar graph for data that can be categorized.

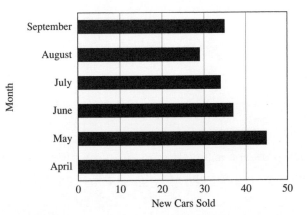

Line Graph

A line graph is used to show change, usually over a specific time period. Time intervals are usually shown on the horizontal axis. The line graph below shows the same new car data.

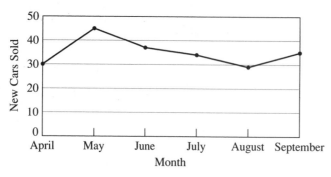

Circle Graph

A circle graph is used to show relative size, usually with percents. You can use a circle graph for data that can be categorized.

Jason brings home $1,000 a month. His budget is shown below.

Monthly Budget

Rent	$300
Food	$250
Gas	$100
Recreation	$150
Savings	$120
Miscellaneous	$ 80

The circle graph below shows the percent of Jason's budget spent on each category.

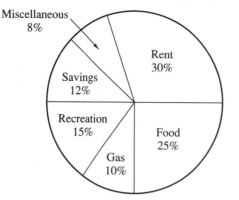

Stem-and-Leaf Diagram

A stem-and-leaf diagram is a way to group data.

Suppose 20 students received the following scores on a math test:

100 89 74 97 100 75 79 94 72 86 65 100 72 85 99 63 78 68 95 83

The stem-and-leaf plot of these data is shown on the right.

The values in the stem represent tens, while the values in the leaves represent ones. We can easily tell from this diagram that there are three scores of 100.

6	3, 5, 8
7	2, 2, 4, 5, 8, 9
8	3, 5, 6, 9
9	4, 5, 7, 9
10	0, 0, 0
Stem	Leaves

Box-and-Whisker Plot

A box-and-whisker plot shows the median and quartiles of a set of scores, along with the maximum score and the minimum score.

Use the data in the stem and leaf plot above. The minimum score is 63 and the maximum score is 100. The first quartile score is 73, the median is 84 and the third quartile is 96.

Draw a number line from 60 to 100 and plot the points above the line.

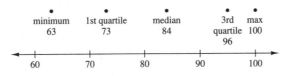

Connect the points using whiskers and boxes as shown below.

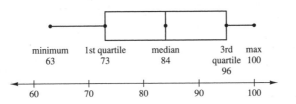

Refer to pages 484–496 in the science section for more information on data collection, representation, and interpretation.

MODEL ACT PROBLEM

A group of 20 students was asked to give the time it takes them, in minutes, to get to school each day. The responses were as follows: 21, 10, 12, 6, 20, 17, 9, 14, 6, 8, 12, 3, 9, 25, 29, 15, 22, 23, 14, 5. Arrange the responses of these 20 students in a stem-and-leaf diagram.

A.
0	1, 2
1	2
2	1, 1, 2
3	0, 2
4	1, 1
5	0, 1, 2
6	0, 0
7	1
8	0
9	0, 0, 2

B.
0	3, 5, 6, 8, 9
1	0, 2, 4, 5, 7
2	1, 2, 3, 5, 9

C.
0	3, 5, 6, 6, 8, 9, 9
1	0, 2, 2, 4, 4, 5, 7
2	0, 1, 2, 3, 5, 9

D.
0	0, 1, 2, 3, 5, 9
1	0, 2, 2, 4, 4, 5, 7
2	3, 5, 6, 6, 8, 9, 9

E.
0	3, 5, 6, 6, 9, 9, 9
1	2, 2, 4, 4, 5, 7, 9
2	0, 2, 3, 5, 9, 0

SOLUTION

The correct answer is C.

Practice

Use these data for questions 1–2.

The weights, in kilograms, of 20 students are given below.

60 43 72 81 76 58 75 53 71 48 57 61 84 55 62 47 88 79 56 82

1. Use the data to create a stem-and-leaf diagram.

2. Use the data to create a box-and-whisker plot.

Use these data for questions 3–4.

Given below is the amount of money Chad spent on his car from April through August.

April	May	June	July	August
$100	$125	$175	$150	$125

3. Create a bar graph.

4. Create a line graph.

Use these data for questions 5–6.

Roman is a car salesman. Given below are the number of cars Roman sold during a six-month period.

Month 1	Month 2	Month 3	Month 4	Month 5	Month 6
13 cars	10 cars	14 cars	12 cars	9 cars	12 cars

5. Create a line graph.

6. Create a bar graph.

7. Using the stem-and-leaf diagram below, write the values for the minimum, first quartile, median, third quartile, and maximum.

```
0 | 1, 3
1 | 2, 5, 7
2 | 1, 4, 6, 6, 7, 8
3 | 3, 3, 4, 5, 7
4 | 1, 2, 4, 8
5 | 2, 3, 9
```

8. Draw a box-and-whisker plot that incorporates the data from problem 7.

9. Below is a bar graph comparing the number of miles 5 friends jog each week. Who jogs the second-farthest each week?

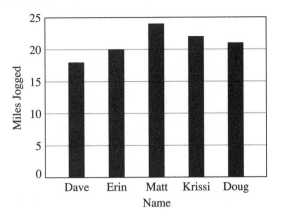

10. The circle graph below displays the amount of time Danielle spends doing a specific activity each day. On which activity does Danielle spend the least amount of time?

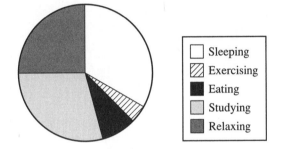

- ☐ Sleeping
- ▨ Exercising
- ■ Eating
- ▨ Studying
- ▨ Relaxing

(Answers on page 343)

ACT-TYPE PROBLEMS

1. Which of the following is greater than 75% of the scores shown in the box-and-whisker plot shown below?

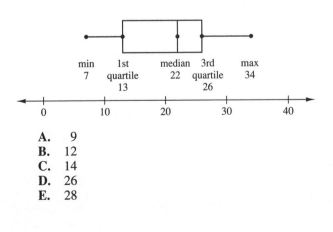

- **A.** 9
- **B.** 12
- **C.** 14
- **D.** 26
- **E.** 28

2. A store manager receives hourly updates about the sales in cash registers throughout the store. The stem-and-leaf display below shows the sales from the most recent hour. What is the median sale?

10	5, 5, 8
11	2, 5, 7
12	1, 4, 7, 8
14	3, 3, 4, 5, 7
15	1, 2, 4, 8
16	0
18	2, 3, 9
20	3

- **F.** 128.5
- **G.** 143
- **H.** 143.5
- **J.** 158
- **K.** 159

3. Alex likes to eat salads for lunch. Alex makes his salads with lettuce, tomatoes, onions, mushrooms, and peppers. Alex uses 2 times as many mushrooms as peppers, the same amount of onions as peppers, 3 times as many tomatoes as peppers, and 4 times as much lettuce as peppers. Which of the following circle graphs best fits this information?

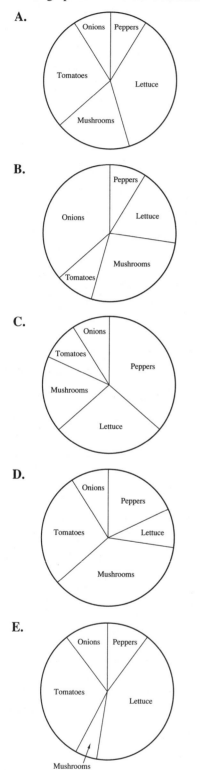

A.

B.

C.

D.

E.

4. The bar graph below shows the scores on a recent math test. What percentage of the class scored in the 80's or higher?

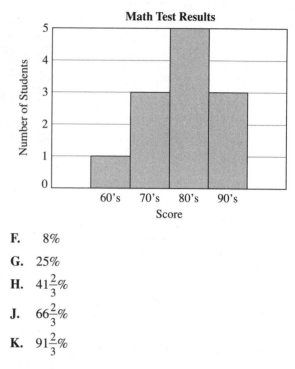

F. 8%

G. 25%

H. $41\frac{2}{3}\%$

J. $66\frac{2}{3}\%$

K. $91\frac{2}{3}\%$

5. During an August heat wave, Wendy kept track of the high temperature every day for a week. She graphed her results on a line graph. To the nearest degree, what was the average high temperature that week?

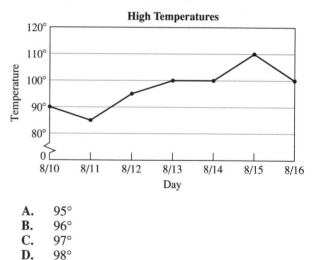

A. 95°
B. 96°
C. 97°
D. 98°
E. 100°

(Answers on page 344)

Probability

The probability of an event is the likelihood that it will occur. If an event will never occur, the probability is 0. If an event will always occur, the probability is 1. All other probabilities fall between 0 and 1. Use a fraction to write the probability of an event.

$$\text{Probability of an event} = \frac{\text{number of favorable outcomes}}{\text{number of possible outcomes}}$$

EXAMPLES

Think of rolling a single fair die.

1. What is the probability of rolling a 3?

> There are six sides and each side is equally likely to land faceup.

$$P(3) = \frac{\text{number of sides with a 3}}{\text{number of sides in all}} \begin{array}{c} \to 1 \\ \to 6 \end{array}$$

The probability of rolling a 3 is $\frac{1}{6}$.

2. What is the probability of rolling an even number?

$$P(\text{even number}) = \frac{\text{number of sides with an even number (2, 4, 6)}}{\text{number of sides in all}} \begin{array}{c} \to \frac{3}{6} = \frac{1}{2} \end{array}$$

The probability of rolling an even number is $\frac{1}{2}$.

3. What is the probability of rolling a 7?

$$P(7) = \frac{\text{number of sides with a 7}}{\text{number of sides in all}} \begin{array}{c} \to \frac{0}{6} = 0 \end{array}$$

The probability of rolling a 7 is 0. This means that the event is impossible.

4. What is the probability of rolling a number less than 10?

$$P(N < 10) = \frac{\text{number of sides with a number less than 10}}{\text{number of sides in all}} \begin{array}{c} \to \frac{6}{6} = 1 \end{array}$$

The probability of rolling a number less than 10 is 1. This means that the event is certain to happen.

Independent and Dependent Events

Independent events — The outcome of one event *does not* affect the probability of the other event.

EXAMPLE

You have a standard deck of 52 cards. The probability of picking the king of hearts is $\frac{1}{52}$. You pick a card without looking and replace it in the deck. There are still 52 cards in the deck. The probability of picking the king of hearts is still $\frac{1}{52}$. A card picked and replaced followed by picking another card are independent events.

Dependent events — The outcome of one event *does* affect the probability of the other event.

EXAMPLE

You have a standard deck of 52 cards. The probability of picking the king of hearts is $\frac{1}{52}$. You pick a card without looking, not the king of hearts, and don't put it back. There are now 51 cards in the deck. The probability of picking the king of hearts is now $\frac{1}{51}$. A card picked and not replaced followed by picking another card are dependent events.

MODEL ACT PROBLEMS

1. In a jar there are 3 blue marbles and 5 red marbles. If you reach into the jar and pick out a marble without looking, what is the probability that you will pick out a blue marble?

   A. $\frac{3}{8}$

   B. $\frac{3}{5}$

   C. $\frac{5}{8}$

   D. $\frac{8}{5}$

   E. $\frac{8}{3}$

   SOLUTION

   $$\text{P(blue marble)} = \frac{\text{number of blue marbles}}{\text{total number of marbles}} = \frac{3}{8}$$

   The probability of picking a blue marble is $\frac{3}{8}$.

   The correct answer is A.

2. What is the probability of choosing, without looking, a red face card from a regular deck of cards? (A face card is a jack, a queen, or a king.)

   F. $\frac{1}{26}$ J. $\frac{1}{4}$

   G. $\frac{1}{13}$ K. $\frac{1}{2}$

   H. $\frac{3}{26}$

   SOLUTION

   There are 6 red face cards in a deck (There are 4 kings, 4 queens, and 4 jacks, or a total of 12 face cards. Half of the face cards are red.)

   Find the probability of getting a red face card.

   $$\text{P(red face card)} = \frac{\text{number of red face cards}}{\text{total number of cards}}$$

   $$= \frac{6}{52} = \frac{3}{26}$$

   The probability of choosing a red face card from the deck is $\frac{3}{26}$.

   The correct answer is H.

Practice

You have a fair penny.

1. What is the probability of flipping a tail?

2. What is the probability of flipping a head?

3. What is the probability of flipping a head or a tail?

4. You flip the penny 10 times and get 10 heads. What is the probability of getting a head on the next flip?

You have six same-size balls numbered 1, 2, 4, 5, 7, and 8 in a box. You pick one without looking. What is the probability of picking

5. an 8?

6. an even number?

7. a number greater than 6?

8. a number divisible by 6?

9. a multiple of 4?

You have a standard deck of 52 cards. What is the probability of picking

10. a 7?

11. a red card?

12. a heart?

13. a non-face card?

14. a 2, 3 , or 4?

15. a 9 or 10?

16. a jack, or a queen, or a king?

You pick a 7 of clubs and don't replace it. What is the probability that the next pick will be

17. a 6 of clubs?

18. a 7 of clubs?

19. a 7?

20. a club?

(Answers on page 344)

1. In a drawer there are 3 brown socks, 2 blue socks, and 5 red socks. You pick a sock without looking. What is the probability of choosing a brown sock?

   A. $\frac{2}{3}$

   B. $\frac{3}{10}$

   C. $\frac{1}{2}$

   D. $\frac{1}{5}$

   E. $\frac{3}{7}$

2. What is the probability of rolling a 1 or a 5 with a single six-sided die?

   F. $\frac{1}{6}$

   G. $\frac{1}{3}$

   H. $\frac{1}{2}$

   J. $\frac{2}{3}$

   K. $\frac{5}{6}$

3. From a regular deck of 52 cards, what is the probability of choosing a card that is not an ace or a face card, but that is divisible by 2?

   A. $\frac{3}{26}$

   B. $\frac{1}{13}$

   C. $\frac{1}{4}$

   D. $\frac{7}{13}$

   E. $\frac{5}{13}$

4. If a king and a jack are removed from a regular 52-card deck, what is the probability of picking a face card?

   F. $\frac{1}{5}$

   G. $\frac{1}{3}$

   H. $\frac{2}{3}$

   J. $\frac{2}{5}$

   K. $\frac{3}{5}$

5. There are 20 colored pencils in a box: 4 red, 5 green, 10 blue, and 1 black. You pick a pencil without looking. What is the sum of the probability of picking a red pencil and the probability of picking a black pencil?

   A. $\frac{7}{10}$

   B. $\frac{1}{5}$

   C. $\frac{1}{20}$

   D. $\frac{1}{4}$

   E. $\frac{1}{2}$

(Answers on page 344)

■ Elementary Counting Techniques

There are some elementary techniques that can help you to count very efficiently. This section reviews three of these techniques: products, permutations, and combinations. Factorials are used to find the number of permutations or combinations.

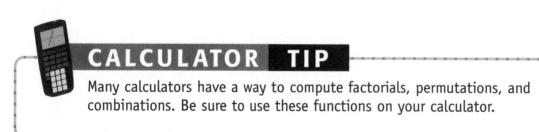

CALCULATOR TIP

Many calculators have a way to compute factorials, permutations, and combinations. Be sure to use these functions on your calculator.

Products

You may just need to multiply. Look at the example below.

You are buying a frozen yogurt cone. The yogurt store has three different types of cones, six different flavors, and eight kinds of toppings. Multiply to find how many types of yogurt cones you can buy.

$$3 \quad \times \quad 6 \quad \times \quad 8 \quad = \quad 144$$

cones   flavors   toppings

There are 144 different types of cones.

Permutations

A permutation is the arrangement of a certain number of items in a specific order.

Three students, Alex, Bonnie, and Charles, line up single file. How many different ways can they line up? You can make a list.

A B C   B A C   C A B

A C B   B C A   C B A

Three students can line up single file six different ways.

You can also use **factorial** to find the number of different permutations. 3 factorial is written 3!.   3! means $3 \times 2 \times 1$, or 6.

Use 6 factorial to find the number of different permutations of 6 items.

$6! = 6 \times 5 \times 4 \times 3 \times 2 \times 1 = 720$

There are 720 different ways to arrange six items.

According to a special rule, $0! = 1$.

Nine students are going to line up single file for movie tickets. In how many different ways can the students line up?

Use $9! = 9 \times 8 \times 7 \times 6 \times 5 \times 4 \times 3 \times 2 \times 1 = 362{,}880$

There are 362,880 ways for the students to line up.

Permutation Formula

You can always use this formula to find the number of permutations, or ways of arranging *n* items in *r* positions.

$$\frac{n!}{(n-r)!}$$

Look at the examples on the next page.

1. Four students in a club are interested in two positions, president and vice president. How many ways can the students be chosen for the two positions?

    Use the permutation formula.

    There are four students and two different positions.

    So $n = 4$ and $r = 2$. Substitute.

    $$\frac{n!}{(n-r)!} = \frac{4!}{2!} = \frac{24}{2} = 12$$

    There are 12 ways of electing four students to two positions.

2. There are five cars and three parking spaces. In how many different ways can the cars be parked in the spaces?

    $n = 5$ and $r = 3$

    $$\frac{n!}{(n-r)!} = \frac{5!}{(5-3)!} = \frac{5!}{2!} = \frac{120}{2} = 60$$

    There are 60 ways to park 5 cars in three spaces.

>
> Order matters here. First—you have 4 to choose from. Second—you have only 3 left to choose from. So $4 \times 3 = 12$ ways.

Combinations

A combination is an arrangement of a certain number of items in which order does not matter. Look at this example.

There are just two parking spaces, but four cars—red, blue, yellow, and green. In how many different ways can two cars be parked? Order doesn't matter.

We are taking four things, two at a time.

Make a list. Remember, order doesn't matter.

RB	BY	YG
RY	BG	
RG		

There are six ways of parking four cars in two spaces.

> Order doesn't matter here. There are fewer ways.
>
> $$\frac{4 \text{ (choices)}}{1 \text{ (1st pick)}} \times \frac{3 \text{ (choices)}}{2 \text{ (2nd pick)}}$$
>
> $$\frac{4 \times 3}{1 \times 2} = \frac{12}{2} = 6 \text{ ways}$$

Combination Formula

You can always use this formula to find the number of combinations that can be made from a group of n items taken r at a time.

$$\frac{n!}{(n-r)!r!}$$

Since we are taking four things two at a time, $n = 4$ and $r = 2$.

Substitute.

$$\frac{n!}{(n-r)!r!} = \frac{4!}{(4-2)!(2)!} = \frac{4!}{2! \times 2!} = \frac{4 \times 3 \times 2 \times 1}{2 \times 2} = \frac{24}{4} = 6$$

There are six ways of parking the cars if order does not matter.

There are three job openings for Level I computer techni-
cians, and six applicants. How many different ways could
these six applicants be chosen to fill the three job openings?

A. 6
B. 20
C. 120
D. 520
E. 720

SOLUTION

The problem asks about choosing, not ranking, applicants. Order does not matter, so this is a combination
problem.

$$\frac{n!}{(n-r)!(r)!} = \frac{6!}{(6-3)!(3)!}$$

$$= \frac{6 \times 5 \times 4 \times \cancel{3} \times \cancel{2} \times \cancel{1}}{3 \times 2 \times 1 \times \cancel{3} \times \cancel{2} \times \cancel{1}} = \frac{120}{6} = 20$$

The correct answer is B.

Practice

1. How many different ways are there to place 5 books on 5 different shelves if you can
   place only one book on each shelf?

2. If Andy has 7 shirts, 6 pants, and 8 ties, how many different outfits can he make with
   the pants, shirts, and ties?

3. There are 4 coach seats left on an airplane and 7 people waiting for those seats. How
   many different ways can you choose people to fill the 4 seats?

4. There are 4 people running for 4 different positions: President, Vice President, Sec-
   retary, and Treasurer. How many different ways are there to place the 4 people into
   these 4 different positions?

5. For breakfast Jane has cereal, orange juice, an apple, and a piece of toast. Jane keeps
   4 different types of cereal, 2 different kinds of orange juice, 3 different types of
   apples, and 2 types of bread. How many different breakfast choices does she have?

6. There are 6 puppies born, but the mother always feeds only 4 puppies at a time. How
   many different groups of puppies can she feed at one time?

7. There are 5 windows in a room, and Rich bought 5 different curtains to go on the
   windows. How many different ways can Rich place the curtains on the windows?

8. Steve knows 5 notes on the guitar and plays 4 notes in a row. How many different
   ways will Steve be able to arrange his notes?

9. There are 9 different things to drink in Lisa's house, but Lisa has only 3 glasses. How
   many different ways are there to put 3 different drinks in the glasses. (You can put
   only one type of drink into each glass.)

10. A 7-digit phone number uses all the digits from 3 to 9. How many different possible
    phone numbers are there?

(Answers on page 344)

1. Eight students try out for two openings on the debate team. In how many different ways can these two openings be filled?

   A. 2
   B. 8
   C. 28
   D. 56
   E. 84

2. A store manager has four different gifts to give to the first four people who enter the store. In how many different ways can he distribute the gifts?

   F. 4
   G. 8
   H. 16
   J. 24
   K. 32

3. A license plate has three letters (A–Z) followed by three digits (0–9). How many different license plates can be produced?

   A. 11,232,000
   B. 12,654,720
   C. 12,812,904
   D. 15,600,000
   E. 17,576,000

4. A newspaper editor needs five stories for the front page of the next issue. There are six reporters available to write the stories. How many ways can they be assigned if each story is written by only one reporter?

   F. 5
   G. 6
   H. 30
   J. 120
   K. 720

5. How many 3-digit area codes can be created if the first digit cannot be zero?

   A. 100
   B. 300
   C. 729
   D. 900
   E. 999

(Answers on page 344)

■ Writing Linear Expressions and Equations

You may have to write an expression or an equation to solve a problem. An **equation** is a statement about variables and numbers that contains an equal sign. An **expression** contains no equal sign. The variables in a **linear equation** are of the first degree. (They have exponents of 1.)

Words or phrases in the problem may help you decide when to write an operation sign or when to write an equal sign. But remember that you can't just use these words and phrases without thinking.

Operation	Words and Phrases
addition	and, more, in all, increased by, sum, total
subtraction	less, decreased by, difference, how many more/less
multiplication	of, product, times
division	per, quotient, shared
equals	is, equal to, equals

1. Write an expression for this statement.

   Three times the sum of x and y.

   Substitute symbols and operation signs for the words. $3 \times (x + y)$

2. Write an equation for this problem.

   The Rooster baseball team scored 8 runs in the first four innings. The Stars scored 3 runs in the first four innings. After four innings, how many more runs have the Roosters scored?

   The variable in this problem is the difference in the number of runs each team has scored. Let d equal the difference.

Write the problem in words.	The difference equals 8 less 3.
Write an equation.	$d = 8 - 3$

3. Write an equation for this problem.

   A magnolia tree on Kiefer's property is 7 feet 3 inches high. If the magnolia tree grows 10 inches a year, how many years will pass before the tree is 16 feet high?

   The unknown quantity in this problem is the number of years it will take the tree to grow to be 16 feet tall. Let y equal the number of years. Then $10y$ is the number of inches the tree will grow in y years.

Write the heights in inches.	now: 7 feet 3 inches = 87 inches
	in y years: 16 feet = 192 inches
Write the problem in words.	87 inches plus 10 times some number of years equals 192 inches.
Write an equation.	$87 + 10y = 192$

MODEL ACT PROBLEM

A newspaper reporter can write w words per hour. Her colleague can write $w - 15$ words per hour. If they work together on a story, which equation below shows the amount of time, t, it would take them to write 1,500 words?

A. $t = w^2 - 15w - 1{,}500$

B. $t = \dfrac{1{,}500}{2w - 15}$

C. $t = 1{,}500(2w - 15)$

D. $t = 1{,}515 - 2w$

E. $t = 100$

SOLUTION

Find the number of words the reporters can write in total in one hour.

$w + (w - 15) = 2w - 15$

Divide the total number of words by the number of words written per hour to find the total number of hours.

total hours $= \dfrac{1{,}500}{2w - 15}$

The correct answer is B.

Practice

Write an expression for each of these statements.

1. The difference of the money in a bank account at the beginning of the year and the amount withdrawn during the year.

2. The average of Scott's grades on 5 quizzes.

3. Laura's pay for one week if she worked 8 hours at her regular rate and 3 hours of overtime at 1.5 times her regular rate.

4. The amount of money in a savings account at the end of a year if P dollars were deposited at the beginning of the year and I% interest was earned on the account.

Write an equation that can be used to solve each problem.

5. The balance of Brynah's account at the Credit Union was $3,155. After she made a withdrawal to pay for car repairs her new balance was $2,855. How much did Brynah withdraw from her account?

6. When Tony turned 9 years old he received an allowance of $4 each week. His parents increased his allowance by $0.75 a week each birthday. What was Tony's allowance when he turned 12?

7. A chemistry researcher has a container filled with 100 ml of hydrochloric acid (HCl). If the researcher does a series of experiments that each use 4 ml of HCl, how much of the acid is left?

8. To design a house, an architect charges a certain fee, f, per square foot of the house. If he gives the Martins a 25% discount off that fee, how much will they pay for the design of an 8,000-square-foot house?

9. Write an equation that could be used to find Marvin's age if 3 more than Kim's age is equal to 29 less than Marvin's age.

10. The tax on a restaurant meal is 8%. If Joe paid a 15% tip on the original (pre-tax) meal cost, write an equation to show how much Joe paid in total for a meal whose original cost was $20.

(Answers on page 345)

ACT-TYPE PROBLEMS

1. A computer virus creates a total of 50 copies of itself on each computer it infects. If n computers at a school are infected with the virus, which equation shows V, the number of copies of the virus at that school?

   A. $V = n^{50}$

   B. $V = 50^n$

   C. $V = 50 + n$

   D. $V = 50n$

   E. $V = \dfrac{50}{n}$

2. A company starts with $500,000 in a savings account. Every 2 years the account grows by $250,000 in regular yearly amounts. Which of the following expressions can be used to represent the amount in dollars in the account after x years?

   F. $500,000 + \dfrac{250,000}{x}$

   G. $500,000 + \left(\dfrac{250,000}{2}\right)x$

   H. $250,000 + 250,000x$

   J. $500,000 + \left(\dfrac{500,000}{2}\right)x$

   K. $250,000 + \left(\dfrac{250,000}{2}\right)x$

3. A bag of 30 chocolate bars is shared equally among 12 children. Which equation could be used to find the number of chocolate bars each child receives?

A. $c = 12 \times 30$
B. $c = 30 - 12$
C. $c = 12 \div 30$
D. $c = 30 \div 12$
E. $c = 12 + 30$

4. A bicycle store pays sales representatives a base salary of $5 per hour worked. In addition, employees receive a commission of 8% of the value of each bicycle sold. Which expression shows Joan's pay for the week if she worked h hours and sold v worth of bicycles?

F. $5h + 0.08v$
G. $5v + 0.08h$
H. $5h + 1.08v$
J. $5.08(h + v)$
K. $5.08hv$

(Answers on page 345)

5. Rachel has an average of 82 on the last four math tests. If she has one more test to take before the end of the semester, which equation could be used to find the score she needs to raise her average to an 85?

A. $\dfrac{82 + T}{5} = 85$

B. $\dfrac{82 + T}{2} = 85$

C. $T = \dfrac{82 + 85}{2}$

D. $\dfrac{4(82) + T}{2} = 85$

E. $\dfrac{4(82) + T}{5} = 85$

▪ Solving Linear Equations

Algebraic symbols are either constants or variables.

A **constant** is one value.

Examples: $2, \dfrac{1}{2}, 0.08, \pi, \sqrt{5}$

A **variable** may have one or more values.

Examples: x, y, a, t

Solving an equation means finding the value of the variable that makes the equation true. Solving an equation is a mechanical process. The idea is to get the variable on one side of the equal sign and the value of the variable (a number) on the other side of the equal sign. You can add or subtract any number on each side of the equal sign, or you can multiply or divide each side of the equation by the same *nonzero* number. Substitute your solution back into the equation to check that the solution is correct.

> When solving equations, first add or subtract, then multiply or divide.

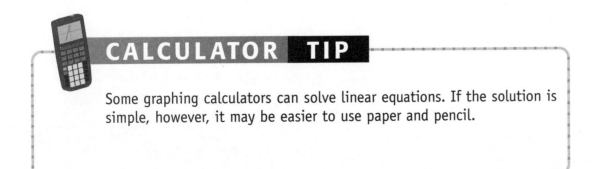

CALCULATOR TIP

Some graphing calculators can solve linear equations. If the solution is simple, however, it may be easier to use paper and pencil.

Using Addition or Subtraction to Solve Equations

1. Solve. $x + 27 = 48$

$$x + 27 = 48$$

Subtract 27. $\dfrac{-27 \quad -27}{x \qquad = 21}$

Check: substitute 21 for x.

$21 + 27 \overset{?}{=} 48$

$\qquad 48 = 48 ✓$

2. Solve. $-47 + x = -63$

$$-47 + x = -63$$

Add 47. $\dfrac{+47 \qquad +47}{x = -16}$

Check: substitute -16 for x.

$-47 + (-16) \overset{?}{=} -63$

$\qquad -63 = -63 ✓$

Using Multiplication or Division to Solve Equations

1. Solve. $\dfrac{y}{4} = 15$

$$\dfrac{y}{4} = 15$$

Multiply by 4. $4\left(\dfrac{y}{4}\right) = 4(15)$

$$y = 60$$

Check: substitute 60 for y.

$$\dfrac{60}{4} \overset{?}{=} 15$$
$$15 = 15 ✓$$

2. Solve. $4k = 61$

$$4k = 61$$

Divide by 4. $\dfrac{4k}{4} = \dfrac{61}{4}$

$$k = 15\dfrac{1}{4}$$

Check: substitute $15\dfrac{1}{4}$ for k.

$$4\left(15\dfrac{1}{4}\right) \overset{?}{=} 61$$
$$61 = 61 ✓$$

Solving Equations in Two or More Steps

You may have to use several steps to solve an equation. Always add or subtract before you multiply or divide.

1. Solve. $5t - 8.5 = 54$

$$5t - 8.5 = 54$$

Add 8.5. $\dfrac{+ 8.5 \quad +8.5}{5t \qquad = 62.5}$

Divide by 5. $\dfrac{5t}{5} = \dfrac{62.5}{5}$

$$t = 12.5$$

Check: substitute 12.5 for t.

$$5(12.5) - 8.5 \overset{?}{=} 54$$
$$62.5 - 8.5 \overset{?}{=} 54$$
$$54 = 54 ✓$$

2. Solve. $\dfrac{k}{8} + 3.4 = -28$

$$\dfrac{k}{8} + 3.4 = -28$$

Subtract 3.4. $\dfrac{-3.4 \qquad -3.4}{\dfrac{k}{8} \qquad = -31.4}$

Multiply by 8. $8\left(\dfrac{k}{8}\right) = 8(-31.4)$

$$k = -251.2$$

Check: substitute -251.2 for k.

$$\dfrac{-251.2}{8} + 3.4 \overset{?}{=} -28$$
$$-31.4 + 3.4 \overset{?}{=} -28$$
$$-28 = -28 ✓$$

3. Solve. $3x - 4 = 5 + x$

$$3x - 4 = 5 + x$$

Subtract x.

$$\underline{\quad -x \qquad\qquad -x \quad}$$
$$2x - 4 = 5$$

Add 4.

$$\underline{\quad +4 \quad +4 \quad}$$
$$2x = 9$$

Divide by 2.

$$\frac{2x}{2} = \frac{9}{2}$$
$$x = 4.5$$

Check: substitute 4.5 for x.

$$3(4.5) - 4 \overset{?}{=} 5 + 4.5$$
$$13.5 - 4 \overset{?}{=} 9.5$$
$$9.5 = 9.5 \checkmark$$

Solving Literal Equations

A **literal equation** is an equation with two or more different variables. For instance, $ax + y = c$ is a literal equation that can be solved for a, x, y, or c.

E X A M P L E

Solve $ax + y = c$ for x.

$$ax + y = c$$

Subtract y.

$$\underline{\quad -y \quad -y \quad}$$
$$ax = c - y$$

Divide by a.

$$\frac{ax}{a} = \frac{c - y}{a}$$
$$x = \frac{c - y}{a}$$

Solving Fractional Equations

Follow these steps to solve fractional equations.

1. Isolate the variable on one side of the equal sign.
2. Multiply by the inverse of the coefficient to solve the problem.

E X A M P L E

Solve. $2\frac{2}{3}n + 31 = -21$

$$2\frac{2}{3}n + 31 = -21$$

Subtract 31.

$$\underline{\qquad\quad -31 \quad -31 \quad}$$
$$2\frac{2}{3}n = -52$$

Write $2\frac{2}{3}$ as a fraction.

$$\frac{8}{3}n = -52$$

Multiply by the reciprocal, $\frac{3}{8}$.

$$\frac{3}{8}\left(\frac{8}{3}n\right) = \frac{3}{8}(-52)$$
$$n = -19\frac{1}{2}$$

1. If $3s + 2 = 8$, what is the value of s?

   A. 2
   B. 3
   C. 4
   D. 5
   E. 6

SOLUTION

Write the equation and solve.

$$3s + 2 = 8$$

Subtract 2. $3s\ \ \ = 6$

Divide by 3. $s\ \ \ \ = 2$

The correct answer is A.

2. Robert's age multiplied by 5, plus 6, equals 41. How old is Robert?

   F. 4 years old
   G. 5 years old
   H. 6 years old
   J. 7 years old
   K. 8 years old

SOLUTION

Write an equation and solve for Robert's age.

Let r = Robert's age.

$$5r + 6 = 41$$

Subtract 6. $5r\ \ \ = 35$

Divide by 5. $r\ \ \ \ = 7$ years old

The correct answer is J.

Practice

Solve each equation. Check each answer.

1. $w + 9 = 25$

2. $-18 = x + 3$

3. $3y = 15$

4. $8z = -28$

5. $30y + 7 = 13$

6. $3y + 8 = -5y + 88$

7. $3v - 9 = -v + 27$

8. $10 - 3t = -5t - 8$

9. $-1.8 = 3m + 3$

10. $2n - 6 = 10 - 14n$

11. $-19 - 6w = -37$

12. $0.5s - 3 = -14$

13. $-7d + 3 = -18$

14. $4g + 6 = 8$

15. $-2k - 3 = 2$

16. Solve $2x + y = c$ for x.

17. Solve $m - 3k = p$ for k.

18. Solve $-x + y = z$ for x.

19. Solve $3k - 2y = -r$ for k.

Solve each equation.

20. $\frac{2}{3}x + 12 = 28$

21. $1\frac{1}{7}y - 9 = 27$

22. $\frac{1}{8}k - 11 = 21$

23. $1\frac{3}{8}a + \frac{1}{4} = \frac{7}{8}$

24. $\frac{2}{5}b + \frac{3}{7} = \frac{11}{5}$

25. $\frac{3}{7}t + \frac{4}{5} = \frac{6}{7}$

26. Jim's weekly pay is half Carl's weekly pay. If Carl is paid $225 a week, how much is Jim paid a week?

27. Henry hit 60 home runs. Ken hit a third fewer home runs than Henry, plus 30. How many home runs did Ken hit?

28. What is the sum of a and b, if $6a + 2 = 14$, and b is 5 more than a?

29. What is three times the value of x if $-4x - 5 = -15$?

30. Julia runs 5 miles a day. Phil runs twice as far each day as Julia, minus 5 miles. How many miles does Phil run each day?

(Answers on page 345)

ACT-TYPE PROBLEMS

1. Solve. $5y - 3y + 13 = 26$

    A. -3
    B.   5
    C.   5.5
    D.   6.5
    E.   13

2. Given the two linear equations $4r - 5 = 17$ and $12 - 2s = 2$, what is the product of r and s?

    F.   5
    G.   5.5
    H.   10.5
    J.   25
    K.   27.5

3. Emily is half Mike's age, plus 7. If Mike is 28 years old, how old is Emily?

    A.   17 years old
    B.   19 years old
    C.   21 years old
    D.   23 years old
    E.   25 years old

(Answers on page 345)

4. Solve for h. $\frac{2}{3}h - 5 = h + 7$

    F. -36
    G.   36
    H. -24
    J.   24
    K.   20

5. The equation $\$2{,}225c - \$2{,}000 = P$ shows the profit (P) a computer store makes selling (c) computers. How many computers must the store sell to make a profit of $\$49{,}175$?

    A.   21 computers
    B.   22 computers
    C.   23 computers
    D.   24 computers
    E.   25 computers

Complete this Cumulative ACT Practice in 10 minutes to reflect real ACT test conditions. This Cumulative ACT Practice gives you an additional opportunity to practice pre-algebra concepts in an ACT format. If you don't know an answer, eliminate and guess. Circle the number of any guessed answer. Then check your answers on page 346. You will also find explanations for the answers and suggestions for further study.

1. If you make x dollars per hour and you work for 8 hours, how much money will you make?

   A. $\dfrac{8}{x}$

   B. $8x$

   C. $16x$

   D. $\dfrac{16}{x}$

   E. $4x$

2. 9 is the answer to which of the following questions?

   F. 15% of 60 is what number?
   G. 2 is 20% of what number?
   H. 40% of what number is 4?
   J. 20% of 180 is what number?
   K. 9% of 50 is what number?

3. An architect is looking at the blueprints of a house and sees that a certain door is 6 centimeters wide on the plans. The architect needs to know the actual width of the door. From the key the architect sees that 1 centimeter represents 6 inches. How wide will the actual door be?

   A. 7 feet
   B. 6 feet
   C. 5 feet
   D. 4 feet
   E. 3 feet

4. A baseball team has three positions open: first base, third base, and right field. Six players are trying out for these positions. In how many different ways can these positions be filled?

   F. 18
   G. 20
   H. 30
   J. 60
   K. 120

5. 80 is the mean of which of the following sets of test scores?

   A. {86, 77, 92, 69, 86, 91, 68}
   B. {85, 76, 91, 68, 84, 89, 67}
   C. {98, 79, 85, 81, 89, 78, 92}
   D. {89, 68, 92, 88, 81, 90, 100}
   E. {75, 98, 86, 83, 88, 94, 99}

6. If you are rolling a pair of dice, which of the following sums has a $\dfrac{1}{9}$ probability of being rolled?

   F. 2
   G. 3
   H. 4
   J. 5
   K. 6

7. Erin puts 10% of her income (I) in a savings account each month. Which of the following expressions could be used to determine how much money Erin has placed into the account after M months?

   A. $0.10 \cdot I + M$
   B. $10 \cdot I + M$
   C. $0.10 \cdot I \cdot M$
   D. $10 \cdot I \cdot M$
   E. $0.10 \cdot I \div M$

8. Jan uses just four budget categories. The circle graph shows the percent of her take-home pay allocated to each category. If her annual take-home pay increases by $900, how much, on average, will her monthly food budget increase?

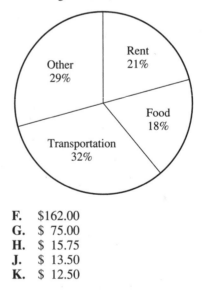

   F. $162.00
   G. $ 75.00
   H. $ 15.75
   J. $ 13.50
   K. $ 12.50

9. Jordan is able to run 3 times as far as Andy. If Jordan can run 18 miles, how many miles can Andy run?

A. 54 miles
B. 27 miles
C. 8 miles
D. 6 miles
E. 4 miles

10. Emily went to a store to buy beads for her necklace. The red (R) beads cost $0.05 per bead, the blue (B) beads cost $0.15 per bead, and the green (G) beads cost $0.25 per bead. How much will it cost if Emily buys 4 red beads, 5 blue beads, and 3 green beads?

F. $2.00
G. $1.90
H. $1.70
J. $1.60
K. $1.50

Subtest Pre-Algebra

Complete this Subtest in 14 minutes to reflect real ACT test conditions. This Subtest has the same number and type of pre-algebra items as are found on the ACT. If you don't know an answer, eliminate and guess. Circle the number of any guessed answer. Then check your answers on page 347. You will also find explanations for the answers and suggestions for further study.

1. Which of the following is a complete list of the factors of 40?

A. 1, 2, 5, 8, 20, 40
B. 10, 20, 40
C. 1, 2, 4, 5, 8, 10, 20, 40
D. 2, 4, 5, 8, 10, 20
E. 1, 2, 4, 5, 8

2. Amazing Computer Company sells three times as many type x computers as type y computers. One year the total number of x and y computers that Amazing Computer Company sold was 10,000. If computer x sells for $1,500 and computer y sells for $1,300, how much more money did the company make from selling type x computers than type y computers?

F. $800,000
G. 8×10^6
H. $600,000
J. 6×106
K. $200,000

3. 6 is 12% of what number?

A. 0.50
B. 0.72
C. 6.72
D. 50
E. 72

4. You are driving from New Jersey to Oregon and you want to know how long you will have to travel. You use a ruler to measure the distance on a map, which is 1 foot. When you look at the map key you see that each inch represents 300 miles. About how far is it from New Jersey to Oregon?

F. 7,200 miles
G. 3,600 miles
H. 3,000 miles
J. 1,800 miles
K. 300 miles

5. There is just room for 5 more people to go on a trip, but 7 people want to go. In how many ways can five people be chosen for the trip?

A. 2
B. 21
C. 35
D. 2,520
E. 5,040

6. What is the mean of the following test scores: 93, 89, 95, 78, 78, 80, 92, 87, 82?

F. 78
G. 80
H. 86
J. 87
K. 93

7. If you roll a pair of fair six-sided dice, what is the probability that their sum will be 7?

A. $\frac{1}{36}$

B. $\frac{1}{18}$

C. $\frac{1}{12}$

D. $\frac{1}{9}$

E. $\frac{1}{6}$

8. Given the numbers $\frac{1}{2}, \frac{2}{8}, \frac{3}{5}, \frac{7}{20}, \frac{11}{40}$, which of the following lists the numbers from smallest to largest?

F. $\frac{3}{5}, \frac{2}{8}, \frac{11}{40}, \frac{7}{20}, \frac{1}{2}$

G. $\frac{2}{8}, \frac{11}{40}, \frac{7}{20}, \frac{1}{2}, \frac{3}{5}$

H. $\frac{1}{2}, \frac{3}{5}, \frac{2}{8}, \frac{7}{20}, \frac{11}{40}$

J. $\frac{11}{40}, \frac{7}{20}, \frac{2}{8}, \frac{3}{5}, \frac{1}{2}$

K. $\frac{7}{20}, \frac{11}{40}, \frac{1}{2}, \frac{2}{8}, \frac{3}{5}$

9. Ed can lift two 50-pound weights and two 25-pound weights on a lifting bar that already weighs 45 pounds. Which of the following choices displays in scientific notation the total weight Ed can lift?

A. 1.2×10^2 pounds

B. 1.5×10^2 pounds

C. 1.2×10^{-2} pounds

D. 1.95×10^2 pounds

E. 1.95×10^{-2} pounds

10. A catering service needs to have 4 tables for every 28 people attending a party. If 224 people are attending the party, how may tables will the catering service need?

F. 8

G. 16

H. 24

J. 32

K. 40

11. What is the value of the following expression rounded to the nearest 100th?

$$\frac{3 \times \frac{1}{4} + 5 + 0.2 - 3.7 + 2 - \frac{1}{3}}{8 - 2.1 + \frac{2}{5} \times 8}$$

A. 0.4

B. 0.43

C. 0.431

D. 0.44

E. 0.5

12. There are three tollbooths open on a toll road. Four cars are headed for the exact-change tollbooth. In how many ways can the cars line up to go through the one exact-change toll?

F. 1

G. 4

H. 12

J. 24

K. 36

13. A large mail order business sells hundreds of thousands of items. The line graph shows the sales levels of three of these items (Item A, Item B, and Item C) from November 19 to November 25. On which date was the sales level of Item A approximately half the sales level of Item C?

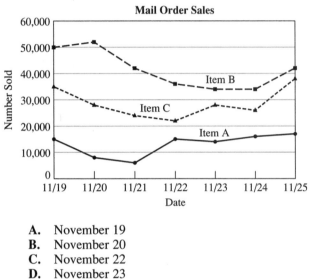

A. November 19

B. November 20

C. November 22

D. November 23

E. November 25

14. The stem-and-leaf plot below shows attendance at the home basketball games this season. What is the difference between the mode and the median attendance at these games?

11	0, 0, 2, 3
13	4, 5, 8, 9
14	5, 7, 9
15	2, 4, 6
16	0, 1, 3

F. 7

G. 29

H. 35

J. 110

K. 145

Chapter 8 ▪

Elementary Algebra

- Ten ACT questions have to do with elementary algebra.
- Easier elementary algebra questions may be about a single skill or concept or may test a combination of pre-algebra and elementary algebra skills.
- More difficult questions will often test a combination of elementary algebra skills and concepts.
- This elementary algebra review covers all the material you need to answer ACT questions.
- Use a calculator for the ACT-Type Problems. Do not use a calculator for the Practice exercises.

▪ Evaluating Formulas and Expressions

Formulas and **expressions** are statements about numbers. To **evaluate** a formula or an expression, substitute values for the variables and then compute. Remember to use the correct order of operations.

> Expressions do not have equal signs. You evaluate formulas and expressions, you don't solve them.

CALCULATOR TIP

Some graphing calculators can evaluate expressions. However, it may be easier to evaluate the expression using paper and pencil.

E X A M P L E S

1. The formula for the area of a square is $A = s^2$, where s is the length of the side of the square.

   Evaluate $A = s^2$ for $s = 4$ centimeters.

   $A = 4^2 = 16$ cm^2

> Use square units for area.

2. The formula for the area of a circle is $A = \pi r^2$, where r is the length of the radius.

   Evaluate $A = \pi r^2$ for $r = 3$ inches. (Use 3.14 for π.)

   $A = (3.14)(3^2) = 3.14 \times 9 = 28.26$ in.^{2}

3. The formula for the area of a trapezoid is $A = \frac{h}{2}(b_1 + b_2)$, where b_1 and b_2 are the lengths of the two parallel bases and h is the height.

Evaluate $A = \frac{h}{2}(b_1 + b_2)$ for $h = 3.5$ meters, $b_1 = 4.4$ meters, and $b_2 = 5.8$ meters.

$$A = \frac{3.5}{2}(4.4 + 5.8) = \frac{3.5}{2}(10.2) = 1.75(10.2) = 17.85 \text{ m}^2$$

4. The formula for the volume of a sphere is $V = \frac{4}{3}\pi r^3$, where r is the radius of the sphere.

Evaluate $V = \frac{4}{3}\pi r^3$ for $r = 6$ centimeters. (Use 3.14 for π.)

$$V = \left(\frac{4}{3}\right)(3.14)(6^3) = 904.32 \text{ m}^3$$

Use cubic units for volume.

5. The formula for simple interest is $I = PRT$.

Simple interest is the amount paid or earned in interest.

P = Principal, the amount borrowed or deposited.

R = Rate, the interest rate.

T = Time, the length of the loan or deposit in years. (For example, for 6 months, $T = \frac{1}{2}$. For 3 months, $T = \frac{1}{4}$.)

Evaluate $I = PRT$ for $P = \$2,800$, $R = 9.8\%$, and $T = 30$ months.

Write the percent as a decimal. $9.8\% = 0.098$

Write T in years. $30 \text{ months} = \frac{30}{12} \text{ years} = 2.5 \text{ years}$

$I = (\$2,800)(0.098)(2.5) = \$686$

The interest is $686.

6. If $y = 3$, $s = 2.8$, and $k = \frac{1}{4}$, what is the value of $3y - 4s + \frac{y}{k}$?

Substitute the numerical value for each variable and then compute.

$3y - 4s + \frac{y}{k}$

$= 3(3) - 4(2.8) + \dfrac{3}{\frac{1}{4}}$ Substitute. Multiply or divide from left to right.

$\left(\text{Remember: } \dfrac{3}{\frac{1}{4}} = \dfrac{3}{1} \div \dfrac{1}{4} = \dfrac{3}{1} \times \dfrac{4}{1} = 12.\right)$

$= 9 - 11.2 + 12$ Add or subtract from left to right.

$= 9.8$

The answer is 9.8.

What is the radius of a circle if the area is 49π in.^{2}?

   A. 24.5 in.
   B. 20 in.
   C. 14 in.
   D. 7 in.
   E. 3.5 in.

SOLUTION

Use the area formula.	A	$=$	πr^2
Substitute 49π for A.	49π	$=$	πr^2
Divide by π.	$\dfrac{49\pi}{\pi}$	$=$	$\dfrac{\pi r^2}{\pi}$
Evaluate.	$\sqrt{49}$	$=$	$\sqrt{r^2}$
	7 in.	$=$	r

The correct answer is D.

Practice

Use these formulas to help you complete the practice exercises that follow.

Geometric Formulas

Triangle	Area $= \dfrac{1}{2}bh$	Circle	Area $= \pi r^2$
Square	Area $= s^2$		Circumference $= 2\pi r$ or πd
Rectangle	Area $= lw$	Cube	Volume $= s^3$
Parallelogram	Area $= bh$	Rectangular Prism	Volume $= lwh$
Trapezoid	Area $= \dfrac{1}{2}h(b_1 + b_2)$	Sphere	Volume $= \dfrac{4}{3}\pi r^3$

For exercises 1–12, all measurements are in centimeters. Use 3.14 for π.

Find the area of each figure.

  1. Triangle: $b = 3$, $h = 8$
  2. Square: $s = 0.7$
  3. Rectangle: $l = 1.5$, $w = 1.2$
  4. Parallelogram: $b = 2.7$, $h = 1.3$
  5. Trapezoid: $b_1 = 4$, $b_2 = 0.5$, $h = 1.3$
  6. Triangle: $b = \dfrac{1}{2}$, $h = \dfrac{1}{4}$

Find the circumference and the area of each circle.

  7. Circle: $r = 2$

  8. Circle: $d = 6$

  9. Circle: $d = 1.8$

Find the volume of each solid.

  10. Cube: $s = 0.9$

  11. Rectangular prism: $l = 4$, $w = 6$, $h = 1.5$

  12. Sphere: $r = 3$

Use the formula $d = rt$, where d = distance, r = rate, and t = time, to find the missing value.

13. $d =$ ___?___

r = 60 mph

t = 2 hours

14. d = 270 feet

$r =$ ___?___

t = 90 seconds

15. d = 420 miles

r = 70 mph

$t =$ ___?___

Use a formula to solve each problem.

16. At an annual rate of 11.25%, what is the simple interest earned in 24 months on an account that has a principal amount of $7,500?

17. What is the circumference of a circle with a diameter of 13 centimeters?

18. What is the average speed of a car that travels 290 miles in 5 hours?

19. The height of a trapezoid is 5 inches. One base is twice the height and the other base is half the height. What is the area of the trapezoid?

20. What is the volume of a sphere with the same radius as a circle with a diameter of 30 millimeters?

21. Belinda has $1,025 in a savings account that pays 4% simple interest. How much will she have in her account after two years if she makes no deposits or withdrawals?

22. Ricardo borrowed $8,500 to buy a car. How much interest will he pay if the loan is for 4 years at 8% simple interest?

(Answers on page 348)

ACT-TYPE PROBLEMS

1. What is the volume of a cube if the area of one face of the cube is 36 m²?

A. 12 m³
B. 18 m³
C. 108 m³
D. 206 m³
E. 216 m³

2. Add the sum of the sides of a square with an area of 30.25 in.² to the sum of the edges of a cube with a volume of 343 in.³.

F. 5.5 in.
G. 7 in.
H. 12.5 in.
J. 64 in.
K. 106 in.

3. What is the area of a circle with a circumference of 24π cm?

A. 12π cm²
B. 24π cm²
C. 120π cm²
D. 144π cm²
E. 240π cm²

4. The area of a rectangle is 30 in.² and the length of the rectangle is 6 in. What is the sum of the length and the width of the rectangle?

F. 11 in.
G. 9 in.
H. 7 in.
J. 5 in.
K. 3 in.

5. John and Eric drove the same distance in separate cars. It took John 6 hours to complete the trip, traveling at an average speed of 65 mph. How much longer would the trip take Eric, if Eric traveled at an average speed of 60 mph?

A. $\frac{1}{4}$ hour

B. $\frac{1}{2}$ hour

C. $\frac{3}{4}$ hour

D. 1 hour

E. 2 hours

(Answers on page 348)

▪ Exponents and Radicals

Exponents

- The **base** shows the factor.
- The **exponent** shows how many times the factor is repeated.

Exponent
↓
Base → $2^6 = 2 \times 2 \times 2 \times 2 \times 2 \times 2 = 64$

The factor is 2.

Use the following rules for exponents.

Rule	Example
$x^0 = 1 \; (x \neq 0)$	$16^0 = 1$
$x^1 = x$	$16^1 = 16$
$(x^n)(x^m) = x^{n+m}$	$7^8 \times 7^5 = 7^{13}$
$\dfrac{x^n}{x^m} = x^{n-m} \; (x \neq 0)$	$\dfrac{7^8}{7^5} = 7^3$
$(x^m)^n = x^{mn}$	$(5^2)^3 = 5^6$
$x^{-m} = \dfrac{1}{x^m} \; (x \neq 0)$	$5^{-4} = \dfrac{1}{5^4}$

Radicals

Write radicals with the radical sign or as a fractional power.

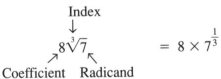

Index
↓
$8\sqrt[3]{7}$ $= 8 \times 7^{\frac{1}{3}}$
Coefficient Radicand

This radical shows *8 times the cube root of 7.*

Simplifying Radicals

Simplify all radicals so that:

- There are no fractions in the radicand.
- All *n*th powers of whole numbers are removed from the radicand.
- There are no radicals in the denominator.
- The index is as low as possible.

$\sqrt[n]{xy} = \sqrt[n]{x} \cdot \sqrt[n]{y}$ and $\sqrt{\dfrac{x}{y}} = \dfrac{\sqrt{x}}{\sqrt{y}}$

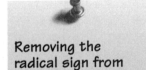

Removing the radical sign from the denominator is called **rationalizing the denominator.**

CALCULATOR TIP

Use the exponent key to find the value of a radical. For example,

$\sqrt[5]{a} = a^{\frac{1}{5}}$.

EXAMPLES

1. Simplify. $\sqrt{75}$

   Look for a perfect square factor. $\sqrt{75} = \sqrt{25 \cdot 3}$

   Extract the square factor. $= \sqrt{25} \cdot \sqrt{3} = 5\sqrt{3}$

2. Simplify. $\sqrt{\frac{7}{16}}$

   Write as separate radicals. $\sqrt{\frac{7}{16}} = \frac{\sqrt{7}}{\sqrt{16}}$

   Find the square root of the denominator. $= \frac{\sqrt{7}}{4}$

3. Simplify. $\sqrt{\frac{3}{7}}$

   Write as separate radicals. $\sqrt{\frac{3}{7}} = \frac{\sqrt{3}}{\sqrt{7}}$

   Multiply the numerator and denominator by $\sqrt{7}$
   to remove the radical from the denominator. $= \frac{\sqrt{3}}{\sqrt{7}} \cdot \frac{\sqrt{7}}{\sqrt{7}} = \frac{\sqrt{21}}{7}$

MODEL ACT PROBLEM

1. What is the sum of the factors of $3^4 \cdot 4^3 \cdot 5^2$?

   A.   9
   B.   12
   C.   24
   D.   34
   E.   60

   SOLUTION

   Write the factors.

   $3^4 \cdot 4^3 \cdot 5^2 = 3 \cdot 3 \cdot 3 \cdot 3 \cdot 4 \cdot 4 \cdot 4 \cdot 5 \cdot 5$

   Add the factors.

   $3 + 3 + 3 + 3 + 4 + 4 + 4 + 5 + 5 = 34$

   The correct answer is D.

2. Simplify $\frac{3^2}{\sqrt{80}}$.

   F. $\frac{\sqrt{80}}{3^{-2}}$

   G. $\frac{4}{9}\sqrt{5}$

   H. $\frac{5}{4}\sqrt{5}$

   J. $\frac{9}{20}\sqrt{5}$

   K. $\frac{9}{4}\sqrt{5}$

   SOLUTION

   Simplify the numerator. Factor and simplify the denominator.

   $\frac{3^2}{\sqrt{80}} = \frac{9}{\sqrt{16}\sqrt{5}} = \frac{9}{4\sqrt{5}}$

   Multiply numerator and denominator by $\sqrt{5}$ to remove the radical from the denominator.

   $\frac{9\sqrt{5}}{4\sqrt{5}\sqrt{5}} = \frac{9\sqrt{5}}{4 \cdot 5} = \frac{9}{20}\sqrt{5}$

   The correct answer is J.

Practice

Compute.

1. $7^8 \times 7^3$
2. $9^{-4} \times 9^7$
3. $2^3 \times (3^2)^2$
4. $3^5 \times 3^{-5}$
5. $3^3 \times 9^2$

6. $5^2 \div 5^5$
7. $4^2 \div 2^4$
8. $12^5 \div 12^{-5}$
9. $4^{\frac{1}{2}} \div 4^1$
10. $6^7 \div 6^2$

Simplify.

11. $\sqrt{25}$
12. $\sqrt{864}$
13. $\sqrt[3]{56}$
14. $\sqrt[3]{576}$
15. $\sqrt{162}$

16. $\dfrac{7}{\sqrt{11}}$
17. $\dfrac{6}{\sqrt{18}}$
18. $\dfrac{19}{\sqrt{19}}$
19. $\dfrac{13}{\sqrt[3]{296}}$
20. $\dfrac{13}{\sqrt{8}}$

(Answers on page 349)

ACT-TYPE PROBLEMS

1. $13^{-2} \times 13^2 = ?$

   A.   2
   B.   1
   C.   0
   D.   −1
   E.   −2

2. $\dfrac{7}{\sqrt{45}} = ?$

   F. $\dfrac{7}{3}$

   G. $\dfrac{7}{9}$

   H. $\dfrac{3}{7}$

   J. $\dfrac{7\sqrt{3}}{9}$

   K. $\dfrac{7\sqrt{5}}{15}$

3. $5^4 \div 25^2 = ?$

   A.   20
   B.   15
   C.   10
   D.   5
   E.   1

4. Which of the following choices is equal to $\dfrac{4^7}{\sqrt{16}}$?

   F.   4,094
   G. $2 \times 2 \times 2 \times 2 \times 2 \times 2 \times 6 \times 6$
   H. 2^{12}
   J. 8^3
   K. $2^6 \times 8^4$

5. All of the following choices are the reciprocal of $\dfrac{\sqrt{72}}{6^5}$ EXCEPT

   A. $(6^2)^2 \times \dfrac{\sqrt{2}}{2}$

   B. $6 \times 6^3 \times \dfrac{\sqrt{2}}{2}$

   C. $6^2 \times 6^2 \times \dfrac{\sqrt{2}}{2}$

   D. $6^8 \div 6^4 \times \dfrac{\sqrt{2}}{2}$

   E. $6^{12} \div 6^3 \times \dfrac{\sqrt{2}}{2}$

(Answers on page 349)

Operations With Radicals

You can add, subtract, multiply, and divide radicals.

Addition and Subtraction

To add or subtract radicals:

- Rewrite so the radicals are the same.
- Add or subtract the coefficients.

> You can add or subtract only when the radicals are the same.

EXAMPLES

1. Add. $5\sqrt{7} + 2\sqrt{112}$

    Rewrite $2\sqrt{112}$. $2\sqrt{112} = 2\sqrt{16 \times 7} = 2 \times 4\sqrt{7} = 8\sqrt{7}$

    Add the coefficients. $5\sqrt{7} + 8\sqrt{7} = 13\sqrt{7}$

    So, $5\sqrt{7} + 2\sqrt{112} = 13\sqrt{7}$.

2. Add. $3\sqrt[3]{162} + 8\sqrt[3]{6}$

    Rewrite $3\sqrt[3]{162}$. $3\sqrt[3]{162} = 3\sqrt[3]{27 \times 6} = 3 \times 3\sqrt[3]{6} = 9\sqrt[3]{6}$

    Add the coefficients. $9\sqrt[3]{6} + 8\sqrt[3]{6} = 17\sqrt[3]{6}$

    So, $3\sqrt[3]{162} + 8\sqrt[3]{6} = 17\sqrt[3]{6}$.

3. Subtract. $3\sqrt{48} - \sqrt{27}$

    Rewrite $3\sqrt{48}$. $3\sqrt{48} = 3\sqrt{16 \times 3} = 12\sqrt{3}$

    Rewrite $\sqrt{27}$. $\sqrt{27} = \sqrt{9 \times 3} = 3\sqrt{3}$

    Subtract the coefficients. $12\sqrt{3} - 3\sqrt{3} = 9\sqrt{3}$

    So, $3\sqrt{48} - \sqrt{27} = 9\sqrt{3}$.

CALCULATOR TIP

If the answers to a question are in radical form, check to see if your calculator represents answers in radical form. Graphing calculators usually don't have this feature.

Multiplication

To multiply radicals:

- Write the factors under one radical sign.
- Multiply.
- Simplify if possible.

> Radicals do NOT need to be the same to multiply.

Multiply. $\sqrt{10} \times \sqrt{15}$

  Write factors under one radical sign. $\sqrt{10} \times \sqrt{15} = \sqrt{10 \times 15}$

  Multiply. $= \sqrt{150}$

  Simplify if possible. $= \sqrt{25 \times 6} = 5\sqrt{6}$

So, $\sqrt{10} \times \sqrt{15} = 5\sqrt{6}$.

Division

To divide radicals:

- Write as a fraction under one radical sign.
- Factor, cancel, and simplify.
- The denominator of the answer must be a whole number.

Divide. $\dfrac{\sqrt{10}}{\sqrt{15}}$

  Rewrite under one radical sign. $\dfrac{\sqrt{10}}{\sqrt{15}} = \sqrt{\dfrac{10}{15}}$

  Factor numerator and denominator. Cancel. $= \sqrt{\dfrac{2 \times \cancel{5}}{3 \times \cancel{5}}} = \sqrt{\dfrac{2}{3}}$

  The denominator is not a whole number.

  Multiply numerator and denominator $= \sqrt{\dfrac{2 \times 3}{3 \times 3}} = \sqrt{\dfrac{6}{9}}$
  by the denominator.

  Simplify. $= \dfrac{\sqrt{6}}{3}$

So, $\dfrac{\sqrt{10}}{\sqrt{15}} = \dfrac{\sqrt{6}}{3}$.

MODEL ACT PROBLEMS

1. $3\sqrt{48} + 11\sqrt{75} = ?$

  A. $14\sqrt{75}$

  B. $33\sqrt{48}$

  C. $67\sqrt{3}$

  D. $72\sqrt{5}$

  E. $81\sqrt{2}$

SOLUTION

Rewrite with radicals that are the same. Then add the coefficients.

$3\sqrt{48} + 11\sqrt{75} = 3\sqrt{16 \times 3} + 11\sqrt{25 \times 3}$

$= 3 \times 4\sqrt{3} + 11 \times 5\sqrt{3} = 12\sqrt{3} + 55\sqrt{3} = 67\sqrt{3}$

The correct answer is C.

2. $\sqrt{14} \div \sqrt{6} = ?$

  F. $\dfrac{1}{7}$

  G. $\dfrac{2}{3}$

  H. $\dfrac{21}{\sqrt{3}}$

  J. $\dfrac{3}{\sqrt{21}}$

  K. $\dfrac{\sqrt{21}}{3}$

SOLUTION

Rewrite the factors under one radical sign. Then simplify.

$\dfrac{\sqrt{14}}{\sqrt{6}} = \sqrt{\dfrac{14}{6}} = \sqrt{\dfrac{2 \cdot 7}{2 \cdot 3}} = \sqrt{\dfrac{7}{3}} = \sqrt{\dfrac{7 \cdot 3}{3 \cdot 3}} = \sqrt{\dfrac{21}{9}} = \dfrac{\sqrt{21}}{3}$

The correct answer is K.

Practice

Add.

1. $5\sqrt{6} + 3\sqrt{216}$ **2.** $6\sqrt[3]{7} + 8\sqrt[3]{875}$ **3.** $9\sqrt{48} + 11\sqrt{75}$

Subtract.

4. $4\sqrt{50} - 12\sqrt{72}$ **5.** $9\sqrt{432} - 4\sqrt{1,323}$ **6.** $11\sqrt[3]{96} - 4\sqrt[3]{324}$

Multiply.

7. $\sqrt{8} \times \sqrt{7}$ **8.** $\sqrt{3} \times \sqrt{39}$ **9.** $\sqrt{8} \times \sqrt{30}$

Divide.

10. $\dfrac{\sqrt{3}}{\sqrt{39}}$ **11.** $\dfrac{\sqrt{72}}{\sqrt{9}}$ **12.** $\dfrac{\sqrt{15}}{\sqrt{135}}$

Simplify.

13. $\sqrt{3} \times \sqrt{6} + \sqrt{72}$ **14.** $\sqrt{13} - \sqrt{52} + \sqrt{208}$ **15.** $-\sqrt{3} + \sqrt{5} \times \sqrt{15}$

16. $\dfrac{\sqrt{13}}{\sqrt{117}} - \dfrac{\sqrt{1}}{\sqrt{16}}$ **17.** $\dfrac{\sqrt{12} + \sqrt{27}}{\sqrt{6}}$ **18.** $\sqrt{6} \times \sqrt{6} - 6 + \sqrt{16}$

19. $\sqrt{32} - \sqrt{8} - 2\sqrt{2}$ **20.** $\sqrt{3} \times \sqrt{5} - \sqrt{375}$

(Answers on page 349)

ACT-TYPE PROBLEMS

1. $\sqrt{22} \times \sqrt{14} = ?$
- **A.** $-4\sqrt{77}$
- **B.** $-2\sqrt{77}$
- **C.** $2\sqrt{77}$
- **D.** $4\sqrt{77}$
- **E.** $8\sqrt{77}$

2. $\sqrt{54} - \sqrt{96} = ?$
- **F.** $-2\sqrt{3}$
- **G.** $-\sqrt{6}$
- **H.** $2\sqrt{3}$
- **J.** $\sqrt{6}$
- **K.** $7\sqrt{6}$

3. $3\sqrt{125} + 6\sqrt{80} = ?$
- **A.** -45
- **B.** $-29\sqrt{5}$
- **C.** $-9\sqrt{5}$
- **D.** $39\sqrt{5}$
- **E.** 145

4. $\sqrt{8}(\sqrt{13} - \sqrt{117}) = ?$
- **F.** $-4\sqrt{26}$
- **G.** $4\sqrt{2}$
- **H.** $2\sqrt{13}$
- **J.** $2\sqrt{26}$
- **K.** $4\sqrt{26}$

5. What is the product of the numerator and denominator in the simplified form of $\dfrac{\sqrt{45}}{\sqrt{10}}$?
- **A.** $\sqrt{6}$
- **B.** $\dfrac{3}{2}\sqrt{6}$
- **C.** $3\sqrt{2}$
- **D.** $6\sqrt{2}$
- **E.** 18

(Answers on page 349)

■ Polynomials

Types of Polynomials

Polynomial is just another name for an expression. Recall that expressions do not contain equal signs. There are different types of polynomials.

A **monomial** can be a constant, a variable, or the product or quotient of constants and variables.

Notice that there are no addition or subtraction signs in monomials.

Examples: $5, z, 4x, y^5, 345, xy^2z^4w^5, \dfrac{3}{8}, -8.6$

A **binomial** is the sum or difference of two monomials. Each monomial is called a term.

Examples: $3x - 2, 5x^3 + 7, -8 + 62x^4y^3, 7z^5y^2 - 8x^3z^2$

A **trinomial** is the sum or difference of three monomials. Each monomial is called a term.

Examples: $9x^5 + 45y^3 - 6y, z^5 - x^3 - y^8, 7x^2 + 5x + 34$

Similar Terms

Terms consist of constants—called coefficients—and variables. In the term $5x$, the coefficient is 5 and the variable is x.

Similar terms have exactly the same variable part. The order of the variables is not important.

Examples: $5xy$ is similar to $15xy$ and $34yx$.

$6x^2y$ is similar to $17yx^2$ and x^2y.

$6x^2y$ is NOT similar to $5xy^2$ or to $7y^2x$.

To combine similar terms, add or subtract the coefficient and keep the variable part.

E X A M P L E

Combine similar terms. $3x + 2xy + 7y - 5x + 3x^2 - 2y$

Group similar terms. $(3x - 5x) + (7y - 2y) + 2xy + 3x^2$

Combine similar terms. $-2x \quad + \quad 5y \quad + 2xy + 3x^2$

Always combine similar terms.

■ Operations on Polynomials

Adding Polynomials

To add polynomials, combine similar terms.

EXAMPLE

Add. $(3x^4 + 5x^2y - x^2 + 8) + (3x^5 - 6x^2y + 7y^2x - 16)$

Remove the parentheses.	$3x^4 + 5x^2y - x^2 + 8 + 3x^5 - 6x^2y + 7y^2x - 16$
Group similar terms.	$3x^5 + 3x^4 + (5x^2y - 6x^2y) - x^2 - 7y^2x + (8 - 16)$
Combine similar terms.	$3x^5 + 3x^4 - x^2y - x^2 - 7y^2x - 8$

Subtracting Polynomials

To subtract polynomials, change the signs in the polynomial being subtracted. Then add the two polynomials and combine similar terms.

EXAMPLE

Subtract. $(5x^2y + 3xy^2 - 7xy + 23) - (-6x^2y^2 - 8x^2y + 3xy + 7)$

Change the signs of the polynomial being subtracted.	$5x^2y + 3xy^2 - 7xy + 23 + 6x^2y^2 + 8x^2y - 3xy - 7$
Group similar terms.	$(5x^2y + 8x^2y) + 3xy^2 + (-7xy - 3xy) + 6x^2y^2 + (23 - 7)$
Combine similar terms.	$13x^2y + 3xy^2 - 10xy + 6x^2y^2 + 16$

Multiplying Polynomials
Polynomial by Monomial

Multiply each term of the polynomial by the monomial.

EXAMPLE

Multiply. $3x^2(7x^5y^3 - 4x + 3y - 2)$

$$7x^5y^3 - 4x + 3y - 2$$
$$\underline{\times \qquad\qquad\qquad 3x^2}$$
$$21x^7y^3 - 12x^3 + 9x^2y - 6x^2$$

Binomial by Binomial

This is the most common form of polynomial multiplication you will encounter. Multiply one binomial by each term of the other binomial. Then add similar terms.

Multiply. $(3x + 4)(6x - 7)$

$$
\begin{array}{r}
6x - 7 \\
\times\ 3x + 4 \\
\hline
24x - 28 \\
18x^2 - 21x \\
\hline
18x^2 + 3x - 28
\end{array}
$$

Multiply by 4.

Multiply by $3x$.

Add similar terms.

Notice that each binomial is a factor of the final product.

The FOIL Method

The diagram below shows how to multiply binomials quickly.

Multiply. $(4x + 8)(-5x - 2)$

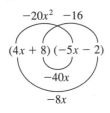

$-20x^2$ -16

$(4x + 8)$ $(-5x - 2)$

$-40x$

$-8x$

Multiply the **F**irst terms. $(4x)(-5x) = -20x^2$

Multiply the **O**uter terms. $(4x)(-2) = -8x$

Multiply the **I**nner terms. $(8)(-5x) = -40x$

Multiply the **L**ast terms. $(8)(-2) = -16$

Combine similar terms.

$(4x + 8)(-5x - 2) = -20x^2 - 8x - 40x - 16 = -20x^2 - 48x - 16$

You can often multiply mentally.

Dividing Polynomials

Dividing polynomials is similar to regular division. The best way to understand division with polynomials is to work through an example.

$(x^3 + 4x^2 + 9x + 10) \div (x + 2) = ?$

Write the terms in each expression in descending order of exponent values.

$x + 2 \overline{)x^3 + 4x^2 + 9x + 10}$

Focus on the leading term in the expression you are dividing by. In this example first ask, "how many times does x go into x^3?" The answer is x^2 times.

$$
\begin{array}{r}
x^2 \\
x + 2 \overline{)x^3 + 4x^2 + 9x + 10}
\end{array}
$$

Multiply $x^2(x + 2)$, write the partial product, and subtract.

Just as in regular division, bring down the next term, $9x$.

$$\begin{array}{r} x^2 \\ x + 2 \overline{)\; x^3 + 4x^2 + 9x + 10} \\ \underline{-\,(x^3 + 2x^2)} \\ 2x^2 + 9x \end{array}$$

Divide $(x + 2)$ into the new term.

What is $2x^2$ divided by x? The answer is $2x$. Write:

$$\begin{array}{r} x^2 + 2x \\ x + 2 \overline{)\; x^3 + 4x^2 + 9x + 10} \\ \underline{-\,(x^3 + 2x^2)} \\ 2x^2 + 9x \end{array}$$

$2x(x + 2) = 2x^2 + 4x$, so subtract and bring down the next term:

$$\begin{array}{r} x^2 + 2x \\ x + 2 \overline{)\; x^3 + 4x^2 + 9x + 10} \\ \underline{-\,(x^3 + 2x^2)} \\ 2x^2 + 9x \\ \underline{-\,(2x^2 + 4x)} \\ 5x + 10 \end{array}$$

Now find $5x$ divided by x. The answer is 5.

Complete the problem.

$$\begin{array}{r} x^2 + 2x + 5 \\ x + 2 \overline{)\; x^3 + 4x^2 + 9x + 10} \\ \underline{-\,(x^3 + 2x^2)} \\ 2x^2 + 9x \\ \underline{-\,(2x^2 + 4x)} \\ 5x + 10 \\ \underline{-(5x + 10)} \\ 0 \end{array}$$

Therefore $(x^3 + 4x^2 + 9x + 10) \div (x + 2) = x^2 + 2x + 5$.

Multiply to check your answer.

$(x + 2)(x^2 + 2x + 5)$

$= x^3 + 2x^2 + 5x + 2x^2 + 4x + 10$

$= x^3 + 4x^2 + 9x + 10$

Dividing always works, but you may be able to:
- write as a fraction
- factor
- simplify

Divide $(x^2 + x - 6) \div (x + 3)$.
Write as a fraction

$$\frac{x^2 + x - 6}{x + 3}$$

Factor.

$$\frac{(x + 3)(x - 2)}{(x + 3)}$$

Simplify.

$$\frac{\cancel{(x + 3)}(x - 2)}{\cancel{(x + 3)}} = x - 2$$

1. $(2x^2 - 3y)(7y - x) = ?$

    A. $2x^2 - 21xy + x$

    B. $-2x^3 + 14x^2y + 3xy - 21y^2$

    C. $-2x^3 + 14x^2y + 24xy$

    D. $7x^2y - 23x^3y^2 + 3y$

    E. $12xy - 21y^2 + 3xy$

SOLUTION

Multiply $(2x^2 - 3y)(7y - x)$.

$$(2x^2 - 3y)(7y - x) = 14x^2y - 2x^3 - 21y^2 + 3xy$$
$$= -2x^3 + 14x^2y + 3xy - 21y^2$$

The correct answer is B.

2. $(3x^3 + 6x) \div 3x = ?$

    F. $x^2 + 2$

    G. $x^2 - 2$

    H. $-x^2 + 2$

    J. $x^2 + 5$

    K. $x^2 - 5x$

SOLUTION

$$
\begin{array}{r}
x^2 + 0 + 2 \\
3x{\overline{\smash{\big)}\,3x^3 + 0x^2 + 6x}} \\
\underline{-(3x^3)} \\
0 + 0x^2 \\
\underline{0} \\
0 + 6x \\
\underline{-(\quad 6x)} \\
0
\end{array}
$$

The correct answer is F.

Practice

Combine similar terms.

1. $3x^2 + 4y + 3x^2y + 6y$

2. $7x + 3x^2y + 17x + 3yx^2 + 7x^2y^2$

3. $17x + 13 - 12xy + 16x^2 - 6y^2x + 4x^2y + 4xy$

4. $2x^2 - 3y + x^2 + 6y^2$

5. $15xy - 7x^2 + 9y^2 - 12xy + 2y^2 + 3y - 7x$

Add.

6. $(3x^5y - 2x^2 + 3y^2 - 12xy) + (7x^5y - 10y^2 - 8xy)$

7. $(5y^4 - 8y^3 + 2x^3 - 4x^2y^2 + 2y) + (12y^3 - 2x^3 + 14x^2y^2 - 19x)$

8. $(7x^2y^4 + 13xy^3 - 18y^2 + 2y) + (-21x^2y^4 - 17xy^3 + 5x^2 - 3y)$

9. $(8x^5 + 3x^2 - 5x^2y + 6xy^2) + (3x^5 - 2x^2 + 4x^3y - x^2y + 3xy^3 - 4y^2)$

10. $(6x^4 - 7x^3 + 2x^2y^2 - 4xy^2) + (6x^5 + 7x^4 + 3x^3 - 2x^2y^3 + 7xy^2)$

Subtract.

11. $(2x^5 - 4y^3 + 15x^2 - 6y) - (3x^5 + 2y^3 - 17y^2 + 6x)$

12. $(4xy^5 - 18x^5y + 3x^4 - 17y^3 + 11) - (5xy^5 - 12x^4 - 3y^3 + 12)$

13. $(11x^4 - 3x^3y^2 + 13xy - 12x) - (15x^4 + 17x^3y^2 - 15x^2y^3 + 19y)$

14. $(15x^8 + 9x^4 - 3x^2y + 5xy^3) - (15x^9 - 15x^4 - 3x^3y + x^2y)$

15. $(6x^2 - 3x + 3xy^2 + 3y^3 - 12) - (6x^3 - 3x^2 + 4xy^2 - 3y^5 - 18)$

Multiply.

16. $-5x^2(4x^3 - 3x + 4y^2 - 1)$

17. $8x^2(3x + 2xy + 4y - 6)$

18. $(3x + 6)(4x + 2)$

19. $(7x - 1)(2x + 5)$

20. $(9x + 4)(3x - 8)$

Divide.

21. $(2x^3 + 10x^2 - 8x) \div 2x$

22. $(3x^3 + 29x^2 + 9x - 6) \div (3x + 2)$

23. $(4x^4 + 8x^3 + 24x^2 + 4x) \div 4x$

24. $(x^3 + 10x^2 + 22x + 12) \div (x + 2)$

25. $(2x^3 - 4x^2 + 16x) \div 2x$

(Answers on page 350)

ACT-TYPE PROBLEMS

1. $(3x^5 - 2x^4y + 9x^2y^2 + 2xy) - (5x^4y^2 - 7x^5 - 18x^2y^2 + 2xy) = ?$

   A. $4x^5 + 7x^4y - 9x^2y^2 + 4xy$

   B. $10x^5 - 3x^4y^2 + 27x^2y^2$

   C. $-5x^4y^2 + 10x^5 - 2x^4y + 27x^2y^2$

   D. $-5x^2y^4 + 4x^5 + 9x^4y^4 + 4x^2y^2$

   E. $-2xy^2 - 9x^5y - 27x^2y^2 + 4xy^6$

2. $5x^2y(2x^3 - 6x^2y^2 + 4xy^2) = ?$

   F. $10x^6y - 30xy^4 + 20x^4y^4$

   G. $10x^5y - 30x^4y^3 + 20x^3y^3$

   H. $5x^5y - 5x^4y^3 + 5x^3y^3$

   J. $2x^6y - 6x^4y^4 + 4x^3y^4$

   K. $10x^6 - 30x^3y^3 + 20x^4y^3$

3. $(2x^2 + 5)(x^2 - 7) = ?$

   A. $2x^4 - 14x^2 - 35$

   B. $2x^4 + 5x^2 - 35$

   C. $2x^4 - 9x^2$

   D. $-9x^2 - 35$

   E. $2x^4 - 9x^2 - 35$

4. $(2x^3 + 7x^2 - 11x - 7) \div (2x + 1) = ?$

   F. $x^2 - 3x + 7$

   G. $x^2 + 3x - 7$

   H. $x^2 - 3x - 7$

   J. $x^2 + 7x - 3$

   K. $x^2 - 7x - 3$

5. $(2y^5x - 7y^3x^2 + 13y^2x - 6x) + (10y^5x + 3y^2x^3 - 21y^2x + 5x) = ?$

   A. $12x^5y - 4y^2x^3 + 8yx^2 + 1$

   B. $8y^5x - 4y^3x^2 + y^2x - x$

   C. $12y^5x - 7y^3x^2 + 3y^2x^3 - 8y^2x - x$

   D. $-12x^5y + 4y^2x^3 - 8yx^2 - 11x$

   E. $-12y^5x - 7y^3x^2 - 3y^2x^3 + 8yx^2 + x$

(Answers on page 350)

Factoring Polynomials

Factors of a polynomial are expressions whose product equals the polynomial.

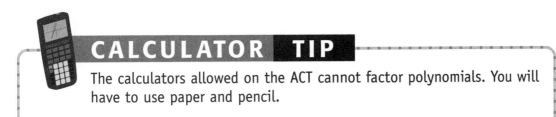

CALCULATOR TIP

The calculators allowed on the ACT cannot factor polynomials. You will have to use paper and pencil.

Factoring Out Common Factors

You may be able to find a common factor in each term of a polynomial.

- Choose the greatest common factor of the coefficients.
- Choose the smallest exponent for each variable.

EXAMPLES

1. Factor. $35x^7y^8 + 14x^2y^5 - 63x^8y^3 + 84x^5y^9$

   7 is the greatest common factor of the coefficients.

   x^2 and y^3 are the smallest powers of the variables.

   Factor out $7x^2y^3$.

   $35x^7y^8 + 14x^2y^5 - 63x^8y^3 + 84x^5y^9 = 7x^2y^3(5x^5y^5 + 2y^2 - 9x^6 + 12x^3y^6)$

2. Factor. $49z^3x - 24z^5x^3 + 10z^2x^6 - z^3x + 14z^2x^6$

   Combine similar terms. $48z^3x - 24z^5x^3 + 24z^2x^6$

   Factor out $24z^2x$. $24z^2x(2z - z^3x^2 + x^5)$

Factoring Completely

You may have to complete several steps before you have factored a polynomial completely.

EXAMPLES

1. Factor. $(4x - 6 - x + 10)(2x - 2) + (3x + 4)(5x - 1)$

   Combine similar terms. $(3x + 4)(2x - 2) + (3x + 4)(5x - 1)$

   Factor out $(3x + 4)$. $(3x + 4)[(2x - 2) + (5x - 1)]$

   Combine similar terms. $(3x + 4)(7x - 3)$

2. Factor. $12w^2 - 4wz + 21wz - 7z^2$

Group to show the two parts to factor.	$(12w^2 - 4wz) + (21wz - 7z^2)$
Factor out $4w$ from one part and $7z$ from the other.	$4w(3w - z) + 7z(3w - z)$
Factor out $(3w - z)$.	$(4w + 7z)(3w - z)$

Special Factors

Memorize these methods of factoring polynomials. The first two are particular favorites of ACT test writers. Remember, these forms may appear either in polynomial form or in factored form. You will be asked to write them in the other form.

Difference of Squares

$x^2 - y^2 = (x + y)(x - y)$

EXAMPLES

1. Factor. $81x^2 - 64y^2$

   $81x^2 - 64y^2 = (9x + 8y)(9x - 8y)$

2. Factor. $144 - 0.25y^2$

   $144 - 0.25y^2 = (12 + 0.5y)(12 - 0.5y)$

> You may have to factor the coefficients to reveal a special factor.
> $3x^2 - 27$
> $= 3(x^2 - 9)$
> $= 3(x + 3)(x - 3)$

Perfect Square

$x^2 + 2xy + y^2 = (x + y)(x + y) = (x + y)^2$

EXAMPLES

1. Factor. $16x^2 + 40xy + 25y^2$

   $16x^2 + 40xy + 25y^2 = (4x + 5y)(4x + 5y) = (4x + 5y)^2$

2. Factor. $4x^2 + 36x + 81$

   $4x^2 + 36x + 81 = (2x + 9)(2x + 9) = (2x + 9)^2$

Sum of Cubes

$x^3 + y^3 = (x + y)(x^2 - xy + y^2)$

EXAMPLES

1. Factor. $8x^3 + 27y^3$

   $8x^3 + 27y^3 = (2x + 3y)(4x^2 - 6xy + 9y^2)$

2. Factor. $64 + 8y^3$

   $64 + 8y^3 = (4 + 2y)(16 - 8y + 4y^2)$

Difference of Cubes

$x^3 - y^3 = (x - y)(x^2 + xy + y^2)$

1. Factor. $125x^3 - y^3$

   $125x^3 - y^3 = (5x - y)(25x^2 + 5xy + y^2)$

2. Factor. $0.008x^3 - 27$

   $0.008x^3 - 27 = (0.2x - 3)(0.04x^2 + 0.6x + 9)$

MODEL ACT PROBLEM

1. Factor. $12x^4y + 18x^3y^2 - 30x^2y$

   A. $4xy(3x^3y + 4x^2y^2 - 7x^2y)$

   B. $6x^2y(2x^2 + 3xy - 5)$

   C. $6xy(2x^2 + 3x^2y - 5y)$

   D. $3x^2y(4x^2 + 6xy - 10)$

   E. $3x^2y(2x^2 + 3x^2y - 5y)$

SOLUTION

Choose the greatest common factor of the coefficients.   6

Choose the smallest exponent of each variable. $x: x^2$

   $y: y$

Factor the polynomial.

   $12x^4y + 18x^3y^2 - 30x^2y = 6x^2y(2x^2 + 3xy - 5)$

The correct answer is B.

2. Factor. $2x^2 - 2x + 3x - 3$

   F. $(x - 1)(x + 3)$

   G. $(x - 3)(2x + 1)$

   H. $(x - 1)(2x + 3)$

   J. $(x + 3)(x - 1)$

   K. $(2x - 1)(x + 3)$

SOLUTION

Group to show the parts to be factored.

   $(2x^2 - 2x) + (3x - 3)$

Factor out $2x$ from the first part and 3 from the second part.

   $2x(x - 1) + 3(x - 1)$

Factor out $(x - 1)$ because it is the common factor in both terms.

   $(x - 1)(2x + 3)$

The correct answer is H.

Practice

Factor each polynomial completely.

1. $4x^4 + 10x^4y^2 - 2x^4y^2 + 2x^4$

2. $4x^2y^3 - 12y^4 + 16xy^4$

3. $12x^3 + 10x^2 - 20x$

4. $z^2 + 8w^2 - 1.49z^2 + 17w^2$

5. $17k^2 + 10y^2 - 2k^2 - 5y^2$

6. $27z^2x^2 - 12z^2x^2 + 9z^2$

7. $5x^3y^2 + 10xy - 15x^2y^2 - 5xy^2$

8. $8x^4y^5 - 16x^2y + 12x^3y^3$

9. $x^2 + 3x - 5x - 15$

10. $72x^4y^2z^4 + 60xyz$

11. $7y^2z - 28yz - 21y^2z + 14yz$

12. $4x^2 - 16y^2$

13. $6x^2 + 15x - 14x - 35$

14. $7x^3z^2 - 13x^2z^2 + 27x^2z$

15. $64x^3 - 27y^3$

16. $16x^4z^3 - 4x^3z^2 - 8x^2z^2 + 20x^5z^3$

17. $2x^2 + 12x - 9x - 54$

18. $(4x - 7 - 2x + 2)(3x - 4) + (2x - 5)(3x + 1)$

19. $125x^3 + 343y^3$

20. $x^2 - x + 2x - 2$

(Answers on page 350)

1. Factor $144x^2y^2 - 169z^2$ completely.

   A. $(12xy - 13z)(12xy + 13z)$

   B. $(12x - 13y)(12x + 13y)$

   C. $(13xz + 12y)(13xz - 12y)$

   D. $(12xz - 13y)(12xz + 13y)$

   E. $(12 - 13z)(12y + 13z)$

2. Factor $6x^2 + 27x - 14x - 63$ completely.

   F. $(7x - 3)(9x + 2)$

   G. $(2x + 9)(3x - 7)$

   H. $(7x + 9)(3x - 2)$

   J. $(7x + 3)(9x - 2)$

   K. $(9x + 3)(7x - 2)$

3. Factor $24x^3y^2 - 12x^2y^2 + 16x^3y^3 - 8x^5y^4$ completely.

   A. $8x^2y^2(3x - 2 + 2xy - x^3y^2)$

   B. $2xy(12x^2y - 6xy + 8x^2y^2 - 4x^4y^3)$

   C. $4x^2y^2(6x - 3 + 4xy - 2x^3y^2)$

   D. $4xy(6x^2y - 3xy + 4x^2y^2 - 2x^4y^3)$

   E. $2x^2y^2(12x - 4 + 8xy - 4x^3y^2)$

4. Factor $(3y - 5 - 2y + 3)(2y + 3) + (y - 2)$.

   F. $(2y - 1)(y + 4)$

   G. $(4y - 1)(y - 1)$

   H. $2(y - 2)(y + 2)$

   J. $2(y^2 - 4)$

   K. $2(y - 1)(y + 2)$

5. When $9x^2 - 16$ is completely factored, what is the sum of the factors?

   A. $3x - 4$

   B. $3x + 4$

   C. $3x$

   D. $6x$

   E. $3x + 8$

(Answers on page 351)

■ Quadratic Equations

Quadratic equations can be written in the form $ax^2 + bx + c = 0$ $(a \neq 0)$. They can also be written in function notation: $f(x) = ax^2 + bx + c$ $(a \neq 0)$.

Factors of the quadratic equation are expressions whose product is $ax^2 + bx + c$.

Factoring Polynomials in the Form $ax^2 + bx + c$ $(a \neq 0)$

Usually, factoring quadratic expressions begins with an educated guess.

EXAMPLES

1. Factor. $x^2 + 4x - 5 = 0$

   The first terms must both be $(x \underline{\quad})(x \underline{\quad})$.

   The second terms must be -5 and 1 or -1 and 5 because the product must be -5.

   Adding -1 and 5 gives 4, the coefficient of the middle term.

   $x^2 + 4x - 5 = (x - 1)(x + 5)$

2. Factor. $10x^2 - 6x = 5x + 6$

Always write the equation as a quadratic equal to 0.

$10x^2 - 6x = 5x + 6 \rightarrow 10x^2 - 11x - 6 = 0$

Then factor the trinomial $10x^2 - 11x - 6$.

The first terms could be $(x \_\_\_)(10x \_\_\_)$ or $(2x \_\_\_)(5x \_\_\_)$.

The product of the second terms must be -6.

So the possible factors are:

$(x \_\_\_)(10x \_\_\_)$		or	$(2x \_\_\_)(5x \_\_\_)$	
-6	1		-6	1
6	-1		6	-1
1	-6		1	-6
-1	6		-1	6
-3	2		-3	2
3	-2		3	-2
-2	3		-2	3
2	-3		2	-3

The sum of the inner and outer products must be $-11x$.

$$\overset{-15x}{\overbrace{(2x - 3)\,(5x + 2)}}_{\underbrace{}_{4x}}$$

$10x^2 - 11x - 6 = 0$

$(2x - 3)(5x + 2) = 0$

Sometimes you may find special factors.

EXAMPLES

1. Factor. $16x^2 + 40x + 25 = 0$

This polynomial is a perfect square.

$16x^2 + 40x + 25 = (4x + 5)(4x + 5) = (4x + 5)^2$

2. Factor. $36x^2 - 49 = 0$

This polynomial is the difference of squares.

$36x^2 - 49 = (6x + 7)(6x - 7)$

Solving Quadratic Equations by Factoring

Follow these steps.

- Write the equation in the form $ax^2 + bx + c = 0$.
- Factor the polynomial $ax^2 + bx + c$.
- Find the solution set.

The solution set contains the roots of the polynomial.

EXAMPLES

1. Factor to solve the quadratic equation. $x^2 + 4x - 5 = 0$

   $x^2 + 4x - 5 = (x - 1)(x + 5)$

   Since $x^2 + 4x - 5 = 0$, $(x - 1)(x + 5) = 0$.

   If $(x - 1)(x + 5) = 0$, then $(x - 1) = 0$ or $(x + 5) = 0$ or they both equal zero.

   Solve the equations $x - 1 = 0$ and $x + 5 = 0$ to find the solution set for the quadratic equation.

   $$x - 1 = 0 \qquad\qquad x + 5 = 0$$
   $$x = 1 \qquad\qquad x = -5$$

   The solution set for the equation $x^2 + 4x - 5 = 0$ is $\{1, -5\}$.

2. Solve the quadratic equation. $10x^2 - 6x = 5x + 6$

   Write in standard form. $10x^2 - 6x = 5x + 6 \rightarrow 10x^2 - 11x - 6 = 0$

   $$10x^2 - 11x - 6 = (2x - 3)(5x + 2)$$

   So $(2x - 3)(5x + 2) = 0$.

   If $(2x - 3)(5x + 2) = 0$, then $(2x - 3) = 0$ or $(5x + 2) = 0$ or they both equal zero.

   Solve the equations $2x - 3 = 0$ and $5x + 2 = 0$ to find the solution set for the quadratic equation.

   $$2x - 3 = 0 \qquad\qquad 5x + 2 = 0$$
   $$2x = 3 \qquad\qquad 5x = -2$$
   $$x = \frac{3}{2} \qquad\qquad x = -\frac{2}{5}$$

   The solution set for the equation $10x^2 - 11x - 6 = 0$ is $\left\{\frac{3}{2}, -\frac{2}{5}\right\}$.

3. Solve the quadratic equation. $16x^2 + 40x + 25 = 0$

   $$16x^2 + 40x + 25 = (4x + 5)^2$$
   $$4x + 5 = 0$$
   $$4x = -5$$
   $$x = -\frac{5}{4}$$

   The solution set is $\left\{-\frac{5}{4}\right\}$.

4. Solve the quadratic equation. $36x^2 - 49 = 0$

$$36x^2 - 49 = (6x + 7)(6x - 7)$$

$6x + 7 = 0$	$6x - 7 = 0$
$6x = -7$	$6x = 7$
$x = -\dfrac{7}{6}$	$x = \dfrac{7}{6}$

The solution set is $\left\{-\dfrac{7}{6}, \dfrac{7}{6}\right\}$.

MODEL ACT PROBLEMS

1. What is the solution set for the quadratic equation $16x^2 - 4 = 0$?

A. $\left\{\dfrac{1}{2}, -\dfrac{1}{2}\right\}$

B. $\{2, -2\}$

C. $\{0\}$

D. $\{1, -1\}$

E. $\left\{\dfrac{1}{3}, -\dfrac{1}{3}\right\}$

SOLUTION

The equation $16x^2 - 4 = 0$ is the difference of two squares.

$$(4x - 2)(4x + 2) = 0$$

Set each factor equal to 0 to find the solution set.

$4x - 2 = 0$	$4x + 2 = 0$
Add 2. $4x = 2$	Subtract 2. $4x = -2$
Divide by 4. $x = \dfrac{1}{2}$	Divide by 4. $x = -\dfrac{1}{2}$

$$x = \left\{\dfrac{1}{2}, -\dfrac{1}{2}\right\}$$

The correct answer is A.

Hint:
You can also substitute answers in the equation. Keep trying answers until you find an answer set that makes the equation correct.

2. What is the solution set to the quadratic equation $2x^2 + 3x = 2$?

F. $\{1, -1\}$

G. $\{2, -2\}$

H. $\left\{\dfrac{1}{2}, -\dfrac{1}{2}\right\}$

J. $\left\{\dfrac{1}{2}, -2\right\}$

K. $\left\{-\dfrac{1}{2}, 2\right\}$

SOLUTION

Rewrite the equation in standard form.
$$2x^2 + 3x = 2 \rightarrow 2x^2 + 3x - 2 = 0$$

The first terms must be: $(2x \underline{\quad})(x \underline{\quad}) = 0$

The product of the outer terms must be -2, so the possible factors are:

$(2x \underline{\quad})(x \underline{\quad}) = 0$

-1	2
2	-1
1	-2
-2	1

The sum of the inner and outer products must be 3, so the correct factorization is $(2x - 1)(x + 2) = 0$.

Solve the quadratic equation.

$2x - 1 = 0$	$x + 2 = 0$
Add 1. $2x = 1$	Subtract 2. $x = -2$
Divide by 2. $x = \dfrac{1}{2}$	

The solution set is $x = \left\{\dfrac{1}{2}, -2\right\}$.

The correct answer is J.

Practice

Factor to solve each quadratic equation.

1. $x^2 + 4x + 3 = 0$
2. $4x^2 - 10x + 6 = 0$
3. $9x^2 + 54x = -81$
4. $6x^2 + x = 15$
5. $7x^2 = 126$
6. $8x^2 + 40x = 0$
7. $x = -2x^2 + 21$
8. $4x^2 = 16$
9. $9x^2 + 12x + 4 = 0$
10. $-19x - 5 = -4x^2$
11. $x^2 - x - 6 = 0$
12. $25x^2 - 36 = 0$
13. $0 = -x^2 + 9$
14. $3x^2 = x + 14$
15. $4x^2 - 20x = -25$
16. $x^2 + 24 = -14x$
17. $14x^2 - 5x - 1 = 0$
18. $2x^2 = -19x - 39$
19. $x^2 + 2x + 1 = 0$
20. $29x = -10x^2 + 21$

(Answers on page 351)

ACT-TYPE PROBLEMS

1. What is the solution set to the quadratic equation $x^2 - 81 = 0$?

   A. $\{1, -1\}$
   B. $\{3, -3\}$
   C. $\{5, -5\}$
   D. $\{7, -7\}$
   E. $\{9, -9\}$

2. What is the solution set to the quadratic equation $4x^2 - 24x = -36$?

   F. $\{3\}$
   G. $\{3, -3\}$
   H. $\{2\}$
   J. $\{2, -2\}$
   K. $\{-3\}$

3. Which of the following cannot be a solution set for a quadratic equation?

   A. $\{15\}$
   B. $\{-1, 1, 2\}$
   C. $\{-5, 5\}$
   D. $\{3\}$
   E. $\{3, 7\}$

(Answers on page 351)

4. What is the sum of the solutions to the quadratic equation $(x - 3)(x + 5) = 0$?

   F. 8
   G. 2
   H. 5
   J. -2
   K. 3

5. What is the product of the solutions to the quadratic equation $10x^2 = -21x + 10$?

   A. -2
   B. -1
   C. 0
   D. 1
   E. 2

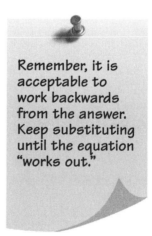

Remember, it is acceptable to work backwards from the answer. Keep substituting until the equation "works out."

Cumulative ACT Practice
Elementary Algebra

Complete this Cumulative ACT Practice in 10 minutes to reflect real ACT test conditions. This Cumulative ACT Practice gives you an additional opportunity to practice elementary algebra concepts in an ACT format. If you don't know an answer, eliminate and guess. Circle the number of any guessed answer. Then check your answers on page 352. You will also find explanations for the answers and suggestions for further study.

1. In the formula $E = IR$, E = Voltage, I = Amperage, and R = Resistance. If the voltage is 12 what must the amperage be so that the resistance is 2.5?

    A. 4
    B. 4.5
    C. 4.8
    D. 5.3
    E. 5.8

2. $x = 2$ and $x = -4$ is the solution set for which of the following quadratic equations?

    F. $x^2 - 2x + 8 = 0$
    G. $x^2 + 2x - 8 = 0$
    H. $x^2 - 8x + 2 = 0$
    J. $x^2 + 8x + 2 = 0$
    K. $x^2 - 8x - 2 = 0$

3. $3\sqrt{15}$ is the simplified form of which expression?

    A. $\dfrac{30\sqrt{3}}{2\sqrt{5}}$
    B. $3\sqrt{10} + 3\sqrt{5}$
    C. $\sqrt{45} - \sqrt{30}$
    D. $\sqrt[3]{5} \cdot \sqrt[3]{3} \cdot 3$
    E. $45 \div 3\sqrt{15}$

4. $5a^2b^3 - 2ab - 3a + 3b$ is formed by adding which two polynomials?

    F. $4a^2b^3 - 5ab + 3a$ and $2a^2b^2 - 3ab - 2b$
    G. $a^2b^3 + 2ab - 3a$ and $4a^2b^3 - 4ab + 3b$
    H. $5a^2b^3 + 2a - 3b$ and $-2ab + 5a$
    J. $a^3b^2 + 2ab + 3a$ and $4a^3b^2 + 4ab - 3b$
    K. $-a^2b^2 + 2ab - 3a$ and $4a^2b^3 - 4ab + 3b$

5. $\dfrac{7x^3 + 9x^2 + 11x - 15}{7x - 5} = ?$

    A. $x^2 + 2x + 3$
    B. $2x^2 - 3x + 5$
    C. $x^2 - 2x - 3$
    D. $2x^2 + 2x + 3$
    E. $x^2 - 3x + 5$

6. What is the sum of the solutions of the equation $x^2 - 5x + 6 = 0$?

    F. -5
    G. -4
    H. 4
    J. 5
    K. 6

7. $2^5 \div \sqrt[3]{64} = ?$

    A. 16
    B. 8
    C. 4
    D. 2.5
    E. 2

8. What is the value of $\dfrac{3^3 \div 3 - 6}{\left(\frac{1}{2}\right)^{-2} - 1}$?

    F. 1
    G. 2
    H. 3
    J. 4
    K. 5

9. In simplest form, $5\sqrt{12} + 7\sqrt{108} =$

    A. $26\sqrt{12}$
    B. $12\sqrt{120}$
    C. $52\sqrt{6}$
    D. $52\sqrt{3}$
    E. $20\sqrt{3}$

10. What is the value of $\dfrac{(\sqrt[3]{8} - 2) + 3}{(-2)^{-1} + 1}$?

    F. 7
    G. 6
    H. 3
    J. 2
    K. 1

Subtest Elementary Algebra

Complete this Subtest in 10 minutes to reflect real ACT test conditions. This Subtest has the same number and type of elementary algebra items as are found on the ACT. If you don't know an answer, eliminate and guess. Circle the number of any guessed answer. Then check your answers on page 353. You will also find explanations for the answers and suggestions for further study.

1. If $a = 2$, $b = -4$, and $c = 2$, what is the value of $\dfrac{-b \pm \sqrt{b^2 - 4ac}}{2a}$?

   A. 0.75
   B. 1
   C. $\sqrt{2}$
   D. 1.5
   E. $\sqrt{2} + 1$

2. $\dfrac{x^3 + 4x^2 + 9x + 10}{x + 2} = $?

   F. $x^2 + 2x + 5$
   G. $x^2 - 5x - 2$
   H. $x^2 + 5x + 2$
   J. $x^2 + 2x - 5$
   K. $x^2 - 2x + 5$

3. Solve the quadratic equation. $2x^2 + 2x - 12 = 0$

   A. $x = -3, x = 2$
   B. $x = 6, x = -2$
   C. $x = 3, x = -2$
   D. $x = 0.5, x = -1$
   E. $x = 2, x = -2$

4. Which of the following is $3x^2 + x\left(2x - \dfrac{3}{x}\right) + 7$ written in its simplest form?

   F. $3x^2 + 2x + 7$
   G. $2x^2 + 7$
   H. $5x^2 + 4$
   J. $5x^2 + 3x + 7$
   K. $3x^2 - 7$

5. For all x and y values, what is the sum of $x^3y + 2y^2x + 3x$ and $3y^2x - 4x^3y - 2x$?

   A. $2x^3y + y^2x + 5x$
   B. $-3x^3y - 5y^2x - x$
   C. $-3x^3y + 5y^2x + x$
   D. $3x^3y - 5y^2x - x$
   E. $5x^3y + (-3y^2x) + x$

6. $3^4 \div \dfrac{1}{\sqrt[4]{81}} = $?

   F. 1
   G. 3
   H. 9
   J. 27
   K. 243

7. What is the product of the solutions of the equation $x^3 - 6x^2 + 5x = 0$?

   A. -6
   B. -4
   C. 0
   D. 5
   E. 6

8. Factor $6x^2 + 13x - 63$ completely.

   F. $(3x - 7)(2x - 9)$
   G. $(3x + 7)(2x - 9)$
   H. $(3x - 7)(2x + 9)$
   J. $(2x - 7)(3x + 9)$
   K. $(2x + 7)(3x - 9)$

9. A car is traveling for 9,000 seconds and travels a distance of 140 miles. What is the speed of the car in miles per hour given the equation distance = rate × time?

   A. 56 miles per hour
   B. 64 miles per hour
   C. 156 miles per hour
   D. 212 miles per hour
   E. 300 miles per hour

10. Factor $24x^3y^2 + 6x^2y^3 - 8xy^2 + 12xy$ completely.

   F. $2xy(12x^2y + 3xy^2 - 4y + 6)$
   G. $4x^2y^2(6x + y - 2 + 3)$
   H. $2xy(12xy + 3xy - 4y + 6x)$
   J. $2x^2y(12xy + 3y^2 - 4y + 6)$
   K. $2xy(12x^2y + 3xy - y + 6)$

Cumulative ACT Practice
Pre-Algebra/Elementary Algebra

Complete this Cumulative ACT Practice in 25 minutes to reflect real ACT test conditions. This Cumulative ACT Practice gives you an additional opportunity to practice pre-algebra and elementary algebra concepts in an ACT format. If you don't know an answer, eliminate and guess. Circle the number of any guessed answer. Then check your answers on page 354. You will also find explanations for the answers and suggestions for further study.

1. There are 6 black marbles, 2 blue marbles, and 4 orange marbles in a jar. If you reach into the jar and pick out a single marble, what is the probability that the marble will be blue?

   A. $\frac{1}{6}$

   B. $\frac{1}{3}$

   C. $\frac{1}{2}$

   D. $\frac{2}{3}$

   E. $\frac{5}{6}$

2. The equation for volume of a right circular cylinder is $V = \pi r^2 h$. The volume can be found if you know the height and radius of the cylinder. Which of the following allows you to find the radius of a right circular cylinder if you know the height and the volume of the cylinder?

   F. $r = \sqrt{\dfrac{V}{\pi h}}$

   G. $r = \dfrac{V}{\pi h}$

   H. $r = \sqrt{V\pi h}$

   J. $r = V\pi h$

   K. $r = \sqrt{\dfrac{\pi h}{V}}$

3. If $x^4 - 3x^3 - 18x^2 = 0$, what is the product of the possible values for x?

   A. -18

   B. -12

   C. -6

   D. -3

   E. 0

4. You roll two fair dice. What is the probability that the sum of the numbers you roll will be divisible by 3?

   F. $\frac{1}{6}$

   G. $\frac{1}{5}$

   H. $\frac{1}{4}$

   J. $\frac{1}{3}$

   K. $\frac{1}{2}$

5. 60 is 40% of what number?

   A. 180
   B. 150
   C. 120
   D. 110
   E. 105

6. $(5x^2y + 2xy + y) - (3x^2y + 3x + 2y) = ?$

   F. $8x^2y + 5xy + 3y$

   G. $2x^2y - 5xy + 3x$

   H. $2x^2y + x - y$

   J. $2x^2y + 2xy - 3x - y$

   K. $-2x^2y - 5xy + 3x - y$

7. How many feet long is a 100-yard football field?

   A. 100 feet
   B. 200 feet
   C. 300 feet
   D. 400 feet
   E. 500 feet

8. Each day, Andy's Ice Cream Shop sells twice as many double-scoop ice cream cones, at $2.50 a cone, as single-scoop ice cream cones, at $1.50 a cone. If Andy's Ice Cream Shop sells 300 double-scoop cones on Monday, how much money did customers spend on single-scoop and double-scoop cones on Monday?

   F. $975
   G. $925
   H. $850
   J. $825
   K. $800

9. Given that $s = 4$ and $t = -3$, evaluate the expression $(s^3 - t^3 + s)(t^2 + \sqrt{s})$.

   A. 1,045
   B. 1,001
   C.   451
   D.   259
   E. −637

10. What is the prime factorization of 24?

   F. $4 \cdot 6$
   G. $2 \cdot 2 \cdot 2 \cdot 2$
   H. $2 \cdot 3 \cdot 3$
   J. $2 \cdot 2 \cdot 2 \cdot 3$
   K. $2 \cdot 2 \cdot 3$

11. If $a = \sqrt{c^2 - b^2}$, and $c = 13$ and $b = 5$, what is the value of a?

   A. 12
   B. 10
   C. 8
   D. 6
   E. 5

12. Laurie jogs 4 miles each day Monday through Friday. The following is a list of her times.

Day of the week	Time
Monday	40 minutes
Tuesday	37 minutes
Wednesday	42 minutes
Thursday	36 minutes
Friday	35 minutes

What is Laurie's average time for the week?

   F. 42 minutes
   G. 40 minutes
   H. 38 minutes
   J. 37 minutes
   K. 36 minutes

13. The initial fee for a cab ride is $2.80. You are also charged $0.15 for every half-mile you travel. How much does it cost for a 27.5-mile cab ride?

   A. $13.85
   B. $11.05
   C. $ 8.25
   D. $ 5.45
   E. $ 2.80

14. What are the roots of the quadratic equation $x^2 - 4x - 12 = 0$?

   F. $x = -4, x = -12$
   G. $x = 4, x = 12$
   H. $x = -1, x = 5$
   J. $x = 3, x = -7$
   K. $x = -2, x = 6$

15. $\dfrac{(5^2 - \sqrt{9} + (-2)^3) \cdot 2^{-2}}{\left(3^{-1} + \frac{2}{3}\right) \cdot 4^{-2}} = ?$

   A. 224
   B. 144
   C. 80
   D. 65
   E. 56

16. $(6x^3 - 25x^2 + 24x - 35) \div (2x - 7) = ?$

   F. $3x^2 + 2x + 5$
   G. $3x^2 - 2x - 5$
   H. $3x^2 + 2x - 5$
   J. $3x^2 - 2x + 5$
   K. $3x^2 - 5x + 5$

17. The area of a circle is $A = \pi r^2$. Which of the following expresses r in terms of A?

   A. $r = \dfrac{A}{\pi}$

   B. $r = \sqrt{A\pi}$

   C. $r = \sqrt{\dfrac{A}{\pi}}$

   D. $r = \dfrac{\pi}{A}$

   E. $r = \sqrt{\dfrac{\pi}{A}}$

18. When Jordan plays basketball, he makes one shot out of every 3 attempts. How many baskets should Jordan make if he takes 39 shots?

   F. 13
   G. 16
   H. 19
   J. 22
   K. 25

19. Which of the following is 3.56×10^5 in standard form?

    A.   0.0000356
    B.   0.000356
    C.   35,600
    D.   356,000
    E.   3,560,000

20. $(5.73 \times 10^2) - (3.74 \times 10^3) = ?$

    F. 1.99×10^5
    G. -3.167×10^3
    H. 1.99×10^{-1}
    J. 3.167×10^{-3}
    K. 1.99×10^1

21. Calvin's Pet Center charges \$10 to wash a dog, and one and a half times as much to cut a dog's hair. On a certain day, Calvin's Pet Center washes x dogs and gives y dog haircuts. Which of the following can be used to determine the amount in dollars they will collect from washes and haircuts that day?

    A. $10x + 1.5y$
    B. $10x + 20y$
    C. $1.5x + 20y$
    D. $5x + 10y$
    E. $10x + 15y$

22. If $r = 4, p = -1$, and $q = 2$, evaluate
    $$\frac{(rp)^2 - pq + 3}{(r - p + q)^2}.$$

    F. $\dfrac{3}{7}$

    G. $\dfrac{7}{3}$

    H.   3
    J.   7
    K.   21

23. What is the sum of $(r^2t + t^2r - r^2 + t)$ and $(3r^2t - 2t^2r + 5t^2 - t)$?

    A. $2r^2t + t^2r - 5t^2r^2 - t$
    B. $4r^2t - t^2r - 5t^2$
    C. $3r^2t + 2t^2r + 5t^2 - r$
    D. $4r^2t - t^2r + 5t^2 - r^2$
    E. $2r^2t - t^2r + 4t^2 - r^2$

24. Given the equation $m = \dfrac{y_2 - y_1}{x_2 - x_1}$, what is the value of m if $y_1 = 3, y_2 = 2, x_1 = 5$, and $x_2 = 7$?

    F.   2

    G. $\dfrac{1}{2}$

    H. $-\dfrac{1}{2}$

    J. -1

    K. -2

25. The area of a triangle equals 38 square meters. The sum of the base and the height is 23 meters. What are the measures of the base and the height?

    A.   11 meters and 12 meters
    B.   10 meters and 13 meters
    C.   8 meters and 15 meters
    D.   4 meters and 19 meters
    E.   3 meters and 20 meters

Subtest Pre-Algebra/Elementary Algebra

Complete this Subtest in 24 minutes to reflect real ACT test conditions. This Subtest has the same number and type of pre-algebra and elementary algebra items as are found on the ACT. On the real ACT, these items make up the EA subscore. If you don't know an answer, eliminate and guess. Circle the number of any guessed answer. Then check your answers on page 356. You will also find explanations for the answers and suggestions for further study.

1. In a drawer there are 5 blue socks, 3 red socks, and 6 black socks. If you reach into the drawer and pull out a sock, what is the probability that you will NOT get a black sock?

   A. $\frac{3}{7}$

   B. $\frac{4}{7}$

   C. $\frac{5}{7}$

   D. $\frac{15}{7}$

   E. 1

2. The equation for the volume of a sphere is $V = \frac{4}{3}\pi r^3$.

   Which of the following expressions shows r written in terms of V?

   F. $r = \sqrt[3]{\frac{V - \frac{4}{3}}{\pi}}$

   G. $r = \sqrt[3]{\frac{4}{3}V\pi}$

   H. $r = \sqrt[3]{\frac{\frac{3}{4}V}{\pi}}$

   J. $r = \left(\frac{4}{3}V\pi\right)^3$

   K. $r = \left(\frac{V - \frac{4}{3}}{\pi}\right)^3$

3. If $x^3 + 2x^2 - 15x = 0$, what is the sum of the possible values for x?

   A. 8
   B. 5
   C. 3
   D. −2
   E. −3

4. What is the prime factorization of 540?

   F. $2 \times 2 \times 2 \times 3 \times 3 \times 3$
   G. $4 \times 5 \times 27$
   H. 20×27
   J. $2 \times 2 \times 3 \times 3 \times 3 \times 5$
   K. $2 \times 2 \times 2 \times 3 \times 3 \times 5$

5. 14% of 60 is?

   A. 0.084
   B. 0.84
   C. 8.4
   D. 84
   E. 840

6. What is the difference of $2ab^2 + 4a^2b - 3a^2b^2$ and $5ab^2 - a^2b + 6a^2b^2$?

   F. $3ab^2 - 5a^2b + 9a^2b^2$

   G. $7ab^2 + 5a^2b - 3a^2b^2$

   H. $-7ab^2 - 5a^2b + 3a^2b^2$

   J. $-3ab^2 + 5a^2b + 9a^2b^2$

   K. $-3ab^2 + 5a^2b - 9a^2b^2$

7. If a 12-foot pole casts a 4.8-foot shadow, how tall is a pole that casts an 8-foot shadow?

   A. 72 feet
   B. 20 feet
   C. 7.2 feet
   D. 3.2 feet
   E. 2 feet

8. The area of a rectangle is equal to 132 square meters. The sum of two of the perpendicular sides of the rectangle is equal to 23 meters. What are the measures of each of these sides?

   F. 7 meters and 16 meters
   G. 8 meters and 15 meters
   H. 11 meters and 12 meters
   J. 9 meters and 14 meters
   K. 10 meters and 13 meters

9. Given that $a = 2$, $b = -4$, and $c = 5$, evaluate the expression $\frac{b^3 - abc - b^a}{c^2 + bc}$.

   A. 24
   B. 6
   C. −8
   D. −24
   E. −30

10. What is the prime factorization of 72?

    F. $2 \cdot 2 \cdot 3 \cdot 3 \cdot 3$
    G. $2 \cdot 4 \cdot 9$
    H. $2 \cdot 2 \cdot 3 \cdot 6$
    J. $2 \cdot 2 \cdot 2 \cdot 3 \cdot 3$
    K. $8 \cdot 9$

11. If $c = \sqrt{a^2 + b^2}$, and $a = -3$ and $b = 4$, what is the value of c?

    A. 1
    B. 2
    C. 3
    D. 4
    E. 5

12. A person reaches into a jar containing 20 marbles and takes marbles out 10 different times. The marbles are returned each time. The following is a list of how many marbles the person removed from the jar: 10, 9, 15, 13, 11, 17, 11, 16, 8, 14. What is the median of this event?

    F.   8
    G. 11
    H. 12
    J. 12.4
    K. 17

13. A car salesperson makes \$7.00 per hour plus \$100.00 for every car sold. Which expression represents the amount in dollars the salesperson makes, before taxes, if x cars are sold in 40 hours?

    A. $100x + 280$
    B. $7x + 100$
    C. $100x + 7$
    D. $280x + 100$
    E. $7x + 280$

14. The quadratic equation $x^2 + 3x - 28$ can be factored in the form $(x + a)(x + b)$, where a and b are integers. Which of the following is the sum of these two factors?

    F. $x - 28$
    G. $x + 28$
    H. $2x - 4$
    J. $2x + 7$
    K. $2x + 3$

15. $\dfrac{2 - 5 \cdot (4 - 7^2) + 5}{(3 - 1) \cdot 2 + 4} = ?$

    A. $\dfrac{-59}{4}$

    B. $\dfrac{-65}{6}$

    C. -10

    D. $\dfrac{101}{4}$

    E. 29

16. The quadratic equation $2x^2 + 12x + 18$ can be factored in the form $(2x + p)(x + q)$, where p and q are whole numbers. Which of the following is $(2x + p) \div (x + q)$ in simplest form?

    F. $\dfrac{2x + 3}{x + 6}$

    G. $\dfrac{2x + 6}{x + 3}$

    H. $\dfrac{2}{3}$

    J.   2

    K. 3

17. The Pythagorean Theorem is $a^2 + b^2 = c^2$. Which of the following expresses b in terms of a and c?

    A. $b = \sqrt{c^2 - a^2}$

    B. $b = c + a$

    C. $b = c^2 - a^2$

    D. $b = \dfrac{c}{a}$

    E. $b = \sqrt{\dfrac{c}{a}}$

18. It is said that every one year of a human's life is equal to 7 years of a dog's life. If a dog is born in the year 2000, how old, in human years, will the dog be in the year 2008?

    F.   42
    G. 46
    H. 48
    J. 54
    K. 56

19. Express 128,000,000 in scientific notation.

    A. 1.28×10^{-8}

    B. 128×10^6

    C. 1.28×10^8

    D. 12.8×10^{-7}

    E. 0.128×10^9

20. $(2.89 \times 10^1) + (5.4 \times 10^2) = ?$

    F. 5.689×10^2

    G. 8.29×10^2

    H. 8.28×10^1

    J. 1.56×10^1

    K. 5.689×10^3

21. Delicious Ice Cream has many delicious flavors such as cookie dough peanut butter crunch. However, their best sellers are still the old favorites: chocolate, vanilla, and strawberry. Delicious Ice Cream uses a gallon of chocolate every 30 minutes, a gallon of vanilla every 0.25 hour, and a gallon of strawberry every hour. If the store is open for 8 hours in a day, how many total gallons of chocolate, vanilla, and strawberry do they use each day?

   A. 64 gallons
   B. 56 gallons
   C. 48 gallons
   D. 40 gallons
   E. 24 gallons

22. If $s = 4$, $t = 2$, and $u = -3$, evaluate $\dfrac{st + ust + 4}{u^4}$.

   F. $\dfrac{5}{27}$

   G. $\dfrac{4}{27}$

   H. $-\dfrac{1}{27}$

   J. $-\dfrac{5}{81}$

   K. $-\dfrac{4}{27}$

23. What is the sum of $m^2n + 2mn + 7n$ and $3m^2n - 5mn + 2m$?

   A. $4m^4n^2 - 3m^2n^2 + 7n + 2m$

   B. $3m^2n - 3mn + 14mn$

   C. $4m^2n + 3mn + 2m + 7n$

   D. $2m^2n + 3mn + 9mn$

   E. $4m^2n - 3mn + 2m + 7n$

24. If $2x^2 - 5x - 12 = 0$, what is the product of the values of x?

   F. -6
   G. -4
   H. 4
   J. 6
   K. 12

Chapter 9 ▪

Intermediate Algebra

- Nine ACT questions have to do with intermediate algebra.
- Easier intermediate algebra questions may be about a single skill or concept or may test a combination of pre-algebra, elementary algebra, and intermediate algebra skills.
- More difficult questions will often test a combination of intermediate algebra skills and concepts.
- This intermediate algebra review covers all the material you need to answer ACT questions.
- Use a calculator for the ACT-Type Problems. Do not use a calculator for the Practice exercises.

▪ Solving Inequalities

Solving inequalities is like solving equations, except that multiplying or dividing by a negative number reverses the inequality sign.

Multiplying or Dividing an Inequality by a Negative Number

EXAMPLES

1. Multiply both sides of $-4x \geq 7$ by -1.

   $$-4x \geq 7 \quad \rightarrow \quad 4x \leq -7$$

2. Multiply both sides of $-\dfrac{x}{6} < -4$ by -6.

   $$-\dfrac{x}{6} < -4 \quad \rightarrow \quad x > 24$$

3. Divide both sides of $9 \leq -3x$ by -3.

   $$9 \leq -3x \quad \rightarrow \quad -3 \geq x$$

4. Divide both sides of $-7x + 42 \leq y$ by -7.

   $$-7x + 42 \leq y \quad \rightarrow \quad x - 6 \geq \dfrac{-y}{7}$$

Reverse the inequality sign when you multiply or divide both sides of the inequality by a negative number.

Solving Inequalities

To solve an inequality, follow the same steps as for solving equations.

Using Addition or Subtraction

E X A M P L E S

1. Solve. $x + 15 < 23$

$$
\begin{array}{r}
x + 15 < 23 \\
-15 \quad -15 \\
\hline
x < 8
\end{array}
$$

Subtract 15.

To check, choose a number for x that is close to but less than 8.

Substitute 7 for x in the inequality. $\quad 7 + 15 < 23$
$$22 < 23 ✓$$

2. Solve. $-19 + x \geq 12$

$$
\begin{array}{r}
-19 + x \geq 12 \\
+19 \quad\quad +19 \\
\hline
x \geq 31
\end{array}
$$

Add 19.

Check: Since x is greater than or equal to 31, choose two values for x.

Substitute 31 for x. \quad Substitute 32 for x.

$-19 + 31 \geq 12 \quad\quad -19 + 32 \geq 12$

$\quad\quad 12 \geq 12 ✓ \quad\quad\quad\quad 13 \geq 12 ✓$

Using Multiplication or Division

E X A M P L E S

1. Solve. $-\dfrac{y}{8} > 7$

Multiply by -8. $\quad\quad -\dfrac{y}{8}(-8) > 7(-8)$

Reverse the inequality. $\quad\quad y < -56$

To check, substitute -57 for y in the original inequality.

$$-\left(\frac{-57}{8}\right) > 7$$

$$-(-7.125) > 7$$

$$7.125 > 7 ✓$$

2. Solve. $-8k \leq 70$

Divide by -8.

$$\frac{-8k}{-8} \leq \frac{70}{-8}$$

Reverse the inequality.

$$k \geq -8\frac{3}{4}$$

Substitute $-8\frac{3}{4}$ for k.   Substitute -8 for k.

$$-8\left(-8\frac{3}{4}\right) \leq 70 \qquad -8(-8) \leq 70$$

$$-8\left(-\frac{35}{4}\right) \leq 70 \qquad\qquad 64 \leq 70 ✓$$

$$70 \leq 70 ✓$$

You may have to use more than one operation to solve some inequalities.

EXAMPLES

1. Solve. $-5t - 8.5 < 54$

$$-5t - 8.5 < 54$$

Add 8.5.

$$\underline{+\ 8.5 \qquad +\ 8.5}$$

Divide by -5.

$$\frac{-5t}{-5} < \frac{62.5}{-5}$$

Reverse the inequality.

$$t > -12.5$$

2. Solve. $k \div 8 + 3.4 > -28$

$$k \div 8 + 3.4 > -28$$

Subtract 3.4.

$$\underline{\qquad -3.4 \qquad -3.4}$$

$$k \div 8 > -31.4$$

Multiply by 8.

$$k \div 8 \times 8 > -31.4 \times 8$$

Do *not* reverse the inequality.

$$k > -251.2$$

3. Solve. $-3x - 4 \geq 5 + x$

$$-3x - 4 \geq 5 + x$$

Subtract x.

$$\underline{-x \qquad\qquad\qquad -x}$$

$$-4x - 4 \geq 5$$

Add 4.

$$\underline{+\ 4 \qquad +\ 4}$$

Divide by -4.

$$\frac{-4x}{-4} \geq \frac{9}{-4}$$

Reverse the inequality.

$$x \leq -2.25$$

> Always add or subtract before you multiply or divide.

1. What is the solution set to the inequality $2x + 5 \geq 13$?

   A. $x \leq 9$
   B. $x \geq 9$
   C. $x \leq 4$
   D. $x \geq 4$
   E. $x \leq -4$

   SOLUTION

   $$2x + 5 \geq 13$$

   Subtract 5. $\qquad\qquad 2x \geq 8$

   Divide by 2. $\qquad\qquad\; x \geq 4$

   The correct answer is D.

2. What is the smallest number in the solution set to the inequality $-4x - 5 \leq -2x + 9$?

   F. -14
   G. -7
   H. 0
   J. 7
   K. 14

   SOLUTION

   $$-4x - 5 \leq -2x + 9$$

   Add 5. $\qquad\qquad -4x \leq -2x + 14$

   Add $2x$. $\qquad\qquad -2x \leq 14$

   Divide by -2. $\qquad\qquad\; x \geq -7$

   The correct answer is G.

Practice

Solve each inequality.

1. $x + 9 > 13$

2. $-13 + x \leq 22$

3. $-3y \leq 7$

4. $-\left(\dfrac{y}{6}\right) > 11$

5. $-4t + 6 \leq 9$

6. $\dfrac{k}{8} - 2.4 > 12$

7. $-5x + 6 \leq 2x - 8$

8. $7 + 4t < 3t - 2$

9. $2x + 5 < 4$

10. $-3x > 9$

11. $2x - 7 \geq 12$

12. $7x - 3 < 3x + 1$

13. $-2x + 9 \leq 13$

14. $16x - 7 \geq 6x + 2$

15. $-4x + 9 \geq 27$

16. $-3x + 7 > 11x - 2$

17. $-x - 13 > 2$

18. $27x - 5 < 35x + 3$

19. $9x - 5 \leq 6x + 2$

20. $x + 3 < 2x - 1$

(Answers on page 359)

1. What is the solution set for the inequality $4x \geq -12$?

   A. $x \geq -4$
   B. $x \leq -3$
   C. $x \geq -3$
   D. $x \geq 3$
   E. $x \leq 4$

2. What is the solution set to the inequality $-3x - 7 < 20$?

   F. $x > 13$
   G. $x < 13$
   H. $x < 9$
   J. $x > 9$
   K. $x > -9$

3. What is the solution set to the inequality $-4x + 17 \le -3$?

    A. $x \le -5$
    B. $x \ge -4$
    C. $x \le 4$
    D. $x \ge 5$
    E. $x \le 5$

4. Which of the following is not in the solution set to the inequality $-5x + 3 > -2x - 12$?

    F. 5
    G. 4
    H. 3
    J. 2
    K. 1

5. Which of the following is the reciprocal of the smallest number in the solution set to the inequality $-9x + 11 \le -4x + 3$?

    A. -40
    B. $-\dfrac{8}{5}$
    C. $-\dfrac{5}{8}$
    D. $\dfrac{5}{8}$
    E. $\dfrac{8}{5}$

(Answers on page 360)

Absolute Value Equations and Inequalities

When solving absolute value equalities and inequalities you must consider two possibilities.

> For example: If $|x| = 7$, then $x = 7$ or $x = -7$.

To solve absolute value equations and inequalities you must solve for each case.

> Case 1: The value is positive. Drop the absolute value and solve.

> Case 2: The value is negative. Drop the absolute value. Use a minus sign to make the expression from inside the absolute value negative and solve.

EXAMPLES

1. $|x - 8| = 5$

    Case 1: $x - 8$ is positive $x - 8 = 5$ $x = 13$

    Case 2: $x - 8$ is negative $-(x - 8) = 5$ $-x + 8 = 5$ $-x = -3$ $x = 3$

    $x = 3$ or $x = 13$

    Check: $|13 - 8| = |5| = 5$

    $|3 - 8| = |-5| = 5$

2. $|x + 4| < 7$

    Case 1: $x + 4 < 7$ $x < 3$

    Case 2: $-(x + 4) < 7$ $-x - 4 < 7$ $-x < 11$ $x > -11$

    Therefore $-11 < x < 3$.

    Check: Check a sample of the values between -11 and 3. Each value makes the original inequality correct.

Which of the following is the largest number that will make the inequality $|2x - 8| \leq 6$ true?

    A. 7
    B. 5
    C. 3
    D. 1
    E. 0

SOLUTION

Solve the inequality.

$$2x - 8 \leq 6 \rightarrow 2x \leq 14 \rightarrow x \leq 7$$

$$-(2x - 8) \leq 6 \rightarrow -2x + 8 \leq 6 \rightarrow -2x \leq -2 \rightarrow -x \leq -1 \rightarrow x \geq 1$$

$$1 \leq x \leq 7$$

The correct answer is A.

Practice

Solve.

1. $|3x - 4| < 14$ 2. $|x - 6| \leq 9$ 3. $|x + 3| = 11$ 4. $|5x - 12| \geq 13$

5. $|2x + 8| > 12$ 6. $|3x - 3| = 12$ 7. $|6x + 5| > 9$ 8. $|x - 5| \geq 2$

9. $|3x - 1| < 5$ 10. $|2x + 4| \leq 6$ 11. $|x - 3| < 1$ 12. $|2x + 7| = 9$

13. $|7x - 3| \geq 7$ 14. $|5x + 4| \leq 4$ 15. $|4x - 9| > 3$ 16. $|3x + 3| = 4$

17. $|8x - 5| \leq 9$ 18. $|2x + 3| < 12$ 19. $|5x - 5| \geq 7$ 20. $|x + 1| > 14$

(Answers on page 360)

ACT-TYPE PROBLEMS

1. 16 and -10 are the solutions to which one of the following equations?

    A. $|x - 2| = 14$
    B. $|x - 2| = 8$
    C. $|x + 2| = 12$
    D. $|x - 3| = 13$
    E. $|x + 3| = 13$

2. Which of the following choices makes the inequality $|2x - 9| < 5$ false?

    F. 2
    G. 3
    H. 4
    J. 5
    K. 6

3. Solve the inequality $|7x - 5| \geq 9$.

    A. $-4 \leq x \leq 2$
    B. $x \leq \dfrac{4}{7}$ or $x \geq 2$
    C. $x \leq -\dfrac{4}{7}$ or $x \geq 2$
    D. $x \leq -\dfrac{4}{7}$ or $x \geq -2$
    E. $-\dfrac{4}{7} \leq x \leq 2$

4. $x = 6$ is the complete solution set to which of the following equations?

    F. $2x = 12$
    G. $|x - 3| = 3$
    H. $|2x - 2| = 10$
    J. $-x = 6$
    K. $|x + 6| = 12$

5. What is the product of the solutions to the equation $|2x - 3| = 15$?

  A.   45
  B.   9
  C. -6
  D. -54
  E. -81

(Answers on page 360)

■ Solving Systems of Linear Equations

A linear equation is any equation in the form $ax + by = c$ $(a \neq 0, b \neq 0)$. The solution to a linear equation is an ordered pair (x,y) that makes the equation true.

A system of linear equations is two or more linear equations that can be solved together. The solution to a system of linear equations must be the solution for all of the equations in the system.

To solve a system of equations, sometimes you can add or subtract the equations to eliminate one of the variables. Other times you will have to change an equation so that when you add or subtract, one of the terms is eliminated.

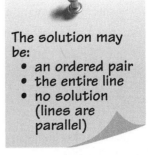

The solution may be:
- an ordered pair
- the entire line
- no solution (lines are parallel)

E X A M P L E S

1. Solve. $3x + 5y = 16$
              $-3x + 3y = 8$

This one is easy. Add the two equations.

$$3x + 5y = 16$$
Add. $$-3x + 3y = 8$$
Solve for y. $$8y = 24$$
$$y = 3$$

Substitute 3 for y in one of the equations. $$3x + 5(3) = 16$$
Solve for x. $$3x + 15 = 16$$
$$3x = 1$$
$$x = \frac{1}{3}$$

The solution to the system is $x = \frac{1}{3}$ and $y = 3$.

The solution as an ordered pair is $\left(\frac{1}{3}, 3 \right)$.

Substitution

You can also:
- Solve one equation
- Substitute the result in the other equation
- Solve that equation

CALCULATOR TIP

Graphing calculators can be used to graph and solve systems of linear equations.

2. Solve. $-4x - 4y = 8$

$2x + 7y = 15$

Multiply the bottom equation by 2. $\quad 2(2x + 7y = 15) \rightarrow 4x + 14y = 30$

$$-4x - 4y = 8$$

Add the equations. $\qquad \underline{4x + 14y = 30}$

Solve for y. $\qquad\qquad\qquad 10y = 38$

$$y = 3.8$$

Substitute 3.8 for y in
one of the equations. $\qquad -4x - 4(3.8) = 8$

Solve for x. $\qquad\qquad\qquad -4x = 23.2$

$$4x = -23.2$$

$$x = -5.8$$

The solution is $x = -5.8$ and $y = 3.8$.

The solution as an ordered pair is $(-5.8, 3.8)$.

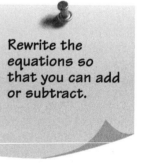

Change the equations so that when you add or subtract the equations, you "get rid" of one of the variables.

3. Solve. $\quad -8y + 5x + 12 = 2$

$6x + 12y = 6$

Rewrite the equations in linear form. $\qquad 5x - 8y = -10$

$$6x + 12y = 6$$

Multiply the top equation by 1.5. $\qquad 1.5(5x - 8y = -10) \rightarrow 7.5x - 12y = -15$

$$7.5x - 12y = -15$$

Add the equations. $\qquad\qquad\qquad \underline{6x + 12y = 6}$

Solve for x. $\qquad\qquad\qquad\qquad 13.5x = -9$

$$x = -\frac{2}{3}$$

Rewrite the equations so that you can add or subtract.

Substitute $-\dfrac{2}{3}$ for x in one
of the equations. $\qquad 6\left(-\dfrac{2}{3}\right) + 12y = 6$

Solve for y. $\qquad\qquad\qquad -4 + 12y = 6$

$$12y = 10$$

$$y = \frac{5}{6}$$

The solution is $x = -\dfrac{2}{3}$ and $y = \dfrac{5}{6}$.

The solution as an ordered pair is $\left(-\dfrac{2}{3}, \dfrac{5}{6}\right)$.

1. What is the solution to the system of linear equations $2x + 5y = 10$ and $2x + 3y = 2$?

    A. $(-5,-4)$
    B. $(-5,4)$
    C. $(-4,-5)$
    D. $(4,-5)$
    E. $(5,4)$

SOLUTION

Subtract one equation from the other.	$2x + 5y = 10$
	$-2x - 3y = -2$
Solve for y.	$2y = 8$
	$y = 4$

Substitute 4 for y in one of the equations.	$2x + 5(4) = 10$
Solve for x.	$2x + 20 = 10$
	$2x = -10$
	$x = -5$

The solution is $x = -5$ and $y = 4$.

The solution as an ordered pair is $(-5,4)$.

The correct answer is B.

2. What is the sum of the solutions to the following system of linear equations?

$$6x - 5y = 15$$
$$-3x + 2y = 10$$

    F. $-61\frac{2}{3}$

    G. -35

    H. $8\frac{1}{3}$

    J. 25

    K. $26\frac{2}{3}$

SOLUTION

Multiply both sides of the second equation by 2. $2(-3x + 2y = 10) \rightarrow -6x + 4y = 20$

	$6x - 5y = 15$
Add the equations.	$-6x + 4y = 20$
	$-y = 35$
	$y = -35$

Substitute -35 for y in one of the equations. $-3x + 2(-35) = 10$

Solve for x. $-3x - 70 = 10$

$$-3x = 80$$

$$x = -\frac{80}{3} = -26\frac{2}{3}$$

The solutions are $x = -26\frac{2}{3}$ and $y = -35$.

Find the sum of the solutions. $-26\frac{2}{3} + (-35) = -61\frac{2}{3}$

The correct answer is F.

Practice

Solve the system of equations.

1. $4x + 5y = 13$

   $4x + 3y = 9$

2. $3x - 2y = 6$

   $9x + 6y = 60$

3. $-5y + 3x = 8\frac{3}{4}$

   $2y + 12x + 13 = 26$

4. $2x + 5y = 6$

   $2x + 4y = 5$

5. $3x + 5y = 7$

   $6x + 5y = 2$

6. $4x - 9y = 8$

   $4x + 9y = 8$

7. $2x + 4y = 9$

   $3x - 4y = 8$

8. $x + 3y = 5$

   $2x + 4y = 6$

9. $4x + 5y = 8$

   $6x + 8y = 7$

10. $5x + 12y = 13$

    $3x + 4y = 5$

11. $12x + 8y = 2$

    $4x + 6y = 10$

12. $-6x + 7y = -13$

    $-12x + 8y = 4$

13. $x - 5y = 10$

    $-2x + 3y = 8$

14. $-12x + 8y = -5$

    $-4x + 6y = -5$

15. $-4x + 9y = 19$

    $-6x + 11y = -20$

16. $-2x + 5y = 6$

    $-3x + 4y = 9$

17. $9x + 7y = 5$

    $8x + 6y = 4$

18. $-16x + 7y = 5$

    $17x - 8y = 2$

19. Holly has 15 dimes and nickels worth $1.05. How many of each type of coin does Holly have?

20. Suresh studied 8 hours for his final exams in math and science. He studied 1.5 hours longer for his math final than for his science final. How many hours did he study for each final?

21. Tia went to a sale at a media store. She bought 8 videos and CDs for $92. If each video cost $16 and each CD cost $10, how many of each did she buy?

(Answers on page 360)

ACT-TYPE PROBLEMS

1. What is the solution to the following system of linear equations?

   $$2x + 5y = 8$$
   $$2x + 4y = 7$$

   A. $(1.5,1)$
   B. $(1,1.5)$
   C. $(1,-1.5)$
   D. $(-1.5,1)$
   E. $(-1.5,-1)$

2. What is the solution to the following system of linear equations?

   $$3x + 5y = 8$$
   $$-3x + 5y = 8$$

   F. $(1.6,0)$
   G. $(0,1.6)$
   H. $(0,1.4)$
   J. $(0,-1.6)$
   K. $(-1.4,0)$

3. What is the product of the solutions to the following system of linear equations?

$$7x + 10y = 12$$

$$5x + 5y = 6$$

   A. -1.2
   B. -1
   C. 0
   D. 1
   E. 1.2

4. You are to find two numbers. When you double the first and triple the second, their sum is 1. When you triple the first and multiply the second by 5, the sum is 2. What are the two numbers?

   F. $(1, 0.5)$
   G. $(1, -1)$
   H. $(-0.5, 1)$
   J. $(-1, 1)$
   K. $(-1, -1)$

(Answers on page 360)

5. There are two paths. In the morning, 6 people walked the first path and 12 people walked the second path. The total distance these people walked was 8 miles. In the afternoon, 9 people walked the first path and 4 people walked the second path. The total distance people walked in the afternoon was 5 miles. How many miles long is each path?

   A. $\left(\frac{1}{2}\text{ mile}, \frac{1}{3}\text{ mile}\right)$

   B. $\left(\frac{1}{3}\text{ mile}, \frac{1}{2}\text{ mile}\right)$

   C. (3 miles, 2 miles)

   D. (2 miles, 3 miles)

   E. (5 miles, 8 miles)

■ Rational and Radical Expressions

Simplifying Expressions

To simplify rational and radical expressions, you may need to use some or all of these equalities.

$\sqrt[n]{x}$ means the *n*th root of *x*.

$$\sqrt[a]{x^b} = x^{\left(\frac{b}{a}\right)}$$

$$x^{(-a)} = \frac{1}{x^a} \quad \left(\frac{1}{x^{(-a)}} = x^a\right)$$

$$\sqrt{x} \cdot \sqrt{x} = x$$

When the bases are the same, use these equalities to multiply and divide exponents.

$$x^b \cdot x^a = x^{(b+a)} \qquad x^b \div x^a = x^{(b-a)}$$

Undefined Expressions

An expression is considered undefined when its denominator is equal to zero, or any time there is division by zero. Otherwise, the expression is defined.

$\dfrac{4\sqrt{3}}{x}$ is defined except when $x = 0$.

$\dfrac{x^{\left(\frac{1}{3}\right)}}{3^x}$ is defined for all values of *x*.

$\dfrac{4(2x + 8)}{x - 8}$ is defined for all values of *x* except $x = 8$.

1. Simplify. $\dfrac{11x}{\sqrt{3x-8}}$

$\dfrac{11x}{\sqrt{3x-8}}$

$= \dfrac{11x}{\sqrt{3x-8}} \cdot \dfrac{\sqrt{3x-8}}{\sqrt{3x-8}}$ Multiply numerator and denominator by $\sqrt{3x-8}$. This removes the radical from the denominator.

$= \dfrac{11x\sqrt{3x-8}}{3x-8}$

2. Simplify. $\dfrac{1}{x^{\left(-\frac{2}{3}\right)}} + \dfrac{\sqrt[3]{x^2}}{x^{\left(-\frac{1}{3}\right)}}$

$\dfrac{1}{x^{\left(-\frac{2}{3}\right)}} + \dfrac{\sqrt[3]{x^2}}{x^{\left(-\frac{1}{3}\right)}}$

$= x^{\left(\frac{2}{3}\right)} + x^{\left(\frac{1}{3}\right)} \cdot \sqrt[3]{x^2}$ Use $x^{(-a)} = \dfrac{1}{x^a}$ $\left(\dfrac{1}{x^{(-a)}} = x^a\right)$

$= x^{\left(\frac{2}{3}\right)} + x^{\left(\frac{1}{3}\right)} \cdot x^{\left(\frac{2}{3}\right)}$ Use $\sqrt[a]{x^b} = x^{\left(\frac{b}{a}\right)}$

$= x^{\left(\frac{2}{3}\right)} + x^{\left(\frac{1}{3} + \frac{2}{3}\right)}$ Use $x^b \cdot x^a = x^{(b+a)}$

$= x^{\left(\frac{2}{3}\right)} + x^{\left(\frac{3}{3}\right)}$ Add fractional exponents with the same base.

$= x^{\left(\frac{2}{3}\right)} + x$ Simplify a fractional exponent. $x^{\left(\frac{3}{3}\right)} = x^1 = x$

3. For which real values of x is $\dfrac{7x}{2^{(4-x)} - 8}$ defined?

Find the values of x for which the expression is *not* defined.

$\dfrac{7x}{2^{(4-x)} - 8}$ is not defined when $2^{(4-x)} - 8 = 0$.

$2^{(4-x)} - 8 = 0$ when $2^{(4-x)} = 8$.

$2^{(4-x)} = 8$ when $x = 1$. ($2^{(4-1)} = 2^3 = 8$)

The expression is *not* defined when $x = 1$.

The expression *is* defined for all real values of x except $x = 1$.

Write the expression $\dfrac{\sqrt{x^3} + x}{\sqrt{x}}$ in simplified form with no radicals and no negative exponents. $(x \neq 0)$

A. x

B. $x^{\left(\frac{5}{2}\right)}$

C. $\dfrac{x + 1}{x^{\left(\frac{1}{2}\right)}}$

D. $x + x^{\left(\frac{1}{2}\right)}$

E. $x^2 + \dfrac{1}{x}$

SOLUTION

$$\frac{\sqrt{x^3} + x}{\sqrt{x}}$$

$$= \frac{x^{\left(\frac{3}{2}\right)} + x}{x^{\left(\frac{1}{2}\right)}}$$

$$= \frac{x^{\left(\frac{3}{2}\right)}}{x^{\left(\frac{1}{2}\right)}} + \frac{x}{x^{\left(\frac{1}{2}\right)}}$$

$$= x^{\left(\frac{2}{2}\right)} + x^{\left(\frac{1}{2}\right)}$$

$$= x + x^{\left(\frac{1}{2}\right)}$$

The correct answer is D.

Practice

Simplify.

1. $\dfrac{x^{-3}}{x^3} + \sqrt[3]{x^2}$

2. $\sqrt[3]{x^3} - \dfrac{x^4}{x^4}$

3. $x^{\left(\frac{2}{5}\right)} - \sqrt[5]{x^3} \div x^{\left(\frac{1}{5}\right)} + 1$

4. $\sqrt[3]{x^2} \cdot \sqrt[2]{x^{-1}} \div \sqrt[6]{x^5} \cdot \sqrt[3]{x^{-2}}$

5. $x^{-5} + \sqrt{x^5} \div x^{\left(\frac{1}{2}\right)} - \sqrt{x^{-10}}$

6. $\sqrt[3]{x^2} + \sqrt[3]{x^3} \cdot \sqrt[3]{x}$

7. $x^{\left(\frac{1}{2}\right)} \cdot \sqrt{x^{-1}} + x^2$

8. $x^4 \cdot x^{\left(\frac{1}{4}\right)} + x^2 \cdot x^{-2}$

9. $\dfrac{1}{x^{-6}} + \sqrt{x^{-3}} \div x^{\left(-\frac{3}{2}\right)} \cdot x$

10. $\sqrt[2]{x^2} + \sqrt[3]{x^2} \cdot \sqrt[4]{x^2} \div \sqrt[5]{x^2}$

Identify the real values of x for which each of the following expressions is defined.

11. $\dfrac{3x^2}{2}$

12. $\dfrac{\sqrt{7y - 9}}{x^2}$

13. $\dfrac{4k - \sqrt{65}}{6x - 2}$

14. $\dfrac{\sqrt{4y - 19}}{3^{(x+2)} - 3}$

15. $\dfrac{x^3 - 6}{x^3 + x^2 + 18}$

(Answers on page 361)

ACT-TYPE PROBLEMS

1. Which of the following expressions is $\dfrac{4x^3y^2 + 4x^2y^3}{x + y}$ expressed in its simplest form?

   A. $\dfrac{8x^5y^5}{x + y}$

   B. $\dfrac{4x^2y^2(x + y)}{x + y}$

   C. $8x^2y^2$

   D. $4x^2y^2$

   E. $4x^2y + 4xy^2$

2. Which of the following is $\dfrac{1}{\sqrt[3]{x}} \cdot \sqrt[3]{x} - \sqrt{x} \cdot \sqrt{\dfrac{1}{x}}$ in simplest form?

   F. 0

   G. 1

   H. $1 - x$

   J. $x^{\left(\frac{1}{3}\right)} - x^{\left(\frac{1}{2}\right)}$

   K. $x^{\left(\frac{2}{3}\right)} - 1$

3. For which real values of x is the expression $\dfrac{7x}{2^{(3x-1)}}$ defined?

   A. All real values

   B. All real values except $\dfrac{1}{3}$

   C. All real values except 0

   D. All real values except 2

   E. All real values except 3

4. What are the real numbers x such that $\dfrac{2x^2 + \sqrt{x}}{x^2 + x - 6}$ is defined?

   F. All real numbers
   G. All real numbers except 2
   H. All real numbers except -3
   J. All non-negative real numbers except 2
   K. All non-negative real numbers except 3

5. Which of the following is $\dfrac{-2x^2 + 2y^2}{x - y}$ in simplest form? ($x - y \neq 0$)

   A. $-2x + 2y$
   B. $-2x - 2y$
   C. $2x - 2y$
   D. $x + y$
   E. $x - y$

(Answers on page 361)

■ Solving Quadratic Equations

Quadratic equations can be written in this standard form: $ax^2 + bx + c = 0$ ($a \neq 0$). Since the largest exponent of x is 2, the equation can have at most two solutions (roots).

It may be difficult to solve a quadratic equation by factoring.

You can always use the quadratic formula to solve a quadratic equation.

$$x = \frac{-b \pm \sqrt{b^2 - 4ac}}{2a}$$

The a, b, and c in the formula are the same as the coefficients a, b, and c in the quadratic equation.

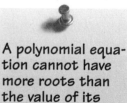

A polynomial equation cannot have more roots than the value of its largest exponent.

Solve. $4x^2 - 2 = 3x$

Write the equation in standard form. $\quad 4x^2 - 3x - 2 = 0$

Identify the values for a, b, and c. $\quad a = 4, b = -3, c = -2$

Substitute the values of a, b, and c into the quadratic formula.

$$x = \frac{-b \pm \sqrt{b^2 - 4ac}}{2a}$$

$$= \frac{-(-3) \pm \sqrt{(-3)^2 - 4(4)(-2)}}{2(4)}$$

$$= \frac{3 \pm \sqrt{9 + 32}}{8}$$

$$= \frac{3 \pm \sqrt{41}}{8}$$

The solutions of the quadratic equation are $\dfrac{3 + \sqrt{41}}{8}$ and $\dfrac{3 - \sqrt{41}}{8}$.

> Write the equation in standard form to identify a, b, and c.

CALCULATOR TIP

If an exact answer is not required and you have an advanced calculator, you don't need to use the quadratic formula. Just use your calculator's method for finding the roots of a polynomial.

MODEL ACT PROBLEMS

1. What is the sum of a, b, and c in the quadratic equation $23x^2 = -13x + 6$?

   A. -4
   B. 4
   C. 16
   D. 30
   E. 42

SOLUTION

Write the equation in standard form. $\qquad 23x^2 = -13x + 6$

Add $13x$ to each side. $\qquad 23x^2 + 13x = 6$

Subtract 6 from each side. $\qquad 23x^2 + 13x - 6 = 0$

Identify the values for a, b, and c. $\qquad a = 23, b = 13, c = -6$

Find the sum of a, b, and c. $\qquad 23 + 13 + (-6) = 36 - 6 = 30$

The correct answer is D.

2. What are the solutions to the quadratic equation
$x^2 - 2x - 35 = 0$?

    F. 5 and 7
    G. 5 and -7
    H. 2 and 12
    J. -2 and 12
    K. -5 and 7

SOLUTION

Identify the values for a, b, and c.
$\qquad a = 1, b = -2, c = -35$

Substitute the values of a, b, and c into the quadratic formula.

$$x = \frac{-b \pm \sqrt{b^2 - 4ac}}{2a}$$

$$= \frac{-(-2) \pm \sqrt{(-2)^2 - 4(1)(-35)}}{2(1)}$$

$$= \frac{2 \pm \sqrt{4 + 140}}{2}$$

$$= \frac{2 \pm \sqrt{144}}{2}$$

$$= \frac{2 \pm 12}{2} = \frac{2 + 12}{2} \text{ or } \frac{2 - 12}{2}$$

$$= 7 \text{ or } -5$$

The solutions are -5 and 7.

The correct answer is K.

Practice

1. Write the quadratic formula.

2. Identify a, b, and c in the equation $3 - 2x^2 = -8x$.

Solve each equation.

3. $x^2 - 3x - 4 = 0$ **4.** $5x^2 - 2 = 4x$ **5.** $5x + 2 = -2x^2$

6. $x^2 = x + 6$ **7.** $3x - 4 = -x^2$ **8.** $2x^2 - 5x - 3 = 0$

9. $x^2 + x = 56$ **10.** $x^2 - 18 = 7x$ **11.** $4x^2 + 9x + 2 = 0$

12. $2x^2 + 4 = -6x$ **13.** $x^2 - 5x = -6$ **14.** $4x^2 - 4x - 15 = 0$

15. $26x = -3x^2 + 9$ **16.** $4x^2 + 25x - 21 = 0$ **17.** $6x^2 = 9x + 15$

18. Mr. Wilson's rectangular garden has an area of 27 square feet. If the length of his garden is three times the width, what are the dimensions of the garden?

19. The area of a rectangular pool is 180 square feet. If the pool is 3 feet longer than it is wide, what are the dimensions of the pool?

20. Chan is framing an 8-inch by 10-inch picture. The area of the picture and the frame is 143 square inches. What is the width of the frame?

(Answers on page 362)

1. What is the product of a, b, and c in the quadratic equation $4x^2 + 2 = -14x$?

   A. -112
   B. -56
   C. -28
   D. 56
   E. 112

2. What are the solutions to the quadratic equation $8x^2 = -44x - 56$?

   F. -4.5 and 3
   G. -3.5 and 1
   H. -3.5 and -2
   J. -3 and 4.5
   K. -2 and 3.5

3. What are the solutions to the quadratic equation $2x^2 + 16x + 24 = 0$?

   A. 2 and 6
   B. 1 and 12
   C. -6 and -2
   D. -12 and 1
   E. -12 and -1

4. What is the sum of the solutions to the quadratic equation $21x^2 - 189 = 0$?

   F. -3
   G. -1
   H. 0
   J. 1
   K. 3

5. What is the product of the solutions to the quadratic equation $-29x - 22 = -6x^2$?

   A. $-\dfrac{11}{3}$
   B. $-\dfrac{3}{11}$
   C. $\dfrac{4}{33}$
   D. $\dfrac{3}{11}$
   E. $\dfrac{33}{4}$

(Answers on page 362)

■ Solving Quadratic Inequalities

Factor or use the quadratic formula to solve quadratic inequalities. Use the same techniques as for quadratic equations. However, you must consider the following cases:

- If the quadratic inequality is written *less than* zero, then the two factors have different signs.

EXAMPLE

Solve. $x^2 - 9 < 0$

$(x + 3)(x - 3) < 0$

$(x + 3)$ and $(x - 3)$ must have different signs. That happens when:

$x + 3 < 0$ and $x - 3 > 0$ OR $x + 3 > 0$ and $x - 3 < 0$

$x < -3$ and $x > 3$ $x > -3$ and $x < 3$

This is impossible. $-3 < x < 3$

The solution is all real numbers between -3 and 3.

- If the quadratic inequality is written *greater than* zero then the two factors must have the same sign.

Solve.

$x^2 - 9 > 0$

$(x + 3)(x - 3) > 0$

$(x + 3)$ and $(x - 3)$ must have the same sign. That happens when:

$x + 3 > 0$ and $x - 3 > 0$ OR $x + 3 < 0$ and $x - 3 < 0$

$x > -3$ and $x > 3$ $x < -3$ and $x < 3$

This means the number must be greater than 3.   This means the number must be less than -3.

$x > 3$ $x < -3$

The solution is all real numbers greater than 3 or less than -3.

MODEL ACT PROBLEM

Which is the solution set of $x^2 + 2x - 8 \geq 0$?

  A. $x \leq 4$
  B. $x \geq 2$
  C. $x \leq -4$ or $x \geq 2$
  D. $-4 \leq x \leq 2$
  E. $-2 \leq x \leq 4$

SOLUTION

$x^2 + 2x - 8 \geq 0$

$(x + 4)(x - 2) \geq 0$

Both factors must have the same sign.

$x + 4 \geq 0$ and $x - 2 \geq 0$ OR $x + 4 \leq 0$ and $x - 2 \leq 0$

$x \geq -4$ and $x \geq 2$ $x \leq -4$ and $x \leq 2$

$x \geq 2$ $x \leq -4$

The correct answer is C.

Practice

Write the solution set for the given inequality.

  1. $x^2 - 16 < 0$ **2.** $x^2 + 7x \leq 0$ **3.** $x^2 - 25 > 0$ **4.** $x^2 - 4x \geq 0$

  5. $3x^2 + 10x \leq 8$ **6.** $4x^2 - 9 \geq 0$ **7.** $2x^2 - 11x + 5 \geq 0$ **8.** $x^2 > 8x + 20$

  9. $x^2 + 27 < 12x$ **10.** $x^2 + 2x < 15$ **11.** $x^2 - 8 > 8$ **12.** $2x^2 - x - 3 > 0$

13. $x^2 + 1 < 2x$ **14.** $3x^2 - 12 \leq 0$ **15.** $x^2 \leq 6x - 5$ **16.** $16x^2 - 32 < 0$

17. $0 > -x^2 + 16$ **18.** $4x^2 + 5x \geq 0$ **19.** $x^2 + 5x \leq -4x$ **20.** $1{,}000{,}000x^2 \geq 0$

(Answers on page 363)

1. Which is the solution set of $x^2 - 8x + 12 < 0$?

   A. $x < 2$ or $x > 6$
   B. $2 < x < 6$
   C. $x < -6$ or $x > -2$
   D. $-6 < x < -2$
   E. $-6 < x < 2$

2. Which quadratic inequality has the solution $-\sqrt{3} < x < \sqrt{3}$?

   F. $x^2 - 3 < 0$
   G. $x^2 - 3 > 0$
   H. $x^2 - 3 \le 0$
   J. $x^2 - 3 \ge 0$
   K. $x^2 + 3 < 0$

3. Which is the solution set of $x^2 + 3x - 4 > 0$?

   A. $-1 < x < 4$ D. $x < -4$ or $x > 1$
   B. $-4 < x < 1$ E. $x < -4$ or $x > -1$
   C. $x < -1$ or $x > 4$

4. Which is the solution set of $x^2 - 3x \le 0$?

   F. $x < 0$ or $x > 3$
   G. $0 < x < 3$
   H. $0 \le x \le 3$
   J. $-3 < x < 0$
   K. $-3 \le x \le 0$

5. Which quadratic inequality has the solution set $-3 \le x \le 8$?

   A. $x^2 + 5x - 24 < 0$
   B. $x^2 - 5x + 24 \le 0$
   C. $x^2 - 5x - 24 \ge 0$
   D. $x^2 - 5x - 24 < 0$
   E. $x^2 - 5x - 24 \le 0$

(Answers on page 363)

■ Complex Numbers

CALCULATOR TIP

Use a calculator that can represent complex numbers to check your work.

You will most frequently encounter complex numbers as you solve quadratic equations.

Imaginary Numbers

We call $\sqrt{-1}$ an imaginary number. It is neither a whole number, a decimal, nor a rational number. We use the symbol i to represent this imaginary number.

$\sqrt{-1} = i$

The Square of *i*

The square of i is -1.

$i^2 = -1$

$24i^2 = -24$

Standard Form

Every complex number has a standard form, $a + bi$, where a and b are real numbers.

Complex Addition

Treat i as a variable when you add complex numbers. Look at these examples.

$(a + bi) + (c + di) = (a + c) + (b + d)i$

$(3 + 4i) + (5 + 6i) = (3 + 5) + (4 + 6)i = 8 + 10i$

Complex Multiplication

Treat i as a variable when you multiply complex numbers, but remember that $i^2 = -1$.
Look at these examples.

$(a + bi)(c + di) = ac + adi + bci + bdi^2 = (ac - bd) + (ad + bc)i$

$(3 + 4i)(5 + 6i) = 15 + 18i + 20i + 24i^2 = (15 - 24) + (18 + 20)i = -9 + 38i$

Complex Division

The expression $(3 + 2i) \div (1 + i)$ can be written as $\dfrac{3 - 2i}{1 + i}$. However, this is not a
complex number in the form $a + bi$. We can simplify the fraction by multiplying by
the *complex conjugate* of the denominator. The **complex conjugate** of $a + bi$ is $a - bi$.
Look at these examples.

$$\frac{3 - 2i}{1 + i} = \frac{3 - 2i}{1 + i} \cdot \frac{1 - i}{1 - i} = \frac{3 - 3i - 2i + 2i^2}{1 - i + i - i^2} = \frac{3 - 5i - 2}{1 + 1} = \frac{1 - 5i}{2} = \frac{1}{2} - \frac{5}{2}i$$

$$\frac{3 - 4i}{2 - 6i} = \frac{3 - 4i}{2 - 6i} \cdot \frac{2 + 6i}{2 + 6i} = \frac{6 + 18i - 8i - 24i^2}{4 + 12i - 12i - 36i^2} = \frac{6 + 10i - 24}{4 + 36}$$

$$= \frac{30 + 10i}{40} = \frac{3}{4} + \frac{1}{4}i$$

Write the value of the number. Use standard form if possible.

MODEL ACT PROBLEMS

1. Which of the following choices represents
$3 + 4i - 7 + 5i$ in standard form?

   A. $7i - 2i$
   B. $-4 + 9i$
   C. $-4 + 9i^2$
   D. $5i$
   E. $4 - i$

SOLUTION

Standard form for a complex number is $a + bi$. Treat i
as a variable when you add complex numbers.

$3 + 4i - 7 + 5i = (3 - 7) + (4i + 5i)$

$\qquad\qquad\qquad = -4 + 9i$

The correct answer is B.

2. In simplest form, $(6 - 4i)(5 + 2i) = ?$

   F. $30 - 8i^2$
   G. $30 - 8i - 8i^2$
   H. $22 - 8i$
   J. $38 - 8i$
   K. $38 + 8i$

SOLUTION

Use FOIL. $\qquad\qquad (6)(5) + (6)(2i) -$
$\qquad\qquad\qquad\qquad\qquad (4i)(5) - (4i)(2i)$

Simplify. $\qquad\qquad 30 + 12i - 20i - 8i^2$

Combine like terms. $\quad (30 - 8i^2) + (12i - 20i)$

Remember $i^2 = -1$. $\quad 38 - 8i$

The correct answer is J.

Practice

1. $18i^2$ 2. i^3 3. $\sqrt{-49}$ 4. $\sqrt{-50}$ 5. $\sqrt{-48}$

Write in standard form.

6. $18 - 6 + 5i$ 7. $3 \times 9 + 3i$ 8. $4i - 16 - 5i$ 9. $2 + 7i - 6 + 5i$ 10. $7(6 - \sqrt{-1})$

Add.

11. $(2 + 9i) + (3 + 7i)$ 12. $(12 + 2i) + (7 + 3i)$ 13. $(13 + 9i) + (-21 + 7i)$

14. $(12.5i^2 + 3i) + (-8 + 6i)$ 15. $\left(\dfrac{1}{2} + \dfrac{3}{4}i\right) + \left(\dfrac{1}{4} + \dfrac{1}{8}i\right)$

Multiply or divide. Express answers in $a + bi$ form.

16. $(2 + 9i)(3 + 7i)$ 17. $(12 + 2i)(7 + 3i)$ 18. $(13 + 9i)(-4 + i)$

19. $(0.5 + 3i)(-8 + 6i)$ 20. $\left(\dfrac{1}{2} + \dfrac{3}{4}i\right) \times \left(\dfrac{1}{4} + \dfrac{2}{3}i\right)$ 21. $4 \div i$

22. $\dfrac{5}{-2 + 6i}$ 23. $(2 - 0.5i) \div (1 + 0.5i)$ 24. $\dfrac{-7 + 2i}{9 + 5i}$

(Answers on page 363)

ACT-TYPE PROBLEMS

1. In standard form, $-3(6 - \sqrt{-25}) = ?$

   A. $-18 + 15i$
   B. $-18 - 15i$
   C. $-18 - 5i$
   D. -3
   E. $-3i$

2. Find the sum of $5 - 2i$ and $-3 + 7i$.

   F. $8 - 9i$
   G. $-2 + 9i$
   H. $2 - 5i$
   J. $2 + 5i$
   K. $2 + 5i^2$

3. Use the quadratic formula to find the roots of $x^2 + 4 = 0$.

   A. -2 only
   B. -2 and 2
   C. $4 + 2i$ and $4 - 2i$
   D. $-8i$ and $8i$
   E. $-2i$ and $2i$

4. Find the quadratic equation whose roots are $3i$ and $-3i$.

   F. $x^2 - 9 = 0$
   G. $x^2 + 9 = 0$
   H. $x^2 - 6ix + 9 = 0$
   J. $x^2 - 6ix - 9 = 0$
   K. $x^2 + 6ix - 9 = 0$

5. Use the quadratic formula to find the roots of $x^2 - 6x + 10 = 0$.

   A. 2 and 4
   B. 2 and 5
   C. $-3i$ and $3i$
   D. $3 + i$ and $3 - i$
   E. $3 + \sqrt{19}$ and $3 - \sqrt{19}$

(Answers on page 363)

■ Patterns, Sequences, and Modeling

Patterns

Pascal's triangle is among the most famous patterns in mathematics.

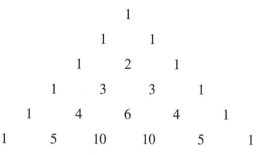

Each term in the triangle is the sum of the two terms immediately above it to the left and to the right. This triangle can be used to model probability events, and each row shows the coefficients of a power of the binomial $x + 1$. $(x + 1)^3 = x^3 + 3x^2 + 3x + 1$.

Sequences

A sequence is a list of numbers that follows a pattern. Sequences frequently model events. The square and triangular numbers below form sequences.

Triangular Numbers

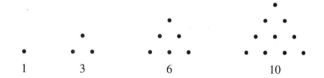

Square Numbers

You can find patterns within sequences. Notice that the sum of two consecutive triangular numbers yields a square number.

The Fibonacci sequence is the most famous mathematics sequence. The first eleven terms of the Fibonacci sequence are shown below. The pattern starts with 1 and then each term is the sum of the two previous terms.

1,   1,   2,   3,   5,   8,   13,   21,   34,   55,   89 …

       +1   +1   +2   +3   +5   +8   +13   +21   +34

Fibonacci numbers model many natural occurrences, including the swirls found in pine cones and the number of petals on sunflowers. The quotient of two consecutive Fibonacci numbers approximates the golden ratio (1.6). This ratio is found often in nature and in architecture.

Arithmetic Sequence

In an arithmetic sequence, consecutive terms differ by the same amount, called the common difference.

3, 5, 7, 9, 11, … is an arithmetic sequence with a common difference of 2.

10, 7, 4, 1, −2, … is an arithmetic sequence with a common difference of −3.

The general term of an arithmetic sequence is given by the formula:

$$a_n = a_1 + (n - 1)d$$

a_n is the nth term.

a_1 is the first term.

n is the position in the sequence.

d is the common difference.

Geometric Sequence

In a geometric sequence, consecutive terms vary from each other by a factor called the common ratio, r.

3, 6, 12, 24, 48, … is a geometric sequence with a common ratio of 2.

36, 12, 4, $\frac{4}{3}$, $\frac{4}{9}$, … is a geometric sequence with a common ratio of $\frac{1}{3}$.

1, −4, 16, −64, 256, … is a geometric sequence with a common ratio of −4.

The general term of a geometric sequence is given by the formula:

$$a_n = a_1 r^{(n-1)}$$

MODEL ACT PROBLEM

Greg went to the duck pond every day for a week. The table shows the number of ducks Greg saw each day.

Day 1	Day 2	Day 3	Day 4	Day 5	Day 6	Day 7
2	4	8	16	32	64	128

If this pattern remains the same, how many ducks will Greg see on the 11th consecutive day he goes to the duck pond?

    A.   136 ducks
    B.   144 ducks
    C.   512 ducks
    D.   1,024 ducks
    E.   2,048 ducks

SOLUTION

The problem shows a geometric sequence with a common ratio of 2. You can solve the problem by just multiplying by 2 until you reach the eleventh term.

You can also use the formula for the nth term of a geometric sequence.

$a_n = a_1 r^{(n-1)}$

$a_{11} = 2(2)^{(11-1)} = 2(2)^{10} = 2 \cdot 1{,}024 = 2{,}048$

Use your calculator.

The correct answer is E.

Practice

Write the next three terms in each sequence.

1. 5, 5, 10, 15, 25, 40, 65, …

2. 1, 3, 5, 7, 9, …

3. 97, 86, 75, 64, 53, …

4. 1, 3, 4, 7, 11, 18, 29, 47, …

5. 3, 8, 13, 18, 23, …

6. 1, 8, 27, 64, 125, …

7. 2,000; 1,000; 500; 250; 125; …

8. 2, 6, 18, 54, 162, 486, …

9. 39, 31, 23, 15, 7, …

10. 2, 3, 5, 7, 11, 13, 17, 19, …

11. 2, 4, 6, 8, 10, …

12. 1, 2, 4, 8, 16, 32, …

13. 3, 6, 12, 24, 48, …

14. 1, 4, 9, 16, 25, …

15. $\frac{1}{2}, \frac{1}{4}, \frac{1}{8}, \frac{1}{16}, \frac{1}{32}, \dots$

16. 15, 11, 7, 3, −1, …

17. 2, 3, 5, 7, 11, …

18. $\frac{1}{9}, \frac{1}{3}, 1, 3, 9, \dots$

19. 1, 3, 7, 15, 31, …

20. $1, \frac{1}{2}, 3, \frac{1}{4}, 5, \frac{1}{6}, 7, \dots$

(Answers on page 364)

ACT-TYPE PROBLEMS

1. Gary swims every week. The table shows the total amount of time Gary swims in five consecutive weeks. If this pattern continues, how long will Gary swim in the 6th week?

Week 1	75 minutes
Week 2	84 minutes
Week 3	93 minutes
Week 4	102 minutes
Week 5	111 minutes

    A. 122 minutes
    B. 120 minutes
    C. 118 minutes
    D. 116 minutes
    E. 114 minutes

2. The first term in a geometric sequence is 3, and the common factor is 2. Which of the following shows the first 5 terms in the sequence?

    F. 3, 5, 7, 8, 9
    G. 3, 6, 12, 24, 48
    H. 2, 5, 8, 11, 14
    J. 2, 6, 18, 54, 162
    K. 3; 18; 108; 648; 3,888

(Answers on page 364)

3. Which of the following choices displays the seventh term in the sequence below?

    4   32   256   2,048

    A. 131,072
    B. 442,368
    C. 838,860
    D. 1,048,576
    E. 8,388,608

4. 81 is the ninth term in which of the following sequences?

    F. 1, 3, 6, 10, 15, …
    G. 11, 21, 31, 41, 51, …
    H. 1, 4, 9, 16, 25, …
    J. 8, 16, 24, 32, 40, …
    K. 5, 10, 15, 20, 25, …

5. What are the next three numbers in the sequence 1, 2, 4, 8, 16, … ?

    A. 24, 36, 52
    B. 32, 64, 128
    C. 48, 96, 192
    D. 32, 48, 64
    E. 48, 64, 136

■ Matrices

Matrix

A matrix is a rectangular array of numbers or variables. The entries in a matrix are called elements. Examples of matrices are shown below.

A. $\begin{bmatrix} 1 & 2 & 3 \\ 4 & 5 & 6 \end{bmatrix}$
    B. $\begin{bmatrix} z & y \\ x & w \\ a & b \end{bmatrix}$
    C. $\begin{bmatrix} 10 & 3 & 5 & 9 \end{bmatrix}$
    D. $\begin{bmatrix} -6 & 5 & 123 \\ 19 & -51 & -1.8 \\ 23 & 2 & -8 \end{bmatrix}$

The horizontal entries are called rows, while the vertical entries are called columns. Notice that there are the same number of elements in each row, and the same number of elements in each column.

Dimension

The dimension of a matrix is the number of rows followed by the number of columns. Here are the dimensions of each matrix shown above:

A. 2×3 B. 3×2 C. 1×4 D. 3×3

Scalar

Scalar is just another name for a number.

Matrix Arithmetic

You are most likely to encounter matrix addition and scalar multiplication on the ACT.

Matrix Addition

You may add matrices that have the same dimension. Just add the corresponding elements of the matrices to form a new matrix. Look at this example.

$$\begin{bmatrix} 3 & 8 \\ -9 & 6 \\ 13 & -6 \end{bmatrix} + \begin{bmatrix} 6 & -5 \\ -3 & -14 \\ 0 & 9 \end{bmatrix} = \begin{bmatrix} 3+6 & 8+(-5) \\ -9+(-3) & 6+(-14) \\ 13+0 & -6+9 \end{bmatrix} = \begin{bmatrix} 9 & 3 \\ -12 & -8 \\ 13 & 3 \end{bmatrix}$$

Scalar Multiplication

Multiply each element in the matrix by a scalar (number). Look at this example.

$$7\begin{bmatrix} 3 & 8 & -9 & 6 \\ 12 & -7 & 5 & 0 \end{bmatrix} = \begin{bmatrix} 7 \times 3 & 7 \times 8 & 7 \times (-9) & 7 \times 6 \\ 7 \times 12 & 7 \times (-7) & 7 \times 5 & 7 \times 0 \end{bmatrix} = \begin{bmatrix} 21 & 56 & -63 & 42 \\ 84 & -49 & 35 & 0 \end{bmatrix}$$

Matrix Multiplication

Multiplying two matrices is more complicated than multiplying a matrix by a scalar. Look at this example.

$$\begin{bmatrix} 5 & -4 & 2 \\ 1 & 3 & -2 \end{bmatrix} \times \begin{bmatrix} 3 & -1 \\ 0 & 5 \\ -2 & 0 \end{bmatrix}$$

Step 1. Find the dimension of the product. This step is easy.
Write the dimension of the first matrix: 2×3
Write the dimension of the second matrix: 3×2
The product will have the same number of rows as the first matrix and the same number of columns as the second matrix.

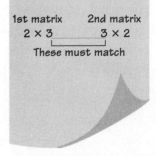

The multiplication cannot be done if the number of columns in the 1st matrix does not equal the number of rows in the 2nd.

1st matrix   2nd matrix
  2 × 3 _____ 3 × 2
  └─────────┘
  These must match

1st matrix   2nd matrix

2×3 3×2

Dimension of the product

The dimension of the product will be 2×2.

Step 2. Multiply the matrices. For each element of the product, multiply a row in the first matrix by a column in the second matrix and add the products of the elements. For example, to find the top left element, multiply the first row in the first matrix by the first column in the second. This is highlighted below:

$$\begin{bmatrix} 5 & -4 & 2 \\ 1 & 3 & -2 \end{bmatrix} \times \begin{bmatrix} 3 & -1 \\ 0 & 5 \\ -2 & 0 \end{bmatrix} = \begin{bmatrix} 5(3) - 4(0) + 2(-2) & ? \\ ? & ? \end{bmatrix} = \begin{bmatrix} 11 & ? \\ ? & ? \end{bmatrix}$$

Repeat this process for all the elements in the product:

$$\begin{bmatrix} 5 & -4 & 2 \\ 1 & 3 & -2 \end{bmatrix} \times \begin{bmatrix} 3 & -1 \\ 0 & 5 \\ -2 & 0 \end{bmatrix} = \begin{bmatrix} 11 & 5(-1) - 4(5) + 2(0) \\ 1(3) + 3(0) - 2(-2) & 1(-1) + 3(5) - 2(0) \end{bmatrix} = \begin{bmatrix} 11 & -25 \\ 7 & 14 \end{bmatrix}$$

CALCULATOR TIP

Some calculators can add, subtract, and multiply matrices. You can always use a calculator to complete or check your calculations.

Problem Solving With Matrices

You can use matrix multiplication to solve one type of problem that may appear on your test. You may see a problem like this on the ACT:

EXAMPLE

Bob and Liz purchased bags of candy, Bags A, B, and C. Bob bought six of Bag A, four of Bag B, and nine of Bag C. Liz bought four of Bag A, eight of Bag B, and five of Bag C. Bags of candy A cost $3, bags of candy B cost $6, and bags of Candy C cost $8. What was the total cost of all the candy in Bags A, B, and C that Bob and Liz bought?

Here's how to solve it using matrix multiplication.

Write a matrix for how many bags of candy each person bought.

Write a matrix for the cost of each bag.

Number of Bags

$$\begin{array}{c} \\ \text{Bob} \\ \text{Liz} \end{array} \begin{array}{cc} \text{A} \ \ \text{B} \ \ \text{C} \\ \begin{bmatrix} 6 & 4 & 9 \\ 4 & 8 & 5 \end{bmatrix} \end{array}$$

Cost

$$\begin{array}{c} \text{A} \\ \text{B} \\ \text{C} \end{array} \begin{bmatrix} \$3 \\ \$6 \\ \$8 \end{bmatrix}$$

Multiply: $\begin{bmatrix} 6 & 4 & 9 \\ 4 & 8 & 5 \end{bmatrix} \times \begin{bmatrix} \$3 \\ \$6 \\ \$8 \end{bmatrix} = \begin{bmatrix} 6 \times \$3 + 4 \times \$6 + 9 \times \$8 \\ 4 \times \$3 + 8 \times \$6 + 5 \times \$8 \end{bmatrix}$

$$= \begin{bmatrix} \$18 + \$24 + \$72 \\ \$12 + \$48 + \$40 \end{bmatrix} = \begin{bmatrix} \$114 \\ \$100 \end{bmatrix}$$

The top entry in the final matrix shows the cost of Bob's candy.

The bottom entry shows the cost of Liz's candy.

Add the entries to find the total cost.

$114 + $100 = $214

Note that you do not need matrices to solve this problem. You can just think the problem through.

Candy A: 6 + 4 = 10 bags, Candy B: 4 + 8 = 12 bags, Candy C: 9 + 5 = 14 bags.

Then multiply by the cost:

10 × $3 + 12 × $6 + 14 × $8 = $30 + $72 + $112 = $214

MODEL ACT PROBLEM

$$\begin{bmatrix} 8 & -8 \\ 0 & -2 \end{bmatrix} + \begin{bmatrix} -2 & -2 \\ -2 & -2 \end{bmatrix} = ?$$

A. $\begin{bmatrix} 6 & 6 \\ -2 & 0 \end{bmatrix}$

B. $\begin{bmatrix} 6 & -10 \\ -2 & -4 \end{bmatrix}$

C. $\begin{bmatrix} 6 & -6 \\ 2 & 4 \end{bmatrix}$

D. $\begin{bmatrix} 10 & -10 \\ -2 & -4 \end{bmatrix}$

E. $\begin{bmatrix} -10 & 10 \\ -2 & 4 \end{bmatrix}$

SOLUTION

Add the corresponding elements.

$$\begin{bmatrix} 8 & -8 \\ 0 & -2 \end{bmatrix} + \begin{bmatrix} -2 & -2 \\ -2 & -2 \end{bmatrix}$$

$$= \begin{bmatrix} 8 + (-2) & -8 + (-2) \\ 0 + (-2) & -2 + (-2) \end{bmatrix}$$

$$= \begin{bmatrix} 6 & -10 \\ -2 & -4 \end{bmatrix}$$

The correct answer is B.

Practice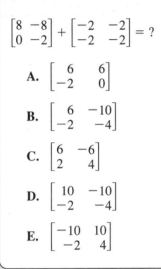

Use the matrices shown below to answer questions 1–15.

A. $\begin{bmatrix} -1 & 1 & -4 \\ 9 & -7 & 0 \\ 12 & -9 & 2 \end{bmatrix}$
B. $\begin{bmatrix} 3 & -8 \\ 6 & 12 \\ 5 & 10 \end{bmatrix}$
C. $\begin{bmatrix} 10 & 3 & 5 & 9 \\ -3 & 6 & 5 & 2 \end{bmatrix}$
D. $\begin{bmatrix} -6 & 5 & 12 \\ 19 & -1 & -2 \\ 23 & 2 & -8 \end{bmatrix}$

E. $\begin{bmatrix} 5 & -9 & 12 & 2 \\ -3 & 7 & -3 & 8 \end{bmatrix}$
F. $\begin{bmatrix} 9 & 18 & -12 \\ 0 & 3 & -6 \\ -2 & 8 & 9 \end{bmatrix}$
G. $\begin{bmatrix} 7 & 2 & -5 & 8 \end{bmatrix}$
H. $\begin{bmatrix} -1 & 3 & 2 \\ 9 & -1 & -8 \\ -4 & 11 & -7 \\ 12 & 3 & 5 \end{bmatrix}$

Write the dimension of the matrix.

1. A
2. C
3. B
4. G
5. H

Add these matrices.

6. C + E
7. A + F
8. D + F
9. F + A
10. D + H

Find the scalar product.

11. $-1 \times G$ **12.** $2 \times F$ **13.** $7 \times C$ **14.** $-2 \times B$ **15.** $0.5 \times H$

Subtract the matrices.

16. $C - E$ **17.** $A - F$ **18.** $D - F$ **19.** $F - A$ **20.** $D - H$

(Answers on page 364)

ACT-TYPE PROBLEMS

1. What are the dimensions of this matrix?

   $[5 \quad 2 \quad -3]$

   A. 1×3
   B. 3×1
   C. 3×0
   D. $5 \times 2 \times -3$
   E. 3

2.

   $$\begin{bmatrix} 5 & 2 & -3 \\ 4 & -7 & 0 \end{bmatrix} + \begin{bmatrix} 5 & 4 \\ 2 & -7 \\ -3 & 0 \end{bmatrix} = \,?$$

   F. $\begin{bmatrix} 10 & 4 & -6 \\ 8 & -14 & 0 \end{bmatrix}$

   G. $\begin{bmatrix} 10 & 8 \\ 4 & -14 \\ -6 & 0 \end{bmatrix}$

   H. $\begin{bmatrix} 10 & 6 & -1 \\ -3 & -10 & 0 \end{bmatrix}$

   J. $\begin{bmatrix} 10 & -3 \\ 6 & -10 \\ -1 & 0 \end{bmatrix}$

   K. The two matrices cannot be added.

3. Multiply.

   $$-3\begin{bmatrix} 4 & -7 & -\frac{1}{2} \end{bmatrix}$$

   A. $\begin{bmatrix} -12 & 21 & -\frac{3}{2} \end{bmatrix}$
   D. $\begin{bmatrix} 1 & -10 & -3\frac{1}{2} \end{bmatrix}$

   B. $\begin{bmatrix} -12 \\ 21 \\ \frac{3}{2} \end{bmatrix}$
   E. $\begin{bmatrix} \frac{21}{2} \end{bmatrix}$

   C. $\begin{bmatrix} -12 & 21 & \frac{3}{2} \end{bmatrix}$

4. $[2 \quad 3 \quad 4] + [-2 \quad -3 \quad -4] = \,?$

   F. $\begin{bmatrix} 2 & 3 & 4 \\ -2 & -3 & -4 \end{bmatrix}$

   G. $[-4 \quad -6 \quad -8]$

   H. $[-4 \quad -9 \quad -16]$

   J. $[0 \quad 0 \quad 0]$

   K. $[0]$

5. Subtract.

   $$\begin{bmatrix} 1 & 3 \\ 5 & 7 \\ 9 & 11 \end{bmatrix} - \begin{bmatrix} -1 & -3 \\ -5 & -7 \\ -9 & -11 \end{bmatrix}$$

   A. $\begin{bmatrix} 0 & 0 \\ 0 & 0 \\ 0 & 0 \end{bmatrix}$

   B. $\begin{bmatrix} 2 & 6 \\ 10 & 14 \\ 18 & 22 \end{bmatrix}$

   C. $\begin{bmatrix} -2 & -6 \\ -10 & -14 \\ -18 & -22 \end{bmatrix}$

   D. $\begin{bmatrix} 8 \\ 24 \\ 40 \end{bmatrix}$

   E. $\begin{bmatrix} 1 & 3 & -1 & -3 \\ 5 & 7 & -5 & -7 \\ 9 & 11 & -9 & -11 \end{bmatrix}$

6. A car dealership had a special sale on Wednesday and another on Thursday. Convertibles (C), midsized cars (M), and SUVs (S) were sold on Wednesday and Thursday. The matrices below show the number of each type of car sold for each day and the bonus the dealer received for each car sold from the manufacturer. What total bonus did the dealership receive from the manufacturer for the convertibles, midsized cars, and SUVs sold on those two sale days?

$$
\begin{array}{c}
\begin{array}{ccc} C & M & S \end{array} \\
\begin{array}{c} \text{Wednesday} \\ \text{Thursday} \end{array}
\begin{bmatrix} 24 & 38 & 13 \\ 32 & 52 & 11 \end{bmatrix}
\end{array}
\qquad
\begin{array}{c}
\text{Bonus} \\
\begin{array}{c} C \\ M \\ S \end{array}
\begin{bmatrix} \$200 \\ \$100 \\ \$300 \end{bmatrix}
\end{array}
$$

F. $10,200

G. $12,500

H. $14,900

J. $27,400

K. $30,800

7. $[450 \quad 700 \quad 900] \times \begin{bmatrix} 28 & 81 & 41 \\ 30 & 36 & 32 \\ 51 & 25 & 29 \end{bmatrix} = ?$

A. $[79{,}500 \quad 84{,}150 \quad 66{,}950]$

B. $[49{,}050 \quad 99{,}400 \quad 91{,}800]$

C. $[230{,}600]$

D. $\begin{bmatrix} 67{,}500 \\ 68{,}600 \\ 94{,}500 \end{bmatrix}$

E. $\begin{bmatrix} 79{,}500 \\ 84{,}150 \\ 66{,}950 \end{bmatrix}$

(Answers on page 364)

8. The Amsco designer shirts were hot sellers. Amsco sells three types of T-shirts: regular T-shirts (R), T-shirts with collars (C), and long sleeve T-shirts (L). The T-shirts are sold in three stores located in New York, Chicago, and Detroit. The matrix below gives the number of shirts sold at each store on opening day.

$$
\begin{array}{c}
\begin{array}{ccc} R & C & L \end{array} \\
\begin{array}{c} \text{New York} \\ \text{Chicago} \\ \text{Detroit} \end{array}
\begin{bmatrix} 120 & 75 & 35 \\ 125 & 60 & 25 \\ 90 & 35 & 40 \end{bmatrix}
\end{array}
$$

The price of each T-shirt is shown in the following matrix.

$$
\begin{array}{c} R \\ C \\ L \end{array}
\begin{bmatrix} \$22.50 \\ \$29.00 \\ \$36.50 \end{bmatrix}
$$

Given these matrices, what were the total sales for opening day?

F. $16,117.50

G. $20,247.50

H. $25,975

J. $27,264

K. $67,145

Cumulative **ACT** Practice

Intermediate Algebra

Complete this Cumulative ACT Practice in 10 minutes to reflect real ACT test conditions. This Cumulative ACT Practice gives you an additional opportunity to practice intermediate algebra concepts in an ACT format. If you don't know an answer, eliminate and guess. Circle the number of any guessed answer. Then check your answers on page 365. You will also find explanations for the answers and suggestions for further study.

1. Which of the following is a solution for the inequality $|9x - 5| \leq 4$?

   A.   2

   B. $\dfrac{1}{8}$

   C. -1

   D. $\dfrac{1}{10}$

   E. $-\dfrac{1}{10}$

2. Given the two linear equations $5x + 3y = 2$ and $6x + 4y = 9$, what is the sum of the x and y values that will make both equations true?

   F. -9.5

   G. -7

   H.   7

   J.   9

   K.   16.5

3. $-4\begin{bmatrix} 4 & -2 \\ 7 & 3 \end{bmatrix} + \begin{bmatrix} -3 & 2 \\ 2 & 5 \end{bmatrix} = ?$

   A. $\begin{bmatrix} -13 & 6 \\ -30 & -17 \end{bmatrix}$

   B. $\begin{bmatrix} -16 & 10 \\ -26 & 7 \end{bmatrix}$

   C. $\begin{bmatrix} -19 & 10 \\ -26 & -7 \end{bmatrix}$

   D. $\begin{bmatrix} -19 & 6 \\ -30 & -17 \end{bmatrix}$

   E. $\begin{bmatrix} -13 & 10 \\ -26 & 7 \end{bmatrix}$

4. Use the quadratic formula to find the roots of $x^2 - 2x + 5 = 0$.

   F.   2 and 3

   G. $2 + i$ and $2 - i$

   H.   1 and 2

   J. $1 + 2i$ and $1 - 2i$

   K. -1 and 3

5. If $|x + 4| \geq 10$ which value is part of the solution set?

   A. -13 D.   5

   B. -5 E.   13

   C.   0

6. Which of the following is NOT a solution to the inequality $2x^2 + 6x - 20 \leq 0$?

   F. -4

   G. -1

   H.   0

   J.   1

   K.   4

7. One morning Pat looked at her rose bushes and saw 3 roses. The second morning Pat saw 5 roses, and the third morning she saw 7 roses. If this pattern continues, how many roses will Pat see on the tenth morning?

   A.   21 roses

   B.   19 roses

   C.   17 roses

   D.   15 roses

   E.   13 roses

8. What is the maximum number of roots that might be found for the function $f(x) = 2x^5 - 3x^4 + 6x^3 - 9x^2 + 5x - 1$?

   F.   2

   G.   3

   H.   4

   J.   5

   K.   6

9. Which choice represents the expression $\left(\sqrt[4]{x}\right)^8 - \dfrac{1}{x^{-1}} - 1 + \dfrac{3}{x^{-2}}$ without radicals and negative exponents?

   A. $4x^2$

   B. $-4x^2 - x$

   C. $4x^2 - x - 1$

   D. $4x^2 + x$

   E. $4x - 1$

10. A chair manufacturer began to make kitchen chairs and rocking chairs in 1990. The function $k(x) = 3{,}000x + 3{,}800$ represents the number of kitchen chairs made x years after 1990. The function $r(x) = 3{,}500x + 3{,}300$ represents the number of rocking chairs made x years after 1990. In what year did the chair manufacturer make the same number of kitchen chairs as rocking chairs?

    F.   1995   J.   1992

    G.   1994   K.   1991

    H.   1993

Subtest

Complete this Subtest in 9 minutes to reflect real ACT test conditions. This Subtest has the same number and type of intermediate algebra items as are found on the ACT. If you don't know an answer, eliminate and guess. Circle the number of any guessed answer. Then check your answers on page 366. You will also find explanations for the answers and suggestions for further study.

1. Which of the following is NOT a solution to the inequality $|-3x + 3| > 6$?

   A. -4
   B. -3
   C. -2
   D. 3
   E. 4

2. Given the linear equations $2x + 5y = 16$ and $4x + 7y = 23$, what is the product of the x and y values that will make both equations true?

   F. 0.5
   G. 1.5
   H. 3
   J. 3.5
   K. 6

3. $2\begin{bmatrix} 4 & 3 \\ 2 & 1 \end{bmatrix} + -1\begin{bmatrix} 1 & 3 \\ 5 & 6 \end{bmatrix} = ?$

   A. $\begin{bmatrix} 7 & 3 \\ 7 & 7 \end{bmatrix}$

   B. $\begin{bmatrix} 5 & 6 \\ 7 & 7 \end{bmatrix}$

   C. $\begin{bmatrix} -5 & -6 \\ -7 & -7 \end{bmatrix}$

   D. $\begin{bmatrix} 7 & 3 \\ -1 & -4 \end{bmatrix}$

   E. $\begin{bmatrix} 9 & 9 \\ 9 & 8 \end{bmatrix}$

4. Use the quadratic formula to find the roots of $2x^2 - 4x = -4$.

   F. $-1 + i$ and $-1 - i$
   G. 0 and 2
   H. $1 + i$ and $1 - i$
   J. $1 + \sqrt{3}$ and $1 - \sqrt{3}$
   K. $4 + i$ and $4 - i$

5. If $|x - 4| \leq 9$, which of the following values is NOT in the solution set?

   A. -5
   B. -4
   C. 0
   D. 13
   E. 14

6. The monthly profit function for Beautiful Boutiques is $P(x) = 20x^2 - 1,000$, where x represents the number of customers. How many customers does Beautiful Boutiques need to have each month to make a profit of at least $1,000?

   F. 100
   G. 50
   H. 25
   J. 20
   K. 10

7. One week Joe and Theresa saw 5 ducks in the lake. One week later they saw 15 ducks in the lake, 2 weeks later they saw 35 ducks in the lake, and 3 weeks later they saw 75 ducks. If this pattern stays consistent, how many weeks after their initial visit to the lake will Joe and Theresa first see more than 315 ducks in the lake?

   A. 8 weeks
   B. 7 weeks
   C. 6 weeks
   D. 5 weeks
   E. 4 weeks

8. What is the maximum number of roots of the function $f(x) = 3x^4 - 2x^3 + 5x - 8$?

   F. 3
   G. 4
   H. 5
   J. 6
   K. 8

9. Write the expression $x^{\frac{1}{2}} \cdot \sqrt[3]{x2} + \frac{1}{\sqrt[4]{x^{-2}}}$ in simplified form without radicals or negative exponents.

   A. $x^{\frac{6}{7}} + x^2$

   B. $2x^{\frac{1}{2}}$

   C. $x^{\frac{5}{3}}$

   D. $x^{\frac{7}{6}} + x^{\frac{1}{2}}$

   E. $x^{\frac{7}{12}}$

Chapter 10 ▪

Coordinate Geometry

- Nine ACT questions have to do with coordinate geometry.
- Easier coordinate geometry questions may be about a single skill or concept, or may test a combination of pre-algebra, elementary algebra, intermediate algebra, and coordinate skills.
- More difficult questions will often test a combination of coordinate geometry skills and concepts.
- This coordinate geometry review covers all the material you need to answer ACT questions.
- Use a calculator for the ACT-Type Problems. Do not use a calculator for the Practice exercises.

▪ Graphing Inequalities on a Number Line

You can graph an inequality on a number line.

- An open circle, ○, shows that a point is not included.
- A closed circle, ●, shows that a point is included.
- An arrow, ← or →, shows that the line continues forever in that direction.

Graphing Inequalities

E X A M P L E S

1. Graph. $x > -3$

   The graph of $x > -3$ should show all points greater than -3 but not including -3.

   Draw an open circle at -3 and an arrow to the right. The open circle shows that -3 is not included. The arrow shows that the graph goes on forever to the right.

2. Graph. $-2 < x < 4$

   The graph of $-2 < x < 4$ should show all points between -2 and 4 but not including -2 or 4. Draw open circles at -2 and 4. Shade the number line between the open circles.

3. Graph. $x \leq 0$

   The graph of $x \leq 0$ should show all points less than 0, including 0.

   Draw a closed circle at 0 and an arrow to the left.

4. Graph. $-4 \leq x < 5$

The graph of $-4 \leq x < 5$ should show all the points between -4 and 5, including -4 but not including 5. Draw a closed circle for -4 and an open circle for 5. Shade the number line between the circles.

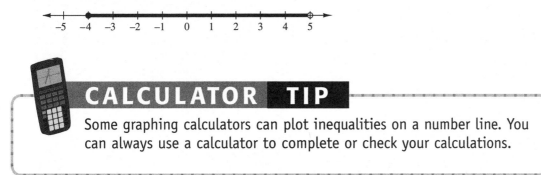

CALCULATOR TIP

Some graphing calculators can plot inequalities on a number line. You can always use a calculator to complete or check your calculations.

5. Graph. $2x + 1 < 3x + 2$

First, solve the inequality.

$$
\begin{array}{rcl}
2x + 1 & < & 3x + 2 \\
-2 & & -2 \\
\hline
2x - 1 & < & 3x \\
-2x & & -2x \\
\hline
-1 & < & x
\end{array}
\qquad (x > -1)
$$

Then, graph the solution.

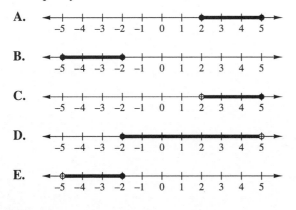

MODEL ACT PROBLEMS

1. Which of the following is the graph of the inequality $-2 \leq x < 5$?

A.

B.

C.

D.

E.

SOLUTION

The inequality $-2 \leq x < 5$ represents a section of the number line from -2 to 5 with -2 included and 5 excluded.

The correct answer is D.

2. Which inequality is graphed below?

F. $x > -3$

G. $3(x + 1) \geq 3(5 - x)$

H. $x + 1 \geq -5\left(1 + \dfrac{x}{5}\right)$

J. $x + 1 \geq 5 + x$

K. $x + 1 \geq -x + 4$

SOLUTION

Since the circle at -3 is closed and the arrow shows that the graph goes on forever to the right, the inequality shown is $x \geq -3$.

Find the inequality with the solution $x \geq -3$.

$$x + 1 \geq -5\left(1 + \dfrac{x}{5}\right)$$

$$x + 1 \geq -5 - x$$

$$2x \geq -6$$

$$x \geq -3$$

The correct answer is H.

Practice

Graph each inequality on a number line.

1. $x < -1$ **2.** $x \geq -2$ **3.** $x \geq -1$

4. $x \geq 3$ **5.** $x < -3$ **6.** $x \geq 4$

7. $x > -5$ **8.** $5 \geq x > -2$ **9.** $-3 \leq x \leq 1$

10. $2 \geq x > -2$ **11.** $-3 < x \leq -1$ **12.** $2 \leq x \leq 4$

13. $-1 < x \leq 3$ **14.** $4 \geq x \geq 1$ **15.** $3x + 1 \geq 2x - 2$

16. $3x - 1 \leq 4x + 3$ **17.** $9x + 7 < 2x + 28$ **18.** $-5x - 2 \geq 3x + 14$

19. $21x + 9 > 14x + 2$ **20.** $x - 15 \leq 20x + 23$

(Answers on page 367)

ACT-TYPE PROBLEMS

1. Which inequality is graphed below?

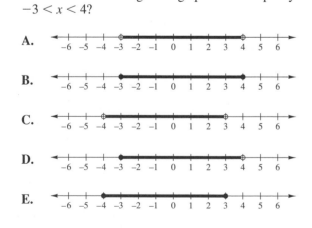

    A. $x < 6$
    B. $x > -6$
    C. $x \leq -6$
    D. $x \geq -6$
    E. $x \leq 6$

2. Which inequality is graphed below?

    F. $x \geq -5$
    G. $-5 \leq x \leq 1$
    H. $x \leq 1$
    J. $-5 \leq x < 1$
    K. $-5 < x \leq 1$

3. Which of the following is the graph of the inequality $-3 < x < 4$?

    A.

    B.

    C.

    D.

    E.

4. Which of the following is the graph of the inequality $3x - 5 \leq 5x + 7$?

    F.

    G.

    H.

    J.

    K.

5. What is the sum of all integers that are solutions of the inequality graphed below?

    A. -11
    B. -12
    C. -13
    D. -14
    E. -15

(Answers on page 367)

Graphing Equations on the Coordinate Plane

A linear equation can be written in the form $ax + by = c$.

Points on the Coordinate Plane

A point on the coordinate plane is named by an ordered pair (x,y). The x refers to the value on the x (horizontal) axis. The y refers to the value on the y (vertical) axis. Look at the points plotted below.

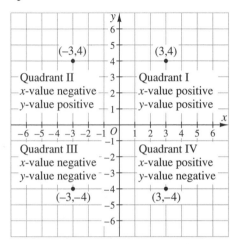

CALCULATOR TIP

Graphing calculators are designed to graph equations and inequalities on the coordinate plane.

EXAMPLES

1. Graph. $3x + 2y = 6$

   Solve the equation for $x = 0$.

   $$0 + 2y = 6$$
   $$y = 3$$

   One ordered pair is $(0,3)$.

   Solve the equation for $y = 0$.

   $$3x + 0 = 6$$
   $$x = 2$$

   A second ordered pair is $(2,0)$.

   Plot the points and connect them.

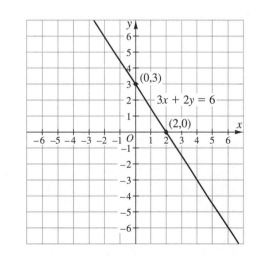

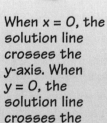

When x = 0, the solution line crosses the y-axis. When y = 0, the solution line crosses the x-axis.

2. Graph. $8x - 3y + 12 = 0$

Write the equation in standard form.

$$8x - 3y = -12$$

Solve the equation for $x = 0$.

$$0 - 3y = -12$$
$$y = 4$$

One ordered pair is (0,4).

Solve the equation for $y = 0$.

$$8x + 0 = -12$$
$$x = -1.5$$

A second ordered pair is (−1.5,0).

Plot the points and connect them.

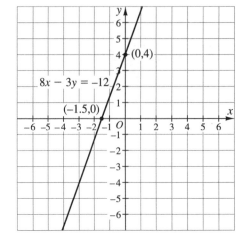

Slope

Slope Formula

In the slope formula, m stands for the slope of a line. To find the slope, identify two points on the line, (x_1, y_1) and (x_2, y_2). The slope is the difference of the y-values divided by the difference of the x-values as long as the x-values are not equal. You should memorize this slope formula.

$$m = \frac{y_2 - y_1}{x_2 - x_1} \text{ with } x_2 - x_1 \neq 0$$

A line with a slope of 0 is parallel to the x-axis. A line with an undefined slope is parallel to the y-axis.

EXAMPLES

1. Find the slope of the line passing through the points (2,6) and (3,5).

Label one ordered pair (x_1, y_1) and the other pair (x_2, y_2).

$$(x_1, y_1) \rightarrow (2,6) \qquad (x_2, y_2) \rightarrow (3,5)$$

Use the formula $m = \dfrac{y_2 - y_1}{x_2 - x_1}$ to find the slope of the line passing through the points.

$$m = \frac{y_2 - y_1}{x_2 - x_1} = \frac{5 - 6}{3 - 2} = \frac{-1}{1} = -1$$

The slope is −1.

2. Find the slope of the line passing through the points (−2,0) and (0,4).

$$m = \frac{y_2 - y_1}{x_2 - x_1} = \frac{4 - 0}{0 - (-2)} = \frac{4}{2} = 2$$

The slope is 2.

Slopes of Lines

- Lines that slope down from left to right have a negative slope.

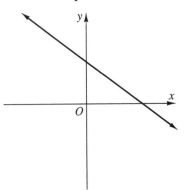

- Lines that slope up from left to right have a positive slope.

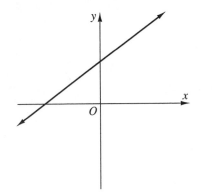

- Vertical lines have no slope. The slope of a vertical line is undefined. All *x*-values are the same.

- Horizontal lines have a slope of 0. All *y*-values are the same.

Slopes of Pairs of Lines

- Parallel lines have the same slope.

- Perpendicular lines have slopes whose product is -1 (except when one of the lines is vertical). The slopes of perpendicular lines are negative reciprocals.

Slope-Intercept Form

Every linear equation can be written in slope-intercept form, which is given below.

$$y = mx + b$$

- *m* is the slope.

- *b* is the *y*-intercept (where the line crosses the *y*-axis and $x = 0$).

You should memorize this equation.

Graph. $8x - 4y = 6$

Write the equation in slope-intercept form.

Solve the equation for y.

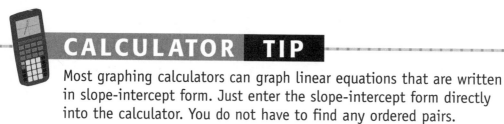

$$8x - 4y = 6$$
$$-4y = -8x + 6$$
$$y = 2x - 1.5$$

From the equation, you can see that the slope is 2 and the y-intercept is -1.5.

To graph the equation, first plot the y-intercept, $(0, -1.5)$. Then use the slope to find another point on the line. From $(0, -1.5)$, go up 2 and to the right 1. Plot the point. Repeat to find another point. Then draw the line.

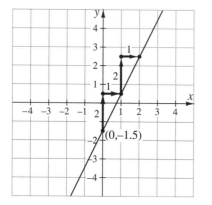

CALCULATOR TIP

Most graphing calculators can graph linear equations that are written in slope-intercept form. Just enter the slope-intercept form directly into the calculator. You do not have to find any ordered pairs.

Graphing to Find the Solution to Systems of Equations

The solution to a system of two equations is the point at which the two graphs intersect. If the two lines are parallel, there is no solution. If the two lines are the same, there are an infinite number of solutions.

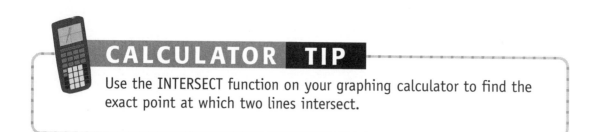

CALCULATOR TIP

Use the INTERSECT function on your graphing calculator to find the exact point at which two lines intersect.

1. Solve the system by graphing.

$$x + y = 4$$
$$2x - y = 5$$

Write each equation in slope-intercept form.

$$x + y = 4$$
$$y = -x + 4$$

$$2x - y = 5$$
$$-y = -2x + 5$$
$$y = 2x - 5$$

The y-intercept is 4.
The slope is -1.

The y-intercept is -5.
The slope is 2.

Graph to find the point of intersection.

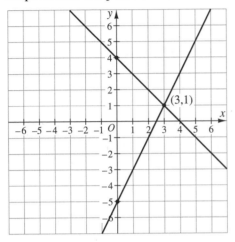

The lines intersect at (3,1). So the solution to the system is $x = 3$ and $y = 1$.

2. Solve the system by graphing.

$$-x + y = 3$$
$$x - y = -6$$

Write each equation in slope-intercept form.

$$-x + y = 3$$
$$y = x + 3$$

$$x - y = -6$$
$$-y = -x - 6$$
$$y = x + 6$$

The y-intercept is 3.
The slope is 1.

The y-intercept is 6.
The slope is 1.

Since the lines have the same slope, their graphs are parallel.

So the system has no solution.

> There is no need to actually graph the equation to find the solution since we can tell from the slopes that the lines are parallel.

3. Solve the system by graphing.

$$-2x - 4y = -6$$
$$x + 2y = 3$$

Write each equation in slope-intercept form.

$$-2x - 4y = -6$$
$$-4y = 2x - 6$$
$$y = -\frac{1}{2}x + \frac{3}{2}$$

$$x + 2y = 3$$
$$2y = -x + 3$$
$$y = -\frac{1}{2}x + \frac{3}{2}$$

Since the lines have the same equation, their graphs are the same.

So the system has an infinite number of solutions.

> There is no need to actually graph the equation to find the solution since we can tell from the slopes and inter-cepts that the lines are the same.

1. What is the slope of a line perpendicular to $y = -3x - 4$?

   A. -3

   B. -4

   C. 4

   D. $\dfrac{1}{4}$

   E. $\dfrac{1}{3}$

SOLUTION

The slopes of perpendicular lines have a product of -1.

The slope of the given line is -3.

Since $-3 \times \dfrac{1}{3} = -1$, the slope of a perpendicular line is $\dfrac{1}{3}$.

The correct answer is E.

2. Which of the following is the graph of the linear equation $x + y = -3$?

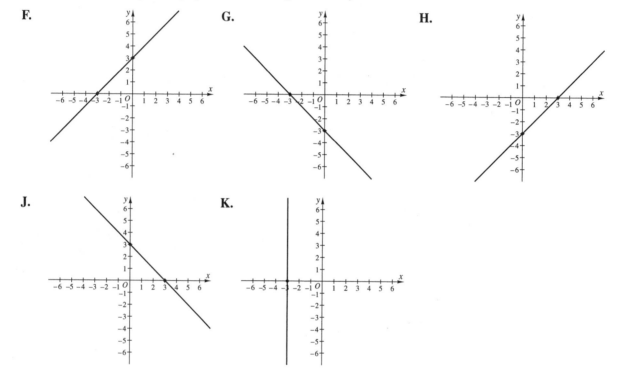

F. **G.** **H.**

J. **K.**

SOLUTION

Find two points that lie on the graph of $x + y = -3$.

Let $x = 0$. Let $y = 0$.

$0 + y = -3$ $x + 0 = -3$

$y = -3$ $x = -3$

One point is $(0, -3)$. Another point is $(-3, 0)$.

The graph in G contains these points.

The correct answer is G.

3. How many solutions are there to the system of equations $y = 3x + 2$ and $12x - 4y = 8$?

A. 0
B. 1
C. 1,000,000
D. 1,000,000,000
E. infinite

SOLUTION

Write each equation in slope-intercept form.

$$y = 3x + 2 \qquad 12x - 4y = 8$$
$$-4y = -12x + 8$$
$$y = 3x - 2$$

The equations represent different lines with the same slope.

When the slopes are the same, the lines are parallel.

Parallel lines never intersect, so there are no solutions.

The correct answer is A.

Practice

1. Plot each point on the coordinate plane. $(-2,5)$, $(4,3)$, $(2,-3)$, $(-3,-1)$

2. What are the coordinates of the point on the y-axis and the coordinates of the point on the x-axis of the graph of $2x + 4y = 4$?

3. Graph the line with equation $3x + 6y = 12$ by plotting points.

4. What is the slope of the line whose equation is given in problem 3?

5. Find the slope and the y-intercept of the line with equation $10x + 5y = 20$.

6. Graph the line with equation $4x + 2y = 8$ using slope-intercept form.

State whether the slope of each line is positive, negative, zero, or undefined.

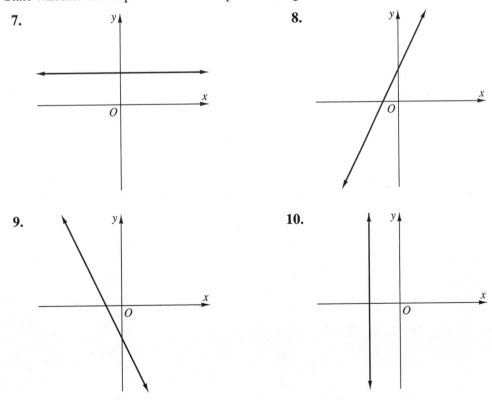

7.

8.

9.

10.

Tell whether each statement is true or false.

11. The lines represented by the following equations are parallel.

$y = -2x + 3$
$y = -3x + 3$

12. The lines represented by the following equations are perpendicular.

$y = \dfrac{1}{4}x + 6$

$y = -4x - 18$

13. The lines represented by the following equations are parallel.

$y = 3x - 4$
$y = 3x + 10$

14. The lines represented by the following equations are perpendicular.

$4y = x - 5$
$y = -4x + 5$

15. The y-intercept of $y = 2x - 19$ is 2.

16. There are an infinite number of solutions to the following system of equations.

$3x - 8y = 6$

$y = \dfrac{3}{8}x - \dfrac{3}{4}$

17. The slope of $2x + 4y = 20$ is 5.

Solve each problem.

18. What is the slope of the line passing through the points with coordinates (1,5) and (3,8)?

19. What is the slope of the line passing through the points with coordinates (4,7) and (7,9)?

20. Are the lines described in problems 18 and 19 parallel, perpendicular, or neither?

(Answers on page 368)

 ACT-TYPE PROBLEMS

1. What is the sum of the y-intercept and the slope of the linear equation $6x - 9y = 3$?

    A. -6

    B. -3

    C. $\dfrac{1}{3}$

    D. $-\dfrac{1}{3}$

    E. 9

2. Which equation is graphed below?

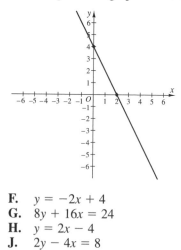

    F. $y = -2x + 4$
    G. $8y + 16x = 24$
    H. $y = 2x - 4$
    J. $2y - 4x = 8$
    K. $y = 4x - 2$

3. What is the slope of a line parallel to the line with equation $12x - 3y = 17$?

    A. $-\dfrac{17}{3}$

    B. 4

    C. $\dfrac{1}{4}$

    D. -4

    E. $\dfrac{3}{17}$

4. Which equation creates an infinite number of solutions when solved in a system with $y = 5x - 7$?

    F. $2y + 10x = -14$

    G. $y = 7x - 5$

    H. $3y - 15x = -28$

    J. $4y - 20x = -28$

    K. $4y + 15x = -21$

5. What is the equation of a line with a y-intercept of -3 that is perpendicular to the line with equation $3y - 4x = 21$?

    A. $y = \dfrac{4}{3}x - 3$

    B. $y = \dfrac{3}{4}x + 3$

    C. $y = -3x - \dfrac{4}{3}$

    D. $y = 3x + \dfrac{3}{4}$

    E. $y = -\dfrac{3}{4}x - 3$

(Answers on page 368)

▪ Distance and Midpoint Formulas

- Use the distance formula to find the distance between two points on a plane.
- Use the midpoint formula to find the midpoint of a line segment.

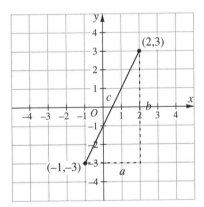

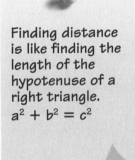

Finding distance is like finding the length of the hypotenuse of a right triangle.
$a^2 + b^2 = c^2$

Distance Formula

Use the following formula to find the distance between two points on the plane.

The distance (d) between (x_1, y_1) and $(x_2, y_2) = \sqrt{(x_2 - x_1)^2 + (y_2 - y_1)^2}$.

Find the distance between the two points $(-1,-3)$ and $(2,3)$ shown in the diagram on the previous page.

$$\begin{aligned} \text{Distance} &= \sqrt{(2-(-1))^2 + (3-(-3))^2} \\ &= \sqrt{(2+1)^2 + (3+3)^2} \\ &= \sqrt{3^2 + 6^2} = \sqrt{9+36} = \sqrt{45} = \sqrt{9 \times 5} = 3\sqrt{5} \end{aligned}$$

The distance between the two points is $3\sqrt{5} \approx 6.71$.

The symbol $\approx$ means "is approximately equal to."

Midpoint Formula

Use this formula to find the midpoint of the line segment between two points in the plane.

The midpoint of the line segment between (x_1,y_1) and $(x_2,y_2) = \left(\dfrac{x_1 + x_2}{2}, \dfrac{y_1 + y_2}{2}\right)$.

Find the midpoint of the line segment between the two points $(-1,-3)$ and $(2,3)$ shown in the diagram on the previous page.

$$\begin{aligned} \text{Midpoint} &= \left(\frac{-1+2}{2}, \frac{-3+3}{2}\right) \\ &= \left(\frac{1}{2}, \frac{0}{2}\right) = \left(\frac{1}{2}, 0\right) \end{aligned}$$

The midpoint is $\left(\dfrac{1}{2}, 0\right)$.

This is like averaging the x-coordinates and then averaging the y-coordinates.

MODEL ACT PROBLEM

1. What is the distance between points $(2,6)$ and $(1,4)$?

   A. $\sqrt{29}$
   B. $\sqrt{13}$
   C. $\sqrt{5}$
   D. 5
   E. 13

   SOLUTION

   Identify each point. $(x_1,y_1) \rightarrow (2,6)$
   $(x_2,y_2) \rightarrow (1,4)$

   Use the distance formula.

   $$\begin{aligned} \text{Distance} &= \sqrt{(1-2)^2 + (4-6)^2} \\ &= \sqrt{(-1)^2 + (-2)^2} = \sqrt{1+4} = \sqrt{5} \end{aligned}$$

   The correct answer is C.

2. What is the midpoint of the line segment between the points $(3,6)$ and $(2,4)$?

   F. $(2.5,5)$
   G. $(4.5,3)$
   H. $(3.5,4)$
   J. $(0.5,1)$
   K. $(1.5,1)$

   SOLUTION

   Identify the points. $(x_1,y_1) \rightarrow (3,6)$
   $(x_2,y_2) \rightarrow (2,4)$

   Use the midpoint formula.

   $$\begin{aligned} \text{Midpoint} &= \left(\frac{3+2}{2}, \frac{6+4}{2}\right) \\ &= \left(\frac{5}{2}, \frac{10}{2}\right) = (2.5,5) \end{aligned}$$

   The correct answer is F.

Calculators do not usually have special functions for finding the midpoint and distance. However, you can still use a calculator to check your answer.

Practice

1. What is the general formula for the distance between two points (x_1,y_1) and (x_2,y_2)?

2. What is the distance between points (3,5) and (1,2)?

3. What is the distance between points $(-2,7)$ and $(4,-6)$?

4. What is the general midpoint formula given two points (x_1,y_1) and (x_2,y_2)?

5. What is the midpoint of the line segment between points (4,2) and (6,4)?

6. What is the midpoint of the line segment between points (9,1) and (2,6)?

7. What is the distance between points $(5,-2)$ and (1,3)?

8. What is the distance between points $(4,-6)$ and $(-5,2)$?

9. What is the distance between points (6,3) and (8,9)?

10. What is the distance between points (0,0) and $(-3,4)$?

11. What is the distance between points $(-3,7)$ and (9,5)?

12. What is the distance between points (1,8) and (7,4)?

13. What is the distance between points $(-3,9)$ and $(-7,2)$?

14. What is the midpoint of the line segment between points $(5,-2)$ and (1,3)?

15. What is the midpoint of the line segment between points $(4,-6)$ and $(-5,2)$?

16. What is the midpoint of the line segment between points (6,3) and (8,9)?

17. What is the midpoint of the line segment between points (0,0) and $(-3,4)$?

18. What is the midpoint of the line segment between points $(-3,7)$ and (9,5)?

19. What is the midpoint of the line segment between points (1,8) and (7,4)?

20. What is the midpoint of the line segment between points $(-3,9)$ and $(-7,2)$?

(Answers on page 369)

1. What is the distance between points $(-4,7)$ and $(5,-3)$?

   A. 19
   B. $\sqrt{181}$
   C. 9
   D. $\sqrt{81}$
   E. 3

2. What is the midpoint of the line segment between points $(2,6)$ and $(3,8)$?

   F. $(2,2.5)$
   G. $(2.5,2)$
   H. $(5,4.5)$
   J. $(4,5.5)$
   K. $(2.5,7)$

3. What is the sum of the length of the line segment between the points $(-1,5)$ and $(3,8)$, and the y-coordinate of the midpoint of that line segment?

   A. 11.5
   B. 12
   C. 12.5
   D. 13
   E. 13.5

4. The midpoint of a line segment is $(3,4)$. One of the endpoints of that line segment is $(7,4)$. What is the other endpoint?

   F. $(1,5)$
   G. $(-4,0)$
   H. $(-1,4)$
   J. $(0,-4)$
   K. $(4,-1)$

5. Which of the following lists the length of each side of $\triangle ABC$ below from shortest to longest?

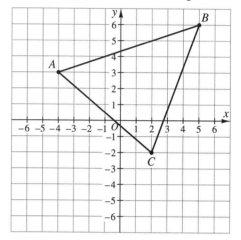

   A. BC, AC, AB
   B. AC, AB, BC
   C. BC, AB, AC
   D. AC, BC, AB
   E. AB, AC, BC

(Answers on page 369)

Graphing Systems of Inequalities on the Coordinate Plane

An inequality is shown as a region on the coordinate plane.

- Inequalities containing $\leq$ or $\geq$ have a solid-line boundary.
- Inequalities containing $<$ or $>$ have a dotted-line boundary.

Graphing an Inequality on the Coordinate Plane

To graph an inequality on the coordinate plane:

- Graph the related equation to show the boundary line.
- Pick a point on one side of the boundary line. If the point satisfies the inequality, shade the region. If the point does NOT satisfy the inequality, shade the region on the other side of the line.

Graph. $3x + 2y \leq 9$

Graph the related equation $3x + 2y = 9$.

Let $x = 0$.

$$3(0) + 2y = 9$$
$$y = 4.5$$

The ordered pair is (0,4.5).

Let $y = 0$.

$$3x + 2(0) = 9$$
$$x = 3$$

The ordered pair is (3,0).

Plot the two points and draw the line.

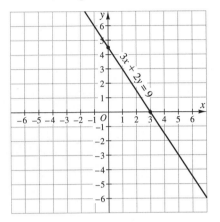

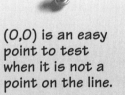

Since the inequality is $\leq$, the boundary is a solid line.

To find the region to shade, test (0,0) in the inequality.

$$3x + 2y \leq 9$$
$$3(0) + 2(0) \leq 9$$
$$0 \leq 9 \qquad \text{True}$$

Shade the region that includes the point (0,0). The shaded region and the boundary line make up the graph of the inequality $3x + 2y \leq 9$.

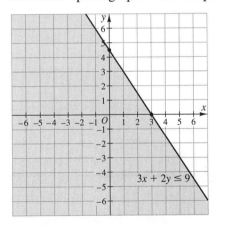

(0,0) is an easy point to test when it is not a point on the line.

Graphing Two Inequalities on the Coordinate Plane

To graph two inequalities on the coordinate plane:

- Graph each inequality on the same coordinate plane.
- The portion of the shaded region common to both inequalities is the graph of the two inequalities.

EXAMPLE

Graph. $3x + 2y \leq 9$ and $2x - y > 1$

> The graph of $3x + 2y \leq 9$ is shown on the previous page.
>
> To graph $2x - y > 1$, begin by graphing the equation $2x - y = 1$.
>
> Two points on this boundary line are $(0, -1)$ and $(0.5, 0)$.

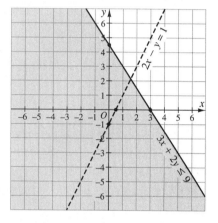

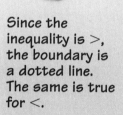

Since the inequality is $>$, the boundary is a dotted line. The same is true for $<$.

> To find out which side of the boundary line to shade, test $(0,0)$ in the inequality $2x - y > 1$.
>
> $$2x - y > 1$$
> $$2(0) - 0 > 1$$
> $$0 > 1 \quad \text{False}$$
>
> The graph of $2x - y > 1$ will *not* include the point $(0,0)$. Shade the region to the right of the boundary line.
>
> The shaded region that is common to both inequalities is the graph of $3x + 2y \leq 9$ and $2x - y > 1$.

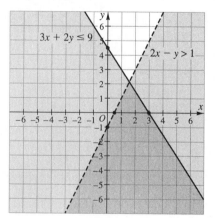

1. Consider the following inequalities.

   I. $2x + 4y > 6$ III. $5x + y < 8$
   II. $3x - 2y \geq 7$ IV. $7x + 9y \leq 3$

   Which choice lists all the inequalities that have graphs with solid boundary lines?

   A. I and III
   B. I and II
   C. III and IV
   D. II and IV
   E. II and III

   SOLUTION

   II and IV have $\geq$ or $\leq$, which result in a solid boundary line.

   The correct answer is D.

2. Which of the following choices shows the graph of the inequality $2x - 4y < 8$?

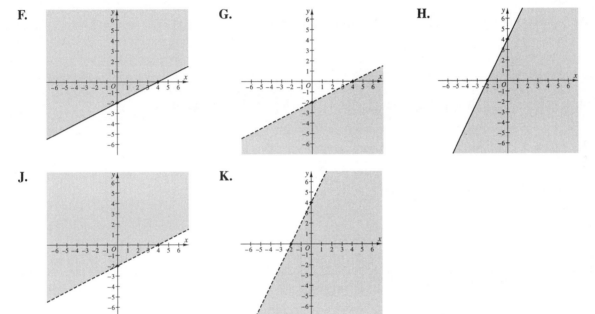

F. G. H.

J. K.

SOLUTION

Graph the line $2x - 4y = 8$ as a dotted boundary line because the inequality is $<$.

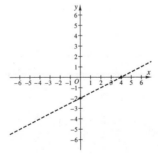

Shade the side of the boundary line that satisfies the inequality. The graph in J shows the correct shading.

The correct answer is J.

Some graphing calculators can plot inequalities. You can, however, use the calculator to graph the equations to find the boundary lines. Then test points to identify mentally the region of the graph that should be shaded in.

Practice

Should a solid line or a dotted line be used to graph each inequality?

1. $6x - 3y < 2$ **2.** $2x + 7y \geq 5$

3. $5x - 9y \leq 15$ **4.** $4x + 13y < 2$

5. When graphing two inequalities simultaneously, which part of the graph do you shade?

Graph each inequality.

6. $2x + y \geq 4$ **7.** $3y - 9x < 18$

8. $-3x - 6y < 12$ **9.** $-18x + 2y \leq 9$

10. $2x - y \geq -6$ **11.** $8x - 2y > 12$

12. $7x + 4y \leq -14$ **13.** $3x + 8y \geq 18$

Graph each system of inequalities.

14. $4x + 2y < 8$ and $-3x + y \geq -5$

15. $2x + 2y \geq 3$ and $-6x + 3y < 9$

16. $x + y < 6$ and $x - 2y \leq -3$

17. $-3x + 6y \geq -18$ and $7x - 7y < -7$

18. $2x - 5y > 10$ and $6x - 2y \geq 6$

19. $-8x - 4y < -16$ and $4x - 2y \leq 8$

20. $-4x - y < -2$ and $-4.5x - 3y \leq -9$

(Answers on page 370)

ACT-TYPE PROBLEMS

1. Which is NOT a solution of the inequality $2y - 6x \leq 13$?

   A. (0,0)
   B. (2,4)
   C. (1,6)
   D. (2,13)
   E. (2,10)

2. Which inequality is graphed below?

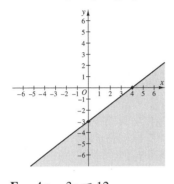

   F. $4x - 3y \leq 12$
   G. $3x + 4y \geq 12$
   H. $3x - 4y \geq 12$
   J. $4x - 3y < 12$
   K. $3x - 4y > 12$

3. Which system of inequalities is graphed below?

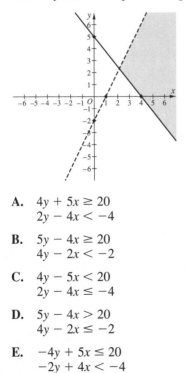

   A. $4y + 5x \geq 20$
      $2y - 4x < -4$

   B. $5y - 4x \geq 20$
      $4y - 2x < -2$

   C. $4y - 5x < 20$
      $2y - 4x \leq -4$

   D. $5y - 4x > 20$
      $4y - 2x \leq -2$

   E. $-4y + 5x \leq 20$
      $-2y + 4x < -4$

4. Which is the graph of the inequality $-2x + 5y \leq 10$?

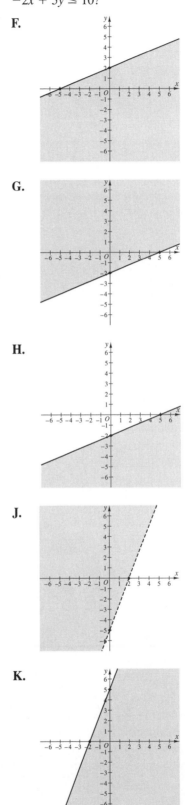

F.

G.

H.

J.

K.

5. Which is the graph of the system of inequalities $8x - 6y \geq -24$ and $-14x - 7y < -28$?

A.

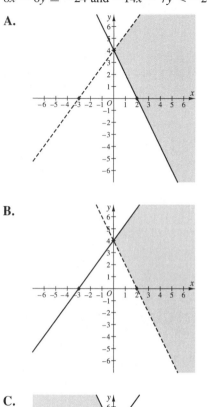

B.

C.

D.

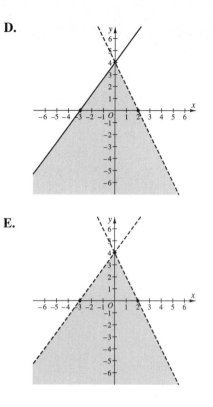

E.

(Answers on page 372)

Graphing Conic Sections

Conic sections are the figures formed by the intersection of a cone or cones and a plane.

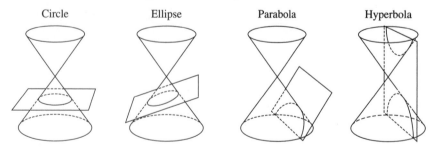

Circle   Ellipse   Parabola   Hyperbola

The equations for a circle, for an ellipse, for a parabola, and for a hyperbola each have a standard form.

CALCULATOR TIP

A graphing calculator can graph conic sections. Be careful when you enter the equation for circles, ellipses, and hyperbolas. These figures have two y-values for each x-value and you may have to use additional keys to display the entire graph.

Circle

The standard form of the equation for a circle is

$$(x - h)^2 + (y - k)^2 = r^2$$

where (h,k) is the center and r is the radius.

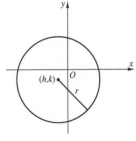

> You may have to simplify or rearrange terms to write an equation in standard form.

EXAMPLE

Graph. $x^2 + y^2 - 6x - 10y + 25 = 0$

To determine if this is an equation of a circle, first write the equation in standard form.

$$x^2 + y^2 - 6x - 10y + 25 = 0$$

Rearrange the terms. $\qquad (x^2 - 6x) + (y^2 - 10y) = -25$

To make each expression a perfect square, add the square of half the coefficient of each variable to both sides of the equation. This is called *completing the square*.

$$x^2 - 6x + \left(\frac{6}{2}\right)^2 + y^2 - 10y + \left(\frac{10}{2}\right)^2 = -25 + \left(\frac{6}{2}\right)^2 + \left(\frac{10}{2}\right)^2$$

$$x^2 - 6x + 9 + y^2 - 10y + 25 = -25 + 9 + 25$$

Write each expression as a perfect square.

$$(x - 3)^2 + (y - 5)^2 = 9$$

This is the equation of a circle with center $(3,5)$ and radius 3.

Graph the equation.

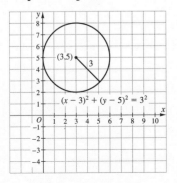

Ellipse

An ellipse is all points such that the sum of the distances from any point to two fixed points called **foci** is constant. The standard form of the equation for an ellipse is

$$\frac{(x - h)^2}{a^2} + \frac{(y - k)^2}{b^2} = 1$$

where (h,k) is the center, a and b are half the lengths of the axes of the ellipse, and the ellipse crosses the axes at $(h + a, k)$, $(h - a, k)$, $(h, k - b)$, and $(h, k + b)$.

If a^2 is the larger denominator, then the major (longer) axis is horizontal.

If b^2 is the larger denominator, then the major (longer) axis is vertical.

The foci are located on the major axis, $\sqrt{|a^2 - b^2|}$ units from the center.

An ellipse looks like a flattened circle.

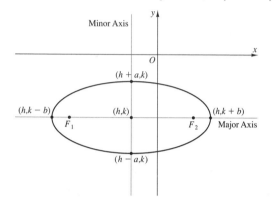

EXAMPLE

Graph. $\dfrac{(x + 3)^2}{36} + \dfrac{(y - 2)^2}{4} + 6 = 7$

First write the equation in standard form.

$$\frac{(x - (-3))^2}{6^2} + \frac{(y - 2)^2}{2^2} = 1$$

The center is $(-3,2)$. a^2 is larger than b^2, so the major axis is horizontal.

The ellipse crosses its axes at $(3,2)$, $(-9,2)$, $(-3,0)$, and $(-3,4)$.

The foci of the ellipse are $\pm\sqrt{|36 - 4|} = \pm\sqrt{32} = \pm4\sqrt{2}$ units along the major axis from the center, or at the points $(-3 + 4\sqrt{2}, 2)$ and $(-3 - 4\sqrt{2}, 2)$.

Use this information to graph the ellipse.

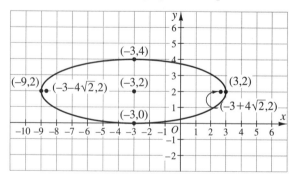

Parabola

There are two standard forms for a parabola.

$$y - k = a(x - h)^2$$

Vertical parabola

Vertex at (h,k)

$$x - h = a(y - k)^2$$

Horizontal parabola

Vertex at (h,k)

> A parabola is somewhat like an open ellipse.

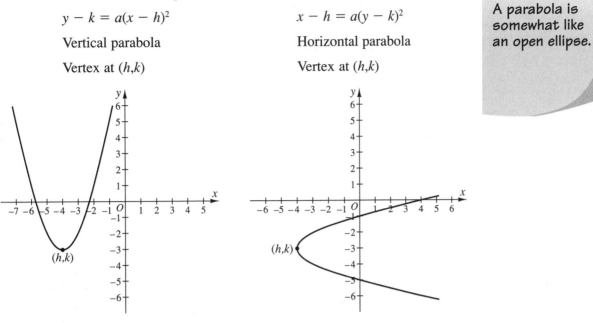

EXAMPLE

Graph the equation. $y = 2 + 5x^2 - 40x + 80$

Write the equation in standard form.

Subtract 2. $\qquad\qquad y - 2 = 2 - 2 + 5x^2 - 40x + 80$

Factor 5 out of the expression on the $\qquad y - 2 = 5(x^2 - 8x + 16)$

right side of the equation.

Write the perfect square. $\qquad\qquad y - 2 = 5(x - 4)^2$

This is the standard form for a vertical parabola.

The equation $y - 2 = 5(x - 4)^2$ is a vertical parabola with a vertex at (4,2). Substitute $x = 3$ and $x = 5$ to find two other points on the parabola. (3,7) and (5,7) are other points on the parabola. Graph the equation $y - 2 = 5(x - 4)^2$.

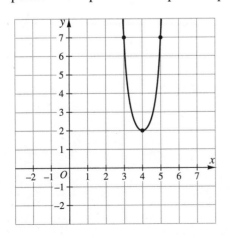

Hyperbola

A hyperbola is the set of all points in the plane for which the difference between the distances to two fixed points, called foci, is constant.

Each hyperbola has two branches. The line segment that connects the two foci intersects the hyperbola at two points, called the vertices. The line segment connecting these vertices is called the transverse axis. The midpoint of the transverse axis is called the center of the hyperbola. See the figure below.

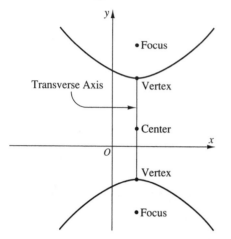

- The general equation of the hyperbola centered at (h,k) is

$$\frac{(x-h)^2}{a^2} - \frac{(y-k)^2}{b^2} = 1 \text{ if the transverse axis is horizontal.}$$

$$\frac{(y-k)^2}{a^2} - \frac{(x-h)^2}{b^2} = 1 \text{ if the transverse axis is vertical.}$$

- If the transverse axis is horizontal, the equation of the axis is $y = k$.
- If the transverse axis is vertical, the equation of the axis is $x = h$.
- The vertices are a units from the center along the transverse axis.
- The distance along the transverse axis from the center to a focus is c, where $c^2 = a^2 + b^2$.
- **Asymptotes** are limit lines approached by the hyperbola.

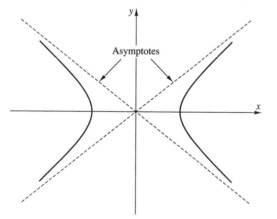

The equation of the asymptotes is:

Horizontal transverse axis: $\quad y = k - \dfrac{b}{a}(x-h) \quad$ and $\quad y = k + \dfrac{b}{a}(x-h)$

Vertical transverse axis: $\quad y = k - \dfrac{a}{b}(x-h) \quad$ and $\quad y = k + \dfrac{a}{b}(x-h)$

1. Given the equation of a circle $(x - 3)^2 + (y + 5)^2 = 16$, what is the sum of the x-coordinate of the center, the y-coordinate of the center, and the radius of the circle?

  A. 24
  B. 18
  C. 12
  D. 6
  E. 2

SOLUTION

Write the equation in standard form.
$(x - h)^2 + (y - k)^2 = r^2$

$(x - 3)^2 + (y + 5)^2 = 16$

$\quad\quad\quad \rightarrow \quad (x - 3)^2 + (y - (-5))^2 = 4^2$

Therefore, $h = 3$, $k = -5$, and $r = 4$.

Find the sum of h, k, and r.

$3 + (-5) + 4 = 2$

The correct answer is E.

2. What are the coordinates of the foci of an ellipse with the equation $\frac{(x + 1)^2}{9} + \frac{(y - 2)^2}{4} = 1$?

  F. $(-1, 2 + \sqrt{5})$ and $(-1, 2 - \sqrt{5})$
  G. $(-1 + \sqrt{5}, 2)$ and $(-1 - \sqrt{5}, 2)$
  H. $(2, -1 + \sqrt{5})$ and $(2, -1 - \sqrt{5})$
  J. $(1, -2 + \sqrt{5})$ and $(1, -2 - \sqrt{5})$
  K. $(1 + \sqrt{5}, -2)$ and $(1 - \sqrt{5}, -2)$

SOLUTION

To identify the major axis and the center, write the equation in standard form,

$\frac{(x - h)^2}{a^2} + \frac{(y - k)^2}{b^2} = 1$

$\frac{(x + 1)^2}{9} + \frac{(y - 2)^2}{4} = 1$

$\frac{(x - (-1))^2}{3^2} + \frac{(y - 2)^2}{2^2} = 1$

So, $a = 3$ and $b = 2$. Since $a > b$, the major axis is horizontal. The center of the ellipse is $(-1, 2)$. The foci of the ellipse are $\pm\sqrt{|3^2 - 2^2|} = \pm\sqrt{9 - 4} = \pm\sqrt{5}$ units along the major axis from the center, or at the points $(-1 + \sqrt{5}, 2)$ and $(-1 - \sqrt{5}, 2)$.

The correct answer is G.

3. What is the vertex of the parabola with equation $y - 5 = 2x^2 - 12x + 18$?

  A. $(-3, 5)$
  B. $(3, -5)$
  C. $(5, 3)$
  D. $(3, 5)$
  E. $(-3, -5)$

SOLUTION

Write the equation in standard form.

$y - k = a(x - h)^2$

$y - 5 = 2x^2 - 12x + 18$

Factor 2 out of the expression on the right side of the equation.

$y - 5 = 2(x^2 - 6x + 9)$

Write the perfect square.

$y - 5 = 2(x - 3)^2$

$h = 3$ and $k = 5$ so the vertex is $(3, 5)$.

The correct answer is D.

4. What is the center of the hyperbola with equation $\frac{x^2}{4} - \frac{y^2}{49} = 1$?

  F. $(-2, 0)$

  G. $(2, 0)$

  H. $(0, 0)$

  J. $(\sqrt{53}, 0)$

  K. $(-\sqrt{53}, 0)$

SOLUTION

The general equation for a hyperbola with its center at (h, k) is $\frac{(x - h)^2}{a^2} - \frac{(y - k)^2}{b^2} = 1$.

In the given equation, $h = 0$ and $k = 0$, therefore the center is $(0, 0)$.

The correct answer is H.

Practice

1. Find the center and the radius of the circle represented by the equation $(x + 2)^2 + (y - 3)^2 = 16$.

2. Complete the square for the equation $x^2 + 4x = 0$.

3. Write the equation for this circle in standard form. $x^2 + y^2 - 10x + 14y - 26 = 0$

4. Graph the circle represented by the equation. $x^2 + y^2 - 4x + 12y + 24 = 0$

The equation of an ellipse is $\dfrac{(x - 2)^2}{4} + \dfrac{(y + 3)^2}{9} = 1$.

5. What is the center of the ellipse?

6. Is the major axis horizontal or vertical?

7. What are the coordinates of the foci of this ellipse?

The equation of an ellipse is $\dfrac{(x + 1)^2}{25} + \dfrac{(y - 2)}{16} = 1$.

8. At what points does this ellipse cross the major and minor axes?

9. What are the coordinates of the foci of this ellipse?

10. Graph the ellipse represented by the equation.

Is each parabola vertical or horizontal?

11. $y - 16 = (x + 6)^2$

12. $x + 25 = 2(y - 7)^2$

13. $x - 34 = 3(y + 14)^2$

14. $y - 8 = 10(x - 12)^2$

The equation of a parabola is $y = 4 + 6x^2 - 36x + 54$.

15. What is the standard equation of the parabola represented by the equation?

16. What is the vertex of the parabola?

17. Determine the vertices of the hyperbola having the equation
$\dfrac{(y + 3)^2}{36} - \dfrac{(x - 1)^2}{9} = 1$.

18. Determine the foci of the hyperbola having the equation
$\dfrac{(x + 3)^2}{8} - \dfrac{(y - 2)^2}{16} = 1$.

19. Graph the hyperbola having the equation $\dfrac{x^2}{25} - \dfrac{y^2}{9} = 1$.

20. Graph the parabola represented by the equation $y + 3 = 2(x - 2)^2$.

21. Graph the parabola represented by the equation $x - 2 = 3(y + 1)^2$.

22. Graph the circle represented by the equation $(x + 5)^2 + (y - 3)^2 = 9$.

23. Graph the ellipse represented by the equation $\dfrac{(x + 1)^2}{16} + \dfrac{(y + 6)^2}{4} = 1$.

24. Graph the hyperbola having the equation

$$\frac{y^2}{4} - \frac{x^2}{16} = 1.$$

(Answers on page 372)

ACT-TYPE PROBLEMS

1. What is the center of a circle with equation $x^2 - 6x + y^2 + 2y + 9 = 0$?

   A. $(3,1)$
   B. $(1,3)$
   C. $(-3,1)$
   D. $(3,-1)$
   E. $(-1,-3)$

2. What is the equation of the major axis of the ellipse with equation $\dfrac{(x-4)^2}{49} + \dfrac{(y+7)^2}{36} = 1$?

   F. $y = -7$
   G. $x = -4$
   H. $y = 4$
   J. $x = 7$
   K. $y = 7$

3. What is the sum of the x-coordinate and the y-coordinate of the vertex of the parabola with equation $y = x^2 - 8x + 10$?

   A. 4
   B. 2
   C. -2
   D. -4
   E. -6

4. Which is the equation of an ellipse?

   F. $8x^2 - 8y^2 = 16$
   G. $(x-3)^2 + (y+7)^2 = 3$
   H. $4x^2 + 9y^2 = 36$
   J. $y - 1 = 5(x+4)^2$
   K. $3x^2 - 4y^2 = 12$

5. Which is the graph of $4y^2 + 8y - 9x^2 + 18x = 41$?

   A.

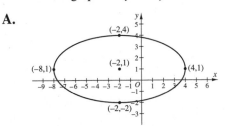

25. Graph the hyperbola having the equation

$$\frac{(x-1)^2}{25} - \frac{(y-2)^2}{36} = 1.$$

   B.

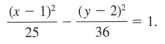

   C.

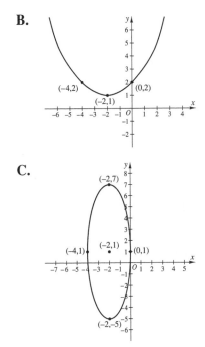

   D.

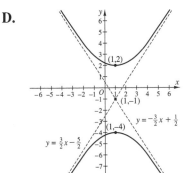

   E.

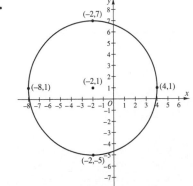

(Answers on page 373)

Coordinate Geometry **215**

Complete this Cumulative ACT Practice in 10 minutes to reflect real ACT test conditions. This Cumulative ACT Practice gives you an additional opportunity to practice coordinate geometry concepts in an ACT format. If you don't know an answer, eliminate and guess. Circle the number of any guessed answers. Then check your answers on page 374. You will also find explanations for the answers and suggestions for further study.

1. Which is the equation for the graph shown below?

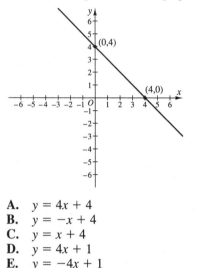

- **A.** $y = 4x + 4$
- **B.** $y = -x + 4$
- **C.** $y = x + 4$
- **D.** $y = 4x + 1$
- **E.** $y = -4x + 1$

2. What is the equation of a line with y-intercept of -3 that is perpendicular to the line with equation $y = 2x + 3$?

- **F.** $y = 2x - 3$
- **G.** $y = 2x + 3$
- **H.** $y = -3x + \frac{1}{2}$
- **J.** $y = -\frac{1}{2}x + 3$
- **K.** $y = -\frac{1}{2}x - 3$

3. What is the equation of the graph below?

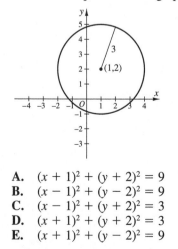

- **A.** $(x + 1)^2 + (y + 2)^2 = 9$
- **B.** $(x - 1)^2 + (y - 2)^2 = 9$
- **C.** $(x - 1)^2 + (y + 2)^2 = 3$
- **D.** $(x + 1)^2 + (y + 2)^2 = 3$
- **E.** $(x + 1)^2 + (y - 2)^2 = 9$

4. The endpoints of a line segment are $(3,5)$ and $(-2,10)$. What is the length of the line segment?

- **F.** $2\sqrt{5}$
- **G.** $5\sqrt{2}$
- **H.** 10
- **J.** 20
- **K.** 25

5. What is the slope of the line whose equation is $y - 5 = 3(x - 2)$?

- **A.** -5
- **B.** -2
- **C.** $-\frac{5}{3}$
- **D.** $-\frac{3}{2}$
- **E.** 3

6. The graph below is formed by which of the following inequalities?

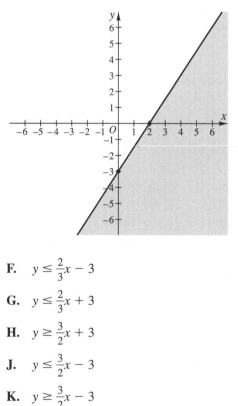

- **F.** $y \leq \frac{2}{3}x - 3$
- **G.** $y \leq \frac{2}{3}x + 3$
- **H.** $y \geq \frac{3}{2}x + 3$
- **J.** $y \leq \frac{3}{2}x - 3$
- **K.** $y \geq \frac{3}{2}x - 3$

7. $\frac{x^2}{4} - \frac{y^2}{9} = 1$ is the equation of

A. a circle
B. an ellipse
C. a hyperbola
D. a line
E. a parabola

8. Which is the equation of a line that passes through the point $(-1,2)$ and is parallel to the line with equation $2y + 6x = 10$?

F. $y = \frac{2}{3}x + 5$

G. $y = -3x + 5$

H. $y = -3x - 1$

J. $y = -\frac{1}{3}x - 1$

K. $y = -\frac{3}{2}x - 1$

9. What is the midpoint of the line segment whose endpoints are $(5,-2)$ and $(-3,7)$?

A. $(-1,2.5)$
B. $(1,2.5)$
C. $(1,-2.5)$
D. $(4,4.5)$
E. $(4.5,4)$

10. In the coordinate plane, what is the distance between the points with coordinates $(9,0)$ and $(4,10)$?

F. $\sqrt{10}$

G. 5

H. $5\sqrt{5}$

J. 10

K. 25

Subtest Coordinate Geometry

Complete this Subtest in 9 minutes to reflect real ACT test conditions. This Subtest has the same number and type of coordinate geometry items as are found on the ACT. If you don't know an answer, eliminate and guess. Circle the number of any guessed answer. Then check your answers on page 375. You will also find explanations for the answers and suggestions for further study.

1. Which is the equation of the line graphed below?

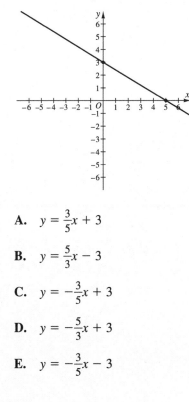

A. $y = \frac{3}{5}x + 3$

B. $y = \frac{5}{3}x - 3$

C. $y = -\frac{3}{5}x + 3$

D. $y = -\frac{5}{3}x + 3$

E. $y = -\frac{3}{5}x - 3$

2. What is the equation of a line parallel to the line with equation $y = 3x + 8$ with a y-intercept of -7?

F. $y = 3x + 8$

G. $y = -3x - 7$

H. $y = 3x + 7$

J. $y = \frac{1}{3}x - 7$

K. $y = 3x - 7$

3. Which is the equation of the graph shown below?

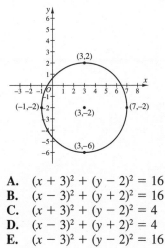

A. $(x + 3)^2 + (y - 2)^2 = 16$
B. $(x - 3)^2 + (y + 2)^2 = 16$
C. $(x + 3)^2 + (y - 2)^2 = 4$
D. $(x - 3)^2 + (y + 2)^2 = 4$
E. $(x - 3)^2 + (y - 2)^2 = 16$

4. What is the length of the longest side of the triangle whose vertices have coordinates $A(-1,-2)$, $B(4,2)$, and $C(-2,4)$?

   F. $\sqrt{45}$

   G. $\sqrt{41}$

   H. $\sqrt{40}$

   J. $\sqrt{39}$

   K. $\sqrt{37}$

5. What is the slope of the line whose equation is $3y + 2x - 9 = 0$?

   A. -5

   B. -3

   C. $-\dfrac{2}{3}$

   D. $\dfrac{2}{3}$

   E. 3

6. Which is the graph of the inequality $-2y + 4x \leq 8$?

F.

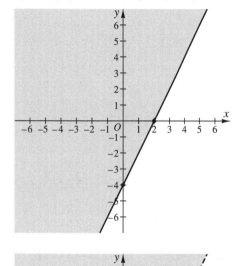

G.

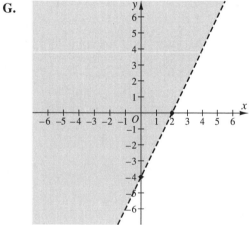

H.

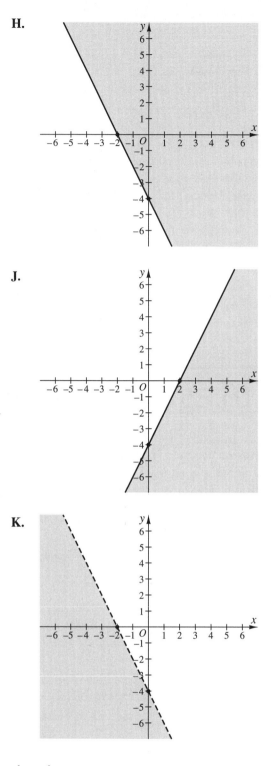

J.

K.

7. $\dfrac{x^2}{9} + \dfrac{y^2}{16} = 1$ is the equation of a(n)

   A. circle

   B. ellipse

   C. hyperbola

   D. line

   E. parabola

8. Which is the equation of a line passing through the point (2,3) and perpendicular to the line with equation $4y - 2x = 12$?

F. $y = -\dfrac{1}{2}x + 4$

G. $y = \dfrac{1}{2}x + 2$

H. $y = 2x - 1$

J. $y = -2x + 7$

K. $y = -2x + 3$

9. What is the midpoint of the line segment whose endpoints are $(-2,4)$ and $(5,-7)$?

A. $(-3.5, -1.5)$

B. $(-3.5, 6)$

C. $(1.5, -1.5)$

D. $(1.5, 6)$

E. $(3.5, -1.5)$

Cumulative **ACT** Practice

Intermediate Algebra / Coordinate Geometry

Complete this Cumulative ACT Practice in 15 minutes to reflect real ACT test conditions. This Cumulative ACT Practice gives you an additional opportunity to practice intermediate algebra and coordinate geometry concepts in an ACT format. If you don't know an answer, eliminate and guess. Circle the number of any guessed answers. Then check your answers on page 376. You will also find explanations for the answers and suggestions for further study.

1. What is the product of the x-values that will make the function $f(x) = \dfrac{1}{x^2 - 4}$ undefined?

A. -4
B. -2
C. 0
D. 2
E. 4

2. Hobb's Pet Store sells dogs and cats. Each cat costs $50 and each dog costs three times as much as a cat. At the end of one day, Hobb's Pet Store sold a total of 25 pets and collected $2,650. How many cats were sold?

F. 25
G. 15
H. 14
J. 12
K. 11

3. What is the y-intercept of the line with equation $3y - x = 6$?

A. $\dfrac{1}{6}$

B. $\dfrac{1}{3}$

C. 2

D. 3

E. 6

4. Which of the following choices would be used to find the distance, d, from 5 to any point, x, on the number line?

F. $|x - 5| = d$
G. $|5 + x| = d$
H. $x - 5 = d$
J. $5 - x = d$
K. $|x + 5| = d$

5. What is the slope of the line passing through the points $(4,-1)$ and $(3,2)$?

A. 4

B. 3

C. 1

D. $\dfrac{1}{3}$

E. -3

6. Which of the following choices shows all the roots of $x^2 - 6x + 10 = 0$?

F. $x = 3i - 1, x = 3i + 1$
G. $x = i, x = 5i$
H. $x = 3 - i, x = 3 + i$
J. $x = 3i - 2, x = 3i + 2$
K. $x = 3 - 2i, x = 3 + 2i$

7. If the slope of a given line is undefined, which could be the equation of a line parallel to the given line?

A. $y = 2x + 5$
B. $x = 2$
C. $y = x$
D. $x = 2y + 4$
E. $y = 3$

8. Which point has an x-coordinate of -4 and is on the graph of the function $f(x) = x^3$?

F. $(16, -4)$
G. $(-4, 16)$
H. $(-4, -16)$
J. $(-4, -64)$
K. $(-64, -4)$

9. Which expression is equal to $(2i + 5) + (-3 - 5i)$?

A. i
B. $5i + 1$
C. $-3i + 2$
D. $2i - 3$
E. $5i + 10$

10. What is the equation of the major axis in the ellipse $\frac{(x-3)^2}{4} + \frac{(y+2)^2}{9} = 1$?

F. $x = -2$
G. $y = 2$
H. $x = 2$
J. $x = 3$
K. $y = -3$

11. $-2\begin{bmatrix} 1 & -2 \\ 3 & 5 \end{bmatrix} + 4\begin{bmatrix} 2 & -3 \\ 6 & 4 \end{bmatrix} = ?$

A. $\begin{bmatrix} 10 & -16 \\ 30 & 26 \end{bmatrix}$

B. $\begin{bmatrix} 6 & -8 \\ 18 & 6 \end{bmatrix}$

C. $\begin{bmatrix} -14 & 12 \\ 10 & 14 \end{bmatrix}$

D. $\begin{bmatrix} 14 & 28 \\ -18 & -2 \end{bmatrix}$

E. $\begin{bmatrix} 22 & 20 \\ 2 & -22 \end{bmatrix}$

12.

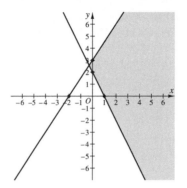

The graph above is the solution set for which system of inequalities?

F. $y \leq \frac{3}{2}x + 3$
   $y \leq -2x + 2$

G. $y \leq \frac{2}{3}x + 3$
   $y \geq 2x - 2$

H. $y \geq \frac{3}{2}x - 3$
   $y \leq x + 2$

J. $y \leq \frac{3}{2}x + 3$
   $y \geq -2x + 2$

K. $y \leq \frac{2}{3}x + 3$
   $y \leq -\frac{1}{2}x + 2$

13. By what factor is the sequence 2, 8, 32, 128, 512, … increasing?

A. 2
B. 3
C. 4
D. 5
E. 6

14. A model rocket is launched off a building 32 feet high. The position function of the rocket is $p(t) = -t^2 + 12t + 32$, where t represents time in seconds. At what time will the rocket be at the highest point?

F. 6 seconds
G. 7 seconds
H. 14 seconds
J. 32 seconds
K. 36 seconds

15. Which is the equation of a circle with a radius of 8 and a center shifted down 6 units and right 3 units from the origin?

A. $(x - 6)^2 + (y + 3)^2 = 8$
B. $(x - 3)^2 + (y + 6)^2 = 64$
C. $(x + 6)^2 + (y - 3)^2 = 8$
D. $(x + 3)^2 + (y - 6)^2 = 64$
E. $(x - 3)^2 + (y - 6)^2 = 64$

Subtest Intermediate Algebra/Coordinate Geometry

Complete this Subtest in 18 minutes to reflect real ACT test conditions. This Subtest has the same number and type of intermediate algebra and coordinate geometry items as are found on the ACT. On the real ACT, these items make up the AG subscore. If you don't know an answer, eliminate and guess. Circle the number of any guessed answer. Then check your answers on page 378. You will also find explanations for the answers and suggestions for further study.

1. If b is any real number, what is the sum of the smallest and largest possible values for b that make the expression $\sqrt{9 - b^2}$ real?

   A. -6
   B. 0
   C. 3
   D. 6
   E. 9

2. The charge for a ticket at a movie theater is $6.50 for adults and $3.00 for children. For the showing of the 8:00 movie, the theater sold a total of 105 tickets, and collected $567.00. How many more adult tickets than children's tickets were sold?

   F. 105 tickets
   G. 72 tickets
   H. 39 tickets
   J. 33 tickets
   K. 25 tickets

3. Which is the equation for the graph shown below?

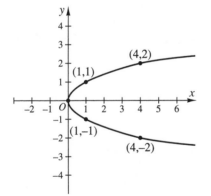

   A. $x = y^2$

   B. $y = x^2$

   C. $y = \sqrt{x}$

   D. $x = \sqrt{y}$

   E. $x = -y^2$

4. Which is the linear equation $2y - 5x + 2 = 0$ written in slope-intercept form?

   F. $2y - 5x = -2$

   G. $2y + 2 = 5x$

   H. $y = \frac{5}{2}x + 2$

   J. $y = \frac{5}{2}x - 1$

   K. $y = \frac{2}{5}x + 1$

5. Which of the following choices would be used to find the distance, d, from -3 to any point, x, on the number line?

   A. $|x + 3| = d$
   B. $|x - 3| = d$
   C. $x - 3 = d$
   D. $|3 - x| = d$
   E. $x + 3 = d$

6. What is the slope of the line that passes through points $(5,2)$ and $(-3,4)$?

   F. -4

   G. -3

   H. $-\frac{1}{4}$

   J. $\frac{1}{3}$

   K. 2

7. How many real roots does the equation $x^2 - 2x + 4 = 0$ have?

   A. 4
   B. 3
   C. 2
   D. 1
   E. 0

8. If the slope of a line is 0, then which could be the equation of a line perpendicular to it?

   F. $x = 5$
   G. $y = 5$
   H. $y = 2x - 4$
   J. $x = -3y - 1$
   K. $y = x$

9. Which of the following is the point on the graph below with a y-coordinate of −8?

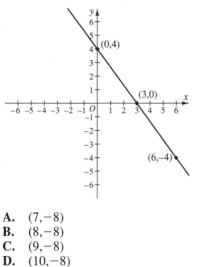

A. (7,−8)
B. (8,−8)
C. (9,−8)
D. (10,−8)
E. (11,−8)

10. Which is equal to $\dfrac{5i + 3}{2i - 4}$ rewritten with no imaginary numbers in the denominator?

F. $3i - 1$

G. $\dfrac{10i + 3}{-4}$

H. $7i + 7$

J. $\dfrac{1 + 13i}{-10}$

K. $\dfrac{5i + 3}{6}$

11. In 1980, an electronics company began manufacturing two different types of stereos, the R1000 and the S1000. The function $R(x) = 6500x + 8000$ represents the number of R1000 stereos manufactured x years after the start of production and $S(x) = 7000x + 6000$ represents the number of S1000 stereos manufactured x years after the start of production. In what year did the total number of R1000 stereos manufactured equal the total number of S1000 stereos manufactured?

A. 1986
B. 1985
C. 1984
D. 1983
E. 1982

12. What is the equation of the transverse axis of the hyperbola with equation $\dfrac{(x - 2)^2}{4} - \dfrac{(y + 1)^2}{9} = 1$?

F. $x = -2$
G. $y = -1$
H. $x = 2$
J. $y = 1$
K. $y = 3$

13. $3\begin{bmatrix} 4 & -2 \\ -7 & 6 \end{bmatrix} - 5\begin{bmatrix} -1 & -3 \\ 5 & 2 \end{bmatrix} = ?$

A. $\begin{bmatrix} 17 & 9 \\ -46 & 8 \end{bmatrix}$

B. $\begin{bmatrix} 7 & -24 \\ -12 & 4 \end{bmatrix}$

C. $\begin{bmatrix} 7 & -24 \\ -6 & 28 \end{bmatrix}$

D. $\begin{bmatrix} 17 & -46 \\ -6 & 8 \end{bmatrix}$

E. $\begin{bmatrix} 7 & -24 \\ -6 & 28 \end{bmatrix}$

14.

The graph above is the solution set for which system of inequalities?

F. $y \leq -\dfrac{4}{5}x + 5$

  $y < \dfrac{3}{2}x + 2$

G. $y \geq -\dfrac{5}{4}x + 5$

  $y \geq \dfrac{3}{2}x + 2$

H. $y \geq -\dfrac{5}{4}x + 5$

  $y < \dfrac{2}{3}x + 2$

J. $y > \dfrac{5}{4}x + 5$

  $y \leq \dfrac{2}{3}x + 2$

K. $y \leq -\dfrac{4}{5}x + 5$

  $y > \dfrac{3}{2}x + 2$

15. What is the factor by which the sequence 120, 96, 76.8, 61.44, 49.152, … is decreasing?

A. $\dfrac{4}{5}$

B. $\dfrac{3}{4}$

C. $\dfrac{2}{3}$

D. $\dfrac{2}{5}$

E. $\dfrac{1}{4}$

16. Given the equation $y = \sqrt{x}$, what is the distance between the two points with x-coordinates of 1 and 9?

F. $2\sqrt{17}$

G. $4\sqrt{5}$

H. $\sqrt{102}$

J. $\sqrt{145}$

K. $9\sqrt{5}$

17. To make a profit, a bicycle manufacturing company needs to produce at least $P(x) = x^2 + 30x$ bicycles an hour, where x represents the number of hours after the start of the day. Which of the following choices represents a production amount where NO profit is made?

A. 140 bicycles 4 hours after the start of the day
B. 64 bicycles 2 hours after the start of the day
C. 175 bicycles 5 hours after the start of the day
D. 250 bicycles 7 hours after the start of the day
E. 105 bicycles 3 hours after the start of the day

18. Which of the following choices is the equation of a circle that has been shifted up 3 units, left 5 units, and has a radius of 5?

F. $(x + 5)^2 + (y - 3)^2 = 25$

G. $(x - 5)^2 + (y + 3)^2 = 25$

H. $(x + 5)^2 + (y - 3)^2 = 5$

J. $(x + 3)^2 + (y - 5)^2 = 25$

K. $(x - 5)^2 + (y + 3)^2 = 5$

▪Chapter 11

Plane Geometry

- Fourteen ACT questions have to do with plane geometry.
- Easier plane geometry questions may be about a single skill or concept or may test a combination of pre-algebra, elementary algebra, intermediate algebra, and plane geometry skills.
- More difficult questions will often test a combination of plane geometry skills and concepts.
- This plane geometry review covers all the material you need to answer ACT questions.

▪ Basic Elements of Plane Geometry

Plane geometry has to do with two dimensions, but three-dimensional concepts tested on the ACT are also reviewed in this chapter. Some important elements and concepts, along with their models and symbols, are given below.

A **point** is a position or location. It has no length or volume.

    $A^{\cdot}$ point A

A **line** is a straight path with an infinite number of points extending infinitely in both directions.

- Two points determine a line.
- Points on the same line are called collinear points.

$\overleftrightarrow{AB}$ is the symbol that represents line AB.

Parallel lines are lines that never intersect.

The notation $\overleftrightarrow{AB} \parallel \overleftrightarrow{CD}$ is read, "line AB is parallel to line CD."

Perpendicular lines intersect to form right angles.

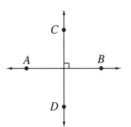

The notation $\overleftrightarrow{AB} \perp \overleftrightarrow{CD}$ is read, "line AB is perpendicular to line CD."

Note: **Skew** lines are lines in three dimensions that are not parallel but do not intersect.

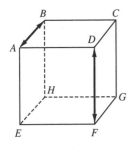

$\overleftrightarrow{AB}$ and $\overleftrightarrow{DF}$ are skew lines.

A **line segment** is a portion of a line with two endpoints.

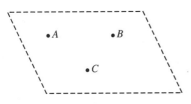

$\overline{AB}$ is the symbol that represents line segment AB. The length of segment AB is represented without the overbar: $AB = 4$ cm.

A **ray** is a portion of a line with one endpoint and extending infinitely in one direction.

When describing a ray, the endpoint is written first. The above ray is written $\overrightarrow{AB}$.

A **plane** is a flat surface that extends infinitely in all directions.

- Three noncollinear points determine a plane.

MODEL **ACT** PROBLEM

Which statement is true about the figure shown below?

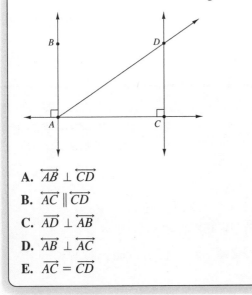

A. $\overleftrightarrow{AB} \perp \overleftrightarrow{CD}$

B. $\overleftrightarrow{AC} \parallel \overleftrightarrow{CD}$

C. $\overrightarrow{AD} \perp \overrightarrow{AB}$

D. $\overleftrightarrow{AB} \perp \overleftrightarrow{AC}$

E. $\overleftrightarrow{AC} = \overleftrightarrow{CD}$

SOLUTION

The small square in the corner of the figure indicates that line $\overleftrightarrow{AB}$ is perpendicular to line $\overleftrightarrow{AC}$.

The correct answer is D.

Practice

1. a. How many points determine a line?

    b. Draw a model of a line AB.

    c. What symbol is used to represent line AB?

2. a. What are parallel lines?

    b. Draw a model of two parallel lines, AB and CD.

    c. Write the symbol for parallel lines.

3. a. What are perpendicular lines?

    b. Draw a model of two perpendicular lines, AB and CD.

    c. Write the symbol for perpendicular lines.

4. a. How many points are needed to determine a plane?

    b. Draw a model of plane ABC.

5. a. What is a line segment?

    b. Draw a model of line segment AB.

    c. Write the symbol for line segment AB.

6. a. What is a ray?

    b. Draw a model of ray AB.

    c. Write the symbol for ray AB.

7. Draw a diagram of four lines that intersect at three points.

8. Explain the difference between parallel lines and skew lines.

9. Draw a diagram of $\overrightarrow{DE} \perp \overrightarrow{FD}$.

10. Segment PQ is 1 cm long. How many points does it contain?

(Answers on page 381)

ACT-TYPE PROBLEMS

1. $\overleftrightarrow{AB} \perp \overleftrightarrow{CD}$ means:

   A. Line AB is perpendicular to line CD.
   B. Ray AB is perpendicular to ray CD.
   C. Segment AB and segment CD are skew.
   D. Segment AB is parallel to segment CD.
   E. Segment AB is perpendicular to segment CD.

2. Lines $\overleftrightarrow{XY}$ and $\overleftrightarrow{WZ}$ are both perpendicular to $\overleftrightarrow{RS}$. Which statement must be true if all three lines lie in the same plane?

   F. $\overleftrightarrow{XY} \perp \overleftrightarrow{WZ}$

   G. $\overleftrightarrow{XY} \parallel \overleftrightarrow{WZ}$

   H. $\overleftrightarrow{XY}, \overleftrightarrow{WZ}$, and $\overleftrightarrow{RS}$ form a triangle.

   J. $\overleftrightarrow{XY}$ and $\overleftrightarrow{WZ}$ are skew.

   K. $\overleftrightarrow{XY}$ and $\overleftrightarrow{WZ}$ intersect at one point, R.

3. Which of the following is true of a ray $\overrightarrow{KL}$?

    A. It extends infinitely in both directions.
    B. Its endpoint is L.
    C. It has length.
    D. It contains exactly two points, K and L.
    E. It is a portion of a line beginning with point K.

4. The opposite edges of a piece of paper are:

    F. Lines
    G. Skew
    H. Parallel
    J. Perpendicular
    K. Rays

(Answers on page 381)

5. The notation FG represents:

    A. The length of a line
    B. The length of a segment
    C. The length of a ray
    D. Two points
    E. A plane

■ Angles

Angles consist of two rays with a common endpoint.

• The common endpoint is called the vertex.

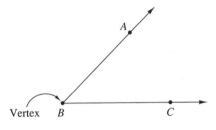

An angle can be described using three points or simply by naming the vertex. The angle above can be named $\angle ABC$ or simply $\angle B$.

The shorthand for the measure of angle B is m$\angle B$. Write m$\angle B = 45°$.

Classifying Angles

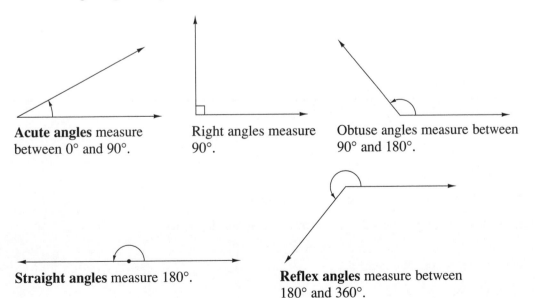

Acute angles measure between 0° and 90°.

Right angles measure 90°.

Obtuse angles measure between 90° and 180°.

Straight angles measure 180°.

Reflex angles measure between 180° and 360°.

Complementary and Supplementary Angles

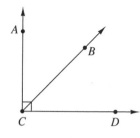

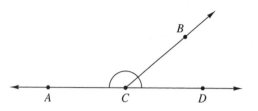

Complementary angles have a total measure of 90°.

∠ACB and ∠BCD are complementary angles.

Supplementary angles have a total measure of 180°.

∠ACB and ∠BCD are supplementary angles.

Congruent Angles

Two angles are **congruent** if they have the same angle measure.

- If ∠ABC and ∠DEF are congruent, then m∠ABC = m∠DEF.
- Right angles measure 90°, therefore all right angles are congruent.

Adjacent, Vertical, Corresponding, Interior, and Exterior Angles

Parallel lines cut by a transversal form vertical, corresponding, interior, and exterior angles.

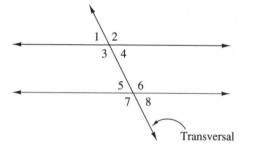

Adjacent angles touch at a common side.

Adjacent angles: ∠1 and ∠2, ∠1 and ∠3, ∠2 and ∠4, ∠3 and ∠4, ∠5 and ∠6, ∠5 and ∠7, ∠6 and ∠8, ∠7 and ∠8

Vertical angles are opposite each other when two lines intersect.

- Pairs of vertical angles are congruent.

Vertical angles: ∠1 and ∠4, ∠2 and ∠3, ∠5 and ∠8, ∠6 and ∠7

Corresponding angles are located in corresponding positions with respect to two parallel lines.

- Pairs of corresponding angles are congruent.

Corresponding angles: ∠1 and ∠5, ∠2 and ∠6, ∠3 and ∠7, ∠4 and ∠8

Alternate interior angles are located on opposite sides of the transversal, inside the parallel lines.

- Pairs of alternate interior angles are congruent.

Alternate interior angles: ∠3 and ∠6, ∠4 and ∠5

Alternate exterior angles are located on opposite sides of the transversal, outside the parallel lines.

- Pairs of alternate exterior angles are congruent.

Alternate exterior angles: ∠1 and ∠8, ∠2 and ∠7

Same-side interior angles are located on the same side of the transversal, inside the parallel lines.

- Pairs of same-side interior angles are supplementary.

Same-side interior angles: ∠3 and ∠5, ∠4 and ∠6

Same-side exterior angles are located on the same side of the transversal, outside the parallel lines.

- Pairs of same-side exterior angles are supplementary.

Same-side exterior angles: ∠1 and ∠7, ∠2 and ∠8

MODEL ACT PROBLEMS

1. In the figure below, $\overleftrightarrow{AB} \parallel \overleftrightarrow{CD}$ and t is a transversal that intersects both $\overleftrightarrow{AB}$ and $\overleftrightarrow{CD}$. Which of the following choices does NOT list a pair of congruent angles?

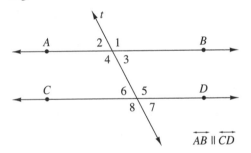

$\overleftrightarrow{AB} \parallel \overleftrightarrow{CD}$

- **A.** ∠1 and ∠4
- **B.** ∠2 and ∠7
- **C.** ∠6 and ∠3
- **D.** ∠3 and ∠5
- **E.** ∠2 and ∠6

SOLUTION

Consider each pair of angles in turn.

A. ∠1 is congruent to ∠4 because they are vertical angles.

B. ∠2 is congruent to ∠7 because they are alternate exterior angles.

C. ∠6 is congruent to ∠3 because they are alternate interior angles.

D. ∠3 is not congruent to ∠5 because they are same-side interior angles.

Same-side interior angles are supplementary and are congruent only when the transversal is perpendicular to the parallel lines.

The correct answer is D.

2. In the figure below, which pairs of angles are adjacent angles?

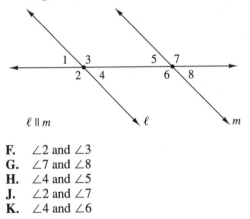

$\ell \parallel m$

- **F.** ∠2 and ∠3
- **G.** ∠7 and ∠8
- **H.** ∠4 and ∠5
- **J.** ∠2 and ∠7
- **K.** ∠4 and ∠6

SOLUTION

Adjacent angles are a pair of angles that share a common side. ∠7 and ∠8 share a common side, therefore they are adjacent angles.

The correct answer is G.

Practice

1. **a.** What is an angle?
    b. Draw a model of an angle *ABC*.
    c. Write the symbol for angle *ABC*.

2. **a.** What type of angle has a measure of 90°?
    b. Draw a model of the angle.

3. **a.** What type of angle has a measure between 0° and 90° ($0° < \theta < 90°$)?
    b. Draw a model of the angle.

4. **a.** What type of angle has a measure between 90° and 180° ($90° < \theta < 180°$)?
    b. Draw a model of the angle.

5. **a.** What type of angle has a measure between 180° and 360° ($180° < \theta < 360°$)?
    b. Draw a model of the angle.

6. **a.** What type of angle has a measure of 180°?
    b. Draw a model of the angle.

7. **a.** What are complementary angles?
    b. Draw a model of complementary angles.

8. **a.** What are supplementary angles?
    b. Draw a model of supplementary angles.

9. What are congruent angles?

10. The measure of two complementary angles and a third angle totals 168°. What is the measure of the third angle?

11. One of two supplementary angles measures 36°. What is the measure of the other angle?

12. Two angles of a triangle are complementary. What is the measure of the third angle?

Questions 13–19 refer to the diagram below.

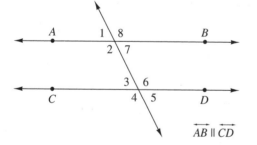

$\overleftrightarrow{AB} \parallel \overleftrightarrow{CD}$

13. **a.** Which pairs of angles in the figure above are adjacent?
    b. Are these angles complementary, supplementary, or congruent?

14. **a.** Which pairs of angles in the figure above are vertical?
    b. Are these angles complementary, supplementary, or congruent?

15. **a.** Which pairs of angles in the figure above are corresponding angles?
    b. Are these angles complementary, supplementary, or congruent?

16. **a.** Which pairs of angles in the figure above are alternate interior angles?
    b. Are these angles complementary, supplementary, or congruent?

17. **a.** Which pairs of angles in the figure above are alternate exterior angles?
    b. Are these angles complementary, supplementary, or congruent?

18. **a.** Which pairs of angles in the figure above are same-side interior angles?
    b. Are these angles complementary, supplementary, or congruent?

19. **a.** Which pairs of angles in the figure above are same-side exterior angles?
    b. Are these angles complementary, supplementary, or congruent?

Questions 20–26 refer to the diagram below.

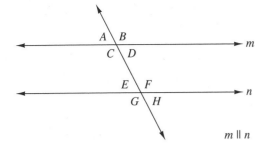

20. m∠C = 105°. What is m∠B?

21. m∠E = 75°. What is m∠G?

22. m∠A = 65°. What is m∠E?

23. m∠D = 70°. What is m∠E?

24. m∠G = 110°. What is m∠B?

25. m∠D = 55°.

   a. m∠A =   **b.** m∠B =   **c.** m∠C =   **d.** m∠E =

   e. m∠F =   **f.** m∠G =   **g.** m∠H =

26. m∠E + m∠H = 120°. What is m∠F? What is m∠A?

(Answers on page 382)

ACT-TYPE PROBLEMS

1. If two parallel lines are cut by a non-perpendicular transversal, which types of angles are NOT congruent?

   A. Same-side exterior angles
   B. Corresponding angles
   C. Alternate interior angles
   D. Vertical angles
   E. Alternate exterior angles

2. In the figure below, $\overleftrightarrow{AB}$ is parallel to $\overleftrightarrow{CD}$ and *t* is a transversal that intersects both $\overleftrightarrow{AB}$ and $\overleftrightarrow{CD}$. If m∠3 = 50°, what is m∠8?

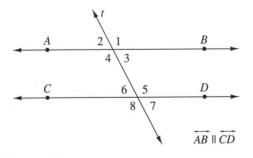

   F. 40°
   G. 50°
   H. 90°
   J. 130°
   K. 180°

3. In the figure below, m∠8 = 120°, $\overleftrightarrow{AB} \parallel \overleftrightarrow{CD}$, and $\overleftrightarrow{AB}$ and $\overleftrightarrow{CD}$ are cut by a transversal, *t*. What is the product of m∠7 and m∠1?

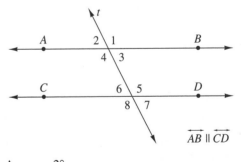

   A. 2°
   B. 180°
   C. 360°
   D. 3,600°
   E. 7,200°

4. In the figure below, m∠6 = 60° and $\overrightarrow{AB} \parallel \overrightarrow{CD}$. What is m∠4?

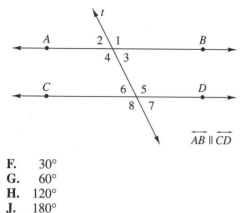

$\overrightarrow{AB} \parallel \overrightarrow{CD}$

F. 30°
G. 60°
H. 120°
J. 180°
K. 360°

5. In the figure below, $p \parallel q$ and $\ell \parallel m$. Which of the following choices is a complete list of all the angles supplementary to ∠13?

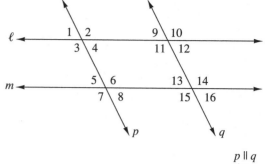

$p \parallel q$
$\ell \parallel m$

A. ∠1, ∠4, ∠5, ∠8, and ∠16
B. ∠2, ∠3, ∠6, ∠7, ∠10, ∠11, ∠14, and ∠15
C. ∠2, ∠3, ∠6, and ∠7
D. ∠1, ∠4, ∠5, ∠8, ∠9, ∠12 and ∠16
E. ∠1, ∠2, ∠3, ∠4, ∠5, ∠6, ∠7, and ∠8

(Answers on page 383)

■ Quadrilaterals

- A **quadrilateral** is a four-sided polygon.

Both of these figures are quadrilaterals.

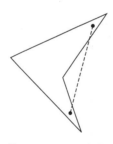

A polygon is any closed figure made up of line segments. A regular polygon has all sides the same length and all angles the same measure.

Convex quadrilateral
A quadrilateral is **convex** if the line segment connecting *any* two points inside the quadrilateral stays inside the quadrilateral.

Concave quadrilateral
A quadrilateral is **concave** if you can find two points inside the figure such that the segment connecting them goes outside the quadrilateral.

Types of Quadrilaterals

Described below are some quadrilaterals with special properties.

Trapezoid

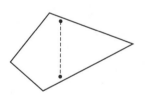

- One and only one pair of opposite sides is parallel.

Parallelogram

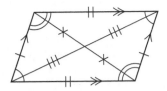

- Both pairs of opposite sides are parallel.
- Opposite sides are congruent.
- Opposite angles are congruent.
- Consecutive angles are supplementary.
- The diagonals bisect each other. That is, their point of intersection is the midpoint of each diagonal.

Rhombus

- Has all the properties of a parallelogram.
- All four sides are congruent.
- Diagonals are perpendicular.

Rectangle

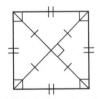

- Has all the properties of a parallelogram.
- All four angles are congruent. (All angles measure 90°.)
- Diagonals are congruent.

Square

- Has all of the properties of a parallelogram, a rhombus, and a rectangle.

Relationships Among Quadrilaterals

Shown below is a partial list of statements describing the relationships among quadrilaterals.

- Every rectangle is a parallelogram.
  Not all parallelograms are rectangles.

- Every rhombus is a parallelogram.
  Not all parallelograms are rhombuses.

- Every square is a rectangle and a rhombus.
  Not all rectangles and rhombuses are squares.

Which of the following statements is NOT true?

A. A square is a rectangle.
B. A square is a rhombus.
C. A rectangle is a parallelogram.
D. Adjacent sides of a rhombus are perpendicular.
E. Opposite sides of a parallelogram are congruent.

SOLUTION

Adjacent sides of a rhombus do not have to be perpendicular. Some rhombuses, called squares, do have perpendicular adjacent sides. But the statement is false because it is not true for all rhombuses.

The correct answer is D.

Practice

1. A line segment is drawn connecting any two points in a quadrilateral. What determines whether the figure is concave or convex?

2. In a quadrilateral, two sides are parallel but are not congruent. What type of quadrilateral must this be?

3. For which types of special quadrilaterals can the sum of two opposite angles be less than 180°?

4. What quadrilateral has diagonals that are perpendicular and congruent?

5. In parallelogram *ABCD*, the measure of ∠*B* is 90°. What type of parallelogram is this?

6. What is the name for a parallelogram that has perpendicular diagonals?

7. What is the name for a parallelogram with congruent diagonals?

8. What properties does a rhombus share with a square?

9. What quadrilateral with a pair of parallel sides is not also a parallelogram?

10. What is the name for a quadrilateral with bisecting diagonals?

(Answers on page 383)

1. An artist wants to draw one geometric shape that has the qualities of two geometric figures. Which of the following shapes could he draw?

   A. A rectangle that is concave
   B. A square that is a trapezoid
   C. A trapezoid that is a parallelogram
   D. A square that is a parallelogram
   E. A parallelogram that is a plane

2. Which of the following is a property of a rectangle?

   F. All four sides are congruent.
   G. Consecutive angles are complementary.
   H. Diagonals are congruent.
   J. Diagonals are perpendicular.
   K. It is a concave quadrilateral.

3. What is the best name for a parallelogram that has four congruent angles and perpendicular diagonals?

   A. Quadrilateral
   B. Parallelogram
   C. Rectangle
   D. Rhombus
   E. Square

4. Consecutive angles in a rhombus must be:

   F. Congruent
   G. Supplementary
   H. Complementary
   J. Right
   K. Acute

5. The perimeter of a figure is 4 times the length of one side. This figure could be:

A. A square
B. A rhombus
C. A trapezoid
D. All of the above
E. None of the above

(Answers on page 383)

■ General Properties of Triangles

A triangle is a three-sided polygon.

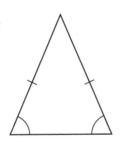

Triangle *ABC*

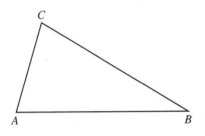

Equilateral triangle

An **equilateral triangle** has three congruent angles and three congruent sides.

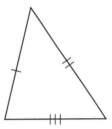

Isosceles triangle

An **isosceles triangle** has two congruent angles, called base angles, and two congruent sides. The third angle is called the vertex angle. The third side is called the base.

Scalene triangle

A **scalene triangle** has all different size angles and all different length sides.

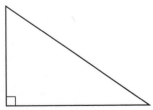

Right triangle
A **right triangle** has one right angle.

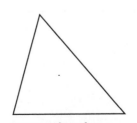

Acute triangle
An **acute triangle** has all acute angles.

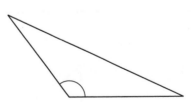

Obtuse triangle
An **obtuse triangle** has one obtuse angle.

Triangle Facts

- The sum of the angles in a triangle is 180°.
- The sum of the lengths of any two sides of a triangle is greater than the length of the third side.
- If two sides of a triangle are congruent, then the angles opposite those sides are congruent.

MODEL ACT PROBLEM

Which of the following statements about triangles is true?

A. The sum of the measures of the angles of a triangle is greater than or equal to 180°.

B. It is possible to have the sum of the lengths of two sides of a triangle be equal to the length of the third side.

C. Triangles always have an angle whose measure is greater than or equal to 90°.

D. A triangle can have more than one right angle.

E. If a triangle has two sides of equal length, then it must have two angles of equal measure.

SOLUTION

Only the final statement is true. All of the others are false.

The correct answer is E.

Practice

1. Indicate whether each triangle is equilateral, isosceles, or scalene.

   a.   b.   c.

   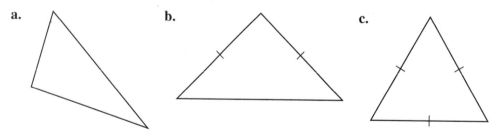

2. Indicate whether each triangle is right, acute, or obtuse.

   a.   b.   c.

   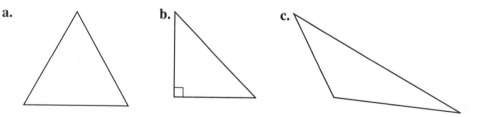

3. What is the measure of each angle in an equilateral triangle?

4. One base angle in an isosceles triangle measures 50°. What are the measures of the other angles?

5. The vertex angle in an isosceles triangle measures 50°. What are the measures of the base angles?

6. If one angle of a scalene triangle measures 10°, then what measure is impossible for either of the other two angles?

7. What is the measure of ∠*a*?

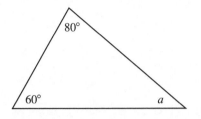

8. Are the measurements of this triangle realistic? If not, explain why.

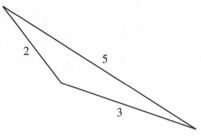

9. What is the measure of ∠*a*?

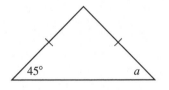

Are the measurements of these triangles realistic? Explain.

10.

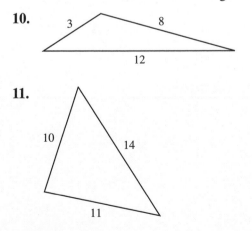

11.

(Answers on page 383)

1. Which of the following statements can NOT be concluded from the information presented in the figure shown below?

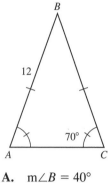

   A. m∠B = 40°
   B. m∠C + m∠B = 100°
   C. AB + BC = 24
   D. m∠A = 70°
   E. BC = 12

2. In the isosceles triangle below, AB = CB. What is the measure of the vertex angle if m∠A = 40°?

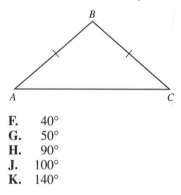

   F. 40°
   G. 50°
   H. 90°
   J. 100°
   K. 140°

(Answers on page 384)

3. Which of the following choices could NOT be the lengths of the sides of a triangle?

   A. 3, 5, 9
   B. 6, 7, 12
   C. 2, 6, 6
   D. 12, 12, 12
   E. 5, 12, 13

4. Which of the following statements is FALSE?

   F. 40°, 60°, and 80° are the angle measures of some scalene triangles.
   G. 60°, 60°, and 60° are the angle measures of some acute triangles.
   H. 60°, 60°, and 60° are the angle measures of equilateral triangles.
   J. 45°, 45°, and 90° are the angle measures of some obtuse triangles.
   K. 45°, 45°, and 90° are the angle measures of some isosceles triangles.

5. Which of the following types of triangle always has at least two angles with equal measures?

   A. Scalene triangle
   B. Obtuse triangle
   C. Equilateral triangle
   D. Right triangle
   E. Acute triangle

■ Right Triangles

The two sides of a right triangle that meet at the right angle are called legs. The third side is called the hypotenuse.

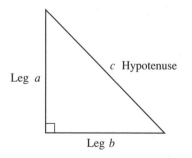

The Pythagorean Theorem

The Pythagorean Theorem describes the relationship among the three sides of a right triangle.

$$a^2 + b^2 = c^2$$

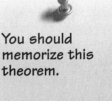
You should memorize this theorem.

EXAMPLES

1. The lengths of the legs of a right triangle are 3 and 4. What is the length of the hypotenuse?

   Use the Pythagorean Theorem.

   $$a^2 + b^2 = c^2$$
   $$3^2 + 4^2 = c^2$$
   $$9 + 16 = c^2$$
   $$25 = c^2$$
   $$5 = c$$

   The length of the hypotenuse is 5.

2. The length of one leg of a right triangle is 4. The length of the hypotenuse is $\sqrt{65}$. How long is the other leg?

   Use the Pythagorean Theorem.

   $$a^2 + b^2 = c^2$$
   $$4^2 + b^2 = (\sqrt{65})^2$$
   $$16 + b^2 = 65$$
   $$b^2 = 65 - 16 = 49$$
   $$b = 7$$

   The length of the other leg is 7.

Whole numbers that satisfy the Pythagorean Theorem are called **Pythagorean triples**. The most famous Pythagorean triple is 3, 4, 5. Memorize it. The multiples of Pythagorean triples are also Pythagorean triples. Here are the first four triples along with the first three multiples of the 3-4-5 triple.

3, 4, 5	5, 12, 13	7, 24, 25	8, 15, 17
6, 8, 10			
9, 12, 15			
12, 16, 20			

Remember: The first two numbers are the lengths of the legs. The third number is the length of the hypotenuse.

Isosceles Right Triangle

Both legs of an isosceles right triangle are the same length. The length of the hypotenuse is $\sqrt{2}$ times the length of a leg. The angle measures for this triangle are 45°, 45°, and 90°.

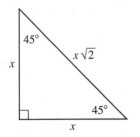

In an isosceles right triangle, you can quickly find the lengths of all the sides once you know the length of any one side.

1. The length of the leg of an isosceles right triangle is 5. How long is the hypotenuse?

   The hypotenuse is $5\sqrt{2}$.

2. In an isosceles right triangle, the length of the hypotenuse is 14. How long are the legs?

   Divide 14 by $\sqrt{2}$: $\dfrac{14}{\sqrt{2}} = \dfrac{14}{\sqrt{2}} \times \dfrac{\sqrt{2}}{\sqrt{2}} = \dfrac{14\sqrt{2}}{2} = 7\sqrt{2}$

   The length of each leg is $7\sqrt{2}$.

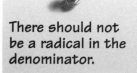

There should not be a radical in the denominator.

30–60–90 Triangle

In this type of triangle, the hypotenuse is twice the length of the shorter leg. The longer leg is $\sqrt{3}$ times the shorter leg.

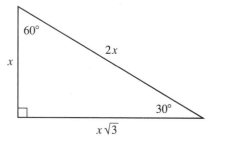

The longer leg of a 30–60–90 triangle is 21. What is the length of the hypotenuse?

Divide 21 by $\sqrt{3}$ to find the length of the shorter leg. $\dfrac{21}{\sqrt{3}} = \dfrac{21}{\sqrt{3}} \times \dfrac{\sqrt{3}}{\sqrt{3}} = \dfrac{21\sqrt{3}}{3} = 7\sqrt{3}$

Multiply the shorter leg by 2 to find the length of the hypotenuse.

$2 \times 7\sqrt{3} = 14\sqrt{3}$

The length of the hypotenuse is $14\sqrt{3}$.

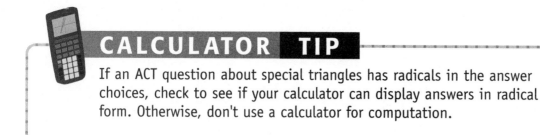

CALCULATOR TIP

If an ACT question about special triangles has radicals in the answer choices, check to see if your calculator can display answers in radical form. Otherwise, don't use a calculator for computation.

What is the length of leg *AB* in the triangle shown below?

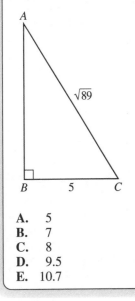

A. 5
B. 7
C. 8
D. 9.5
E. 10.7

SOLUTION

Use the Pythagorean Theorem to find the measure of *AB*.

$$(\text{leg}_1)^2 + (\text{leg}_2)^2 = (\text{hypotenuse})^2$$

$$a^2 + b^2 = c^2$$

$$5^2 + b^2 = \left(\sqrt{89}\right)^2$$

$$25 + b^2 = 89$$

$$b^2 = 64$$

$$b = 8$$

The correct answer is C.

Practice

1. What are the measures of the angles in an isosceles right triangle?

2. Use the Pythagorean Theorem to find the measure of the hypotenuse.

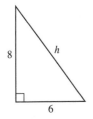

3. What is the length of the missing leg?

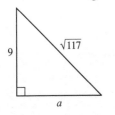

4. This is an isosceles right triangle. What are the lengths of the missing leg and the hypotenuse?

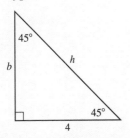

5. This is a 30–60–90 triangle. What are the lengths of legs *a* and *b*?

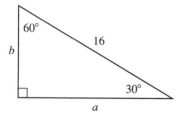

6. Use the Pythagorean Theorem to find the length of the hypotenuse.

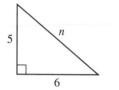

In 7 and 8, use the Pythagorean Theorem to find the lengths of the missing legs.

7.

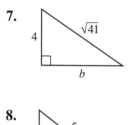

8.

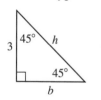

9. The figure below is an isosceles right triangle. What are the lengths of the missing leg and the hypotenuse?

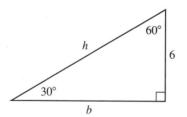

10. The figure below is a 30–60–90 triangle. What are the lengths of the missing leg and the hypotenuse?

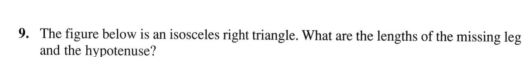

(Answers on page 384)

1. What is the length of the hypotenuse of the triangle shown below?

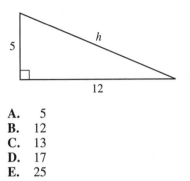

 A. 5
 B. 12
 C. 13
 D. 17
 E. 25

2. What is the product of b and h in the triangle shown below?

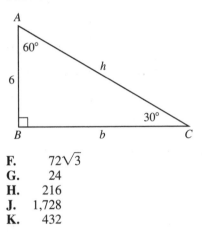

 F. $72\sqrt{3}$
 G.   24
 H.   216
 J.   1,728
 K.   432

3. What is the sum of a and h in the figure shown below?

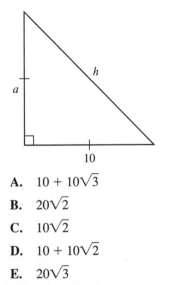

 A. $10 + 10\sqrt{3}$
 B. $20\sqrt{2}$
 C. $10\sqrt{2}$
 D. $10 + 10\sqrt{2}$
 E. $20\sqrt{3}$

(Answers on page 384)

4. The base of a ladder is placed 10 feet from a house, and the top of the ladder touches the house 14 feet above the ground. If the house creates a 90° angle with the ground, how long is the ladder, rounded to the nearest tenth?

 F.   9.8 feet
 G.   17.2 feet
 H.   19.6 feet
 J.   21.0 feet
 K.   24.0 feet

5. The triangle below is a 30–60–90 triangle. What is the sum of the lengths of side $\overline{AB}$ and side $\overline{BC}$?

 A.   4 feet
 B. $4 + 4\sqrt{3}$ feet
 C. $4\sqrt{3}$ feet
 D. $12 + \sqrt{3}$ feet
 E.   12 feet

■ Similar Triangles

Similar figures have the same shape but not necessarily the same size.

- The lengths of corresponding sides of similar triangles are proportional.
- Corresponding angles of similar triangles are congruent.
- The vertices of similar triangles are listed in the same order. In other words, if triangle *ABC* is similar to triangle *DEF*, then ∠*A* corresponds to ∠*D*, ∠*B* corresponds to ∠*E*, and ∠*C* corresponds to ∠*F*. This is true for other congruent figures as well.

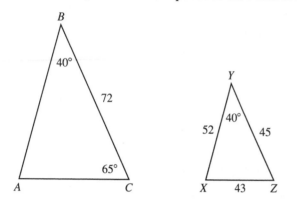

Triangles *ABC* and *XYZ* are similar. This can be written △*ABC* ~ △*XYZ*.

Angles

∠*B* and ∠*Y* are corresponding angles. Corresponding angles have equal measures. So m∠*B* = m∠*Y* = 40°. Now we can find the measure of ∠*A*.

The sum of the measures of ∠*B* and ∠*C* is 105°.

The sum of all the angle measures in a triangle is 180°.

The measure of ∠*A* = 180° − 105° = 75°.

By visual inspection of these triangles, we see that the following angles are corresponding angles.

Use the similarity of these triangles to find the missing measurements.

∠*B* and ∠*Y*	m∠*B* = m∠*Y* = 40°
∠*A* and ∠*X*	m∠*A* = m∠*X* = 75°
∠*C* and ∠*Z*	m∠*C* = m∠*Z* = 65°

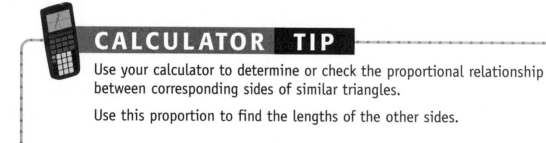

CALCULATOR TIP

Use your calculator to determine or check the proportional relationship between corresponding sides of similar triangles.

Use this proportion to find the lengths of the other sides.

Sides

Corresponding sides are opposite corresponding angles. So $\overline{BC}$ and $\overline{YZ}$ are corresponding sides.

The lengths of corresponding sides are proportional.

$$\frac{BC}{YZ} = \frac{72}{45} = 1.6$$

The length of a side of $\triangle ABC$ will be 1.6 times the length of the corresponding side in $\triangle XYZ$.

Therefore we can find the missing lengths of the other corresponding sides.

XZ and AC.	The length of $XZ = 43$.
	$1.6 \times 43 = 68.8$. The length of AC is 68.8.
XY and AB.	The length of $XY = 52$.
	$1.6 \times 52 = 83.2$. The length of AB is 83.2.

MODEL ACT PROBLEMS

1. In the figure below, $\triangle ABC$ and $\triangle DEF$ are similar triangles. How many units long is $\overline{DE}$?

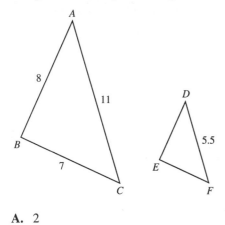

   A. 2
   B. 3
   C. 4
   D. 5
   E. 6

SOLUTION

Since $\triangle ABC$ is similar to $\triangle DEF$, corresponding sides are proportional.

Therefore, $\dfrac{DE}{AB} = \dfrac{DF}{AC}$.

Substitute known values. $\dfrac{DE}{8} = \dfrac{5.5}{11}$

Multiply by 8. $DE = \dfrac{5.5}{11} \times 8$

$$DE = \frac{1}{2} \times 8 = 4$$

The correct answer is C.

2. $\triangle XRK$ is similar to $\triangle STV$, and both triangles are scalene. Which of the following statements is NOT true?

   F. $\angle X \cong \angle S$

   G. $\angle R \cong \angle V$

   H. $\overline{XR}$ and $\overline{ST}$ are corresponding sides.

   J. $\dfrac{KX}{RK} = \dfrac{VS}{TV}$

   K. $\angle X + \angle K = \angle V + \angle S$

SOLUTION

$\angle R$ and $\angle V$ are not corresponding angles, therefore they are not congruent.

The correct answer is G.

Practice

1. What is true about corresponding angles of similar triangles?

2. What is true about corresponding sides of similar triangles?

3. Are these two triangles similar? Explain why or why not.

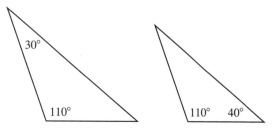

Given that △ABC is similar to △XYZ, find the missing measurements of each triangle.

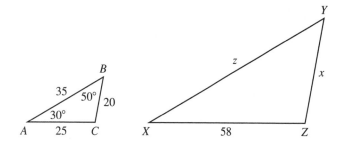

4. What is the length of z?

5. What is the length of x?

6. What is m∠C?

7. What is m∠X?

8. What is m∠Z?

9. What is m∠Y?

In the diagram below, △QRS is similar to △DEF.

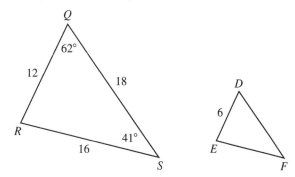

10. What is the measure of ∠R?

11. What is the measure of ∠E?

12. What is the measure of ∠D?

13. What is the measure of ∠F?

14. What is the length of EF?

15. What is the length of DF?

In the diagram below, △KLM is similar to △TUV.

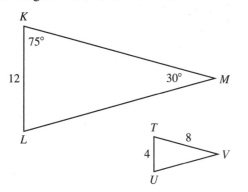

16. What is the measure of ∠L?

17. What is the measure of ∠T?

18. What is the measure of ∠U?

19. What is the measure of ∠V?

20. What is the length of KM?

21. What is the length of LM?

22. If △ABC below is similar to △DEF, how much longer is the perimeter of △DEF than the perimeter of △ABC?

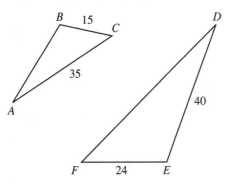

23. If △PQR below is similar to △YXR, what are the perimeter and area of each triangle?

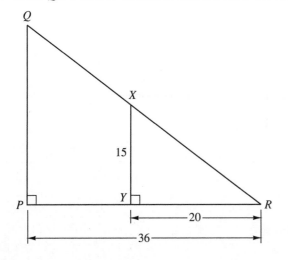

(Answers on page 384)

ACT-TYPE PROBLEMS

1. In the figure below, △ABC is similar to △DEF. What is the length of DF?

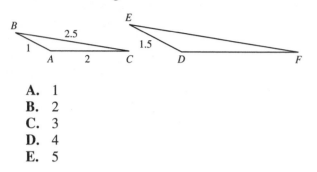

    A. 1
    B. 2
    C. 3
    D. 4
    E. 5

2. If △QRS is similar to △XYZ, which choice gives corresponding sides?

    F. $\overline{QR}$ and $\overline{RQ}$

    G. $\overline{RS}$ and $\overline{YZ}$

    H. $\overline{QS}$ and $\overline{XY}$

    J. $\overline{SR}$ and $\overline{SQ}$

    K. $\overline{RQ}$ and $\overline{XZ}$

3. In the figure below, △ABC is similar to △QRS. What is the sum of the measures of ∠B and ∠Q?

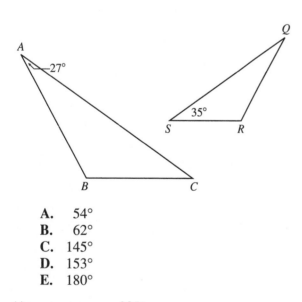

    A.   54°
    B.   62°
    C. 145°
    D. 153°
    E. 180°

(Answers on page 385)

4. In the figure below, △DEF is similar to △XYZ. Which of the following statements is FALSE?

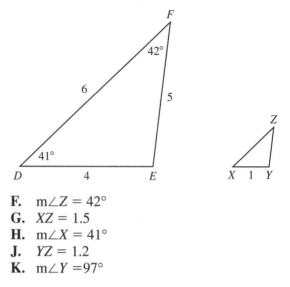

    F. m∠Z = 42°
    G. XZ = 1.5
    H. m∠X = 41°
    J. YZ = 1.2
    K. m∠Y = 97°

5. In the figure below, △ABE is similar to △CDE. What is the sum of AC and BD?

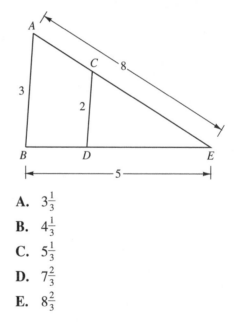

    A. $3\frac{1}{3}$

    B. $4\frac{1}{3}$

    C. $5\frac{1}{3}$

    D. $7\frac{2}{3}$

    E. $8\frac{2}{3}$

■ Concept of Proof and Proof Techniques

Proof means using what is known in a logically convincing way to establish that a hypothesis is true or false. Much of geometry involves proving or disproving hypotheses. Consider this simple example.

EXAMPLE

Given: One angle of a triangle measures 90°, while another angle measures 50°. The sum of the measures of the angles in a triangle is 180°.

Hypothesis: The third angle in the triangle measures 40°.

Proof: $90° + 50° = 140°$ $180° - 140° = 40°$

The proof uses only what is known and is logical.

Congruent Triangles

Many proofs involve showing that two triangles are congruent. We use Side Side Side (SSS), Angle Side Angle (ASA), and Side Angle Side (SAS) congruence relations between triangles to prove that triangles are congruent.

- SSS: If each side of one triangle is congruent to the corresponding side of another triangle, then the two triangles are congruent.

- ASA: In a triangle, if two angles and the side between them are congruent to the corresponding angles and side of another triangle, then the two triangles are congruent.

- SAS: In a triangle, if two sides and the angle between them are congruent to the corresponding sides and angle of a second triangle, then the two triangles are congruent.

Here is a traditional example.

EXAMPLE

Given: 1. The SSS congruence relation between two triangles: If each side of one triangle is congruent to the corresponding side of another triangle, then the two triangles are congruent.

2. The dimensions of the two right triangles below:

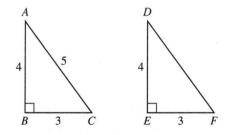

3. The Pythagorean Theorem: In a right triangle, the sum of the squares of the legs (a and b) equals the square of the hypotenuse (c). ($a^2 + b^2 = c^2$)

Hypothesis: Triangle ABC is congruent to triangle DEF ($\triangle ABC \cong \triangle DEF$).

Proof: From the diagram: $\overline{AB} \cong \overline{DE}$ and $\overline{BC} \cong \overline{EF}$

Use the Pythagorean Theorem to find the length of side DF.

$$a^2 + b^2 = c^2$$
$$(EF)^2 + (DE)^2 = (DF)^2$$
$$3^2 + 4^2 = (DF)^2$$
$$9 + 16 = (DF)^2$$
$$25 = (DF)^2$$
$$5 = DF$$

$\overline{DF}$ and $\overline{AC}$ are both 5 units long, so $\overline{DF} \cong \overline{AC}$.

$\triangle ABC \cong \triangle DEF$ by SSS.

The proof uses only what is known and logically shows that the two triangles are congruent by SSS. (*Note:* The proof could also have been done using SAS.)

Here is a more formal way to present this proof. It presents each step in the proof and gives the reason for that step.

EXAMPLE

Hypothesis: Triangle ABC is congruent to triangle DEF ($\triangle ABC \cong \triangle DEF$).

Statement	Reason
1. $AB = DE = 4$ $BC = EF = 3$ $AC = 5$	1. Given
2. $(EF)^2 + (DE)^2 = (DF)^2$ $3^2 + 4^2 = (DF)^2$ $9 + 16 = (DF)^2$ $25 = (DF)^2$ $5 = DF$	2. Pythagorean Theorem
3. $AC = DF$	3. These two sides are both 5 units long.
4. $\overline{AB} \cong \overline{DE}, \overline{BC} \cong \overline{EF}$, and $\overline{AC} \cong \overline{DF}$	4. Segments of equal length are congruent.
5. $\triangle ABC \cong \triangle DEF$	5. SSS congruence theorem

Look at the figure below.

Line ℓ and line m are parallel, and line t is a transversal that cuts through ℓ and m.

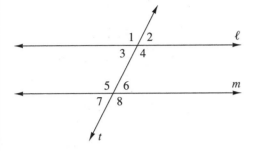

Which of the following statements is FALSE?

A. $\angle 6 \cong \angle 7$

B. $\angle 6 \cong \angle 3$

C. $\angle 8 \cong \angle 4$

D. $\angle 1 \cong \angle 7$

E. $\angle 2 \cong \angle 7$

SOLUTION

We know quite a few things about parallel lines cut by a transversal.

Corresponding angles are congruent. The corresponding angles in this figure are: $\angle 1$ and $\angle 5$, $\angle 2$ and $\angle 6$, $\angle 3$ and $\angle 7$, $\angle 4$ and $\angle 8$.

Alternate interior angles are congruent. The pairs of alternate interior angles in this figure are: $\angle 3$ and $\angle 6$, $\angle 4$ and $\angle 5$.

Alternate exterior angles are congruent. The pairs of alternate exterior angles in this figure are: $\angle 1$ and $\angle 8$, $\angle 2$ and $\angle 7$.

Vertical angles are congruent. The pairs of vertical angles in this figure are: $\angle 1$ and $\angle 4$, $\angle 2$ and $\angle 3$, $\angle 5$ and $\angle 8$, $\angle 6$ and $\angle 7$.

Statement		Reason
A. $\angle 6 \cong \angle 7$	True	vertical angles
B. $\angle 6 \cong \angle 3$	True	alternate interior angles
C. $\angle 8 \cong \angle 4$	True	corresponding angles
D. $\angle 1 \cong \angle 7$	?	does not match known information
E. $\angle 2 \cong \angle 7$	True	alternate exterior angles

All the other statements were proved true, so statement D must be false.

$\angle 1$ is not congruent to $\angle 7$.

The correct answer is D.

Practice

1. Line t is a transversal intersecting lines ℓ and m. Given that $\angle 2 \cong \angle 6$, write a two-column proof that shows that $\angle 3 \cong \angle 6$.

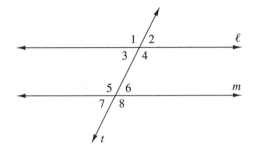

2. In the diagram below, $\overline{AB} \cong \overline{DE}$, $\overline{AC} \cong \overline{DF}$, and $\angle A \cong \angle D$. Is $\triangle ABC \cong \triangle DEF$? If so, how would you prove them congruent?

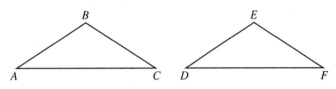

3. Given that $m\angle BCA = 70°$ in the diagram below, write a two-column proof that proves $m\angle BCD = m\angle A + m\angle B$.

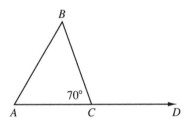

4. In $\triangle ABC$ and $\triangle DEF$ below, $\angle C \cong \angle F$. What other piece(s) of information would you need to prove that $\triangle ABC \cong \triangle DEF$ by ASA congruence?

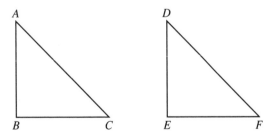

5. In the figure below, name the congruent triangles.

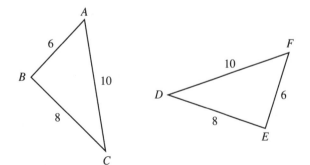

6. Given two triangles, $\triangle ABC$ and $\triangle DEF$, where $\angle A \cong \angle D$, $\angle B \cong \angle E$, and $\angle C \cong \angle F$. Is $\triangle ABC \cong \triangle DEF$? Explain why or why not.

7. In the figure shown below, is $\triangle ABC \cong \triangle EDC$? Explain why.

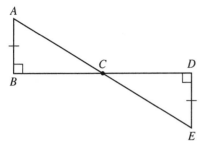

8. Given: $\overline{AB} \cong \overline{DE}$, $\angle B \cong \angle E$, and $\overline{BC} \cong \overline{EF}$ in the figure shown.

How would you explain that $\angle A \cong \angle D$?

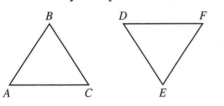

9. Which corresponding angles or corresponding sides in the triangles below could be congruent but not help to prove that the triangles are congruent?

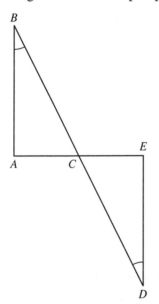

10. Given: $\angle A \cong \angle F$, $\overline{AC} \cong \overline{DF}$, and $\angle C \cong \angle D$. Are triangles ABC and FED in the figure below congruent? Prove they are congruent or explain why they are not congruent.

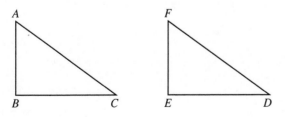

(Answers on page 385)

1.

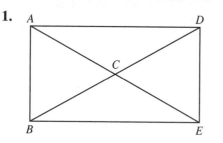

Below is a two-column proof that
m∠CBA + m∠ACB = m∠CDE + m∠DCE
in the figure above.

Statement	Reason
1. ?	1. Given
2. m∠ACB = m∠DCE	2. Vertical angles have equal measures.
3. m∠CBA + m∠ACB = m∠CDE + m∠DCE	3. When equals are added to equals, the sums are equal.

Which of the following statements must be included in Statement 1 so that the proof is complete?

    A. m∠ABE = m∠DEB

    B. m∠CBA = m∠CDE

    C. m∠ECB = m∠ACD

    D. m∠EAB = m∠BDE

    E. m∠ECA = m∠BCD

2. If you are given that △ABC ≅ △EDF, which of the following statements is NOT known?

    F. $\overline{AC} \cong \overline{EF}$

    G. ∠ABC ≅ ∠FDE

    H. $\overline{CA} \cong \overline{FD}$

    J. ∠C ≅ ∠F

    K. △BCA ≅ △DFE

3.

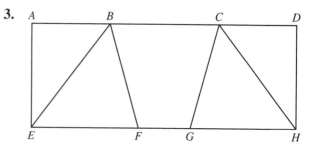

Given: AB = DC, EB = HC, BF = CG, and ∠EAB and ∠HDC are right angles.

Which of the following can NOT be proved about the figure above?

    A. AC = BD

    B. △ABE ≅ △DCH

    C. AE = DH

    D. EF = GH

    E. m∠ABE = m∠DCH

4. Given two triangles where △ABC ≅ △DEF, which of the following could NOT be proved congruent?

    F. ∠B ≅ ∠E

    G. $\overline{AC} \cong \overline{FD}$

    H. $\overline{AB} \cong \overline{EF}$

    J. ∠C ≅ ∠F

    K. $\overline{DE} \cong \overline{ED}$

5. Given △ABC ≅ △DEF, what is the measure of ∠C in the figure below?

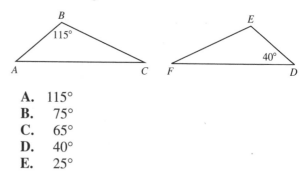

    A. 115°
    B. 75°
    C. 65°
    D. 40°
    E. 25°

(Answers on page 386)

Circles

A **circle** is all the points a set distance from a fixed point called the center. The diagram below shows a circle along with its center, a radius, and a diameter. A **diameter** is a line segment that passes through the center and has its endpoints on the circle. A **radius** is a line segment with the center for one endpoint and the other endpoint on the circle. A circle is named by its center. The circle below is circle O.

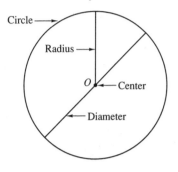

The circle is just the points equidistant from the center. The circle does not include any interior points.

Chords, Secants, and Tangents

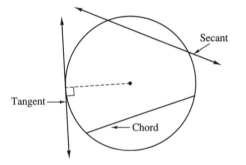

A **chord** is a line segment with endpoints on the circle. The diameter is the longest chord.

A **secant** is a line that contains a chord.

A **tangent** is a line that touches exactly one point on the circle.

The tangent is perpendicular to the radius at the point of tangency.

Arcs and Angles

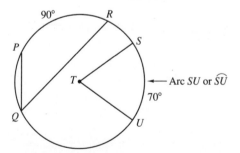

An **arc** is any portion of the circle. $\overset{\frown}{SU}$ is an arc.

A **central angle** is an angle formed by two radii. $\angle STU$ is a central angle. $\angle STU$ intercepts $\overset{\frown}{SU}$.

An **inscribed angle** is an angle formed by two chords with the vertex on the circle. $\angle PQR$ is an inscribed angle. $\angle PQR$ intercepts $\overset{\frown}{PR}$.

Angle and Arc Measures

The total degree measure of a circle is 360°.

- The measure of a central angle equals the measure of the intercepted arc.

    In the previous figure, m$\widehat{SU}$ = 70°, m∠STU = 70°.

- The measure of an inscribed angle equals $\frac{1}{2}$ the measure of the intercepted arc.

    In the previous figure, m$\widehat{PR}$ = 90°, m∠PQR = 45°.

MODEL ACT PROBLEMS

1. In the figure below, the measure of $\widehat{AB}$ is 180°. What is the measure of the inscribed angle θ?

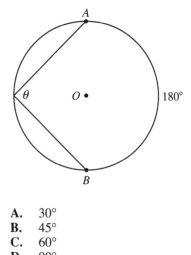

    A. 30°
    B. 45°
    C. 60°
    D. 90°
    E. 180°

SOLUTION

An inscribed angle is half the measure of the intercepted arc. Therefore:

$$m\angle\theta = \frac{1}{2}(m\widehat{AB})$$

$$m\angle\theta = \frac{1}{2}(180°)$$

$$m\angle\theta = 90°$$

The correct answer is D.

2. In the circle below with center O, what is the measure of $\widehat{BD}$?

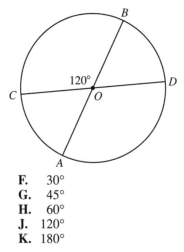

    F. 30°
    G. 45°
    H. 60°
    J. 120°
    K. 180°

SOLUTION

Segment $\overline{CD}$ is a diameter.

m∠BOC + m∠BOD = 180°

m∠BOD = 180° − 120°

m∠BOD = 60°

The measure of a central angle is equal to the measure of the arc that it intercepts.

Therefore m∠BOD = m$\widehat{BD}$.

    60° = m$\widehat{BD}$

The correct answer is H.

Practice

1. What is a diameter?

2. What is a radius?

3. In the diagram below, identify the center, radius, and diameter.

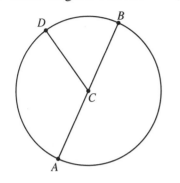

4. What is a chord?

5. What is a secant?

6. What is a tangent to a circle?

7. In the diagram below, identify a chord, secant, and tangent.

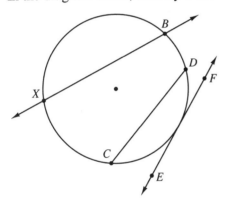

8. What is a central angle of a circle?

9. What is an inscribed angle?

Questions 10 and 11 refer to the figure below.

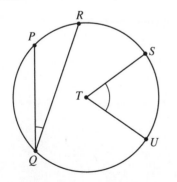

10. Identify the central angle and the inscribed angle.

11. What arcs are intercepted by the inscribed angle and the central angle?

Questions 12 and 13 refer to the figure below.

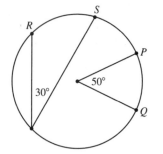

12. What is the measure of $\widehat{PQ}$?

13. What is the measure of $\widehat{RS}$?

Questions 14 and 15 refer to the figure below.

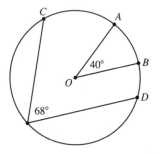

14. What is the measure of $\widehat{AB}$?

15. What is the measure of $\widehat{CD}$?

Questions 16 and 17 refer to the figure below.

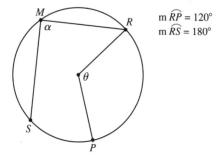

m $\widehat{RP}$ = 120°
m $\widehat{RS}$ = 180°

16. What is the measure of $\angle\,\theta$?

17. What is the measure of $\angle\,\alpha$?

Questions 18–20 refer to the figure below.

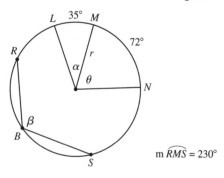

m $\widehat{RMS}$ = 230°

18. What is the measure of ∠θ?

19. What is the measure of ∠α?

20. What is the measure of ∠β?

21. Circle *E* has a circumference of 6 units.

m∠*DEF* = 80°   m∠*CEF* = 55°   m∠*ABD* = 60°

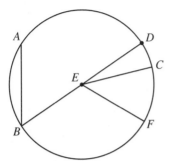

What is the shortest distance around the circle from:

a. Point *A* to point *D*

b. Point *D* to point *F*

c. Point *B* to point *C*

(Answers on page 386)

ACT-TYPE PROBLEMS

1. Which of the figures below shows a chord?

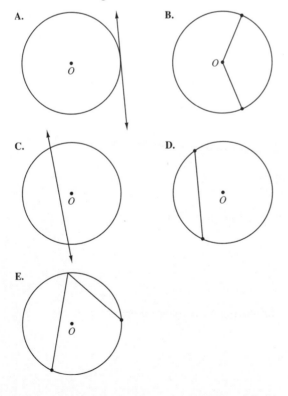

2. Which of the figures below shows an inscribed angle?

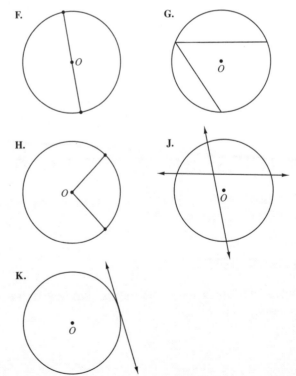

3. In the figure below, what is the measure of arc *RS*?

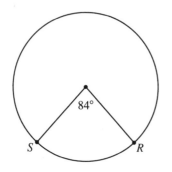

A. 16°
B. 36°
C. 42°
D. 84°
E. 168°

4. In the figure below, $\widehat{AB}$ measures 180° and both ∠θ and ∠α intercept this arc. What is the sum of m∠θ and m∠α?

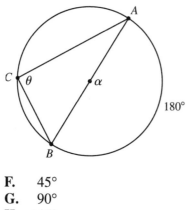

F. 45°
G. 90°
H. 145°
J. 180°
K. 270°

(Answers on page 387)

5. In the figure below, the measure of $\widehat{AC}$ is equal to twice the measure of $\widehat{BC}$. What is the length of segment *AC* in △*ABC*?

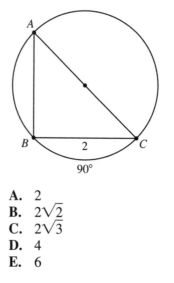

A. 2
B. 2√2
C. 2√3
D. 4
E. 6

■ Transformations in the Plane

A **transformation** in the plane means shifting a figure by sliding, flipping, or rotating it from one location to another.

Translation

A **translation** in the plane means sliding a figure from its original position to another position.

Rectangle A is the original figure. Sliding rectangle A 3 units up and 3 units to the right formed rectangle B.

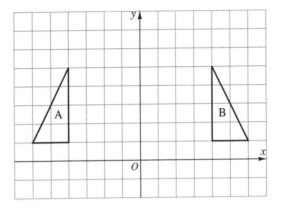

Reflection

A **reflection** in the plane means flipping a figure across a line such as the x-axis or the y-axis.

Triangle A is the original triangle. Triangle B is formed by flipping triangle A across the y-axis.

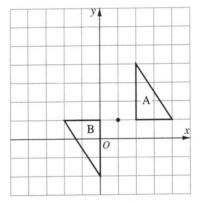

Rotation

A **rotation** in the plane means rotating a figure about a point in the plane.

Triangle A is the original triangle. Triangle B is formed by rotating triangle A 180° about the point (1,1).

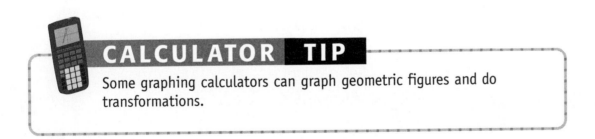

CALCULATOR TIP

Some graphing calculators can graph geometric figures and do transformations.

Given square A in the graph, which of the following choices shows that square B is obtained by sliding 3 units left and 3 units up?

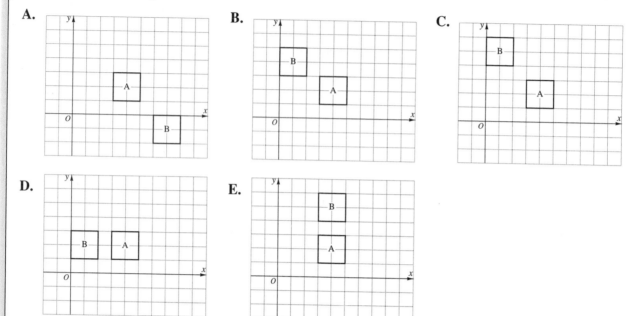

SOLUTION

In choice C, square B is formed from square A by a horizontal translation of 3 to the left and a vertical translation of 3 up.

The correct answer is C.

Practice

1. Using triangle A as your starting triangle, create a triangle B by flipping triangle A across the *x*-axis.

2. Using square A as your starting square, rotate the figure 180° about the origin to create square B.

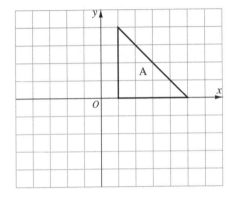

3. Using triangle A as your starting triangle, create triangle B by sliding triangle A 2 units down and 3 units to the left.

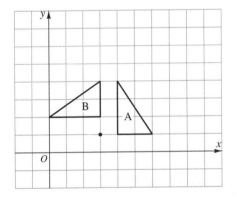

4. Look at the diagram below. How is triangle B obtained from triangle A?

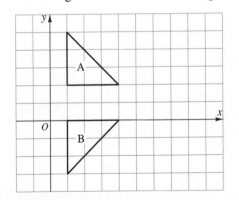

5. In the diagram below, how is triangle B obtained from triangle A?

6. Draw triangle C as a reflection of triangle D across the line $x = 2$.

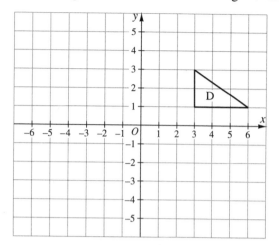

7. Draw rectangle C as a rotation of rectangle D by 180° about the point $(-2,1)$.

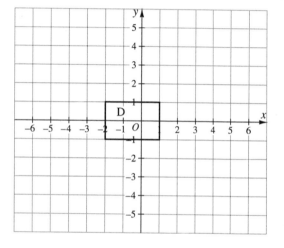

8. Draw triangle C by translating triangle D down 2 units and then reflecting it across the y-axis.

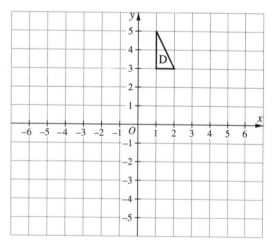

9. What transformation creates triangle C from triangle D in the diagram below?

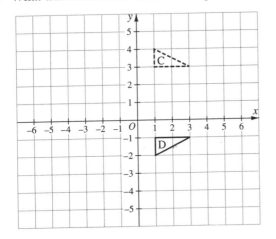

10. What transformation creates rectangle C from rectangle D in the diagram below?

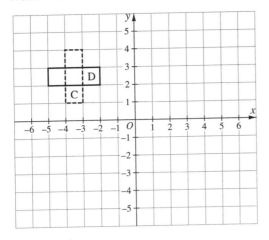

(Answers on page 387)

ACT-TYPE PROBLEMS

1. Which of the following translations produced triangle B from triangle A in the diagram below?

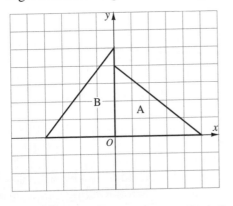

   A. Horizontal shift 4 units to the left
   B. Flip across the *y*-axis
   C. 90° rotation about the origin
   D. Flip across the *x*-axis
   E. 60° rotation about the origin

2. If the center of a circle is the origin, which of the following translations will NOT place the circle on top of itself?

   F. Horizontal translation to the left 1 unit, followed by a horizontal translation to the right 1 unit
   G. Any rotation about the origin
   H. A reflection across the *x*-axis
   J. A vertical translation up 2 units, followed by a vertical translation up 2 units
   K. A reflection across the *y*-axis

3. Which of the following creates a square B by flipping square A across the line $x = 1$?

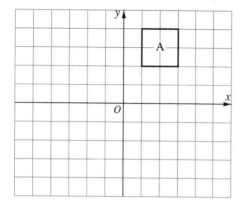

A.

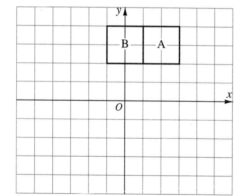

B.

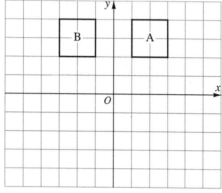

C.

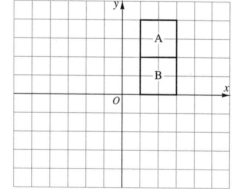

D.

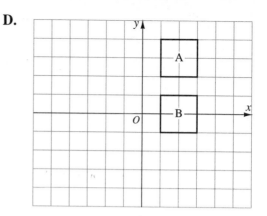

E.

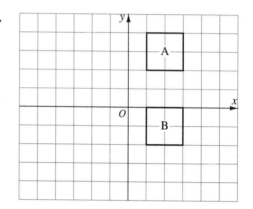

4. Which of the following transformations does NOT move triangle A on top of triangle B?

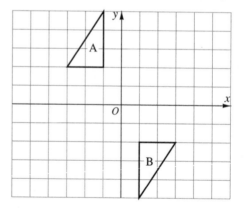

F. 180° rotation about the origin

G. Flip across the x-axis followed by a flip across the y-axis

H. Flip across the y-axis followed by a flip across the x-axis

J. Slide 2 units right, followed by a flip across the line $x = 2$, followed by a flip across the x-axis

K. Slide 2 units left, followed by a flip across the line $x = -1$, followed by a flip across the x-axis

5. In which of the following choices can rectangle B NOT be created from rectangle A by reflections across the y-axis and/or reflections across the x-axis?

A.

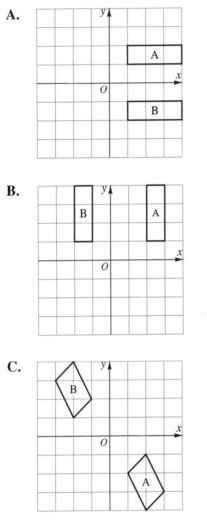

D.

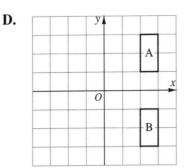

B.

E.

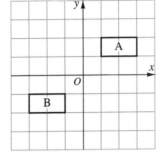

C.

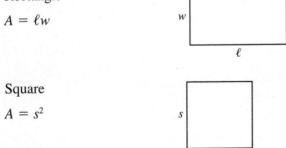

(Answers on page 388)

■ Geometric Formulas

The ACT frequently gives formulas needed to answer a question. This section summarizes area and circumference formulas for reference.

Rectangle

$A = \ell w$

w

ℓ

Square

$A = s^2$

s

Parallelogram

$A = bh$

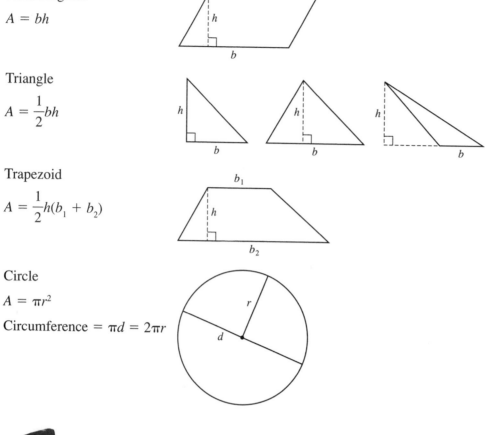

The height must be perpendicular to the base. Look for the right angle symbol to identify the height.

Triangle

$A = \frac{1}{2}bh$

Trapezoid

$A = \frac{1}{2}h(b_1 + b_2)$

Circle

$A = \pi r^2$

Circumference $= \pi d = 2\pi r$

CALCULATOR TIP

If an ACT question has answer choices in pi form (for example, 6π), check to see if your calculator can display answers in pi form. Otherwise, don't use a calculator.

MODEL ACT PROBLEM

What is the area of a circle that has a diameter equal to 6?

A. 36π
B. 18π
C. 12π
D. 9π
E. 6π

SOLUTION

Find the radius of the circle.

$r \text{ (radius)} = \frac{1}{2} \times d \text{ (diameter)}$

$r = \frac{1}{2} \times 6 = 3$

Find the area.

$A = \pi r^2$

$A = \pi(3)^2$

$A = \pi \times 9$

$A = 9\pi$

The correct answer is D.

Practice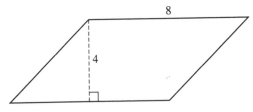

Where appropriate, use 3.14 for π.

1. What is the area of the parallelogram below in square units?

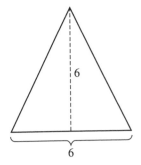

2. What is the height of a parallelogram if the base is 4 cm and the area is 20 cm²?

3. What is the area of this triangle in square units?

4. What is the base of a triangle if the area is equal to 24 in.² and the height is 6 in.?

5. What is the width of the rectangle?

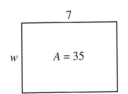

6. What is the area of a rectangle if the length is 9 units and the width is two units less than the length?

7. What is the area of the square in square units?

8. What is the length of a side of a square if the area is 121 square units?

9. What are the area and the circumference of a circle with radius equal to 5 cm?

10. What is the radius of a circle if the area is equal to 16π m²?

11. What is the height of this trapezoid?

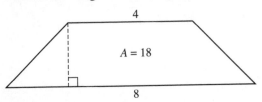

12. What is the area of a trapezoid with height of 4 inches, one base equal to 6 inches, and the other base equal to twice the length of b_1?

13. Find the area of the rectangular deck with a semicircular end shown below.

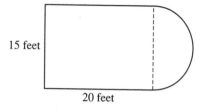

15 feet

20 feet

14. If the height of a triangle is 3.2 meters and the area is 6.4 square meters, what is the length of the base?

15. If the bases of a trapezoid measure 1.4 and 3.4 meters and the area is 19.2 square meters, what is the height of the trapezoid?

16. What is the radius of a circle with an area of 36π square units?

17. What is the area of the figure below, in pi form? It is a rectangle with a triangle at one end and a semicircle at the other end. The length of the rectangle is 10. The radius of the circle is 3. The height of the triangle is equal to its base.

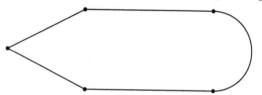

18. What is the area of the figure below?

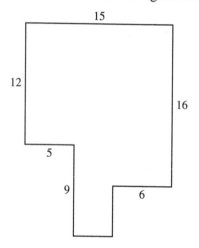

15

12

16

5

9

6

19. What is the area of the shaded region in the figure below in pi form?

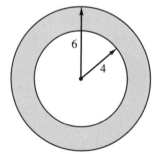

6

4

20. The side of a square is 8 units long. The square is cut and pieced together to form a rectangle with a width of 4 units. What is the length of this rectangle?

(Answers on page 388)

ACT-TYPE PROBLEMS

1. What is the area of the trapezoid shown below?

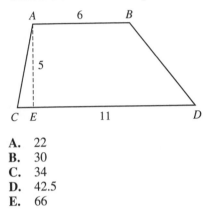

  A. 22
  B. 30
  C. 34
  D. 42.5
  E. 66

2. What is the circumference of a circle that has an area equal to 49π?

  F. 7π
  G. 14π
  H. 21π
  J. 28π
  K. 49π

3. What is the area of a square whose diagonal has a length of 4 cm?

  A. 2 cm^2

  B. $2\sqrt{2}$ cm^2

  C. 4 cm^2

  D. 8 cm^2

  E. 16 cm^2

(Answers on page 388)

4. In the figure below, what is the area of the shaded region?

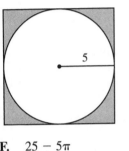

  F. $25 - 5\pi$
  G. $100 - 25\pi$
  H. 10π
  J. $100 - 10\pi$
  K. 25π

5. The figure below is made up of a 45–45–90 triangle and a semicircle. Find the area of the entire figure.

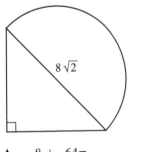

  A. $8 + 64\pi$
  B. $8 + 128\pi$
  C. $32 + 16\pi$
  D. $32 + 32\pi$
  E. $32 + 64\pi$

■ Geometry in Three Dimensions

Space occupies three dimensions and extends infinitely in all directions. Volume is a measure of how much space a three-dimensional figure takes up. The surface area of a three-dimensional figure is the total area occupied by the surface of the figure. Volume is expressed in cubic units. Surface area is expressed in square units.

Given on the next page are some three-dimensional figures with formulas for surface area and volume.

Rectangular Prism

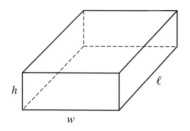

A rectangular prism has 6 rectangular faces: two faces with dimensions ℓh, two faces with dimensions ℓw, and two faces with dimensions hw.

The surface area of a rectangular prism is the sum of the areas of the sides.

$SA = 2(\ell h + \ell w + hw)$

The volume of a rectangular prism is the product of the length, width, and height.

$V = \ell wh$

EXAMPLE

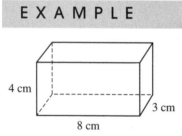

Find the surface area and volume of the rectangular prism above.

Write the dimensions.

$\ell = 8$ cm, $h = 4$ cm, $w = 3$ cm

Use the formulas.

Surface area

$SA = 2(\ell h + \ell w + hw)$

$SA = 2[(8 \text{ cm} \cdot 4 \text{ cm}) + (8 \text{ cm} \cdot 3 \text{ cm}) + (4 \text{ cm} \cdot 3 \text{ cm})]$

$SA = 2(32 \text{ cm}^2 + 24 \text{ cm}^2 + 12 \text{ cm}^2) = 136 \text{ cm}^2$

Volume

$V = \ell wh$

$8 \text{ cm} \cdot 3 \text{ cm} \cdot 4 \text{ cm} = 96 \text{ cm}^3$

Remember: Volume is measured in units^3.

Cube

A cube is a special rectangular prism whose faces are all identical squares. A cube has six faces. The area of each face is s^2.

The surface area of the cube is 6 times the area of a face.

$SA = 6s^2$

The length, width, and height are all equal, so the volume of the cube is the cube of one side.

$V = s^3$

EXAMPLE

What are the surface area and the volume of a cube with side length 5 cm?

$SA = 6s^2$

$SA = 6 \cdot 25 = 150 \text{ cm}^2$

The volume of a cube is the cube of one side.

$V = s^3$

$V = 5^3 = 125 \text{ cm}^3$

Cylinder

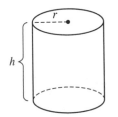

A cylinder is formed by two congruent, parallel, circular bases with radius r connected by a rectangular face with a height h.

The surface area is the sum of the areas of the two circular bases, plus the area of the rectangular side. The length of the rectangle is equal to the circumference of one of the bases; the height is h.

$SA = 2\pi r^2 + 2\pi rh$

The volume of the cylinder is the area of the base multiplied by the height.

$V = \pi r^2 h$

Sphere

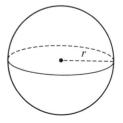

A sphere is all points a given distance (the radius, r) from a center point.

$SA = 4\pi r^2$

$V = \dfrac{4}{3}\pi r^3$

EXAMPLE

To the nearest tenth, what are the surface area and volume of a sphere with a radius of 4 cm? Use 3.14 for π.

$SA \approx 4 \cdot 3.14 \cdot 4^2 \approx 201.0 \text{ cm}^2$

$V \approx \dfrac{4}{3} \cdot 3.14 \cdot 4^3 \approx 267.9 \text{ cm}^3$

Cone

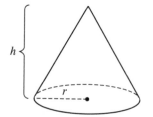

A cone consists of a circular base with a radius (r) and a single vertex a fixed height from the center of the base.

$V = \dfrac{1}{3}\pi r^2 h$

EXAMPLE

To the nearest tenth, what is the volume of a cone with a radius of 3 cm and a height of 6 cm? Use 3.14 for π.

$V \approx \dfrac{1}{3} \cdot 3.14 \cdot 3^2 \cdot 6 = 56.5 \text{ cm}^3$, to the nearest tenth.

Square Pyramid

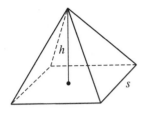

A square pyramid has a square base and 4 congruent triangles as sides.

SA = the sum of the areas of the faces of the pyramid. Note that h above does not provide the heights of the triangles. Those measurements must be taken along the outside of the pyramid.

$V = \dfrac{1}{3}Bh$, where B is the area of the square base and h is the height from the base to the vertex.

MODEL ACT PROBLEMS

1. If the surface area of a cube is 96 cm², what is the length of each side of the cube?

   A. 3 cm
   B. 4 cm
   C. 5 cm
   D. 6 cm
   E. 7 cm

   SOLUTION

   Each face of a cube has the same area. Divide 96 cm² by 6, the number of faces, to find the area of one face of the cube. (96 cm² ÷ 6 = 16 cm²)

   Take the square root of the area of the face to find the length of the side.

   $\sqrt{16 \text{ cm}^2} = 4$ cm. The length of each side of the cube is 4 cm.

   The correct answer is B.

2. What is the radius of a sphere with a volume of $\dfrac{500\pi}{3}$?

   F. 125π
   G. 125
   H. 25
   J. 5π
   K. 5

   Formula: $V = \dfrac{4}{3}\pi r^3$

   Solve for r. $\quad r^3 = \dfrac{V}{\dfrac{4}{3}\pi}$

   $r^3 = \dfrac{\dfrac{500\pi}{3}}{\dfrac{4\pi}{3}} = \dfrac{500\pi}{4\pi} = \dfrac{500}{4} = 125$

   $r = \sqrt[3]{125}$

   $r = 5$

   The correct answer is K.

Practice

Use 3.14 for π, and round your answer to the nearest tenth.

1. What is the surface area of a sphere with a radius of 5 cm?

2. What is the length of each side of a cube if the volume is 729 cm³?

3. What is the length of a rectangular prism with a volume of 576 cm³, a height of 6 cm, and a width of 8 cm?

4. What is the volume of a right circular cone if the height is 15 cm and the radius is 3 cm?

5. What is the radius of a right circular cylinder with a height of 5 cm and a volume of 62.8 cm³?

6. What is the volume of a cube that has a side equal to 4 units?

7. What is the length of a side of a cube whose volume is 216 cubic units?

8. What is the height of a rectangular prism whose volume is 350 cm³, length is 10 cm, and width is 7 cm?

9. What is the volume of a rectangular prism whose length is 4 units, width is 7 units, and height is 2 units?

10. What is the volume of a sphere that has a radius equal to 3 units?

11. What is the radius of a sphere that has a volume of 2,304π cubic centimeters?

12. What is the volume of a cylinder in terms of π if the radius is 3 in. and the height is 9 in.?

13. What is the height of a cylinder that has a volume of 768π cubic units and a radius equal to 8 units?

14. If a cylinder has a volume of 251.2 cm³ and a height of 5 cm, what is its diameter?

15. What is the volume of each storage box after assembly?

   a.

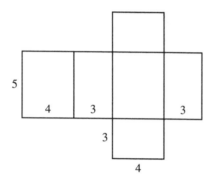

   b.

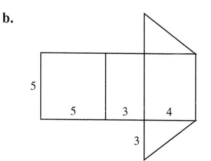

16. A cylindrical can has a radius of 5 and a height of 10. What is the surface area of the can?

17. What is the volume of the sphere shown below?

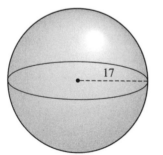

18. A sphere has a radius of 12 units. A cylinder has a height of 9 units and the same volume as the sphere. What is the radius of the cylinder?

19. What is the volume of a cube if the length of a side is 10 inches?

20. The volume of a cube is 216 cm³. What is the surface area of the cube?

(Answers on page 389)

ACT-TYPE PROBLEMS

1. What is the surface area of the rectangular prism below?

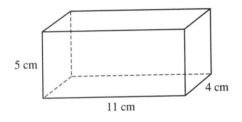

5 cm

4 cm

11 cm

   A. 119 cm²
   B. 220 cm²
   C. 238 cm²
   D. 357 cm²
   E. 440 cm²

2. What is the area of each triangle in a square pyramid if the length of each side of the square base is 6 cm, and the total surface area is 132 cm²?

   F. 20 cm²
   G. 21 cm²
   H. 22 cm²
   J. 23 cm²
   K. 24 cm²

(Answers on page 389)

3. If the volume of a sphere is 3,052.08 cm³, what is the radius of the sphere? (Use 3.14 for π.)

   A. 7 cm
   B. 8 cm
   C. 9 cm
   D. 10 cm
   E. 11 cm

4. What is the volume of a cylinder with a height of 16 and a base area of 25π?

   F. 5π
   G. 25π
   H. 50π
   J. 200π
   K. 400π

5. What is the area of a circle if it has the same radius as a sphere whose volume is $7,776\pi$?

   A. 18π
   B. 36π
   C. 81π
   D. 324π
   E. $1,296\pi$

Cumulative **ACT** Practice
Plane Geometry

Complete this Cumulative ACT Practice in 15 minutes to reflect real ACT test conditions. This Cumulative ACT Practice gives you an additional opportunity to practice plane geometry concepts in an ACT format. If you don't know an answer, eliminate and guess. Circle the number of any guessed answer. Then check your answers on page 390. You will also find explanations for the answers and suggestions for further study.

1. $\angle\theta$ and $\angle\alpha$ are vertical angles. If $\angle\theta = 30°$, what is the measure of $\angle\alpha$?

   A.   30°
   B.   60°
   C.   90°
   D.   120°
   E.   150°

2. What is the circumference of a circle that has an area of 49π?

   F. 3.5π
   G. 7π
   H. 14π
   J. 28π
   K. 49π

3. $\triangle QRS$ is similar to $\triangle LMN$. What is the length of line segment $\overline{LM}$?

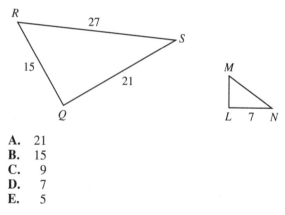

   A.   21
   B.   15
   C.   9
   D.   7
   E.   5

4. Which of the following triangle postulates could be used to prove that $\triangle ABC \cong \triangle DEF$?

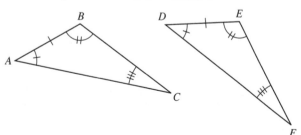

   F.   SSS Postulate
   G.   AAA Postulate
   H.   ASA Postulate
   J.   SSA Postulate
   K.   SAS Postulate

5. $\triangle EFG$ is an isosceles triangle. If the measure of $\angle E = 50°$ what is the measure of $\angle F$?

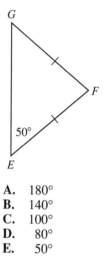

   A.   180°
   B.   140°
   C.   100°
   D.   80°
   E.   50°

6. Which single transformation could be used to place line segment $\overline{AB}$ directly on top of line segment $\overline{CD}$?

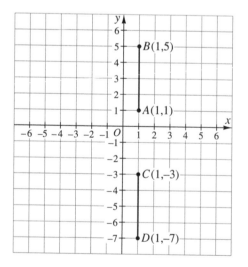

   F.   A 90° rotation about the origin
   G.   A 180° rotation about the point $(1,-1)$
   H.   A vertical slide down 4 units
   J.   A 90° rotation about the point $(1,-1)$
   K.   A vertical slide down 9 units

7. What is the radius of a sphere that has a volume of 288π?

   A. 3
   B. 4
   C. 5
   D. 6
   E. 7

8. $p \| q$, and both are cut by transversal t. If the measure of $\angle 4 = 40°$, what is the measure of $\angle 8$?

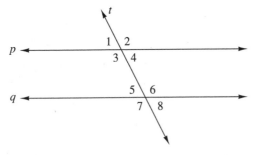

   F. 40°
   G. 80°
   H. 90°
   J. 140°
   K. 150°

9. $\ell \| m$ and s and t are transversals that cross ℓ and m and intersect with line m at point C. If the measure of $\angle BAC = 70°$ and the measure of $\angle 1 = 30°$ what is the measure of $\angle 2$?

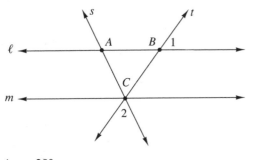

   A. 30°
   B. 70°
   C. 80°
   D. 130°
   E. 150°

10. $\triangle ABC$ is a 30–60–90 triangle. What is the sum of the lengths of sides AC and BC?

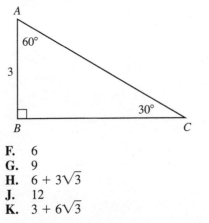

   F. 6
   G. 9
   H. $6 + 3\sqrt{3}$
   J. 12
   K. $3 + 6\sqrt{3}$

11. Which of the following choices properly identifies two congruent triangles?

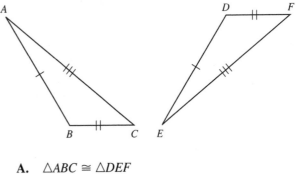

   A. $\triangle ABC \cong \triangle DEF$
   B. $\triangle CBA \cong \triangle FDE$
   C. $\triangle BAC \cong \triangle FDE$
   D. $\triangle CAB \cong \triangle DFE$
   E. $\triangle BCA \cong \triangle DEF$

12. What is the volume of a right circular cylinder whose base is a circle with a diameter of 10 and whose height is equal to the radius?

   F. 5π
   G. 25π
   H. 75π
   J. 125π
   K. 250π

13. $\triangle DEF$ is an isosceles triangle. The measure of $\angle DEF = 120°$ and the altitude of the triangle is 4. What is the area of $\triangle DEF$?

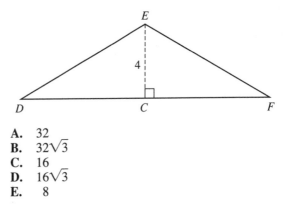

   A. 32
   B. $32\sqrt{3}$
   C. 16
   D. $16\sqrt{3}$
   E. 8

14. Which of the following transformations would NOT place △*ABC* on △*DEF*?

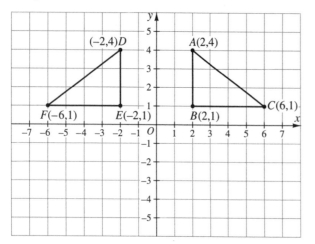

F. Reflection across the *y*-axis
G. Horizontal slide left 2 units; reflection across the line $x = -1$
H. Reflection across the line $x = 2$; horizontal slide left 4 units
J. Horizontal slide left 4 units; reflection across the line $x = -2$
K. Horizontal slide left 3 units; reflection across the line $x = -2$

15. What is the area of the right triangle below if *AB* = 6 ft?

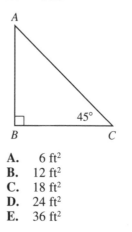

A. 6 ft²
B. 12 ft²
C. 18 ft²
D. 24 ft²
E. 36 ft²

Subtest Plane Geometry

Complete this Subtest in 14 minutes to reflect real ACT test conditions. This Subtest has the same number and type of plane geometry items as are found on the ACT. If you don't know an answer, eliminate and guess. Circle the number of any guessed answer. Then check your answers on page 391. You will also find explanations for the answers and suggestions for further study.

1. ∠θ and ∠α form a linear pair. If ∠θ is twice as large as ∠α, what is m∠θ?

A. 30°
B. 60°
C. 90°
D. 120°
E. 180°

2. What is the area of a circle that has a circumference of 24π?

F. 12π
G. 24π
H. 36π
J. 72π
K. 144π

3. △*ABC* and △*DEF* are similar triangles. What is the length of the longest side of △*DEF*?

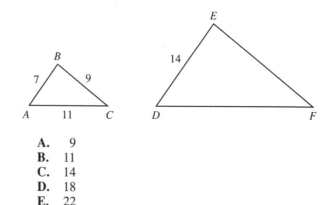

A. 9
B. 11
C. 14
D. 18
E. 22

4. Which of the following is a postulate of congruent triangles that could be used to prove that △ABC ≅ △DEF?

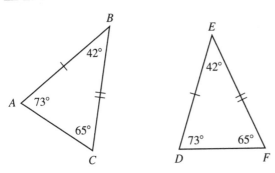

F. AAS Postulate
G. SSA Postulate
H. SAA Postulate
J. AAA Postulate
K. SAS Postulate

5. ABCD is an isosceles trapezoid, $\overline{BE} \perp \overline{AD}$, and $\overrightarrow{FG}$ is parallel to $\overline{CD}$. Given m∠CFG = 50°, what is the measure of ∠ABE?

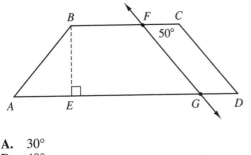

A. 30°
B. 40°
C. 50°
D. 60°
E. 90°

6. A single transformation can be used to move △ABC directly on top of △DEF. What is the transformation?

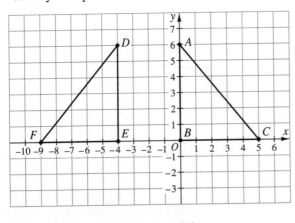

F. A 90° rotation about the origin
G. A reflection across the y-axis
H. A −270° rotation about the origin
J. A reflection across the line x = −2
K. A 90° rotation about the point (−2, 0)

7. What is the volume of a right circular cylinder of height 14 cm whose base has an area of 49π cm²?

A. 53π cm²
B. 686π cm²
C. $\frac{1,372\pi}{3}$ cm²
D. 53π cm³
E. 686π cm³

8. ℓ∥m, and both lines are cut by the transversal t. If m∠1 = 115°, then m∠5 = ?

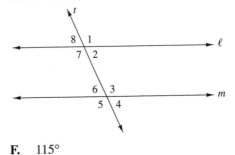

F. 115°
G. 85°
H. 75°
J. 65°
K. 30°

9. ℓ∥m and s and t are transversals that cross ℓ and m and intersect line ℓ at point A. What is the measure of ∠CAB?

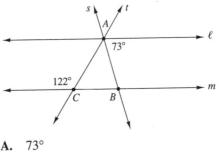

A. 73°
B. 64°
C. 58°
D. 49°
E. 34°

10. △ABC is an isosceles right triangle, BC = 3, and AD = DC. What is the length of BD?

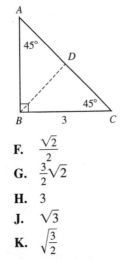

F. $\frac{\sqrt{2}}{2}$
G. $\frac{3}{2}\sqrt{2}$
H. 3
J. $\sqrt{3}$
K. $\sqrt{\frac{3}{2}}$

11. In the figure below, $\overline{BD}$ and $\overline{AE}$ bisect one another. Which of the following choices properly identifies two congruent triangles?

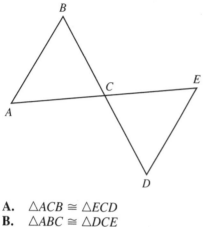

- **A.** $\triangle ACB \cong \triangle ECD$
- **B.** $\triangle ABC \cong \triangle DCE$
- **C.** $\triangle ACB \cong \triangle DCE$
- **D.** $\triangle BCA \cong \triangle ECD$
- **E.** $\triangle CAB \cong \triangle CDE$

12. What is the volume of a sphere that has the same radius as the circle inscribed in the square below?

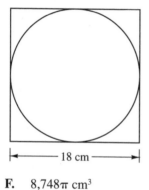

- **F.** $8{,}748\pi$ cm³
- **G.** $7{,}776\pi$ cm³
- **H.** 972π cm³
- **J.** $7{,}776\pi$ cm²
- **K.** 972π cm²

13. $\triangle ABC$ is an equilateral triangle, each side having a length of 8 in. $\triangle QRS$ is formed by joining together the midpoints of $\overline{AB}$, $\overline{BC}$, and $\overline{CA}$. What is the area of the shaded region?

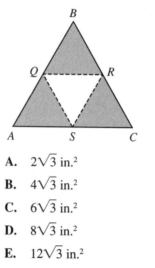

- **A.** $2\sqrt{3}$ in.²
- **B.** $4\sqrt{3}$ in.²
- **C.** $6\sqrt{3}$ in.²
- **D.** $8\sqrt{3}$ in.²
- **E.** $12\sqrt{3}$ in.²

14. Which of the following transformations would place $\triangle ABC$ on top of $\triangle DEF$?

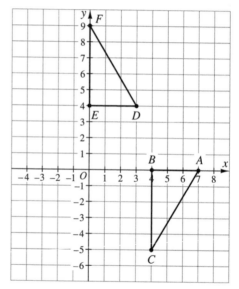

- **F.** Horizontal slide -4 units, reflection across the line $y = 2$
- **G.** Reflection across the line $x = 2$, vertical slide 8 units
- **H.** Reflection across the line $y = 2$, horizontal slide -3 units
- **J.** Reflection across the line $y = x$
- **K.** Vertical slide 4 units up, rotation 180° about the point (4,4)

Chapter 12

Trigonometry

- There are only four trigonometry problems on the entire ACT.
- Typically the ACT questions contain only right triangle or graph identification problems.
- The questions do not require an extensive knowledge of trigonometry.
- This review covers the material you will need to answer the trigonometry questions on the ACT.

▪ Right Triangle Trigonometry

Right triangle trigonometry deals with angles less than 90°. The trigonometric functions are ratios of the sides of right triangles.

Here is the familiar right triangle with the names of the legs labeled.

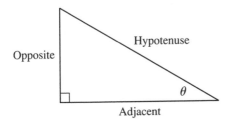

The "opposite" side is opposite the angle θ. The "adjacent" side is adjacent to angle θ.

Trigonometric Functions

The basic trigonometric functions are given below. You should memorize these functions.

$$\text{sine of } \theta = \frac{\text{length of the opposite side}}{\text{length of the hypotenuse}} \qquad \sin \theta = \frac{\text{opp}}{\text{hyp}}$$

$$\text{cosine of } \theta = \frac{\text{length of the adjacent side}}{\text{length of the hypotenuse}} \qquad \cos \theta = \frac{\text{adj}}{\text{hyp}}$$

$$\text{tangent of } \theta = \frac{\text{length of the opposite side}}{\text{length of the adjacent side}} \qquad \tan \theta = \frac{\text{opp}}{\text{adj}}$$

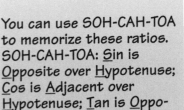

You can use SOH-CAH-TOA to memorize these ratios. SOH-CAH-TOA: Sin is Opposite over Hypotenuse; Cos is Adjacent over Hypotenuse; Tan is Opposite over Adjacent.

CALCULATOR TIP

Scientific and graphing calculators allow you to compute trigonometric functions directly. However, the vast majority of the ACT trigonometry questions ask about your understanding of trigonometric relationships. The calculator will not help you answer these questions.

Inverse Trigonometric Functions

These three trigonometric functions are inverses of the previously mentioned three functions. You should memorize them.

cosecant of $\theta = \dfrac{1}{\text{sine of } \theta}$ $\qquad$ $\csc \theta = \dfrac{\text{hyp}}{\text{opp}}$

secant of $\theta = \dfrac{1}{\text{cosine of } \theta}$ $\qquad$ $\sec \theta = \dfrac{\text{hyp}}{\text{adj}}$

cotangent of $\theta = \dfrac{1}{\text{tangent of } \theta}$ $\qquad$ $\cot \theta = \dfrac{\text{adj}}{\text{opp}}$

Recall that

$\dfrac{1}{\frac{a}{b}} = \dfrac{b}{a}$

Angles can be measured using degrees or radians. The relationship between degrees (D) and radians (R) is $R = \dfrac{\pi}{180} \times D$.

It is helpful to memorize these common values of trigonometric functions.

	Degrees	Radians	Degrees	Radians	Degrees	Radians	Degrees	Radians	Degrees	Radians
θ	$0°$	0	$30°$	$\dfrac{\pi}{6}$	$45°$	$\dfrac{\pi}{4}$	$60°$	$\dfrac{\pi}{3}$	$90°$	$\dfrac{\pi}{2}$
$\sin \theta$	0		$\dfrac{1}{2}$		$\dfrac{\sqrt{2}}{2}$		$\dfrac{\sqrt{3}}{2}$		1	
$\cos \theta$	1		$\dfrac{\sqrt{3}}{2}$		$\dfrac{\sqrt{2}}{2}$		$\dfrac{1}{2}$		0	
$\tan \theta$	0		$\dfrac{\sqrt{3}}{3}$		1		$\sqrt{3}$		undefined	

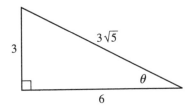

1. Find the sin, cos, and tan of $\angle\theta$. Remember to simplify your answers.

$$\sin\theta = \frac{\text{opp}}{\text{hyp}} = \frac{3}{3\sqrt{5}} = \frac{1}{\sqrt{5}} = \frac{1}{\sqrt{5}} \cdot \frac{\sqrt{5}}{\sqrt{5}} = \frac{\sqrt{5}}{5}$$

$$\cos\theta = \frac{\text{adj}}{\text{hyp}} = \frac{6}{3\sqrt{5}} = \frac{2}{\sqrt{5}} = \frac{2}{\sqrt{5}} \cdot \frac{\sqrt{5}}{\sqrt{5}} = \frac{2\sqrt{5}}{5}$$

$$\tan\theta = \frac{\text{opp}}{\text{adj}} = \frac{3}{6} = \frac{1}{2}$$

2. Find the csc, sec, and cot of $\angle\theta$ in the triangle above.

$$\sin\theta = \frac{\sqrt{5}}{5}, \text{ so } \csc\theta = \frac{5}{\sqrt{5}} = \frac{5\sqrt{5}}{\sqrt{5}\cdot\sqrt{5}} = \frac{5\sqrt{5}}{5} = \sqrt{5}$$

$$\cos\theta = \frac{2\sqrt{5}}{5}, \text{ so } \sec\theta = \frac{5}{2\sqrt{5}} = \frac{5\sqrt{5}}{2\sqrt{5}\cdot\sqrt{5}} = \frac{5\sqrt{5}}{2\cdot5} = \frac{\sqrt{5}}{2}$$

$$\tan\theta = \frac{1}{2}, \text{ so } \cot\theta = 2$$

MODEL ACT PROBLEMS

1. In the triangle below, what is $\sin\theta = $?

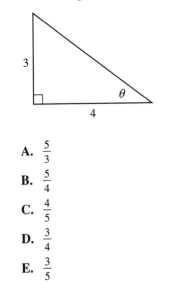

A. $\frac{5}{3}$

B. $\frac{5}{4}$

C. $\frac{4}{5}$

D. $\frac{3}{4}$

E. $\frac{3}{5}$

SOLUTION

Find the length of the hypotenuse.

You may notice that this is a 3–4–5 triangle, with the hypotenuse equal to 5.

Find $\sin\theta$.

$$\sin\theta = \frac{\text{opposite}}{\text{hypotenuse}} = \frac{3}{5}$$

The correct answer is E.

2. In the figure below, the base of the ladder is 3 feet from the house. If the ladder forms a 79° angle with the ground, what is the length of the ladder rounded to the nearest hundredth?

(*Note:* sin 79° ≈ .981, cos 79° ≈ .191, tan 79° ≈ 5.145)

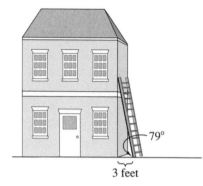

3 feet

F. 3.06 feet

G. 6.72 feet

H. 9.23 feet

J. 13.61 feet

K. 15.71 feet

SOLUTION

We know the length of the side adjacent to the angle and the measure of the angle. We want to find the length of the ladder, which is the hypotenuse.

$\theta = 79°$

adjacent = 3

hypotenuse = ?

$$\cos \theta = \frac{\text{adjacent}}{\text{hypotenuse}}$$

$$\cos 79° = \frac{3}{\text{hypotenuse}}$$

Multiply both sides by the hypotenuse. $\quad \text{hyp} \times \cos 79° = 3$

Divide both sides by cos 79°. $\quad \text{hyp} = \dfrac{3}{\cos 79°} \approx \dfrac{3}{.191}$

Use a calculator. $\quad \text{hyp} \approx 15.71 \text{ feet}$

The correct answer is K.

Practice

In 1–6, find the trigonometric ratios for $\angle \theta$ in the triangle below.

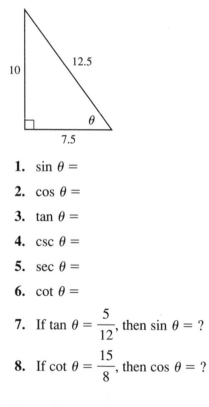

1. sin $\theta =$

2. cos $\theta =$

3. tan $\theta =$

4. csc $\theta =$

5. sec $\theta =$

6. cot $\theta =$

7. If tan $\theta = \dfrac{5}{12}$, then sin $\theta = $?

8. If cot $\theta = \dfrac{15}{8}$, then cos $\theta = $?

9. If $\sin \theta = \dfrac{12}{13}$, then $\csc \theta = ?$

10. If $\sec \theta = \dfrac{3}{2}$, then $\sin \theta = ?$

(Answers on page 393)

ACT-TYPE PROBLEMS

1. In the triangle below, $\sec \theta = ?$

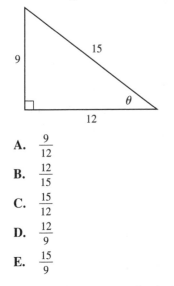

   A. $\dfrac{9}{12}$

   B. $\dfrac{12}{15}$

   C. $\dfrac{15}{12}$

   D. $\dfrac{12}{9}$

   E. $\dfrac{15}{9}$

2. In the triangle below, what is the length of side a rounded to the nearest hundredth?

   (Note: $\sin 40° \approx .643,$
   $\cos 40° \approx .766,$
   $\tan 40° \approx .839)$

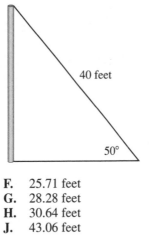

   F. 5.38
   G. 5.03
   H. 4.60
   J. 3.86
   K. 3.05

3. If $\sin \theta = \dfrac{7}{25}$, then $\csc \theta = ?$

   A. $\dfrac{7}{24}$

   B. $\dfrac{24}{25}$

   C. $\dfrac{25}{24}$

   D. $\dfrac{24}{7}$

   E. $\dfrac{25}{7}$

(Answers on page 393)

4. In the figure below, a rope is run from the top of the pole to the ground. The rope is 40 feet long and it forms a 50° angle with the ground. What is the height of the pole rounded to the nearest hundredth?

   (Note: $\sin 50° \approx .766$, $\cos 50° \approx .643$,
   $\tan 50° \approx 1.192)$

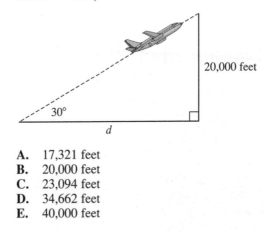

   F. 25.71 feet
   G. 28.28 feet
   H. 30.64 feet
   J. 43.06 feet
   K. 47.68 feet

5. An airplane takes off and climbs at a 30° angle to an altitude of 20,000 feet. To the nearest foot, what ground distance will the plane have flown when it reaches 20,000 feet?

   (Note: $\sin 30° = .5$, $\cos 30° \approx .866$,
   $\tan 30° \approx .577)$

   A. 17,321 feet
   B. 20,000 feet
   C. 23,094 feet
   D. 34,662 feet
   E. 40,000 feet

■ Trigonometric Identities

You can use these identities to find the values of other trigonometric ratios or to rewrite the ratios in equivalent forms. You should memorize them.

Reciprocal Identities

$\sec \theta = \dfrac{1}{\cos \theta}$ or $\sec \theta \cdot \cos \theta = 1$ or $\cos \theta = \dfrac{1}{\sec \theta}$

$\csc \theta = \dfrac{1}{\sin \theta}$ or $\csc \theta \cdot \sin \theta = 1$ or $\sin \theta = \dfrac{1}{\csc \theta}$

$\cot \theta = \dfrac{1}{\tan \theta}$ or $\cot \theta \cdot \tan \theta = 1$ or $\tan \theta = \dfrac{1}{\cot \theta}$

Quotient Identities

$\tan \theta = \dfrac{\sin \theta}{\cos \theta}$

$\cot \theta = \dfrac{\cos \theta}{\sin \theta}$

Pythagorean Identities

$\sin^2 \theta + \cos^2 \theta = 1$

$1 + \tan^2 \theta = \sec^2 \theta$

$\cot^2 \theta + 1 = \csc^2 \theta$

Double-Angle Identities

$\sin 2\theta = 2 \sin \theta \cdot \cos \theta$

$\cos 2\theta = \cos^2 \theta - \sin^2 \theta = 2 \cos^2 \theta - 1 = 1 - 2 \sin^2 \theta$

$\tan 2\theta = \dfrac{2 \tan \theta}{1 - \tan^2 \theta}$

Half-Angle Identities

$\sin \dfrac{\theta}{2} = \pm \sqrt{\dfrac{1 - \cos \theta}{2}}$

$\cos \dfrac{\theta}{2} = \pm \sqrt{\dfrac{1 + \cos \theta}{2}}$

$\tan \dfrac{\theta}{2} = \pm \sqrt{\dfrac{1 - \cos \theta}{1 + \cos \theta}}$

1. The sine of θ is $\sqrt{\dfrac{3}{4}}$. What is the cosine of θ?

   Use $\sin^2 \theta + \cos^2 \theta = 1$.

   Substitute. $\left(\sqrt{\dfrac{3}{4}}\right)^2 + \cos^2 \theta = 1$

   Solve. $\dfrac{3}{4} + \cos^2 \theta = 1$

   $\cos^2 \theta = \dfrac{1}{4}$

   $\cos \theta = \sqrt{\dfrac{1}{4}} = \pm\dfrac{1}{2}$

   The cosine of θ is $\pm\dfrac{1}{2}$.

2. The secant of θ is $\dfrac{5}{\sqrt{5}}$. What is the sine of θ?

   $\sec \theta = \dfrac{5}{\sqrt{5}}$, so $\cos \theta = \dfrac{\sqrt{5}}{5}$. (Secant and cosine are inverse functions.)

   Use $\sin^2 \theta + \cos^2 \theta = 1$.

   Substitute. $\sin^2 \theta + \left(\dfrac{\sqrt{5}}{5}\right)^2 = 1$

   Solve. $\sin^2 \theta + \dfrac{5}{25} = 1$

   $\sin^2 \theta + \dfrac{1}{5} = 1$

   $\sin^2 \theta = \dfrac{4}{5}$

   $\sin \theta = \sqrt{\dfrac{4}{5}}$

   $\sin \theta = \pm 2\sqrt{\dfrac{1}{5}}$

   The sine of θ is $\pm 2\sqrt{\dfrac{1}{5}}$.

3. Demonstrate that $2(1 - \sin^2 \theta) - 1 = \cos 2\theta$.

   Use trigonometric identities to simplify the left side.

   $2(1 - \sin^2 \theta) - 1$

   $= 2(\cos^2 \theta) - 1$ Use the Pythagorean identity.

   $= 2 \cos^2 \theta - 1$ Remove parentheses.

   $= \cos 2\theta$ Use the double-angle identity.

1. $\dfrac{\sin 2\theta}{\tan \theta} - 1 =$

   A. $\sin 2\theta$

   B. $\cos 2\theta$

   C. $\tan 2\theta$

   D. $\sin \dfrac{\theta}{2}$

   E. $\cos \dfrac{\theta}{2}$

   SOLUTION

   $$\dfrac{\sin 2\theta}{\tan \theta} - 1 = \dfrac{2 \sin \theta \cos \theta}{\frac{\sin \theta}{\cos \theta}} - 1$$

   $$= 2 \sin \theta \cos \theta \times \dfrac{\cos \theta}{\sin \theta} - 1$$

   $$= 2 \cos^2 \theta - 1 = \cos 2\theta$$

   The correct answer is B.

2. Which of the following statements is FALSE?

   F. $\cos^2 \dfrac{\theta}{2} = \dfrac{1 + \cos \theta}{2}$

   G. $\sec^2 \theta - \tan^2 \theta = 1$

   H. $(\csc^2 \theta - \cot^2 \theta)(\sin^2 \theta + \cos^2 \theta) = 1$

   J. $\tan 2\theta - \tan 2\theta \tan^2 \theta = 2 \tan \theta$

   K. $\cos 2\theta = 2 \sin^2 \theta - 1$

   SOLUTION

   The identity for $\cos 2\theta$ is: $\cos 2\theta = 1 - 2\sin^2 \theta$.

   Choice K, $\cos 2\theta = 2 \sin^2 \theta - 1$, cannot be derived from that identity. All of the other choices can be derived from an identity.

   The correct answer is K.

Practice

1. If $\cos \theta = \dfrac{\sqrt{3}}{2}$, find the sine of θ using trigonometric identities.

2. If $\csc \theta = \sqrt{2}$, find the cosine of θ using trigonometric identities.

3. $\dfrac{1}{\tan \theta} \cdot \cot \theta + 1 = ?$

4. $\cos \theta \left(\dfrac{-\sin^2 \theta}{\cos \theta} + \cos \theta \right) = ?$

5. $\pm\sqrt{\dfrac{1 + \cos 2\theta}{2}} = ?$

6. $(2 \cos^2 \theta - 1)(1 - 2 \sin^2 \theta) = ?$

7. $\dfrac{\sin^2 \theta}{\cos^2 \theta} + 1 = ?$

8. $\pi(\sin^2 \theta + \cos^2 \theta) = ?$

9. $\dfrac{1 - \cos \theta}{1 + \cos \theta} = ?$

10. $\dfrac{\sec^2 \theta}{1 + \tan^2 \theta} = ?$

(Answers on page 394)

1. $\dfrac{\frac{1}{1 + \tan^2 \theta}}{\sec \theta} \cdot 2\sin \theta = ?$

   A. $\csc^2 \theta$
   B. $\tan 2\theta$
   C. $\sin 2\theta$
   D. $\sin \dfrac{\theta}{2}$
   E. $\sec^2 \theta$

2. $\dfrac{\sin 2\theta}{2 \cos \theta} \cdot \sin \theta + \cos^2 \theta = ?$

   F. 1
   G. $\sec^2 \theta$
   H. π
   J. $\tan 2\theta$
   K. $\tan^2 \theta$

3. Which of the following statements is FALSE?

   A. $\sin^2 \theta + \cos^2 \theta = \csc^2 \theta - \cot^2 \theta$
   B. $\cos^2 \theta - \sin^2 \theta = \cos^2 \theta + \sin^2 \theta$
   C. $2\cos^2 \theta - 1 = 1 - 2\sin^2 \theta$
   D. $\dfrac{\tan^2 \theta}{\sec^2 \theta - 1} = 1$
   E. $\sec^2 \theta - \csc^2 \theta = \tan^2 \theta - \cot^2 \theta$

   (Answers on page 394)

4. $1 - \dfrac{\sin^2 \theta + \cos^2 \theta}{\csc^2 \theta} = ?$

   F. $\sin^2 \theta$
   G. $\tan^2 \theta$
   H. $\sec^2 \theta$
   J. $\cos^2 \theta$
   K. $\cot^2 \theta$

5. $(2 \cos^2 \theta - 1) - (1 - 2 \sin^2 \theta) = ?$

   A. 0
   B. $\tan 2\theta$
   C. 1
   D. $\tan \dfrac{\theta}{2}$
   E. $\tan^2 \theta$

■ Unit-Circle Trigonometry

The right triangle is used as a reference for the trigonometric ratios of acute angles, angles less than 90°. A circle is used as a reference for trigonometric ratios of angles greater than or equal to 90°.

Draw a circle on the coordinate plane with the center at the origin and a radius of 1.

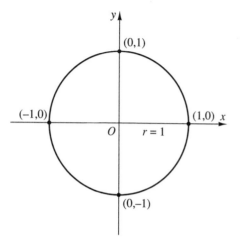

Draw an angle with one side on the x-axis. Start at point (1,0) and go around the circle counterclockwise to place the other side of the angle. You can use the coordinates on the unit circle to find trigonometric values.

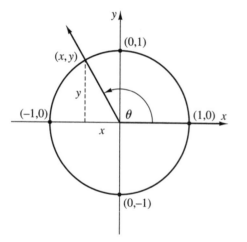

Notice that x and y will always be between -1 and $+1$. For the angle θ use the triangle with sides x and y to find the values of trigonometric ratios. The hypotenuse of this triangle always measures 1.

$$\sin \theta = y \qquad\qquad \cos \theta = x \qquad\qquad \tan \theta = \frac{y}{x} \ (x \neq 0)$$

$$\csc \theta = \frac{1}{y} \ (y \neq 0) \qquad \sec \theta = \frac{1}{x} \ (x \neq 0) \qquad \cot \theta = \frac{x}{y} \ (y \neq 0)$$

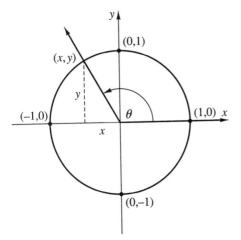

Find the sine, cosine, and tangent of angle θ in the unit circle above. Use the coordinates of the point $\left(x, \dfrac{1}{3}\right)$ on the circle that describes the angle.

We know that $\sin \theta = y$. In this example, $\sin \theta = \dfrac{1}{3}$. To find the cosine, use the trigonometric identity $\sin^2 \theta + \cos^2 \theta = 1$.

$$\left(\frac{1}{3}\right)^2 + \cos^2 \theta = 1$$

$$\frac{1}{9} + \cos^2 \theta = 1$$

$$\cos^2 \theta = \frac{8}{9}$$

$$\cos = \pm \sqrt{\frac{8}{9}}$$

$$\cos = \pm \sqrt{\frac{4 \times 2}{9}} = \pm \frac{2}{3}\sqrt{2}$$

$$x = -\frac{2}{3}\sqrt{2} \quad \text{(Take the negative root since } x \text{ is negative in the diagram.)}$$

$$\tan \theta = \frac{y}{x}$$

$$= \frac{\dfrac{1}{3}}{-\dfrac{2}{3}\sqrt{2}}$$

$$= \frac{1}{-2\sqrt{2}} \times \frac{\sqrt{2}}{\sqrt{2}} = -\frac{\sqrt{2}}{4}$$

Thus, $\sin \theta = \dfrac{1}{3}$, $\cos \theta = -\dfrac{2}{3}\sqrt{2}$, and $\tan \theta = -\dfrac{\sqrt{2}}{4}$.

When working with angles greater than 90° $\left(\theta > 90° \text{ or } \theta > \dfrac{\pi}{2}\right)$ it is necessary to consider whether the trigonometric ratios are positive or negative. The coordinate grid below shows the quadrants in which the ratios are positive.

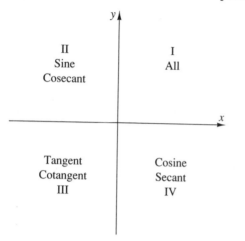

Quadrant I: All ratios are positive.

Quadrant II: Sine and cosecant are positive. All others are negative.

Quadrant III: Tangent and cotangent are positive. All others are negative.

Quadrant IV: Cosine and secant are positive. All others are negative.

A reference angle is an angle in Quadrant I whose trigonometric ratios can be used to determine those of an angle in any other quadrant $\left(\text{angles greater than } 90° \text{ or } \dfrac{\pi}{2}\right)$. The reference angle can be found by calculating the difference between the given angle and 180° (π) or 360° (2π). Sometimes a problem may ask for a trigonometric function within a particular range of angle values. For example, $\dfrac{\pi}{2} < \theta < \pi$ means to find the function when θ falls in Quadrant II $\left(\text{between } \dfrac{\pi}{2} \text{ and } \pi\right)$.

EXAMPLES

1. The reference angle for 135° is 180° − 135°, or 45°.

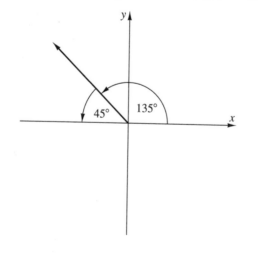

2. The reference angle for 300° is 360° − 300°, or 60°.

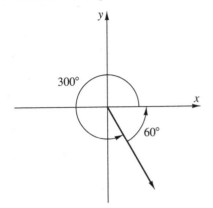

MODEL ACT PROBLEMS

1. What is the reference angle of a 210° angle?

    A. 30°
    B. 45°
    C. 60°
    D. 90°
    E. 180°

SOLUTION

The number of degrees in the reference angle is the difference between the number of degrees in the given angle and the *x*-axis.

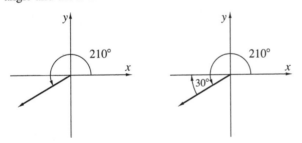

The 210° angle is 30° away from 180°. Therefore 30° is the reference angle for 210°.

The correct answer is A.

2. $\cos 120° = ?$

    F. $-\dfrac{\sqrt{3}}{2}$

    G. $-\dfrac{1}{2}$

    H. $\dfrac{1}{2}$

    J. $\dfrac{\sqrt{3}}{2}$

    K. 2

SOLUTION

Find the reference angle.

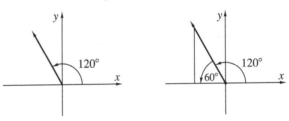

120° is 60° away from 180°. Therefore the reference angle is 60°. Find cos 60°.

This is a 30–60–90 triangle so we know that

$$\cos 60° = \frac{\text{adjacent}}{\text{hypotenuse}} = \frac{1}{2}$$

Find cos 120°.

Since 120° is in the second quadrant, the cosine of the angle is negative.

$$\cos 120° = -\frac{1}{2}$$

The correct answer is G.

Practice

What is the reference angle for each angle?

1. 125°   **2.** 315°   **3.** 425°   **4.** 156°

For 5–10, find the trigonometric ratios for angle θ in the circle below.

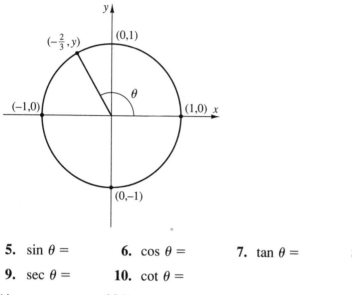

5. $\sin \theta =$ **6.** $\cos \theta =$ **7.** $\tan \theta =$ **8.** $\csc \theta =$

9. $\sec \theta =$ **10.** $\cot \theta =$

(Answers on page 394)

ACT-TYPE PROBLEMS

1. What is the reference angle of a 245° angle?

    A. 25°
    B. 35°
    C. 45°
    D. 55°
    E. 65°

2. What is the reference angle of a 396° angle?

    F. 26°
    G. 36°
    H. 46°
    J. 56°
    K. 66°

3. $\sin 300° = ?$

    A. $\sqrt{3}$

    B. $\dfrac{\sqrt{3}}{2}$

    C. $\dfrac{1}{2}$

    D. $-\dfrac{1}{2}$

    E. $-\dfrac{\sqrt{3}}{2}$

4. $\cos 225° = ?$

    F. $\dfrac{\sqrt{3}}{2}$

    G. $\dfrac{1}{2}$

    H. $\dfrac{\sqrt{2}}{3}$

    J. $-\dfrac{\sqrt{2}}{2}$

    K. $-\dfrac{\sqrt{3}}{2}$

5. If $\sin \theta = -\dfrac{3}{5}$, and $\pi < \theta < \dfrac{3\pi}{2}$, then $\tan \theta = ?$

    A. $-\dfrac{4}{5}$

    B. $\dfrac{4}{5}$

    C. $\dfrac{4}{3}$

    D. $-\dfrac{3}{4}$

    E. $\dfrac{3}{4}$

(Answers on page 394)

Graphs of Trigonometric Functions

Trigonometric functions repeat themselves. The **period** of a trigonometric function is the distance required to show one full cycle. You should be able to recognize the graphs of trigonometric functions. Those shown below are for one period of each function.

$y = \sin x$

$y = \cos x$

$y = \tan x$

$y = \csc x$

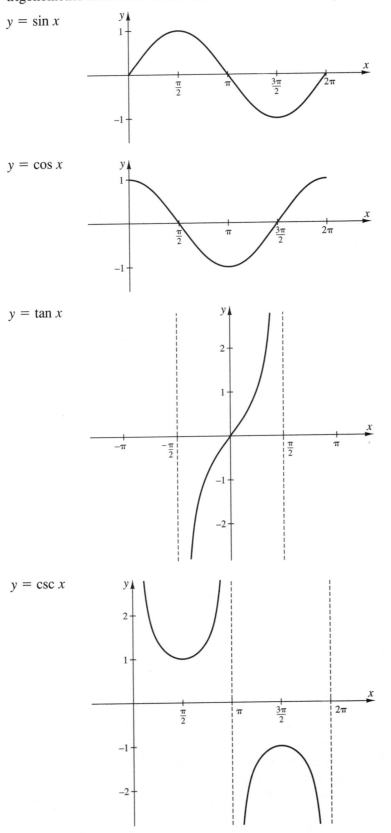

$y = \sec x$

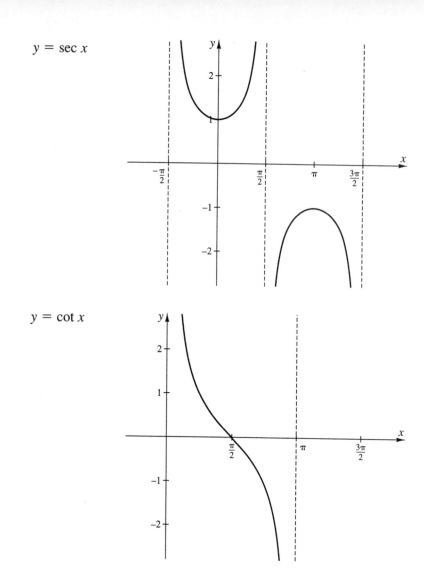

$y = \cot x$

The period of a trigonometric graph changes depending on the coefficient of x.

E X A M P L E

The function $y = \sin 2x$ has a period that is *half* as long as the period of $y = \sin x$.

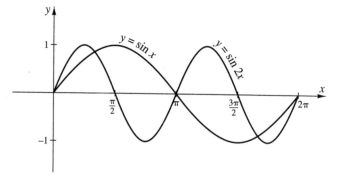

The **amplitude** of a sine or cosine curve is half the distance between the smallest and largest y-values for the function. The amplitudes of the sine and cosine graphs change depending on the coefficient of sine or cosine. The amplitude for tangent, cotangent, secant, and cosecant is undefined.

EXAMPLE

The function $y = 2\sin x$ has an amplitude that is twice the amplitude of $y = \sin x$.

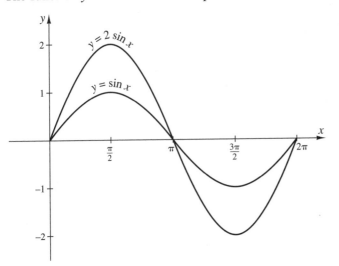

MODEL ACT PROBLEMS

1. What is the period of the graph $y = 4\tan 2x$?

A. 4π

B. 4

C. π

D. 2

E. $\dfrac{\pi}{2}$

SOLUTION

The period of $y = 4\tan 2x$ is $\dfrac{\pi}{2}$.

Sketch the graph to convince yourself that the entire graph repeats after the interval $0 \le x \le \dfrac{\pi}{2}$.

The correct answer is E.

2. Which graph below shows one period of $y = 2\sin x$?

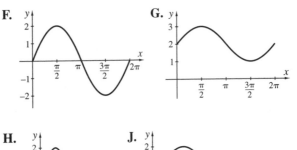

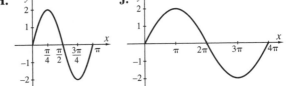

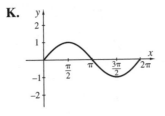

SOLUTION

The graph of $y = 2\sin x$ looks just like the graph of $y = \sin x$, except that the amplitude is 2.

The correct answer is F.

Practice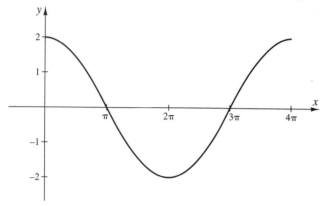

Given below is one full period of the graph of a trigonometric function.

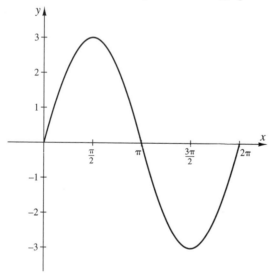

1. What is the amplitude of the graph?

2. What is the period of the graph?

3. What is the equation of the graph?

For 4–6, use the equation $y = \frac{1}{2} \csc x$.

4. What is the amplitude of the graph of this equation?

5. What is the period of the graph?

6. Draw a graph of one period of this function.

Given below is one full period of the graph of a trigonometric function.

7. What is the amplitude of the graph?

8. What is the period of the graph?

9. What is the equation of the graph?

10. Draw a graph of one period of the function $y = -\cos 2x$.

(Answers on page 395)

1. Which of the following is a graph of the function $y = \frac{1}{2} \sin 2x$?

   A.
   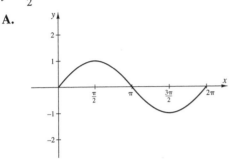

   B.
   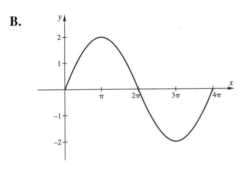

   C.
   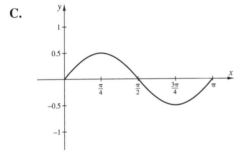

   D.
   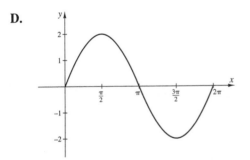

   E.
   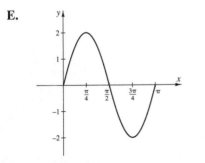

2. What is the product of the amplitude and period of the graph of the equation $y = 3\cos 2x$?

   F. 15π
   G. 6π
   H. 3π
   J. -3π
   K. -6π

3. What is the period of the graph of the equation $y = -2\cot \pi x$?

   A. $\dfrac{\pi}{2}$
   B. 1
   C. -1
   D. $-\dfrac{\pi}{2}$
   E. $-\pi$

4. Which of the following choices is the equation of the graph shown below?

   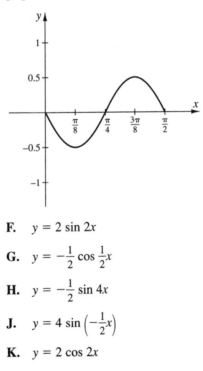

   F. $y = 2 \sin 2x$
   G. $y = -\dfrac{1}{2} \cos \dfrac{1}{2}x$
   H. $y = -\dfrac{1}{2} \sin 4x$
   J. $y = 4 \sin \left(-\dfrac{1}{2}x\right)$
   K. $y = 2 \cos 2x$

5. Which of the following is the graph of one period of $y = \tan 2x$?

A.

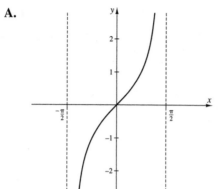

B.

C.

D.

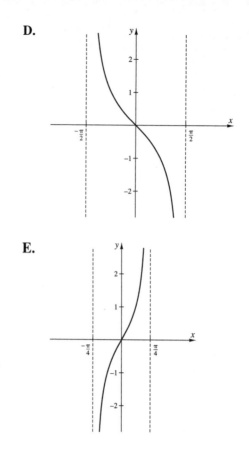

E.

(Answers on page 395)

Complete this Cumulative ACT Practice in 5 minutes to reflect real ACT test conditions. This Cumulative ACT Practice gives you an additional opportunity to practice trigonometry concepts in an ACT format. If you don't know an answer, eliminate and guess. Circle the number of any guessed answers. Then check your answers on page 395. You will also find explanations for the answers and suggestions for further study.

1. $\tan \theta = ?$

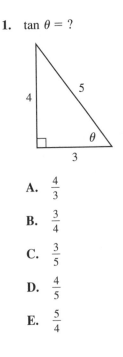

A. $\dfrac{4}{3}$

B. $\dfrac{3}{4}$

C. $\dfrac{3}{5}$

D. $\dfrac{4}{5}$

E. $\dfrac{5}{4}$

2. The graph below shows one period of which of the following trigonometric functions?

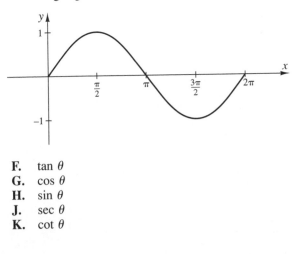

F. $\tan \theta$
G. $\cos \theta$
H. $\sin \theta$
J. $\sec \theta$
K. $\cot \theta$

3. A 140-foot rope is tied taut to the top of a flagpole and staked into the ground forming a 37° angle with the ground. Which of the following choices could be used to find the length from the base of the flagpole to where the rope is staked into the ground?

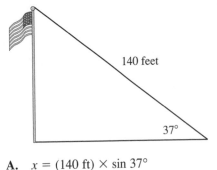

A. $x = (140 \text{ ft}) \times \sin 37°$
B. $x = (140 \text{ ft}) \times \cos 37°$
C. $x = (140 \text{ ft}) \times \sec 37°$
D. $x = (140 \text{ ft}) \times \csc 37°$
E. $x = (140 \text{ ft}) \times \tan 37°$

4. $(\cos^2 \theta + \sin^2 \theta) \times \dfrac{\sin \theta}{\cos \theta}$ is equivalent to which of the following?

F. $\sec \theta$
G. $\cot \theta$
H. $\cos \theta$
J. $\sin \theta$
K. $\tan \theta$

5. A 15-foot ladder is laid against a wall forming a 30° angle with the ground. How high up does the ladder touch the wall?

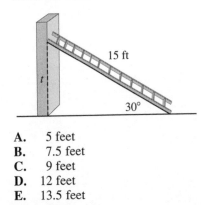

A. 5 feet
B. 7.5 feet
C. 9 feet
D. 12 feet
E. 13.5 feet

Complete this Subtest in 8 minutes to reflect real ACT test conditions. This Subtest has eight of the types of trigonometry items found on the ACT. The actual ACT will have four trigonometry items. If you don't know an answer, eliminate and guess. Circle the number of any guessed answer. Then check your answers on page 396. You will find explanations for the answers and suggestions for further study.

1. $\sin \theta = ?$

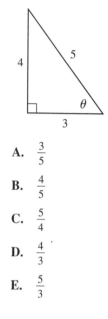

A. $\dfrac{3}{5}$

B. $\dfrac{4}{5}$

C. $\dfrac{5}{4}$

D. $\dfrac{4}{3}$

E. $\dfrac{5}{3}$

2. The graph below is one period of which trigonometric function?

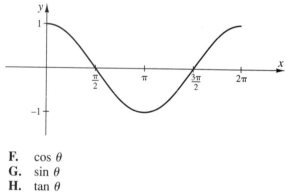

F. $\cos \theta$
G. $\sin \theta$
H. $\tan \theta$
J. $\sec \theta$
K. $\csc \theta$

3. Given that $0° \le \theta < 360°$, what is the value of θ if $\cos \theta \times \sin^2 \theta + \cos^3 \theta = 1$?

A. $\theta = 270°$
B. $\theta = 180°$
C. $\theta = 135°$
D. $\theta = \ 90°$
E. $\theta = \ \ \ 0°$

4. $(\sec^2 \theta - 1) \times \csc^2 \theta$ is equivalent to which of the following?

F. $\tan^2 \theta$
G. $\cos^2 \theta$
H. $\sec^2 \theta$
J. $\cot^2 \theta$
K. $\csc^2 \theta$

5. An airplane leaves a runway at an angle of θ. When the airplane has reached an altitude of 5,000 feet it has traveled 10,000 feet in the direction it took off. At what angle (θ) did the airplane leave the runway?

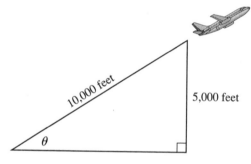

A. 15°
B. 30°
C. 45°
D. 60°
E. 75°

6. If $\sec \theta = \sqrt{2}$, then the measure of angle $\theta = ?$

F. 30°
G. 45°
H. 60°
J. $\dfrac{3\pi}{4}$
K. 2π

7. The base of a ladder is placed 5 feet from a house. If the ladder creates a 46° angle with the ground, which of the following would be used to find out how high up the wall the ladder touches the house?

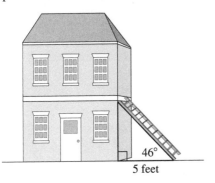

5 feet

A. 5 ft × sin 46°

B. $\dfrac{\cos 46°}{5 \text{ ft}}$

C. 5 ft × tan 46°

D. $\dfrac{5 \text{ ft}}{\sin 46°}$

E. $\dfrac{\tan 46°}{5 \text{ ft}}$

8. If $\cot \theta = x^2$, then $\sec \theta = ?$

F. $\dfrac{x^2}{\sqrt{x^4 + 1}}$

G. $\dfrac{1}{\sqrt{x^4 + 1}}$

H. $\sqrt{x^4 + 1}$

J. $\dfrac{1}{x^2}$

K. $\dfrac{\sqrt{x^4 + 1}}{x^2}$

Cumulative **ACT** Practice
Plane Geometry/Trigonometry

Complete this Cumulative ACT Practice in 16 minutes to reflect real ACT test conditions. This Cumulative ACT Practice gives you an additional opportunity to practice plane geometry and trigonometry concepts in an ACT format. If you don't know an answer, eliminate and guess. Circle the number of any guessed answer. Then check your answers on page 397. You will also find explanations for the answers and suggestions for further study.

1. Triangle *ABC* is an isosceles triangle whose base angles are *A* and *C*. *AC* = 12 inches and *BC* = 10 inches. If the height of the triangle, $\overline{BD}$, is four-fifths the length of $\overline{AB}$, what is the area of triangle *ABC*?

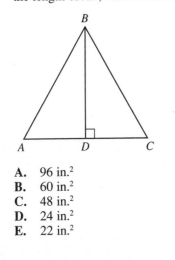

A. 96 in.²
B. 60 in.²
C. 48 in.²
D. 24 in.²
E. 22 in.²

2. Given the triangle below, what is the measure of ∠*FGE*?

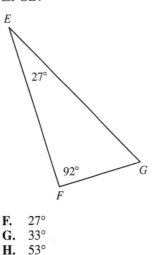

F. 27°
G. 33°
H. 53°
J. 61°
K. 90°

3. Which of the following choices is NOT a list of the lengths of the sides of a right triangle?

   A. 12, 16, 20
   B. 4, 5, 6
   C. 5, 12, 13
   D. 7, 24, 25
   E. 8, 15, 17

4. The radius of a sphere is 6 meters. What is the volume of the sphere?

   F. 12π m³
   G. 36π m³
   H. 184π m³
   J. 288π m³
   K. 366π m³

5. Which of the following postulates could be used to prove that $\triangle ABC \cong \triangle DEF$ in the figure below?

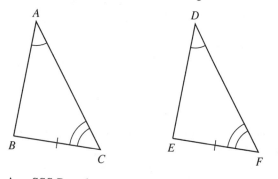

   A. SSS Postulate
   B. ASA Postulate
   C. SAS Postulate
   D. AAA Postulate
   E. SSA Postulate

6. What will the coordinates of the endpoints of $\overline{PQ}$ be if the segment is flipped across the line $y = x$?

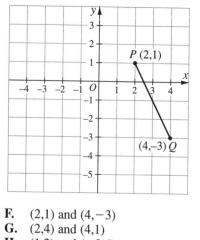

   F. (2,1) and (4,−3)
   G. (2,4) and (4,1)
   H. (1,2) and (−3,4)
   J. (2,−3) and (4,1)
   K. (4,1) and (2,−3)

7. In the figure below, $\ell \parallel m$ and both lines are cut by the transversal t. Which of the following is a complete list of all the angles congruent to $\angle 8$?

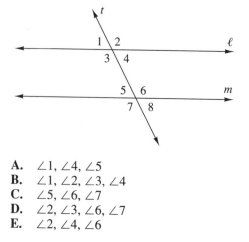

   A. $\angle 1, \angle 4, \angle 5$
   B. $\angle 1, \angle 2, \angle 3, \angle 4$
   C. $\angle 5, \angle 6, \angle 7$
   D. $\angle 2, \angle 3, \angle 6, \angle 7$
   E. $\angle 2, \angle 4, \angle 6$

8. Given that $\triangle ABC \sim \triangle DEF$, what is the area of $\triangle DEF$ if the area of $\triangle ABC$ is 36?

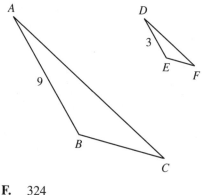

   F. 324
   G. 108
   H. 24
   J. 12
   K. 4

9. In the rectangle $ACDF$ below, $\overline{AC}$ is three times as long as $\overline{BC}$. What is the area of $ABEF$?

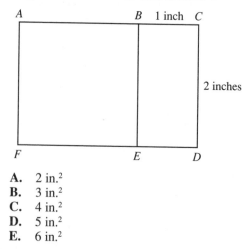

   A. 2 in.²
   B. 3 in.²
   C. 4 in.²
   D. 5 in.²
   E. 6 in.²

10. If the circumference of the circle below is 30 inches and $\angle ABC$ is a right angle, what is the length of $\overset{\frown}{AC}$?

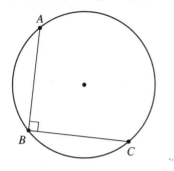

- **F.** 5 in.
- **G.** 10 in.
- **H.** 15 in.
- **J.** 20 in.
- **K.** 25 in.

11. A sphere with a radius of 3 has the same volume as a right circular cylinder with a height of 4. What is the radius of the cylinder?

- **A.** 3
- **B.** 5
- **C.** 7
- **D.** 8
- **E.** 9

12. Which of the following transformations could NOT be used to place triangle *ABC* on top of triangle *DEF*?

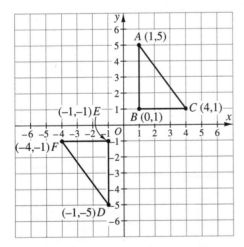

- **F.** Reflection over the *y*-axis, slide 6 units up, flip over the line $y = 3$
- **G.** Rotate 270° about the origin
- **H.** Reflection across the *y*-axis, reflection across the *x*-axis
- **J.** Rotate about the origin 180°
- **K.** Rotate about the origin −540°

13. In the graph $y = \cos\theta$, which of the following values of θ will make the function equal to 0?

- **A.** 0
- **B.** $\dfrac{\pi}{4}$
- **C.** $\dfrac{\pi}{2}$
- **D.** π
- **E.** 2π

14. If $\tan\theta = \dfrac{3}{4}$ and θ lies within the third quadrant, what will be the value of $\sin\theta$?

- **F.** $\dfrac{4}{3}$
- **G.** $\dfrac{3}{5}$
- **H.** $-\dfrac{3}{5}$
- **J.** $-\dfrac{4}{5}$
- **K.** $-\dfrac{4}{3}$

15. $\dfrac{1 - \cos^2\theta}{(\sin^2\theta)(\cot^2\theta)} = ?$

- **A.** $\cot^2\theta$
- **B.** $\sin^2\theta$
- **C.** $\sec^2\theta$
- **D.** $\cos^2\theta$
- **E.** $\tan^2\theta$

16. If $\sin 30° = \dfrac{x}{15}$, what is the value of x?

- **F.** 5
- **G.** 7.5
- **H.** 10
- **J.** 12.5
- **K.** 15

Subtest Plane Geometry/Trigonometry

Complete this Cumulative ACT Practice in 18 minutes to reflect real ACT test conditions. This Subtest has the same number and type of plane geometry/trigonometry items found on the ACT. These items make up the GT subscore on the ACT. If you don't know an answer, eliminate and guess. Circle the number of any item that you have guessed the answer. Check your answers on page 398.

1. $ABDC$ is a parallelogram. $\overline{BC}$ is $\frac{2}{3}$ the size of $\overline{AD}$. If $AE = 9$ how long is $\overline{BE}$?

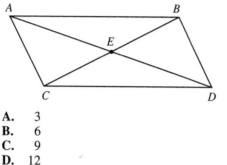

   A.   3
   B.   6
   C.   9
   D. 12
   E. 18

2. Given the figure below with $\overrightarrow{BE} \parallel \overrightarrow{AD}$, what is m$\angle CBE$?

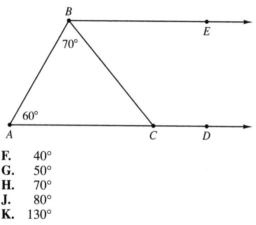

   F.   40°
   G.   50°
   H.   70°
   J.   80°
   K. 130°

3. Given circle O with diameter $\overline{AB}$ and inscribed angle ABC with measure 30°, what is the measure of arc $\overset{\frown}{BC}$?

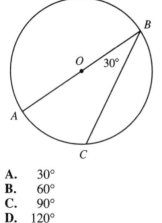

   A.   30°
   B.   60°
   C.   90°
   D. 120°
   E. 180°

4. The volume of a cone is 32π m³. If the area of the base is 16π m², what is the height of the cone?

   F.   4 m
   G.   6 m
   H.   8 m
   J.   9 m
   K. 10 m

5. Which set of information could NOT be used to prove that $\triangle ABC \cong \triangle DEF$?

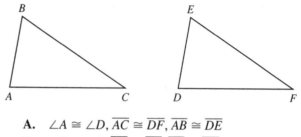

   A. $\angle A \cong \angle D, \overline{AC} \cong \overline{DF}, \overline{AB} \cong \overline{DE}$
   B. $\angle B \cong \angle E, \overline{BC} \cong \overline{EF}, \overline{AB} \cong \overline{DE}$
   C. $\overline{AC} \cong \overline{DF}, \overline{CB} \cong \overline{FE}, \overline{BA} \cong \overline{ED}$
   D. $\angle B \cong \angle E, \angle C \cong \angle F, \overline{CB} \cong \overline{EF}$
   E. $\angle A \cong \angle D, \angle B \cong \angle E, \angle C \cong \angle F$

6. Which of the following will be the vertices of triangle ABC if it is flipped across the y-axis and then the x-axis?

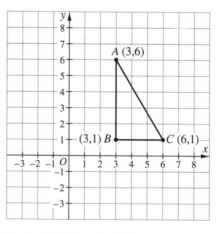

   F. $(-3,6), (3,-1), (6,1)$
   G. $(3,6), (3,1), (6,1)$
   H. $(-3,-6), (-3,-1), (-6,-1)$
   J. $(3,-6), (-3,1), (-6,-1)$
   K. $(-3,-6), (3,1), (-6,-1)$

7. What is the perimeter of triangle *ABC* below if the length of $\overline{BC}$ is $5\sqrt{3}$ centimeters?

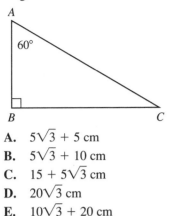

A. $5\sqrt{3} + 5$ cm
B. $5\sqrt{3} + 10$ cm
C. $15 + 5\sqrt{3}$ cm
D. $20\sqrt{3}$ cm
E. $10\sqrt{3} + 20$ cm

8. Which of the following is a complete list of all the angles that are supplementary to ∠11 given that $\ell \parallel m$ and $p \parallel q$?

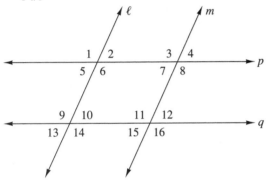

F. ∠2, ∠4, ∠5, ∠7, ∠10, ∠12, ∠13, ∠15
G. ∠9, ∠10, ∠12, ∠13, ∠14, ∠15, ∠16
H. ∠1, ∠3, ∠6, ∠8, ∠9, ∠14, ∠16
J. ∠1, ∠2, ∠3, ∠4, ∠5, ∠6, ∠7, ∠8
K. None of the angles shown are supplementary to ∠11.

9. Given that $\overline{AB}$ and $\overline{CD}$ are perpendicular bisectors of each other, which of the following are congruent triangles?

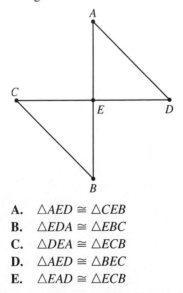

A. △AED ≅ △CEB
B. △EDA ≅ △EBC
C. △DEA ≅ △ECB
D. △AED ≅ △BEC
E. △EAD ≅ △ECB

10. Given that △*ABC* is similar to △*DEF*, what is the area of △*DEF*?

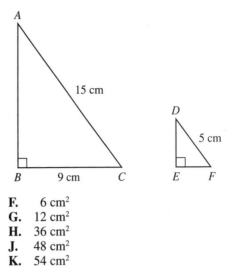

F. 6 cm²
G. 12 cm²
H. 36 cm²
J. 48 cm²
K. 54 cm²

11. *ABFG* is a rectangle, *BCDF* is a square, *BC* = 6 inches, and *E* is the midpoint of $\overline{FD}$. The length of $\overline{AB}$ is twice the length of $\overline{BC}$. What is the area of the trapezoid *ABEG*?

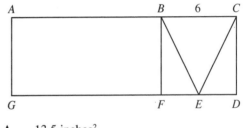

A. 13.5 inches²
B. 45 inches²
C. 72 inches²
D. 81 inches²
E. 162 inches²

12. If *D* is the midpoint of $\overline{AB}$ and *C* is the center of the circle, which of the following statements is NOT true?

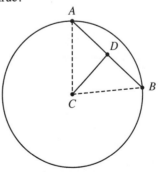

F. $CD \perp AB$
G. △ADC ≅ △BDC
H. $\overline{CA} \cong \overline{CB}$
J. m∠ACD = m∠BCD
K. ∠DCB and ∠DBC are supplementary

13. A right circular cylinder with a height of 12 meters has the same volume and radius as a sphere. What is the radius?

A. 6 meters
B. 7 meters
C. 8 meters
D. 9 meters
E. 10 meters

14. Which of the following transformations could be used to place △ABC on top of △FED?

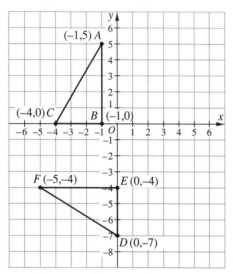

F. Vertical shift −4 and horizontal shift −1

G. Counterclockwise rotation of 90° about the origin and vertical shift −3

H. Counterclockwise rotation of 90° about the origin and vertical shift 4

J. Horizontal shift 1 and vertical shift −7

K. Counterclockwise rotation of 90° about the point (−1,0) and vertical shift −4

15. In the graph of $y = \sec \theta$, for which of the following values of θ is the function undefined?

A. $\dfrac{\pi}{4}$

B. $\dfrac{\pi}{2}$

C. $\dfrac{3\pi}{4}$

D. π

E. 2π

16. If $\sin \theta = \dfrac{5}{13}$, what are the possible values of $\cot \theta$?

F. $\dfrac{5}{12}$

G. $\dfrac{5}{12}$ and $-\dfrac{5}{12}$

H. $\dfrac{13}{12}$

J. $\dfrac{12}{5}$

K. $-\dfrac{12}{5}$ and $\dfrac{12}{5}$

17. $\dfrac{1 + \cot^2 \theta}{1 + \tan^2 \theta}$ is equivalent to which of the following trigonometric expressions?

A. $\cot^2 \theta$
B. $\tan^2 \theta$
C. $\csc^2 \theta$
D. $\sec^2 \theta$
E. $\sin^2 \theta$

18. A rope is tied to the top of a tree to keep the tree perpendicular to the ground. The length of the rope from the top of the tree to the ground is 30 feet and the rope forms a 70° angle with the ground. Which of the following choices gives the height of the tree rounded to the nearest hundredth?

(*Note:* sin 70° ≈ .9396926208, cos 70° ≈ .3420201433, tan 70° ≈ 2.747477419)

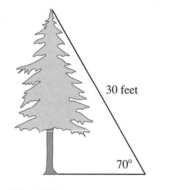

F. 30 feet
G. 28.2 feet
H. 28.19 feet
J. 10.3 feet
K. 10.26 feet

Chapter 13

Diagnostic Mathematics ACT: Model Mathematics **ACT I**

Take this Diagnostic Mathematics ACT after you complete the Mathematics review. This Diagnostic ACT is just like a real ACT, and it is the first of four Mathematics ACTs in this book. But this Diagnostic ACT is different. It is specially designed to help you decide which parts of the Mathematics section to review in more detail.

Take the Diagnostic Mathematics ACT under simulated test conditions. Allow 60 minutes to answer the 60 test questions. Use a pencil to mark the answer sheet on page 313, and answer the questions in the Test 2 (Mathematics) section.

Use the Diagnostic Mathematics Checklist on pages 323–324 to mark the answer sheet. Review the answer explanations on pages 325–333. The Checklist directs you to the Mathematics skills you should review in more detail.

The test scoring chart on page 464 shows you how to convert the number correct to an ACT scale score. The chart on pages 466–467 shows you how to find the Pre-Algebra/ Elementary Algebra (EA), Intermediate Algebra/Coordinate Geometry (AG), and Plane Geometry/Trigonometry (GT) Subscores.

DO NOT leave any answers blank. There is no penalty for guessing on the ACT. Remember that the test is yours. You may mark up, write on, or draw on the test. You may use a calculator.

When you are ready, note the time and turn to the Diagnostic Mathematics ACT. Stop in exactly 60 minutes.

ANSWER SHEET

The ACT answer sheet looks something like this one. Use a No. 2 pencil
to completely fill the circle corresponding to the correct answer.
If you erase, erase completely; incomplete erasures may be read as answers.

TEST 1—English

1 Ⓐ Ⓑ Ⓒ Ⓓ	11 Ⓐ Ⓑ Ⓒ Ⓓ	21 Ⓐ Ⓑ Ⓒ Ⓓ	31 Ⓐ Ⓑ Ⓒ Ⓓ	41 Ⓐ Ⓑ Ⓒ Ⓓ	51 Ⓐ Ⓑ Ⓒ Ⓓ	61 Ⓐ Ⓑ Ⓒ Ⓓ	71 Ⓐ Ⓑ Ⓒ Ⓓ
2 Ⓕ Ⓖ Ⓗ Ⓙ	12 Ⓕ Ⓖ Ⓗ Ⓙ	22 Ⓕ Ⓖ Ⓗ Ⓙ	32 Ⓕ Ⓖ Ⓗ Ⓙ	42 Ⓕ Ⓖ Ⓗ Ⓙ	52 Ⓕ Ⓖ Ⓗ Ⓙ	62 Ⓕ Ⓖ Ⓗ Ⓙ	72 Ⓕ Ⓖ Ⓗ Ⓙ
3 Ⓐ Ⓑ Ⓒ Ⓓ	13 Ⓐ Ⓑ Ⓒ Ⓓ	23 Ⓐ Ⓑ Ⓒ Ⓓ	33 Ⓐ Ⓑ Ⓒ Ⓓ	43 Ⓐ Ⓑ Ⓒ Ⓓ	53 Ⓐ Ⓑ Ⓒ Ⓓ	63 Ⓐ Ⓑ Ⓒ Ⓓ	73 Ⓐ Ⓑ Ⓒ Ⓓ
4 Ⓕ Ⓖ Ⓗ Ⓙ	14 Ⓕ Ⓖ Ⓗ Ⓙ	24 Ⓕ Ⓖ Ⓗ Ⓙ	34 Ⓕ Ⓖ Ⓗ Ⓙ	44 Ⓕ Ⓖ Ⓗ Ⓙ	54 Ⓕ Ⓖ Ⓗ Ⓙ	64 Ⓕ Ⓖ Ⓗ Ⓙ	74 Ⓕ Ⓖ Ⓗ Ⓙ
5 Ⓐ Ⓑ Ⓒ Ⓓ	15 Ⓐ Ⓑ Ⓒ Ⓓ	25 Ⓐ Ⓑ Ⓒ Ⓓ	35 Ⓐ Ⓑ Ⓒ Ⓓ	45 Ⓐ Ⓑ Ⓒ Ⓓ	55 Ⓐ Ⓑ Ⓒ Ⓓ	65 Ⓐ Ⓑ Ⓒ Ⓓ	75 Ⓐ Ⓑ Ⓒ Ⓓ
6 Ⓕ Ⓖ Ⓗ Ⓙ	16 Ⓕ Ⓖ Ⓗ Ⓙ	26 Ⓕ Ⓖ Ⓗ Ⓙ	36 Ⓕ Ⓖ Ⓗ Ⓙ	46 Ⓕ Ⓖ Ⓗ Ⓙ	56 Ⓕ Ⓖ Ⓗ Ⓙ	66 Ⓕ Ⓖ Ⓗ Ⓙ	
7 Ⓐ Ⓑ Ⓒ Ⓓ	17 Ⓐ Ⓑ Ⓒ Ⓓ	27 Ⓐ Ⓑ Ⓒ Ⓓ	37 Ⓐ Ⓑ Ⓒ Ⓓ	47 Ⓐ Ⓑ Ⓒ Ⓓ	57 Ⓐ Ⓑ Ⓒ Ⓓ	67 Ⓐ Ⓑ Ⓒ Ⓓ	
8 Ⓕ Ⓖ Ⓗ Ⓙ	18 Ⓕ Ⓖ Ⓗ Ⓙ	28 Ⓕ Ⓖ Ⓗ Ⓙ	38 Ⓕ Ⓖ Ⓗ Ⓙ	48 Ⓕ Ⓖ Ⓗ Ⓙ	58 Ⓕ Ⓖ Ⓗ Ⓙ	68 Ⓕ Ⓖ Ⓗ Ⓙ	
9 Ⓐ Ⓑ Ⓒ Ⓓ	19 Ⓐ Ⓑ Ⓒ Ⓓ	29 Ⓐ Ⓑ Ⓒ Ⓓ	39 Ⓐ Ⓑ Ⓒ Ⓓ	49 Ⓐ Ⓑ Ⓒ Ⓓ	59 Ⓐ Ⓑ Ⓒ Ⓓ	69 Ⓐ Ⓑ Ⓒ Ⓓ	
10 Ⓕ Ⓖ Ⓗ Ⓙ	20 Ⓕ Ⓖ Ⓗ Ⓙ	30 Ⓕ Ⓖ Ⓗ Ⓙ	40 Ⓕ Ⓖ Ⓗ Ⓙ	50 Ⓕ Ⓖ Ⓗ Ⓙ	60 Ⓕ Ⓖ Ⓗ Ⓙ	70 Ⓕ Ⓖ Ⓗ Ⓙ	

TEST 2—Mathematics

1 Ⓐ Ⓑ Ⓒ Ⓓ Ⓔ	9 Ⓐ Ⓑ Ⓒ Ⓓ Ⓔ	17 Ⓐ Ⓑ Ⓒ Ⓓ Ⓔ	25 Ⓐ Ⓑ Ⓒ Ⓓ Ⓔ	33 Ⓐ Ⓑ Ⓒ Ⓓ Ⓔ	41 Ⓐ Ⓑ Ⓒ Ⓓ Ⓔ	49 Ⓐ Ⓑ Ⓒ Ⓓ Ⓔ	57 Ⓐ Ⓑ Ⓒ Ⓓ Ⓔ
2 Ⓕ Ⓖ Ⓗ Ⓙ Ⓚ	10 Ⓕ Ⓖ Ⓗ Ⓙ Ⓚ	18 Ⓕ Ⓖ Ⓗ Ⓙ Ⓚ	26 Ⓕ Ⓖ Ⓗ Ⓙ Ⓚ	34 Ⓕ Ⓖ Ⓗ Ⓙ Ⓚ	42 Ⓕ Ⓖ Ⓗ Ⓙ Ⓚ	50 Ⓕ Ⓖ Ⓗ Ⓙ Ⓚ	58 Ⓕ Ⓖ Ⓗ Ⓙ Ⓚ
3 Ⓐ Ⓑ Ⓒ Ⓓ Ⓔ	11 Ⓐ Ⓑ Ⓒ Ⓓ Ⓔ	19 Ⓐ Ⓑ Ⓒ Ⓓ Ⓔ	27 Ⓐ Ⓑ Ⓒ Ⓓ Ⓔ	35 Ⓐ Ⓑ Ⓒ Ⓓ Ⓔ	43 Ⓐ Ⓑ Ⓒ Ⓓ Ⓔ	51 Ⓐ Ⓑ Ⓒ Ⓓ Ⓔ	59 Ⓐ Ⓑ Ⓒ Ⓓ Ⓔ
4 Ⓕ Ⓖ Ⓗ Ⓙ Ⓚ	12 Ⓕ Ⓖ Ⓗ Ⓙ Ⓚ	20 Ⓕ Ⓖ Ⓗ Ⓙ Ⓚ	28 Ⓕ Ⓖ Ⓗ Ⓙ Ⓚ	36 Ⓕ Ⓖ Ⓗ Ⓙ Ⓚ	44 Ⓕ Ⓖ Ⓗ Ⓙ Ⓚ	52 Ⓕ Ⓖ Ⓗ Ⓙ Ⓚ	60 Ⓕ Ⓖ Ⓗ Ⓙ Ⓚ
5 Ⓐ Ⓑ Ⓒ Ⓓ Ⓔ	13 Ⓐ Ⓑ Ⓒ Ⓓ Ⓔ	21 Ⓐ Ⓑ Ⓒ Ⓓ Ⓔ	29 Ⓐ Ⓑ Ⓒ Ⓓ Ⓔ	37 Ⓐ Ⓑ Ⓒ Ⓓ Ⓔ	45 Ⓐ Ⓑ Ⓒ Ⓓ Ⓔ	53 Ⓐ Ⓑ Ⓒ Ⓓ Ⓔ	
6 Ⓕ Ⓖ Ⓗ Ⓙ Ⓚ	14 Ⓕ Ⓖ Ⓗ Ⓙ Ⓚ	22 Ⓕ Ⓖ Ⓗ Ⓙ Ⓚ	30 Ⓕ Ⓖ Ⓗ Ⓙ Ⓚ	38 Ⓕ Ⓖ Ⓗ Ⓙ Ⓚ	46 Ⓕ Ⓖ Ⓗ Ⓙ Ⓚ	54 Ⓕ Ⓖ Ⓗ Ⓙ Ⓚ	
7 Ⓐ Ⓑ Ⓒ Ⓓ Ⓔ	15 Ⓐ Ⓑ Ⓒ Ⓓ Ⓔ	23 Ⓐ Ⓑ Ⓒ Ⓓ Ⓔ	31 Ⓐ Ⓑ Ⓒ Ⓓ Ⓔ	39 Ⓐ Ⓑ Ⓒ Ⓓ Ⓔ	47 Ⓐ Ⓑ Ⓒ Ⓓ Ⓔ	55 Ⓐ Ⓑ Ⓒ Ⓓ Ⓔ	
8 Ⓕ Ⓖ Ⓗ Ⓙ Ⓚ	16 Ⓕ Ⓖ Ⓗ Ⓙ Ⓚ	24 Ⓕ Ⓖ Ⓗ Ⓙ Ⓚ	32 Ⓕ Ⓖ Ⓗ Ⓙ Ⓚ	40 Ⓕ Ⓖ Ⓗ Ⓙ Ⓚ	48 Ⓕ Ⓖ Ⓗ Ⓙ Ⓚ	56 Ⓕ Ⓖ Ⓗ Ⓙ Ⓚ	

TEST 3—Reading

1 Ⓐ Ⓑ Ⓒ Ⓓ	6 Ⓕ Ⓖ Ⓗ Ⓙ	11 Ⓐ Ⓑ Ⓒ Ⓓ	16 Ⓕ Ⓖ Ⓗ Ⓙ	21 Ⓐ Ⓑ Ⓒ Ⓓ	26 Ⓕ Ⓖ Ⓗ Ⓙ	31 Ⓐ Ⓑ Ⓒ Ⓓ	36 Ⓕ Ⓖ Ⓗ Ⓙ
2 Ⓕ Ⓖ Ⓗ Ⓙ	7 Ⓐ Ⓑ Ⓒ Ⓓ	12 Ⓕ Ⓖ Ⓗ Ⓙ	17 Ⓐ Ⓑ Ⓒ Ⓓ	22 Ⓕ Ⓖ Ⓗ Ⓙ	27 Ⓐ Ⓑ Ⓒ Ⓓ	32 Ⓕ Ⓖ Ⓗ Ⓙ	37 Ⓐ Ⓑ Ⓒ Ⓓ
3 Ⓐ Ⓑ Ⓒ Ⓓ	8 Ⓕ Ⓖ Ⓗ Ⓙ	13 Ⓐ Ⓑ Ⓒ Ⓓ	18 Ⓕ Ⓖ Ⓗ Ⓙ	23 Ⓐ Ⓑ Ⓒ Ⓓ	28 Ⓕ Ⓖ Ⓗ Ⓙ	33 Ⓐ Ⓑ Ⓒ Ⓓ	38 Ⓕ Ⓖ Ⓗ Ⓙ
4 Ⓕ Ⓖ Ⓗ Ⓙ	9 Ⓐ Ⓑ Ⓒ Ⓓ	14 Ⓕ Ⓖ Ⓗ Ⓙ	19 Ⓐ Ⓑ Ⓒ Ⓓ	24 Ⓕ Ⓖ Ⓗ Ⓙ	29 Ⓐ Ⓑ Ⓒ Ⓓ	34 Ⓕ Ⓖ Ⓗ Ⓙ	39 Ⓐ Ⓑ Ⓒ Ⓓ
5 Ⓐ Ⓑ Ⓒ Ⓓ	10 Ⓕ Ⓖ Ⓗ Ⓙ	15 Ⓐ Ⓑ Ⓒ Ⓓ	20 Ⓕ Ⓖ Ⓗ Ⓙ	25 Ⓐ Ⓑ Ⓒ Ⓓ	30 Ⓕ Ⓖ Ⓗ Ⓙ	35 Ⓐ Ⓑ Ⓒ Ⓓ	40 Ⓕ Ⓖ Ⓗ Ⓙ

TEST 4—Science Reasoning

1 Ⓐ Ⓑ Ⓒ Ⓓ	6 Ⓕ Ⓖ Ⓗ Ⓙ	11 Ⓐ Ⓑ Ⓒ Ⓓ	16 Ⓕ Ⓖ Ⓗ Ⓙ	21 Ⓐ Ⓑ Ⓒ Ⓓ	26 Ⓕ Ⓖ Ⓗ Ⓙ	31 Ⓐ Ⓑ Ⓒ Ⓓ	36 Ⓕ Ⓖ Ⓗ Ⓙ
2 Ⓕ Ⓖ Ⓗ Ⓙ	7 Ⓐ Ⓑ Ⓒ Ⓓ	12 Ⓕ Ⓖ Ⓗ Ⓙ	17 Ⓐ Ⓑ Ⓒ Ⓓ	22 Ⓕ Ⓖ Ⓗ Ⓙ	27 Ⓐ Ⓑ Ⓒ Ⓓ	32 Ⓕ Ⓖ Ⓗ Ⓙ	37 Ⓐ Ⓑ Ⓒ Ⓓ
3 Ⓐ Ⓑ Ⓒ Ⓓ	8 Ⓕ Ⓖ Ⓗ Ⓙ	13 Ⓐ Ⓑ Ⓒ Ⓓ	18 Ⓕ Ⓖ Ⓗ Ⓙ	23 Ⓐ Ⓑ Ⓒ Ⓓ	28 Ⓕ Ⓖ Ⓗ Ⓙ	33 Ⓐ Ⓑ Ⓒ Ⓓ	38 Ⓕ Ⓖ Ⓗ Ⓙ
4 Ⓕ Ⓖ Ⓗ Ⓙ	9 Ⓐ Ⓑ Ⓒ Ⓓ	14 Ⓕ Ⓖ Ⓗ Ⓙ	19 Ⓐ Ⓑ Ⓒ Ⓓ	24 Ⓕ Ⓖ Ⓗ Ⓙ	29 Ⓐ Ⓑ Ⓒ Ⓓ	34 Ⓕ Ⓖ Ⓗ Ⓙ	39 Ⓐ Ⓑ Ⓒ Ⓓ
5 Ⓐ Ⓑ Ⓒ Ⓓ	10 Ⓕ Ⓖ Ⓗ Ⓙ	15 Ⓐ Ⓑ Ⓒ Ⓓ	20 Ⓕ Ⓖ Ⓗ Ⓙ	25 Ⓐ Ⓑ Ⓒ Ⓓ	30 Ⓕ Ⓖ Ⓗ Ⓙ	35 Ⓐ Ⓑ Ⓒ Ⓓ	40 Ⓕ Ⓖ Ⓗ Ⓙ

Diagnostic Mathematics **ACT**

60 Questions – 60 Minutes

INSTRUCTIONS: Find the solution to each problem, choose the correct answer choice, then darken the appropriate oval on your answer sheet.

Do not spend too much time on any one problem. Answer as many problems as you can easily and then work on the remaining problems within the time limit for this test.

Unless the problem indicates otherwise:

- figures are NOT necessarily drawn to scale
- geometric figures are plane figures
- a *line* is a straight line
- an *average* is the arithmetic mean

1. In $\triangle XYZ$, $\angle Y$ is obtuse. Which of the following statements is true about the measures of $\angle X$ and $\angle Z$?

   A. The sum of the measures of the two angles is equal to 180°.

   B. One and only one of the angles is obtuse.

   C. The sum of the measures of the two angles is less than 90°.

   D. One and only one of the angles is acute.

   E. The sum of the measures of the two angles is equal to 90°.

2. The Matos family decided to go on a 3-day trip. The first day they drove 125.5 miles, the second day they drove $80\frac{2}{5}$ miles, and on the third day they traveled $95\frac{3}{4}$ miles. What was the total distance that they traveled during the 3 days?

   F. 30,165 miles
   G. 3,016.5 miles
   H. 301.65 miles
   J. 300.165 miles
   K. 3.0165 miles

3. Three different items (a, b, and c) are purchased at a store. The sales tax on these items can be found by multiplying the price of each item by 5% and adding the results. Which of the following equations represents the total cost of all three items including tax?

   A. $0.5(a + b + c)$
   B. $1.5a + 1.5b + 1.5c$
   C. $2a + 2b + 2c$
   D. $a + b + c$
   E. $1.05a + 1.05b + 1.05c$

4. Which of the following is equivalent to $(2x - 6y) + (8x - 5y)$?

   F. $6x + y$
   G. $10x - 11y$
   H. $7x + 2y$
   J. $8x - 5y$
   K. $2x + 6y$

5. Which of the following sets of numbers contains all prime numbers?

   A. $\{1, 3, 5, 7, 9, 11\}$
   B. $\{1, 3, 5, 8, 10, 11\}$
   C. $\{2, 5, 7, 13, 17\}$
   D. $\{1, 2, 3, 5, 7, 9\}$
   E. $\{2, 7, 9, 11, 13\}$

GO ON TO THE NEXT PAGE.

6. In the diagram below, what is the length of the line segment joining the two points located in the standard coordinate plane?

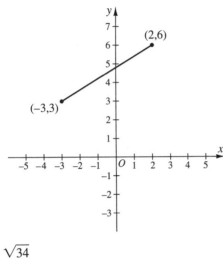

F. $\sqrt{34}$
G. 8
H. $2\sqrt{6}$
J. 5
K. $\sqrt{82}$

7. In the figure below, $\triangle ABC$ is an equilateral triangle. What is the degree measure of $\angle 1$?

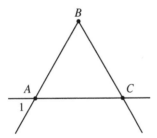

A. 120°
B. 100°
C. 90°
D. 60°
E. 30°

8. An athlete spends 2 hours (120 minutes) exercising each day. Of that time, the athlete spends 20 minutes stretching and warming up, 50 minutes jogging, and 35 minutes lifting weights. If the athlete spends the rest of the exercise time cooling down, what percent of the exercise time is devoted to cooling down?

F. 20%
G. 17.5%
H. 15%
J. 13%
K. 12.5%

9. $4\frac{1}{4} + 5\frac{1}{3} = ?$

A. $1\frac{1}{2}$
B. 9
C. $9\frac{2}{7}$
D. $9\frac{7}{12}$
E. $12\frac{3}{4}$

10. Simplify the expression
$(3a^2 + b^2 - 1) - (a^2 - b^2 - 1)$.

F. $2a^2$
G. $a^2 + 2b^2$
H. $2a^2 + b^2 + 1$
J. $a^2 + 2b^2 - 1$
K. $2(a^2 + b^2)$

11. Ben, Jim, Alice, Erin, and Rick are friends. Ben is 20 years old, Jim is 17 years old, Alice is 21 years old, and Erin is 19 years old. The average age of the five friends is 20. How old is Rick?

A. 19
B. 20
C. 21
D. 22
E. 23

12. Which of the following linear equations has a slope of $-\frac{5}{3}$?

F. $3x + 4y - 5 = 0$
G. $5x + 3y + 2 = 0$
H. $2x - 5y = -3$
J. $6y = 2x + 1$
K. $3x + 5y + 4 = 0$

13. The intersection of which of the following always forms a point?

A. Two separate planes
B. The circumference of a circle and a plane
C. Two non-parallel lines in a plane
D. The surface of a cube and the circumference of a circle
E. The circumferences of two circles

14. The expression $(w + 1)^3 + 3$ is equivalent to

F. $w^2 - 2w + 3$
G. $w^3 + 3w^2 + 3w + 4$
H. $w^3 - w^2 + 4$
J. $w^2 + 4w + 4$
K. $w^3 - 3w^2 + 4$

GO ON TO THE NEXT PAGE.

15. A wheel with a circumference of 1 meter rolls over a flat road for 20 kilometers. Then the same wheel rolls over a flat road for 15 kilometers and uphill for 5 kilometers. How far did the wheel roll in all? (1 kilometer = 1,000 meters)

   A. $40\pi d$ kilometers

   B. 40,000 meters

   C. $35 + \sqrt{29}$ meters

   D. 42 kilometers

   E. 38 kilometers

16. $4\sqrt[3]{135} - \sqrt[3]{320} = ?$

   F. $\sqrt[3]{4}$

   G. $-4\sqrt[3]{5}$

   H. $3\sqrt[3]{4}$

   J. $8\sqrt[3]{5}$

   K. $12\sqrt[3]{5}$

17. In $\triangle ABC$ and $\triangle XYZ$ shown below, $\angle A \cong \angle X$ and $\angle C \cong \angle Z$. The length of side $YZ = 3$ and the length of side $XY = 8$. If the length of side $CB = 14$, what is the measure of side AB?

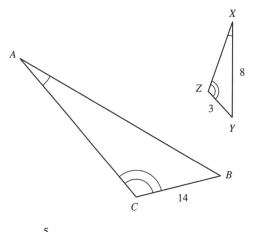

   A. $1\frac{5}{7}$

   B. $5\frac{1}{4}$

   C. 28

   D. $37\frac{1}{3}$

   E. 42

18. If a right triangle has legs of length 4 and 6, what is the length of the hypotenuse?

   F. $2\sqrt{13}$

   G. 10

   H. $4\sqrt{7}$

   J. 13

   K. 26

19. Factor $4x^3 + 32$.

   A. $4(x - 4)(x + 4)$

   B. $(x - 4)(x^2 + 2x - 8)$

   C. $(x - 8)(x + 8)$

   D. $4(x + 2)(x^2 - 2x + 4)$

   E. $(x - 2)^2(x + 2)$

20. Mary rides her bike a distance s, and Victor rides his bike $\frac{1}{2}$ Mary's distance. Which of the following represents the total distance that Mary and Victor rode?

   F. $\dfrac{3s}{2}$

   G. $\dfrac{s}{2}$

   H. $2s$

   J. $\dfrac{2s}{2}$

   K. s^2

21. Which one of the following number lines represents the solutions for the variable a in the inequality

$$7 + 5a(2 - a) > -5a^2 + a\left(12 + \frac{11}{a}\right)?$$

   A. (number line, closed dot at -2, shaded to the right)

   B. (number line, open dot at -1, shaded to the left)

   C. (number line, open dot at -1, closed dot at 1, shaded between)

   D. (number line, open dot at -2, shaded to the left)

   E. (number line, closed dot at 1, shaded to the right)

22. $\sqrt{4}\left(\sqrt{4} + \sqrt{2}\right) - \left(\sqrt{5}\right)\left(\sqrt{20}\right) = ?$

   F. $16 + 3\sqrt{4}$

   G. $2 - 5\sqrt{3}$

   H. $2\sqrt{2} - 6$

   J. $\sqrt{2} - 3$

   K. $3\sqrt{4} + 5$

GO ON TO THE NEXT PAGE.

23. In the right triangle *XYZ* below, side *XZ* is $\sqrt{18}$ units long. What is the length of side *YZ*?

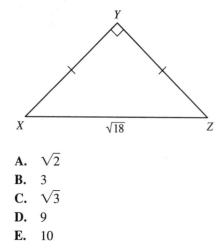

A. $\sqrt{2}$

B. 3

C. $\sqrt{3}$

D. 9

E. 10

24. What is the value of *x* in the equation $3\frac{1}{4} + x = \frac{x}{2} + \frac{3}{8}$?

F. -4

G. $4\frac{1}{4}$

H. -5

J. $-5\frac{3}{4}$

K. 6

25. What is the slope of the line below?

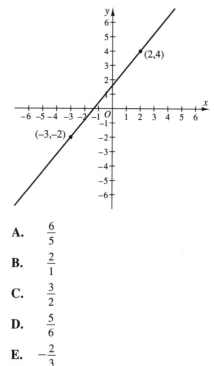

A. $\frac{6}{5}$

B. $\frac{2}{1}$

C. $\frac{3}{2}$

D. $\frac{5}{6}$

E. $-\frac{2}{3}$

26. A line segment in a standard coordinate plane with endpoints at $(-3,2)$ and $(7,-12)$ is reflected across the *x*-axis and then across the *y*-axis. What are the coordinates of one of the endpoints of the new line segment?

F. $(-3,2)$

G. $(-3,-2)$

H. $(-7,12)$

J. $(-7,-12)$

K. $(3,2)$

27. In the diagram below, what is the midpoint of segment *RT*, which is located in a standard coordinate plane?

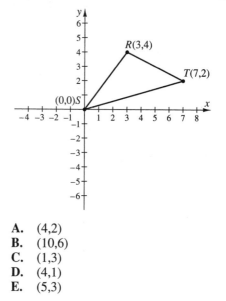

A. $(4,2)$

B. $(10,6)$

C. $(1,3)$

D. $(4,1)$

E. $(5,3)$

28. If two lines are parallel, you can be sure that:

F. The slopes of the two lines are inverses of each other.

G. The *y*-intercepts of the two lines are inverses of each other.

H. The slopes of the two lines are equal to each other.

J. The *y*-intercepts of the two lines are equal to each other.

K. The slope of one line is the negative value of the slope of the other.

29. Seven less than the product of 4 and a certain number *x* is equal to 3 more than the product of 2 and the same number *x*. Which of the following equations shows this relationship?

A. $7 - 4x = 2x - 3$

B. $7x - 4 = 3x + 2$

C. $4x - 7 = 3 + 2x$

D. $4x + 7 = 3 - 2x$

E. $7 - 4x = 3 - 2x$

GO ON TO THE NEXT PAGE.

30. The area of a certain triangle with a base of 8 units is 16 square units. A rectangle has an area equal to the area of that triangle. If the length of the rectangle is twice as long as the height of the triangle, what is the width of the rectangle?

F. 2
G. 3
H. 4
J. 5
K. 6

31. The area of a triangle is 12 square units and the height is 6 units. What is the length of the base of the triangle?

A. 2
B. 3
C. 4
D. 5
E. 6

32. The measure of the inscribed angle ACB in the circle O below is 90°. If the circle has a radius of 5 units, what is the length of arc $\overset{\frown}{AB}$?

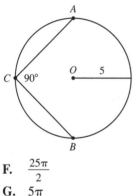

F. $\dfrac{25\pi}{2}$
G. 5π
H. 180°
J. 90°
K. 5

33. If the area of a circle is 25π square units, what is the radius of the circle?

A. 5 units
B. 12.5 units
C. 15 units
D. 23 units
E. 25 units

34. A sphere $(V = \frac{4}{3}\pi r^3)$ has a radius of 3. A right circular cylinder $(V = \pi r^2 h)$ has a radius of 6. What is the height of the cylinder if the cylinder has the same volume as the sphere?

F. 5
G. 4
H. 3
J. 2
K. 1

35. Which of the following is NOT a rational number?

A. $\sqrt{\dfrac{1}{49}}$
B. $2\dfrac{1}{3}$
C. $\sqrt{27}$
D. $\sqrt{324}$
E. 0.6

36. Given the equation $|-2 - x| = 6$, which of the following could NOT be true?

F. $-2 - x = 8$
G. $x = -8$
H. $x = 4$
J. $-2 - x = 6$
K. $-2 - x = -6$

37. $\dfrac{3}{2 + \sqrt{3}} = ?$

A. $\dfrac{\sqrt{3} - 1}{\sqrt{3}}$
B. $\dfrac{3}{2 - \sqrt{3}}$
C. $2 + \sqrt{3}$
D. $6 - 3\sqrt{3}$
E. $2 - 3\sqrt{3}$

38. Let $E = IR$ be the formula for voltage in a circuit, where $E =$ volts, $I =$ amperes, and $R =$ ohms. If a circuit has 3 amperes and 12 volts, how many ohms are in the circuit?

F. 3
G. 4
H. 9
J. 12
K. 16

39. If $\cos \theta = \dfrac{\sqrt{2}}{2}$, what are the possible values of $\sin \theta$?

I. $\dfrac{\sqrt{2}}{2}$
II. $-\dfrac{\sqrt{2}}{2}$
III. 1

A. I only
B. III only
C. I and II
D. II only
E. I and III

GO ON TO THE NEXT PAGE.

40. Which of the following answers gives all the solutions for x in the equation $x^2 - 4 < 0$?

   F. $x > -2$
   G. $x < 2$
   H. $-4 < x < 4$
   J. $2 < x < -2$
   K. $-2 < x < 2$

41. Which of the following expressions is equivalent to $(-b)^{-2} - a^2$?

   A. $b^2 - a^2$

   B. $\dfrac{1 - a^2}{b^2}$

   C. $\dfrac{b^2 - a}{b^2}$

   D. $\dfrac{1 - a^2b^2}{b^2}$

   E. $\dfrac{b^2 - a^2b}{b^2}$

42. For what real values of t is $\dfrac{t^2(3 + t)}{(t^2 - 4)}$ defined?

   F. All values of t except -3
   G. All values of t except 2
   H. All values of t except 2 and -2
   J. All values of t except 3 and -2
   K. All values of t except -2

43. A line segment AB, with a midpoint at $(3,5)$, is located in the standard (x,y) coordinate plane. If point A is $(2,7)$, what are the coordinates of point B?

   A. $(4,3)$
   B. $(1,-2)$
   C. $(3,4)$
   D. $(-2,7)$
   E. $(7,3)$

44. The graph below shows one period of which of the following trigonometric functions?

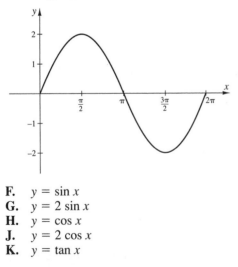

   F. $y = \sin x$
   G. $y = 2 \sin x$
   H. $y = \cos x$
   J. $y = 2 \cos x$
   K. $y = \tan x$

45. In rectangle $WXYZ$ shown below, $\overline{WX}$ is 8 units long. U is the midpoint of $\overline{WX}$. How long is $\overline{ZU}$ if the measure of $\angle ZUW$ is $45°$?

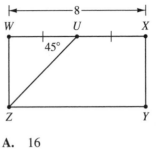

   A. 16
   B. 8
   C. $4\sqrt{3}$
   D. $4\sqrt{2}$
   E. 4

46. $27^{\frac{1}{3}} = ?$

   F. 3
   G. $2\sqrt{3}$
   H. $\sqrt[3]{(27)^2}$
   J. 18
   K. 27

47. For all positive values of a and b, the expression $\dfrac{2\sqrt{a} - 3\sqrt{b}}{\sqrt{a} + \sqrt{b}}$ simplifies to:

   A. $\dfrac{3a + 2b}{a - b}$

   B. $\dfrac{2a - 5\sqrt{ab}}{a + b}$

   C. $\dfrac{2a - 5\sqrt{ab} + 3b}{a - b}$

   D. $\dfrac{2b - 3\sqrt{ab} + 5a}{a - b}$

   E. $\dfrac{3ab - 2\sqrt{a} + 5b}{a + b}$

48. A town planner designs a town based on the standard (x,y) plane. How long is a street that runs from coordinate $(5,18)$ to coordinate $(21,6)$?

   F. 72
   G. 20
   H. $\sqrt{216}$
   J. $\sqrt{12}$
   K. 3

GO ON TO THE NEXT PAGE.

49. Which of the following is equivalent to $\dfrac{\frac{2}{a} + 1}{4 + \frac{5}{a}}$?

    A. $\dfrac{4a^2 + 13a + 10}{a^2}$

    B. $\dfrac{2 + a}{4a + 5}$

    C. $\dfrac{2}{9}$

    D. $\dfrac{2 + a}{4a}$

    E. $\dfrac{a + 4}{2}$

50. What is $\csc \theta$ if $\sin \theta = \dfrac{3}{5}$?

    F. $\dfrac{5}{3}$

    G. $\dfrac{12}{13}$

    H. $\dfrac{3}{4}$

    J. $\dfrac{5}{12}$

    K. $\dfrac{9}{25}$

51. $4^{-2} + 3^2 - 5^0 = ?$

    A. $8\dfrac{1}{16}$

    B. $9\dfrac{1}{2}$

    C. 11

    D. 25

    E. 26

52. You have two routes to drive. If you drive the first route twice and the second route five times, you travel nine miles. If you drive the first route twice and the second route four times you travel eight miles. What is the sum of the lengths of the two routes?

    F. 2
    G. 3
    H. 4
    J. 5
    K. 6

53. What is the circumference of a circle whose radius is 4 cm long?

    A. 16π m
    B. 8π cm
    C. 6π cm
    D. 4π cm
    E. 2π cm

54. Which of the following graphs, located in the standard coordinate plane, shows the points found on the graph of the equation $y = 2 - \sqrt{x}$ for $x = 4$ and $x = 9$?

    F.

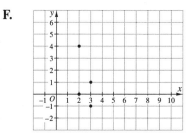

    G.

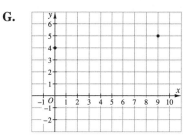

    H.

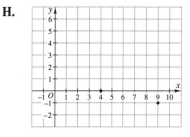

    J.

    K.

GO ON TO THE NEXT PAGE.

55. Given that $\triangle XYZ$ is similar to $\triangle ABC$, what is the sum of the lengths of sides $\overline{XY}$ and $\overline{XZ}$ in $\triangle XYZ$ below?

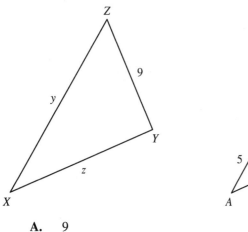

- **A.** 9
- **B.** 12
- **C.** 15
- **D.** 24
- **E.** 27

56. A clay sphere has a volume of 36π cm^3 ($V = \frac{4}{3}\pi r^3$). If someone cuts straight through the sphere, the cross section is a circle. What is the area of the largest circular cross section?

- **F.** 48π cm^2
- **G.** 36π cm^2
- **H.** 9π cm^2
- **J.** 6π cm^2
- **K.** 4π cm^2

57. If the roots of an equation are 4 and $-\frac{1}{2}$, which of the following could be the equation?

- I. $12x^2 - 12x - 4 = 30x + 20$
- II. $(x - 4)(2x + 4) = 0$
- III. $(2x - 8)(6x + 3) = 0$

- **A.** II only
- **B.** I and III
- **C.** I, II, and III
- **D.** III only
- **E.** I and II

58. If $|3x + 7| \geq 5$, which of the following is the largest negative number included in the solution set?

- **F.** -3
- **G.** -4
- **H.** -5
- **J.** -7
- **K.** -12

59. $\dfrac{\frac{1}{\csc \theta}}{\cos^2 \theta + \sin^2 \theta} = ?$

- **A.** $\cos \theta$
- **B.** $\sin \theta$
- **C.** $\tan \theta$
- **D.** $\cot \theta$
- **E.** $\csc \theta$

60. Find the lowest value of $0.4p$ that meets the requirement for p and q in all of the inequalities shown below.

$3 \leq q \leq 6$
$p \leq 8$
$pq \geq 15$

- **F.** 1
- **G.** 2.5
- **H.** 3
- **J.** 5
- **K.** 6

END OF TEST 2

Diagnostic Mathematics Checklist

Answer	Check if missed	Review this section	Pages
1. C	☐	General Properties of Triangles	235–238
2. H	☐	Decimal Computation; Fractions; Addition and Subtraction of Fractions and Mixed Numbers	55–58, 64–67 68–72
3. E	☐	Writing Linear Expressions and Equations	114–117
4. G	☐	Polynomials; Operations on Polynomials	135, 135–140
5. C	☐	Factors, Divisibility, and Primes	58–61
6. F	☐	Distance and Midpoint Formulas	199–202
7. D	☐	Angles; General Properties of Triangles	227–231, 235–238
8. K	☐	Percent; Percent Problems	86–89, 89–94
9. D	☐	Addition and Subtraction of Fractions and Mixed Numbers	68–72
10. K	☐	Operations on Polynomials	135–140
11. E	☐	Statistics—Mean, Median, and Mode; Whole Number Computation	98–101 48–51
12. G	☐	Graphing Equations on the Coordinate Plane	191–199
13. C	☐	Basic Elements of Plane Geometry	224–227
14. G	☐	Operations on Polynomials	135–140
15. B	☐	Geometric Formulas	267–271
16. J	☐	Square Roots, Exponents, and Scientific Notation	61–64
17. D	☐	Similar Triangles	244–248
18. F	☐	Right Triangles	238–243
19. D	☐	Factoring Polynomials	141–144
20. F	☐	Writing Linear Expressions and Equations	114–117
21. D	☐	Graphing Inequalities on a Number Line	188–190
22. H	☐	Square Roots, Exponents, and Scientific Notation	61–64
23. B	☐	Right Triangles	238–243
24. J	☐	Addition and Subtraction of Fractions and Mixed Numbers	68–72
25. A	☐	Graphing Equations on the Coordinate Plane	191–199
26. H	☐	Transformations in the Plane	260–267
27. E	☐	Distance and Midpoint Formulas	199–202
28. H	☐	Graphing Equations on the Coordinate Plane	191–199
29. C	☐	Writing Linear Expressions and Equations	114–117
30. F	☐	Geometric Formulas	267–271
31. C	☐	Evaluating Formulas and Expressions	125–128
32. G	☐	Circles	255–260
33. A	☐	Evaluating Formulas and Expressions	125–128
34. K	☐	Geometry in Three Dimensions	271–277
35. C	☐	Square Roots, Exponents, and Scientific Notation	61–64
36. F	☐	Absolute Value Equations and Inequalities	161–163
37. D	☐	Exponents and Radicals; Operations with Radicals	129–131, 132–134

Answer	Check if missed	Review this section	Pages
38. G	☐	Evaluating Formulas and Expressions	125–128
39. C	☐	Unit Circle Trigonometry	292–296
40. K	☐	Rational and Radical Expressions	167–170
41. D	☐	Polynomials	135
42. H	☐	Rational and Radical Expressions	167–170
43. A	☐	Distance and Midpoint Formulas	199–202
44. G	☐	Graphs of Trigonometric Functions	297–301
45. D	☐	Right Triangles	238–243
46. F	☐	Exponents and Radicals	129–131
47. C	☐	Rational and Radical Expressions	167–170
48. G	☐	Distance and Midpoint Formulas	199–202
49. B	☐	Rational and Radical Expressions	167–170
50. F	☐	Right Triangle Trigonometry	283–287
51. A	☐	Square Roots, Exponents, and Scientific Notation	61–64
52. G	☐	Solving Systems of Linear Equations	163–167
53. B	☐	Evaluating Formulas and Expressions	125–128
54. H	☐	Graphing Equations on the Coordinate Plane	191–199
55. E	☐	Similar Triangles	244–248
56. H	☐	Geometry in Three Dimensions; Geometric Formulas	271–277, 267–271
57. B	☐	Solving Quadratic Equations	170–173
58. G	☐	Absolute Value Equations and Inequalities	161–163
59. B	☐	Trigonometric Identities	288–291
60. F	☐	Solving Inequalities	157–161

1. **C** The sum of the measures of the angles of a triangle is 180°. We know that one of the angle measures is greater than 90°, so the sum of the measures of the other two angles must be less than 90°.

2. **H** Change the fractions to decimals. Then add all the distances.

    Day 1: $\qquad\qquad\qquad\qquad$ 125.50

    Day 2: $\quad 80\frac{2}{5} \rightarrow 5\overline{)2.00} \rightarrow \overset{0.40}{\phantom{5\overline{)2.00}}}$ $\quad$ 80.40

    Day 3: $\quad 95\frac{3}{4} \rightarrow 4\overline{)3.00} \rightarrow \overset{0.75}{\phantom{4\overline{)3.00}}}$ $\quad \underline{+95.75}$
    $\qquad\qquad\qquad\qquad\qquad\qquad\qquad$ 301.65

3. **E** $\quad$ Total cost $=$ Cost of the items $+$ sales tax
    $$= a + b + c + 0.05(a + b + c)$$
    $$= a + b + c + 0.05a + 0.05b + 0.05c$$
    $$= 1.05a + 1.05b + 1.05c$$

4. **G** $\quad (2x - 6y) + (8x - 5y) = 2x + 8x - 6y - 5y = 10x - 11y$

5. **C** A prime number has only one pair of factors, itself and 1. The number 1 is neither prime nor composite. Eliminate answer choices A, B, and D. In answer choice E, 9 can be factored as 1×9 or 3×3. Therefore, 9 is not a prime.

6. **F** $\quad d = \sqrt{(x_2 - x_1)^2 + (y_2 - y_1)^2}$
    $$d = \sqrt{(2 - (-3))^2 + (6 - 3)^2}$$
    $$= \sqrt{5^2 + 3^2}$$
    $$= \sqrt{25 + 9}$$
    $$= \sqrt{34}$$

7. **D** $\angle 1$ and $\angle CAB$ are vertical angles, which means that $\angle 1$ and $\angle CAB$ have the same measure. Since $\triangle ABC$ is equilateral, m$\angle CAB$ is 60°, so m$\angle 1$ is also 60°.

8. **K** First find the total number of minutes before the athlete cools down.

    $\qquad 20 + 50 + 35 = 105$ minutes

    Subtract to find the time the athlete spends cooling down.

    $\qquad 120 - 105 = 15$ minutes

    To find the percent, divide the time spent cooling down by the total time.

    $\qquad \frac{15}{120} = 0.125 = 12.5\%$

9. **D** Before you add, write the fractions with common denominators.

    $$4\frac{1}{4} = 4\frac{3}{12}$$
    $$\underline{+ \ 5\frac{1}{3} = 5\frac{4}{12}}$$
    $$9\frac{7}{12}$$

10. K $(3a^2 + b^2 - 1) - (a^2 - b^2 - 1) = 3a^2 + b^2 - 1 - a^2 + b^2 + 1$

$$= 3a^2 - a^2 + b^2 + b^2 - 1 + 1$$
$$= 2a^2 + 2b^2$$
$$= 2(a^2 + b^2)$$

11. E Average $= \dfrac{\text{Sum of the ages}}{\text{Number of friends}}$

$20 = \dfrac{20 + 17 + 21 + 19 + R}{5}$ where R is Rick's age

Solve for R.

$$20 = \frac{77 + R}{5}$$
$$100 = 77 + R$$
$$23 = R$$

12. G The slope-intercept form of the equation of a line is $y = mx + b$ where m represents the slope. Rewrite the equations in the answer choices in slope-intercept form. Look for a slope of $-\dfrac{5}{3}$. A linear equation containing $5x + 3y$ is most likely to have this slope.

Try G.

$$5x + 3y + 2 = 0$$
$$5x + 3y = -2$$
$$3y = -5x - 2$$
$$y = -\frac{5}{3}x - \frac{2}{3} \quad \text{Correct. The slope is } -\frac{5}{3}.$$

13. C Look at the answer choices.

A. Two separate planes: the intersection is a line.

B. The circumference of a circle and a plane: the intersection is at least one point but could be two points.

C. Two non-parallel lines: the intersection is a point.

14. G

$$(w + 1)^3 + 3 = (w + 1)(w + 1)(w + 1) + 3$$
$$= (w^2 + 2w + 1)(w + 1) + 3$$
$$= w^3 + 3w^2 + 3w + 1 + 3$$
$$= w^3 + 3w^2 + 3w + 4$$

15. B The wheel rolled 40 kilometers. Neither the circumference of the wheel, nor the incline of the road, affects the distance traveled. There are 1,000 meters in a kilometer, so the wheel rolled 40,000 meters.

16. J Think of perfect cubes:

$$2^3 = 8 \qquad 3^3 = 27 \qquad 4^3 = 64 \qquad 5^3 = 125$$

Find perfect cube factors of each radicand and simplify.

17. D $\triangle ABC$ is similar to $\triangle XYZ$ because two pairs of corresponding angles are congruent. So the ratio of corresponding sides is proportional. Use a proportion to find the length of side AB.

$$\frac{YZ}{BC} = \frac{XY}{AB}$$
$$\frac{3}{14} = \frac{8}{AB}$$
$$3(AB) = 112$$
$$AB = 37\frac{1}{3}$$

18. F Use the Pythagorean Theorem.

$$c^2 = a^2 + b^2$$
$$c^2 = 4^2 + 6^2$$
$$c^2 = 16 + 36$$
$$c^2 = 52$$
$$c = \sqrt{52}$$
$$c = 2\sqrt{13}$$

19. D You can factor out a 4.

$$4x^3 + 32 = 4(x^3 + 8)$$

Look at the answer choices. The answer will either be A or D. Notice that $x^3 + 8$ is the sum of the cubes. If you don't remember how to factor the sum of cubes, work backwards from the answer choices. Compare the products to $4x^3 + 32$.

Try A.

$$4(x - 4)(x + 4) = 4(x^2 - 16)$$
$$= 4x^2 - 64 \quad \text{No.}$$

Try D.

$$4(x + 2)(x^2 - 2x + 4) = 4(x^3 + 8)$$
$$= 4x^3 + 32 \quad \text{Correct.}$$

20. F Total distance = Mary's distance + Victor's distance

$$= s + \frac{1}{2}s$$
$$= 1\frac{1}{2}s = \frac{3s}{2}$$

21. D

$$7 + 5a(2 - a) > -5a^2 + a\left(12 + \frac{11}{a}\right)$$
$$7 + 10a - 5a^2 > -5a^2 + 12a + 11$$
$$7 + 10a > 12a + 11$$
$$10a > 12a + 4$$
$$-2a > 4$$
$$a < -2$$

22. H

$$\sqrt{4}\left(\sqrt{4} + \sqrt{2}\right) - \left(\sqrt{5}\right)\left(\sqrt{20}\right) = 2\left(2 + \sqrt{2}\right) - \sqrt{100}$$
$$= 4 + 2\sqrt{2} - 10 = 2\sqrt{2} - 6$$

23. B The triangle is a 45–45–90 triangle. The relationship among the lengths of the sides of this type of triangle is shown below.

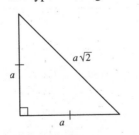

In triangle XYZ,

$$a\sqrt{2} = \sqrt{18} = \sqrt{9 \times 2} = 3\sqrt{2}$$
$$a = 3$$

The length of each leg is 3.

24. J You may substitute the answer choices for x and look for a true statement or you may solve the equation. We will solve the equation.

$$3\tfrac{1}{4} + x = \tfrac{x}{2} + \tfrac{3}{8}$$

Change $3\tfrac{1}{4}$ to $\tfrac{13}{4}$.

$$\tfrac{13}{4} + x = \tfrac{x}{2} + \tfrac{3}{8}$$

To eliminate fractions with denominators 2, 4, and 8, multiply each term by 8.

$$\left(8 \times \tfrac{13}{4}\right) + 8x = \left(8 \times \tfrac{x}{2}\right) + \left(8 \times \tfrac{3}{8}\right)$$

Solve this simple equation.

$$26 + 8x = 4x + 3$$
$$26 + 4x = 3$$
$$4x = -23$$
$$x = -\tfrac{23}{4} = -5\tfrac{3}{4}$$

25. A Slope of a line $= \dfrac{\text{rise}}{\text{run}} = \dfrac{y_2 - y_1}{x_2 - x_1} = \dfrac{4 - (-2)}{2 - (-3)} = \dfrac{4 + 2}{2 + 3} = \dfrac{6}{5}$

26. H When (a,b) is reflected across the x-axis, it becomes $(a,-b)$. When (a,b) is reflected across the y-axis, it becomes $(-a,b)$. So $(-3,2)$ and $(7,-12)$ reflected across the x-axis become $(-3,-2)$ and $(7,12)$. $(-3,-2)$ and $(7,12)$ reflected across the y-axis become $(3,-2)$ and $(-7,12)$.

27. E Midpoint $= \left(\dfrac{x_1 + x_2}{2}, \dfrac{y_1 + y_2}{2}\right) = \left(\dfrac{3 + 7}{2}, \dfrac{4 + 2}{2}\right) = (5,3)$

28. H Parallel lines have the same slope.

29. C Translate the words into an equation.

seven less than   is equal to   the product of 2 and x.
↓   ↓
$4x - 7$ $=$ $3 + 2x$
↑   ↑
the product of 4 and x 3 more than

30. F Find the height of the triangle.

$$A = \tfrac{1}{2}bh$$
$$16 = \tfrac{1}{2}(8)h$$
$$16 = 4h$$
$$4 = h$$

Second, find the length of the rectangle.

$$l = 2h$$
$$l = 2(4)$$
$$l = 8$$

Finally, find the width of the rectangle. The area of the rectangle is equal to the area of the triangle.

$$lw = A$$
$$lw = 16$$
$$8w = 16$$
$$w = 2$$

31. C

$$A = \frac{1}{2}bh$$

$$12 = \frac{1}{2}b(6)$$

$$12 = 3b$$

$$4 = b$$

32. G The measure of the circumference of a circle is $2\pi r$. The radius of circle O is 5 units, so the circumference is 10π.

The measure of an inscribed angle of a circle is equal to one-half the measure of its intercepted arc.

$$m\angle ABC = \frac{1}{2}m\,\widehat{AB}$$

$$90 = \frac{1}{2}m\,\widehat{AB}$$

$$180 = m\widehat{AB} \qquad \text{half the circle}$$

$$m\,\widehat{AB} = \frac{1}{2}(10\pi) = 5\pi$$

33. A The formula for the area of a circle is $A = \pi r^2$.

The area of this circle is 25π. $\qquad 25\pi = \pi r^2$

$$25 = r^2$$

$$r = 5$$

34. K Find the volume of the sphere in π form.

$$V = \frac{4}{3}\pi r^3$$

$$= \frac{4}{3}\pi(3)^3$$

$$= \frac{4}{3}\pi(27)$$

$$= 36\pi$$

Then use the volume of the sphere for the volume of the cylinder.

$$V = \pi r^2 h$$

$$36\pi = \pi(6^2)h$$

$$36\pi = \pi(36)h$$

$$1 = h$$

35. C Every rational number can be represented by a fraction $\frac{a}{b}$ (a and b are whole numbers and $b \neq 0$) or by a decimal that terminates or repeats. Otherwise, the number is not rational.

Look at the answer choices. Both B and E are clearly rational numbers.

For choice A, $\sqrt{\frac{1}{49}} = \frac{1}{7}$, a rational number

For choice D, $\sqrt{324} = 18$, a rational number

But choice C, $\sqrt{27} = 3\sqrt{3}$, and $\sqrt{3}$ is not a rational number.

36. F $|-2 - x| = 6$ means that $-2 - x = \pm 6$. So $x = 4$ or $x = -8$.

The only answer choice that is not possible is $-2 - x = 8$.

37. **D** Simplify the expression by eliminating the radical in the denominator.

$$\frac{3}{2 + \sqrt{3}} = \frac{3(2 - \sqrt{3})}{(2 + \sqrt{3})(2 - \sqrt{3})}$$

$$= \frac{6 - 3\sqrt{3}}{4 - 2\sqrt{3} + 2\sqrt{3} - 3}$$

$$= \frac{6 - 3\sqrt{3}}{4 - 3}$$

$$= \frac{6 - 3\sqrt{3}}{1}$$

$$= 6 - 3\sqrt{3}$$

38. **G** $E = IR$

$12 = 3R$

$R = 4$

The reading for the circuit is 4 ohms.

39. **C** Use the identity $\sin^2 \theta + \cos^2 \theta = 1$.

$$\sin^2 \theta + \left(\frac{\sqrt{2}}{2}\right)^2 = 1$$

$$\sin^2 \theta + \frac{2}{4} = 1$$

$$\sin^2 \theta = 1 - \frac{2}{4}$$

$$\sin^2 \theta = \frac{1}{2}$$

$$\sin \theta = \pm\sqrt{\frac{1}{2}} = \frac{\pm\sqrt{2}}{2}$$

40. **K** Solve.

$$x^2 - 4 < 0$$

$$(x + 2)(x - 2) < 0$$

The signs of $x + 2$ and $x - 2$ must be different.

So $-2 < x < 2$.

41. **D** $(-b)^{-2} - a^2 = \dfrac{1}{b^2} - \dfrac{a^2b^2}{b^2} = \dfrac{1 - a^2b^2}{b^2}$

42. **H** When either 2 or -2 is substituted for t, the denominator becomes 0 since $(2)^2 - 4 = 0$ and $(-2)^2 - 4 = 0$.

43. **A** Use the midpoint formula.

$$\text{Midpoint} = \left(\frac{x_1 + x_2}{2}, \frac{y_1 + y_2}{2}\right)$$

$$(3,5) = \left(\frac{x_1 + 2}{2}, \frac{y_1 + 7}{2}\right)$$

$3 = \dfrac{x + 2}{2}$ and $5 = \dfrac{y + 7}{2}$

$6 = x + 2 \qquad 10 = y + 7$

$4 = x \qquad\qquad 3 = y$

The coordinates of point B are (4,3).

44. **G** The graph shows a sine function. Since the amplitude is 2, the function is $y = 2 \sin x$.

45. **D** The length of the hypotenuse of a 45–45–90 triangle is $a\sqrt{2}$, where a is the length of a leg. Since $\triangle ZUW$ is a 45–45–90 triangle, and the length of one of the legs (WU) is 4, the length of the hypotenuse (ZU) is $4\sqrt{2}$.

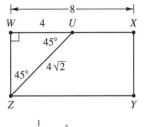

46. **F** $27^{\frac{1}{3}} = \sqrt[3]{27} = 3$

47. **C** Solve to get $a + b$ or $a - b$ in the denominator.

$$\frac{2\sqrt{a} - 3\sqrt{b}}{\sqrt{a} + \sqrt{b}} = \frac{(2\sqrt{a} - 3\sqrt{b})}{(\sqrt{a} + \sqrt{b})} \times \frac{(\sqrt{a} - \sqrt{b})}{(\sqrt{a} - \sqrt{b})} = \frac{2a - 2\sqrt{ab} - 3\sqrt{ab} + 3b}{a - \sqrt{ab} + \sqrt{ab} - b} = \frac{2a - 5\sqrt{ab} + 3b}{a - b}$$

48. **G** Use the distance formula.

$$d = \sqrt{(x_1 - x_2)^2 + (y_1 - y_2)^2}$$
$$d = \sqrt{(5 - 21)^2 + (18 - 6)^2}$$
$$d = \sqrt{(-16)^2 + 12^2}$$
$$d = \sqrt{256 + 144}$$
$$d = \sqrt{400}$$
$$d = 20$$

49. **B** Manipulate the expression to eliminate the fractions in the numerator and denominator.

$$\frac{\frac{2}{a} + 1}{4 + \frac{5}{a}} = \frac{\frac{2}{a} + \frac{a}{a}}{\frac{4a}{a} + \frac{5}{a}} = \frac{\frac{2 + a}{a}}{\frac{4a + 5}{a}} = \frac{2 + a}{a} \cdot \frac{a}{4a + 5} = \frac{2 + a}{4a + 5}$$

50. **F** $\csc \theta = \dfrac{1}{\sin \theta} = \dfrac{1}{\frac{3}{5}} = \dfrac{5}{3}$

51. **A** $4^{-2} + 3^2 - 5^0 = \dfrac{1}{16} + 9 - 1 = \dfrac{1}{16} + 8 = 8\dfrac{1}{16}$

52. **G** Write the equations.

$$2x + 5y = 9$$
$$\underline{2x + 4y = 8}$$

Solve simultaneously. $y = 1$

Substitute to find x. $x = 2$

Add the distances. $2 + 1 = 3$

53. **B** Circumference $= C = 2\pi r = 2\pi(4) = 8\pi$

54. **H** For $x = 4$:

$y = 2 - \sqrt{x}$

$y = 2 - \sqrt{4}$

$\quad = 2 - 2$

$y = 0$

The point is $(4,0)$.

For $x = 9$:

$y = 2 - \sqrt{x}$

$y = 2 - \sqrt{9}$

$\quad = 2 - 3$

$y = -1$

The point is $(9,-1)$.

55. **E**   Since the two triangles are similar, the ratios of corresponding sides are proportional.

$$\frac{BC}{YZ} = \frac{AB}{XY} \quad \text{and} \quad \frac{BC}{YZ} = \frac{AC}{XZ}$$

$$\frac{3}{9} = \frac{4}{z} \qquad\qquad \frac{3}{9} = \frac{5}{y}$$

$$3z = 36 \qquad\qquad 3y = 45$$

$$z = 12 \qquad\qquad y = 15$$

$XY + XZ = 12 + 15 = 27$

56. **H**   To form the largest circular cross section, cut through the thickest part of the sphere, the center. The radius of this circle will be the radius of the sphere.

First, find the radius of the sphere.

$$V = \frac{4}{3}\pi r^3$$

$$36\pi = \frac{4}{3}\pi r^3$$

$$36 = \frac{4}{3}r^3$$

$$36 \times \frac{3}{4} = \frac{4}{3}r^3 \times \frac{3}{4}$$

$$27 = r^3$$

$$3 = r$$

Then find the area of the circle.

$A = \pi r^2$

$A = \pi(3)^2$

$A = 9\pi \text{ cm}^2$

57. **B**   Choices II and III are easier to work with. Try them first.

<u>Choice II</u>: $(x - 4)(2x + 4) = 0$

Substitute 4 for x. When $x = 4$, $x - 4 = 0$. So 4 **is** a root of this equation.

Substitute $-\frac{1}{2}$ for x. When $x = -\frac{1}{2}$, $x - 4 \neq 0$ and $2x + 4 \neq 0$. So $-\frac{1}{2}$ is **not** a root of this equation.

Answer choices A, C, and E are eliminated.

<u>Choice III</u>: $(2x - 8)(6x + 3) = 0$

Substitute 4 for x. When $x = 4$, $2x - 8 = 0$. So 4 **is** a root of this equation.

Substitute $-\frac{1}{2}$ for x. When $x = -\frac{1}{2}$, $6x + 3 = 0$. So $-\frac{1}{2}$ **is** a root of this equation.

<u>Choice I</u>: $12x^2 - 12x - 4 = 30x + 20$

$12x^2 - 42x - 24 = 0$

Factor: $(2x - 8)(6x + 3) = 0$

This equation is the same as the equation in Choice III. So Choice I is also correct.

58. **G**   Of the choices given, the largest negative solution occurs when $-(3x + 7) \geq 5$.

Solve. $-3x - 7 \geq 5$

$$-3x \geq 12$$

$$x \leq -4$$

Of the choices given, the largest negative number that is included in the solution set is -4.

59. **B** $\quad \dfrac{\dfrac{1}{\csc \theta}}{\cos^2 \theta + \sin^2 \theta} = \dfrac{\sin \theta}{1} = \sin \theta$

60. **F** Begin by finding the highest value for q because the higher q is, the lower p can be. Since $3 \le q \le 6$, 6 is the greatest value for q.

Now find the lowest value of p. Substitute 6 for q in the inequality $pq \ge 15$. $6p \ge 15$, so $p \ge 2.5$. The lowest value of p is 2.5.

Check to make sure that $p < 8$. Yes, $2.5 < 8$.

All conditions of the inequalities have been satisfied.

Find $0.4p$. $\qquad p = 2.5 \qquad 0.4p = (0.4)(2.5) = 1$

Mathematics Answers

Chapter 6 Pre-Algebra I

Whole Numbers

Practice (p. 47)

1. $16 < 24$
2. $31 < 91$
3. $1,031 > 976$
4. $1,891 < 2,001$
5. $18 < 24 < 36$
6. $58 > 40 > 38$
7. 3 three million
   4 forty
   6 six billion
   5 five thousand
   0 zero ten millions

   1 one hundred
   8 eight hundred million
   2 two
   9 ninety thousand
   7 seven hundred thousand
8. $10,037 > 9,869$
9. $23,568 < 23,602$
10. $728,315 > 728,298$
11. $101,834,561 > 98,898,789$
12. $83 < 196 < 234$ or $234 > 196 > 83$
13. $108 < 562 < 1,174 < 1,961$ or $1,961 > 1,174 > 562 > 108$
14. 3,000,000
15. 3,058,000
16. 3,057,810
17. 3,057,800
18. 3,100,000
19. 17,600
20. 17,650

ACT-Type Problems (p. 47)

1. **D** The 8 is in the thousands place. The number to the right is 7. Since 7 is greater than 5, round up. $28,750 rounded to the thousands place is $29,000.
2. **G** In all of the choices except for G the symbol > will make the statement true.
3. **A** Identify the digits: hundred-thousands place → 8, ones place → 4, ten-thousands place → 2. Add the digits: $8 + 4 + 2 = 14$.
4. **H** The digit 3 is in the thousands place. The total value of the 3 is three thousand $(3 \times 1,000)$.

5. **D** Find the digit that appears in the ten thousand place in each choice.

   A. 4 **B.** 6 **C.** 8 **D.** 9 **E.** 5

   Choice D has the largest value.

Whole Number Computation

Practice (p. 51)

Possible estimates are shown in parentheses. Estimates may differ depending on estimation techniques.

1. 434 (500)
2. 21,276 (21,000)
3. 230,001 (230,000)
4. 984,904 (1,000,000)
5. 379 (400)
6. 68,989 (69,000)
7. 106,097 (100,000)
8. 586,867 (600,000)
9. 943 (800)
10. 15,238 (16,000)
11. 51,207 (50,000)
12. 624,519 (600,000)
13. 28 (30)
14. 314 (300)
15. 78 (80)
16. 302 (300)
17. 71,202 (70,000)
18. 4,127 (4,000)
19. 15,960 (16,000)
20. 621 (600)

ACT-Type Problems (p. 51)

1. **D** Multiply. $11 \times 12 = 132$. There are a total of 132 players on the 12 teams.
2. **G** Divide. $36 \div 4 = 9$. They need 9 cars.
3. **E** First find Drew's age. Andy's Age $- 5 =$ Drew's Age. $27 - 5 = 22$.

   Add Andy's age and Drew's age.
   Andy's Age $+$ Drew's Age $=$ Sum of Ages

   $$27 \quad + \quad 22 \quad = \quad 49$$
4. **H** Multiply. $8 \times 24 = 192$
5. **C** Subtract: 26 points $-$ 19 points $=$ 7 points

Decimals

Practice (p. 54)

1. $0.32 > 0.289$

2. $1.91 < 9.1$

3. $20.347 < 20.351$

4. $387.9036 > 38.79039$

5. $408.246 > 389.978$

6. $0.73846 < 0.73921$

7. $16.4 < 16.43 < 16.432$

8. $0.56893 < 0.56981 < 0.5699$

9. 1 10 thousand   6 6 thousand
  2 2 hundredths   7 7 hundred-thousandths
  3 thirty   8 8 thousandths
  4 4 hundred   9 nine
  5 5 ten-thousandths   0 0 tenths

10. $18.097 < 18.598 < 18.61$ or
$18.61 > 18.598 > 18.097$

11. $402.5163 < 402.5191 < 403.8799 < 403.8917$ or
$403.8917 > 403.8799 > 402.5191 > 402.5163$

12. 3,245.605

13. 3,000

14. 3,250

15. 3,245.61

16. 3,245.6

17. 3,246

18. 3,245.6054

19. 3,200

20. thousandths place

ACT-Type Problems (p. 54)

1. **E** Identify the digits: tenths place → 9, ten-thousandths place → 7. Add the digits. $9 + 7 = 16$

2. **F** The $<$ symbol makes choice F true. $345.982 < 345.992$

3. **B** The digit 6 is in the thousandths place.

4. **H** The digit in the hundredths place is 6 and the digit to the right is 4. The digit in the hundredths place will remain 6 because 4 is less than 5. The rounded number is 3,482.96.

5. **D** Look at the digits just after the decimal point. Choice D has the largest digit, 9, in the tenths place.

Decimal Computation

Practice (p. 57)

Possible estimates are shown in parentheses. Estimates may differ depending on estimation techniques.

1. 19.863 (20)

2. 560.803 (560)

3. 18.455 (19)

4. 48.0039 (48)

5. 24.19 (24)

6. 0.0119 (0.01)

7. 290.74 (290)

8. 223.95124 (224)

9. 1,002.982 (1,000)

10. 4.688 (4.8)

11. 901.8123 (900)

12. 21.15 (20)

13. 891 (1,000)

14. 1,242.966752 (1,215)

15. 0.673 (0.7)

16. 52.68421053 (50)

17. 398.8

18. 27.04

19. 33

20. 0

ACT-Type Problems (p. 58)

1. **B** Subtract $92.6° - 77.8° = 14.8°$

2. **H** Add the test scores: $87.7 + 79.4 = 167.1$. Divide the sum by 2: $167.1 \div 2 = 83.55$

3. **A** Multiply to find the length of the 1.5-inch nails and the length of the 0.75-inch nails.

50×1.5 inches = 50×0.75 inches =
75 inches   37.5 inches

Add to find the total length:
  75 inches + 37.5 inches = 112.5 inches.

4. **J** Find how many pounds Gene lost after 5 days: $5 \times 0.8 = 4$. Subtract the amount Gene lost from Gene's original weight: $170 - 4 = 166$.

5. **B**

Step 1. Add the mileage from the table:
5.6 miles + 10.4 miles + 7.8 miles + 11.2 miles + 22.7 miles = 57.7 miles.

Step 2. Add this sum to the original odometer reading: 2004.7 miles + 57.7 miles = 2062.4 miles.

Factors, Divisibility, and Primes

Practice (p. 60)

1. 1, 3, 5, 7

2. 1, 3

3. 1, 2, 3, 4, 6, 8

4. 1, 2, 4, 8

5. 1, 2, 3, 6, 9, 18, 27, 54

6. 1, 67

7. 1, 3, 29, 87

8. 1, 2, 3, 4, 5, 6, 8, 10, 12, 15, 20, 24, 30, 40, 60, 120

9. Neither. 1 has only one factor and is neither prime nor composite.

10. Composite. The number is divisible by 1, 5, 41, and 205.

11. Prime. Try factors up to the square root of the number. The square root of 41 is between 6 and 7. So try dividing by 2, 3, 4, 5, 6, 7. None of these are factors, so 41 is prime.

12. Composite. The sum of the digits is 9, so 3 and 9 are factors.

13. Prime. Try dividing by 2, 3, 4, 5, 6, 7, 8, 9, 10, 11, 12. Some you can eliminate immediately. None are factors of 127 so 127 is prime.

14. Composite. The sum of the digits is 12 so the number is divisible by 3.

15. Composite. Try dividing by 2, 3, 4, 5, 6, 7, 8, 9, 11. 7 is a factor of 119, so 119 is composite.

16. Composite. The sum of the digits is 21, so the number is divisible by 3.

ACT-Type Problems (p. 60)

1. **E** List all the factors of 20: 1, 2, 4, 5, 10, 20. Add the factors: $1 + 2 + 4 + 5 + 10 + 20 = 42$.

2. **F** Is 473,124 divisible by:

   1 Yes. Every number is divisible by 1.
   2 Yes. It is even.
   3 Yes. The sum of the digits is divisible by 3.
   4 Yes. The last two digits are divisible by 4.
   5 No. It doesn't end in 0 or 5.
   6 Yes. It is divisible by 2 and 3.
   7 No. You have to work this out.
   8 No. The last three digits are not divisible by 8.
   9 No. The sum of the digits is not divisible by 9.

   Multiply: $1 \times 2 \times 3 \times 4 \times 6 = 144$

3. **A** All even numbers greater than 2 are composite. That eliminates choices B and D. Choice C (81) is composite. It has factors 1, 3, 9, 27, and 81.

   That leaves choice A and choice E. Choice E is 117. Find the sum of the digits: $1 + 1 + 7 = 9$. This means that 117 has both 3 and 9 as factors. Choice A is the only non-composite number.

4. **H** No even numbers between 30 and 40 are prime.

   31 Prime.
   33 Composite. $3 \times 11 = 33$
   35 Composite. $5 \times 7 = 35$
   37 Prime.
   39 Composite. $13 \times 3 = 39$

   The prime numbers between 30 and 40 are 31 and 37.

   Find their sum: $31 + 37 = 68$.

5. **B** The last two digits of 235,783,784 are evenly divisible by 4. So 4 is a factor of this number.

Square Roots, Exponents, and Scientific Notation

Practice (p. 63)

1. 20
2. 0.5
3. 18
4. 1.7
5. 1,000
6. 64

7. 1
8. 7
9. $\frac{1}{2}$ or 0.5
10. $\frac{1}{36}$
11. $\frac{1}{1,000}$ or 0.001
12. $\frac{1}{4}$ or 0.25
13. 2.325×10^9
14. 1.75×10^{-2}
15. 2.75×10^0
16. 5.659×10^3
17. 6.2×10^{-2}
18. 5.079×10^4
19. 2.57×10^{-3}
20. 6.89459×10^6
21. 420,000
22. 5,010
23. 0.0235
24. 660
25. 0.00909
26. 67,000
27. 0.000405
28. 19,000

ACT-Type Problems (p. 64)

1. **C** The area of a square is side $\times$ side, or s^2. So the side of a square is the square root of the area.

   $\sqrt{5.76} = 2.4$

2. **G** Consider the answer choices. The number can't be even because the product is not even, so F, H, and K are eliminated. Try G.

   $7^4 = 7 \times 7 \times 7 \times 7 = 2,401$

   That's our answer. (You could also get the answer by finding $\sqrt[4]{2,401}$. This method is discussed in Chapter 7.)

3. **D** Count to see that there are 7 places to the right of the decimal point. Use that number as a negative power of 10: 10^{-7}.

4. **J** $15^{\frac{1}{2}} = \sqrt{15}$. Use a calculator to find that $\sqrt{15}$ is approximately equal to 3.87298. . . . Answer choice H and answer choice J are closest to 3.87298.

   The difference between choice H, 3.8, and 3.87298 is about 0.7.

   The difference between choice J, 3.9, and 3.87298 is about 0.03.

   Choice J, 3.9, is the closer approximation.

5. **B** Six billion is written 6,000,000,000. Use scientific notation. Move the decimal point nine places to the left. Use 10^9.

   $6,000,000,000 = 6 \times 10^9$

Fractions

Practice (p. 66)

1. True
2. False
3. True
4. True
5. False
6. True
7. False
8. True
9. $\frac{1}{3}$

10. $\dfrac{2}{3}$

11. $\dfrac{3}{5} > \dfrac{4}{7}$

12. $\dfrac{2}{9} < \dfrac{5}{12}$

13. $\dfrac{2}{3} > \dfrac{1}{2} > \dfrac{3}{8}$

14. $\dfrac{3}{4} < \dfrac{6}{7} < \dfrac{7}{8}$

15. $\dfrac{5}{6}, \dfrac{7}{8}$

16. $\dfrac{3}{5}, \dfrac{61}{100}$

17. $\dfrac{2}{7}, \dfrac{4}{9}$

18. $\dfrac{11}{20}, \dfrac{4}{7}, \dfrac{3}{5}$

19. $\dfrac{5}{8}, \dfrac{7}{10}, \dfrac{3}{4}$

20. $\dfrac{7}{16}, \dfrac{1}{2}, \dfrac{5}{9}, \dfrac{4}{5}$

ACT-Type Problems (p. 67)

1. **B** Fractions equivalent to $\dfrac{4}{7}$ can be created by multiplying or dividing the numerator and denominator of $\dfrac{4}{7}$ by the same number. This is not true for choice B. $\dfrac{4}{7} \times \dfrac{3}{5} = \dfrac{12}{35}$

2. **H** Write the fractions with a common denominator. All the denominators are factors of 30 so use 30 for the common denominator.

$\dfrac{3}{5} = \dfrac{18}{30}$ $\dfrac{5}{6} = \dfrac{25}{30}$ $\dfrac{3}{10} = \dfrac{9}{30}$ $\dfrac{1}{3} = \dfrac{10}{30}$ $\dfrac{8}{15} = \dfrac{16}{30}$

$\dfrac{3}{10} = \dfrac{9}{30}$ $\dfrac{1}{3} = \dfrac{10}{30}$ $\dfrac{8}{15} = \dfrac{16}{30}$ $\dfrac{3}{5} = \dfrac{18}{30}$ $\dfrac{5}{6} = \dfrac{25}{30}$

3. **B** Look for equivalent fractions: $\dfrac{7}{13} \times \dfrac{3}{3} = \dfrac{21}{39}$.

4. **H** Use your calculator to find which fraction is the largest: $\dfrac{5}{7}$.

5. **B** Use your calculator to change each fraction to a decimal.

$\dfrac{7}{15} \approx 0.467, \dfrac{8}{13} \approx 0.615, \dfrac{9}{17} \approx 0.529, \dfrac{4}{9} \approx 0.444,$

$\dfrac{5}{12} \approx 0.417$

Compare the decimal values. You can then see that $\dfrac{8}{13}$ is the greatest. Alex walks the furthest.

Addition and Subtraction of Fractions and Mixed Numbers

Practice (p. 71)

Possible estimates are shown in parentheses. Estimates may differ depending on estimation techniques.

1. $1\dfrac{2}{21}\left(1\dfrac{1}{2}\right)$

2. $\dfrac{13}{18}$ (1)

3. $\dfrac{31}{39}$ (1)

4. $1\dfrac{1}{4}\left(1\dfrac{1}{2}\right)$

5. $1\dfrac{7}{24}$ (1 to 2)

6. $1\dfrac{7}{24}$ (1 to 2)

7. $1\dfrac{13}{20}$ (1 to 2)

8. 1 (near 1)

9. $14\dfrac{16}{21}$ (14 to 15)

10. $16\dfrac{8}{21}$ (16 to 17)

11. $21\dfrac{5}{8}$ (21 to 22)

12. $42\dfrac{7}{12}$ (near 42)

13. $\dfrac{1}{34}$ (0)

14. $\dfrac{83}{342}\left(\dfrac{1}{3}\right)$

15. $\dfrac{19}{70}\left(\dfrac{2}{10}\right)$

16. $\dfrac{2}{15}\left(\dfrac{1}{10}\right)$

17. $\dfrac{17}{30}\left(\dfrac{1}{3} \text{ to } \dfrac{2}{3}\right)$

18. $2\dfrac{1}{24}$ (near 2)

19. $16\dfrac{1}{5}$ (16 to 17)

20. $3\dfrac{23}{24}$ (near 4)

21. $4\dfrac{17}{20}$ (near 5)

22. $7\dfrac{7}{15}$ (7 to 8)

23. $14\dfrac{1}{4}$ (14 to 15)

24. $4\dfrac{13}{24}$ (4 to 5)

ACT-Type Problems (p. 71)

1. **B** Use common denominators.

$\dfrac{3}{5} = \dfrac{18}{30}, 1\dfrac{4}{15} = 1\dfrac{8}{30}, \dfrac{23}{30}$

Add. $\dfrac{18}{30}$

$\quad 1\dfrac{8}{30}$

$\quad + \dfrac{23}{30}$

$\quad 1\dfrac{49}{30} = 1 + 1\dfrac{19}{30} = 2\dfrac{19}{30} = 2\dfrac{19}{30}$ pounds

Multiply the price per pound by the number of pounds.

$2\dfrac{19}{30} \times \$6 = \$15.80$

2. **J** Rewrite the fractions with common denominators.

$2\dfrac{1}{2} = 2\dfrac{10}{20}$ $\dfrac{3}{4} = \dfrac{15}{20}$ $1\dfrac{2}{5} = 1\dfrac{8}{20}$ $5\dfrac{7}{10} = 5\dfrac{14}{20}$

Add. $2\dfrac{10}{20} + \dfrac{15}{20} + 1\dfrac{8}{20} + 5\dfrac{14}{20} = 8\dfrac{47}{20}$

$= 8 + 2\dfrac{7}{20} = 10\dfrac{7}{20}$

3. **C** Add the two fractions: $4\dfrac{2}{5}$ and $3\dfrac{3}{4}$.

Use common denominators. $4\dfrac{2}{5} = 4\dfrac{8}{20}$ $3\dfrac{3}{4} = 3\dfrac{15}{20}$

Add. $4\dfrac{8}{20} + 3\dfrac{15}{20} = 7\dfrac{23}{20} = 7 + 1\dfrac{3}{20} = 8\dfrac{3}{20}$

Subtract $\dfrac{1}{4}$. $\dfrac{1}{4} = \dfrac{5}{20}$

$8\dfrac{3}{20} - \dfrac{5}{20} = 7\dfrac{23}{20} - \dfrac{5}{20} = 7\dfrac{18}{20} = 7\dfrac{9}{10}$

$7\dfrac{9}{10} < 8$

4. **H** Rewrite the fractions with common denominators.

$$\frac{3}{8} = \frac{9}{24} \quad \frac{20}{24} \quad \frac{1}{6} = \frac{4}{24}$$

Compute. $\frac{9}{24} + \frac{20}{24} - \frac{4}{24} = \frac{29}{24} - \frac{4}{24} = \frac{25}{24} \neq 1$

5. **D** Subtract the number of pieces that have been eaten from 8.

$$8 - 2 - 1 - 3 = 2$$

There were 2 of 8 pieces or $\frac{1}{4}$ of the pie left for Chad to eat. Simplify the answer: $\frac{2}{8} = \frac{1}{4}$.

Multiplication and Division of Fractions and Mixed Numbers

Practice (p. 74)

1. $\frac{2}{15}$ 2. $\frac{7}{18}$ 3. 12

4. $2\frac{11}{32}$ 5. $51\frac{33}{50}$ 6. $22\frac{1}{7}$

7. $13\frac{1}{8}$ 8. $4\frac{1}{2}$ 9. $1\frac{1}{9}$

10. 2   11. $\frac{1}{12}$ 12. 64

13. $1\frac{7}{9}$ 14. 2   15. $1\frac{11}{65}$

16. $1\frac{7}{25}$ 17. $\frac{1}{9}$ 18. $1\frac{1}{3}$

19. $1\frac{2}{9}$ 20. $1\frac{1}{3}$

ACT-Type Problems (p. 75)

1. **D** $\frac{1}{2} \times \frac{2}{3}$ mile $= \frac{1}{3}$ mile

2. **H** 225 pounds $\times \frac{1}{3} = 75$ pounds

3. **D** $\frac{2}{5} + \frac{25}{9} \div \frac{1}{3} = \frac{2}{5} + \frac{25}{9} \times \frac{3}{1} = \frac{2}{5} + \frac{25}{3}$

$$= \frac{6}{15} + \frac{125}{15} = \frac{131}{15} = 8\frac{11}{15}$$

4. **F** 160 exams $\times \frac{1}{2} = 80 \times \frac{1}{4} = 20$ exams

5. **B** $\frac{5}{7} \div \frac{15}{3} \times \frac{11}{13} =$

$$\frac{5}{7} \times \frac{3}{15} \times \frac{11}{13} =$$

$$\frac{15}{105} \times \frac{11}{13} =$$

$$\frac{1}{7} \times \frac{11}{13} = \frac{11}{91}$$

Positive and Negative Numbers

Practice (p. 76)

1. 8   2. 17
3. 25   4. 17
5. 34   6. 54
7. 0.5   8. 39,000
9. $-8 < 3$ 10. $54.6 > 39.897$
11. $-\frac{1}{6} > -\frac{5}{6}$ 12. $7\frac{1}{4} > 5\frac{8}{9}$
13. $-20.876 < -20.8759$ 14. $-\frac{7}{16} > -\frac{37}{79}$
15. $-15 > -17$ 16. $-0.05 < 0.05$
17. $|-45| < 46$ 18. $-20 < |-19|$
19. $|74 - 32| > 33$ 20. $65 > |-64|$

ACT-Type Problems (p. 77)

1. **C** $-34 < 2$ The $>$ symbol will make the statement false.

2. **G** The only true statement is $14 > -16$.

3. **E** $5 + |-3| - 8 + 3 =$
   $5 + 3 - 8 + 3 =$
   $8 - 8 + 3 =$
   $0 + 3 = 3$

4. **J** Rewrite all the numbers without absolute value symbols.

   $\{-3, 9, 4, -5, 10\}$

   Write the numbers in order from least to greatest.
   $\{-5, -3, 4, 9, 10\}$

5. **B** $|-120| = 120$, the largest number in the list.

Computation with Positive and Negative Numbers

Practice (p. 80)

1. -126 2. 14   3. -3
4. $-23\frac{3}{5}$ 5. $-1,403$ 6. -18.9
7. 568   8. $10\frac{3}{32}$ 9. -18
10. 18   11. -180 12. -6.5
13. 179.4   14. 0   15. -4.76
16. 9,100   17. -5 18. -9
19. 38   20. 29

ACT-Type Problems (p. 81)

1. **C** $1,364 - 1,200 - 64 + 100 = 164 - 64 + 100 = 100 + 100 = 200$

2. **H** $3 + 5 \times (-2) + 8 = 3 - 10 + 8 = -7 + 8 = 1$, which is not equal to -5.

3. **B** $16 \times \$7 + 10 \times \$9 = \$202$

4. **J** $-5 \times 2 + 6 \div (-2) + 5 = -10 + (-3) + 5 = -13 + 5 = -8$

5. **C** $2.5 \times 14 - 30 + 2 \times 5 \div (-2) = 35 - 30 + 10 \div (-2) = 35 - 30 - 5 = 5 - 5 = 0$

Order of Operations

Practice (p. 83)

1. 28	2. 45	3. 51
4. 29	5. 8	6. 8.5
7. -3	8. 23	9. 3
10. -21	11. -4	12. 0
13. 18	14. 0	15. 8
16. 6	17. -3	18. 7
19. 0	20. -5	

ACT-Type Problems (p. 83)

1. **D** $(3 - 2) \times 15 - 4 \div 3$
   $= 1 \times 15 - 4 \div 3$
   $= 15 - 4 \div 3$
   $= 15 - \frac{4}{3}$
   $= 15 - 1\frac{1}{3}$
   $= 14\frac{3}{3} - 1\frac{1}{3} = 13\frac{2}{3}$

2. **G** $3 \times 4 + (7 - 4) + 3$
   $= 3 \times 4 + 3 + 3$
   $= 12 + 3 + 3 = 18$

3. **A** $(16 - 4) \div 3 + 7 - 2 \times 5$
   $= 12 \div 3 + 7 - 2 \times 5$
   $= 4 + 7 - 10$
   $= 11 - 10 = 1$

4. **K** $7 \times (5 + 7) - 8 \times 2$
   $= 7 \times 12 - 8 \times 2$
   $= 84 - 16 = 68$

5. **C** $(21 - 7) \times (25 - 17) - 4 \times 3$
   $= 14 \times 8 - 4 \times 3$
   $= 112 - 12 = 100$

Cumulative ACT Practice
Pre-Algebra I (p. 84)

Correct your Pre-Algebra I Cumulative ACT Practice answers and work through the answer explanations. If you missed an item or guessed at the answer, circle the number of the item below and review the pages indicated.

1. **D** Topic: Whole Numbers (pp. 44–48)

   I is not correct because $534,916 > 534,619$.

   II is correct because 4 is in the thousands place for both numbers and the digit to the right of 4 is 5 or greater in both cases.

   III is correct because the 4 is in the thousands place for both numbers.

2. **G** Topic: Whole Numbers (pp. 44–48)

   Look at the thousands position for each number to see that 3,861,932 has a 1 in the thousands place.

3. **B** Topic: Whole Number Computation (pp. 48–51)

   Twelve full tables plus 7 more people is $12 \cdot 12 + 7 = 144 + 7 = 151$.

4. **K** Topic: Whole Number Computation (pp. 48–51)

   Jim has already traveled 178 out of the 300 miles. He can travel an additional $300 - 178 = 122$ miles.

5. **C** Topic: Decimals (pp. 52–55)

   2,593.7365: There is a 7 in the tenths place and a 5 is in the hundreds place. $7 + 5 = 12$

6. **H** Topic: Decimals (pp. 52–55)

   78,616.874825: There is a 4 in the thousandths place and an 8 to the right of the 4. Round up to 78,616.875.

7. **D** Topic: Multiplication and Division of Fractions and Mixed Numbers (pp. 72–75)

   $\frac{3}{5} \div \frac{6}{5} \cdot \frac{1}{3} = \frac{3}{5} \cdot \frac{5}{6} \cdot \frac{1}{3} = \frac{1}{2} \cdot \frac{1}{3} = \frac{1}{6}$

8. **F** Topic: Decimal Computation (pp. 55–58)

   Add to find the amount of money that Anita has donated.

   $0.10 + 0.09 + 0.15 + 0.07 + 0.20 = 0.61$

   Subtract to find the amount of money remaining in Anita's pocket.

   $1.00 - 0.61 = 0.39$. There is $0.39 remaining.

9. **B** Topic: Factors, Divisibility, and Primes (pp. 58–61)

   The complete list of factors of 60 is: $\{1, 2, 3, 4, 5, 6, 10, 12, 15, 20, 30, 60\}$.

10. **K** Topic: Fractions (pp. 64–67)

One way to find the answer is to rewrite each fraction with a common denominator.

F. $\frac{15}{30}$ **G.** $\frac{20}{30}$ **H.** $\frac{21}{30}$ **J.** $\frac{15}{30}$ **K.** $\frac{24}{30}$

$\frac{24}{30} = \frac{4}{5}$ is the largest number.

It might be faster to use your calculator to find the decimal equivalent of each fraction and then compare the decimals.

11. **D** Topic: Addition and Subtraction of Fractions and Mixed Numbers (pp. 68–72)

A. $\frac{1}{2} + \frac{3}{2} + \frac{2}{2} = \frac{6}{2} = 3$

B. $\frac{4}{6} + \frac{5}{6} = \frac{9}{6} = \frac{3}{2} = 1.5$

C. $\frac{30}{24} + \frac{12}{24} + \frac{6}{24} = \frac{48}{24} = 2$

D. $\frac{77}{14} - \frac{7}{14} = \frac{70}{14} = 5$

E. $\frac{40}{36} + \frac{12}{36} + \frac{54}{36} = \frac{106}{36} = \frac{53}{18} = 2\frac{17}{18}$

12. **G** Topic: Addition and Subtraction of Fractions and Mixed Numbers (pp. 68–72)

Subtract. $\frac{4}{5} - \frac{2}{3} = \frac{12}{15} - \frac{10}{15} = \frac{2}{15}$

13. **E** Topic: Multiplication and Division of Fractions and Mixed Numbers (pp. 72–75)

Amount Joni withdrew: $\$60 \cdot \left(\frac{5}{6}\right) = \$50$

Total amount withdrawn: $\$60 + \$50 = \$110$

14. **H** Topics: Addition and Subtraction of Fractions and Mixed Numbers (pp. 68–72); Multiplication and Division of Fractions and Mixed Numbers (pp. 72–75)

Add the fractions.

$\frac{6}{16} = \frac{3}{8}$ $\frac{1}{8} = \frac{1}{8}$ $\frac{1}{4} = \frac{2}{8}$

Fraction of money spent: $\frac{3}{8} + \frac{1}{8} + \frac{2}{8} + \frac{1}{8} = \frac{7}{8}$

Fraction of money left: $\frac{8}{8} - \frac{7}{8} = \frac{1}{8}$

Amount of money left: $\frac{1}{8} \cdot \$1,600 = \$200$ left to put into savings account

15. **D** Topic: Computation with Positive and Negative Numbers (pp. 77–81)

Write each transaction as a positive or negative number:

In bank to start	+ 50
Paycheck	+450
Loan	−175
Groceries	−150
Rent	−200

Add up the transactions:
$50 + 450 - 175 - 150 - 200 = -25$

Janette needs $25 back from her brother in order to have enough money to pay the rent.

Chapter 7 Pre-Algebra II

Percent

Practice (p. 88)

Fraction	Decimal	Percent
1. $\frac{9}{50}$	**2.** 0.18	18%
3. $\frac{9}{250}$	0.036	**4.** 3.6%
$\frac{5}{8}$	**5.** 0.625	**6.** 62.5%
7. $\frac{21}{2,500}$	**8.** 0.0084	0.84%
9. $\frac{1}{5,000}$	0.0002	**10.** 0.02%
$\frac{5}{6}$	**11.** 0.8333…	**12.** $83\frac{1}{3}\%$

13. 37.5% **14.** $\frac{3}{4}$

15. 25% **16.** $\frac{3}{10}$

17. $66\frac{2}{3}\%$ **18.** 80%

19. $\frac{1}{2}$ **20.** $\frac{4}{5}$

ACT-Type Problems (p. 88)

1. **C** Write the given information as a fraction. The total number of people interviewed is the denominator. The number of females is the numerator. Write the fraction as a percent.

   $\frac{2}{5} = 0.40 = 40\%$

2. **J** Find the percentage of the people under 18.

   $100\% - 65\% = 35\%$ of the people are under 18

   Change the percent to a fraction. $35\% = \frac{35}{100} = \frac{7}{20}$

3. **E** Write the information as a fraction. The denominator is the total number of days in four weeks. The numerator is the total number of times Francine cooked stir-fry. Write the fraction as a percent.

$$\frac{21}{28} = 0.75 = 75\%$$

4. **G** Write both test results as fractions and find the difference. $84\% = \frac{84}{100}$ $\frac{6}{15} = \frac{40}{100}$

$$\frac{84}{100} - \frac{40}{100} = \frac{44}{100} = \frac{11}{25}$$

5. **D** Write a fraction. The number of chairs is the denominator and the number of chairs with arm rests is the numerator.

Rewrite the fraction as a percent. $\frac{10}{16} = 0.625 = 62.5\%$

Percent Problems

Practice (p. 93)

1. $75\% \times \$99 = \textbf{\$74.25}$

2. $\frac{5}{8} \times 640 = \textbf{400}$

3. $0.5\% \times 2.5 = \textbf{0.0125}$

4. $90\% \times \textbf{200} = 180$

5. $\textbf{75\%} \times 120 = 90$

6. $\textbf{66}\frac{2}{3}\textbf{\%} \times 39 = 26$

7. $\textbf{15\%} \times 60 = 9$

8. $\textbf{40\%} \times 50 = 20$

9. $19\% \times \textbf{720} = 136.8$

10. $75\% \times \textbf{100} = 75$

11. $60\% \times \textbf{25,000} = 15,000$

12. $\$18.30 - 15 = \$3.30$
    $\$3.30 \div 15 = 0.22 = 22\%$

13. $\$82 - \$50.84 = \$31.16$
    $\$31.16 \div \$82 = 0.38 = 38\%$

14. $4,850 - 3,104 = 1,746$
    $1,746 \div 4,850 = 0.36 = 36\%$

15. $19,008 - 13,500 = 5,508$
    $5,508 \div 13,500 = 0.408 = 40.8\%$

16. $0.08 \times \$28.50 = \$2.28$
    $\$28.50 + \$2.28 = \$30.78$

17. $0.25 \times \$510 = \$127.50$

18. $\$510 - \$127.50 = \$382.50$

19. $\$2,000 \times 1.065 = \$2,130$

20. $\$2,000 \times 0.08 = \$160$

ACT-Type Problems (p. 93)

1. **D** Find the base.
    percent × base = percentage
    $0.4 \times \text{base} = 21$
    $\text{base} = \frac{21}{0.4} = 52.5$

2. **H** Find 75% of $4,000.
    percent × base = percentage
    $0.75 \times \$4,000 = \$3,000$

3. **C** Find out how much of the rent money Joe has.
    percent × base = percentage
    $0.80 \times \$600 = \$480$
    Find out how much money Joe needs to make to pay the rent.
    $\$600 - \$480 = \$120$
    Find out how many hours Joe has to work to collect $120.
    $\$120 \div \$6 \text{ per hour} = 20 \text{ hours}$

4. **G** Find out how far Ellen has traveled when she stops for breakfast.
    percent × base = percentage
    $0.6 \times 0.75 = 0.45 \text{ mile}$

5. **E** Find out what percent of Dan's average he scores during this game.
    percent × base = percentage
    $\text{percent} \times 20 = 18$
    $\text{percent} = \frac{18}{20} = 0.9 = 90\%$

6. **G** Find out what percent of the team's goals Chad scored.
    percent × base = percentage
    $\text{percent} \times 6 = 2$
    $\text{percent} = \frac{2}{6} = \frac{1}{3} = 33\frac{1}{3}\%$

7. **C** Find out how much the 9% tax will add to the cost of the house.
    percent × base = percentage
    $0.09 \times \$150,000 = \$13,500$
    Add to find the total cost of the house.
    $\$150,000 + \$13,500 = \$163,500$

8. **H** Find the price of the grill after the 10% reduction.
    percent × base = percentage
    $0.10 \times \$120 = \$12$
    There is a $12 reduction from the original price.
    $\$120 - \$12 = \$108$
    Find the amount of the 6% sales tax.
    percent × base = percentage
    $0.06 \times \$108 = \$6.48$
    Find the final price of the grill by adding the tax to the sale price.
    $\$108 + \$6.48 = \$114.48$

9. C Find the percent of decrease.

$$\text{Percent of decrease} = \frac{\text{Original price} - \text{New price}}{\text{Original price}}$$

$$\text{Percent of decrease} = \frac{\$225 - \$180}{\$225} = \frac{\$45}{\$225} = 0.2$$

$$= 20\%$$

10. K Find the percent of increase.

$$\text{Percent of increase} = \frac{\text{New amount} - \text{Original amount}}{\text{Original amount}}$$

$$\text{Percent of increase} = \frac{5 \text{ pounds} - 3 \text{ pounds}}{3 \text{ pounds}} = \frac{2 \text{ pounds}}{3 \text{ pounds}}$$

$$= 66\frac{2}{3}\%$$

Ratio and Proportion

Practice (p. 96)

1. 5:7 and 5 to 7
2. $\frac{3}{13}$ and 3 to 13
3. $\frac{4}{9}$ and 4:9
4. 2:5 and 2 to 5
5. 6
6. 0.6
7. 5.2
8. 21
9. 14
10. 20
11. 55
12. 5
13. $x = 52$
14. 4 to 9 and 4:9
15. $250
16. 30 scoops
17. 84 eggs
18. 450 tiles are needed to cover 120 square feet of the roof.
19. The person casts a 9.6-foot shadow.
20. 20 apples

ACT-Type Problems (p. 97)

1. D Find the number of people who are eating dinner.

17 friends + John = 18 people eating dinner

Set up a proportion.

$$\frac{3 \text{ people}}{5 \text{ tomatoes}} = \frac{18 \text{ people}}{x \text{ tomatoes}}$$

Cross multiply and solve.

$$3x = 18 \times 5$$
$$3x = 90$$
$$x = 30$$

John will use 30 tomatoes.

2. J Set up a proportion and solve.

$$\frac{1 \text{ person}}{8 \text{ glasses}} = \frac{45 \text{ people}}{x \text{ glasses}}$$

$$x = 45 \times 8$$
$$x = 360 \text{ glasses}$$

They will drink 360 glasses of water.

3. D Rewrite the ratio $\frac{5}{6}$ as 5:6 and 5 to 6.

4. G Set up a proportion.

$$\frac{6 \text{ books}}{\$13} = \frac{x \text{ books}}{\$32.50}$$

Cross multiply and solve.

$$6 \times 32.5 = 13x$$
$$195 = 13x$$
$$15 = x$$

Jeff could buy 15 books.

5. B Find the number of players that are getting taped.

percent × base = percentage

$$0.8 \times 15 = 12$$

Set up a ratio.

$$\frac{1 \text{ player}}{0.75 \text{ roll of tape}} = \frac{12 \text{ players}}{x \text{ rolls of tape}}$$

Cross multiply and solve.

$$x = 12 \times 0.75$$
$$x = 9$$

Nine rolls of tape will be needed.

Statistics—Mean, Median, and Mode

Practice (p. 100)

1. 6
2. 6
3. 6
4. 80
5. 79.5
6. 94
7. 407
8. 458
9. 467

	Mean	Median	Mode
10.	5.7	5	4, 5, and 9
11.	4.4	4.9	None
12.	421.5	475	475
13.	30.1	31.3	37

14. $25.6 \times 12 = 307.2$ The sum of the items is 307.2.
15. $761.8 \div 58.6 = 13$ There are 13 items.
16. $6 \times 47 = 282$ $9 \times 49 = 441$
    $282 + 441 = 723$ $723 \div 15 = 48.2 \approx 48$ bushels
17. $(6 \times 61) + (5 \times 60) + [(15 - 11) \times 59] = 902$
    $902 \div 15 = 60.13 \approx 60$ inches
18. The average grade is 80.5.
19. 194.6 miles
20. 1.36 hours

ACT-Type Problems (p. 100)

1. **C** Write the numbers in a list from lowest to highest.

   1, 2, 3, 4, 4, 7, 9, 10

   Find the mean.

   $(1 + 2 + 3 + 4 + 4 + 7 + 9 + 10) \div 8 = 5$

   Find the median.

   $(4 + 4) \div 2 = 4$

   Find the mode.

   4

   Add the mean, median, and mode.

   $5 + 4 + 4 = 13$

2. **J** Find the mean.

   $(3 + 5 + 2 + 6 + 8) \div 5 = 4.8$

   Round to the ones place. 5 glasses.

3. **B** t = total number of points scored = 93

   a = average number of points scored = 18.6

   n = number of games played = ?

   $$\frac{t}{n} = a$$

   $$\frac{93}{n} = 18.6$$

   $$n \times 18.6 = 93$$

   $$n = 5$$

4. **H** Find the mean.

   $(4 \times 172) + (6 \times 134) \div 10 = 149.2$

5. **E** t = total number of push-ups completed = ?

   a = average number of push-ups = 56.5

   n = number of attempts = 16

   $$\frac{t}{n} = a$$

   $$\frac{t}{16} = 56.5$$

   $$t = 56.5 \times 16 = 904$$

Data Collection, Representation, and Interpretation

Practice (p. 104)

1.

4	3, 7, 8
5	3, 5, 6, 7, 8
6	0, 1, 2
7	1, 2, 5, 6, 9
8	1, 2, 4, 8

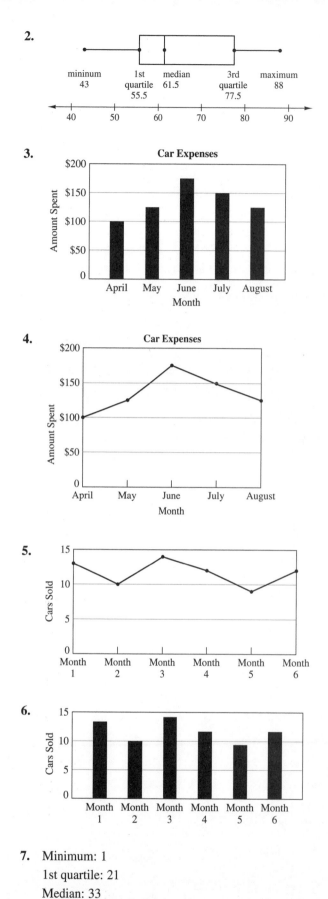

2.

3. Car Expenses

4. Car Expenses

5.

6.

7. Minimum: 1

   1st quartile: 21

   Median: 33

   3rd quartile: 42

   Maximum: 59

8.

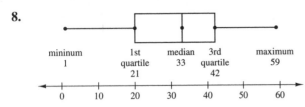

minimum
1

1st
quartile
21

median
33

3rd
quartile
42

maximum
59

9. Krissi

10. Exercising

ACT-Type Problems (p. 105)

1. **E** The third quartile score is 26. That means 75% of the scores are at or below 26. A score greater than 75% of the scores will be greater than 26.

2. **H** There are 24 sales figures. The median sale is between the 12th sales figure (143) and 13th sales figure (144).

3. **A**

4. **J** Add the lengths of all the bars to find how many students are in the class: $1 + 3 + 5 + 3 = 12$ students total. Five students scored in the 80's; three students scored in the 90's. So 8 out of 12 students scored in the 80's or higher. Convert to a percentage:
$\frac{8}{12} = 0.\overline{66} = 66\frac{2}{3}\%$

5. **C** Add the temperature for each day:
$90 + 85 + 95 + 100 + 100 + 110 + 100 = 680$. Divide by the number of days, 7: $680 \div 7 \approx 97.1$. To the nearest degree, the average temperature is 97°.

Probability

Practice (p. 109)

1. $\frac{1}{2}$
2. $\frac{1}{2}$
3. 1
4. $\frac{1}{2}$
5. $\frac{1}{6}$
6. $\frac{3}{6} = \frac{1}{2}$
7. $\frac{2}{6} = \frac{1}{3}$
8. $\frac{0}{6} = 0$
9. $\frac{2}{6} = \frac{1}{3}$
10. $\frac{4}{52} = \frac{1}{13}$
11. $\frac{26}{52} = \frac{1}{2}$
12. $\frac{1}{4}$
13. $\frac{40}{52} = \frac{10}{13}$
14. $\frac{12}{52} = \frac{3}{13}$
15. $\frac{8}{52} = \frac{2}{13}$
16. $\frac{12}{52} = \frac{3}{13}$
17. $\frac{1}{51}$
18. $\frac{0}{51} = 0$
19. $\frac{3}{51}$
20. $\frac{12}{51} = \frac{4}{17}$

ACT-Type Problems (p. 110)

1. **B** P(brown sock) $= \frac{\text{number of brown socks}}{\text{total number of socks}} = \frac{3}{10}$

2. **G** P(1 or 5) $= \frac{\text{number of sides with a 1 or 5}}{\text{total number of sides}} = \frac{2}{6} = \frac{1}{3}$

3. **E** Step 1: Find the number of cards divisible by 2.
2, 4, 6, 8, and 10 are all divisible by 2. In a deck of cards there are 4 of each, so there are 20 cards divisible by 2.

Step 2: Find the probability.

P(card divisible by 2) $= \frac{\text{number of cards divisible by 2}}{\text{total number of cards in deck}}$

$= \frac{20}{52} = \frac{5}{13}$

4. **F** Step 1: Find the number of face cards left.
There are 12 face cards. If 2 are removed, there are 10 face cards left.

Step 2: Find the number of cards left in the deck.
There are 52 cards in the deck. If two are removed, there are 50 cards left.

Step 3: Find the probability.

P(picking face card after removal)
$= \frac{\text{number of face cards after removal}}{\text{number of cards in deck after removal}}$
$= \frac{10}{50} = \frac{1}{5}$

5. **D** Step 1: Find the probability of picking a red pencil (do not simplify).
P(red pencil) $= \frac{\text{number of red pencils}}{\text{total number of pencils}} = \frac{4}{20}$

Step 2: Find the probability of picking a black pencil (do not simplify).

P(black pencil) $= \frac{\text{number of black pencils}}{\text{total number of pencils}} = \frac{1}{20}$

Step 3: Add the probabilities.
$\frac{4}{20} + \frac{1}{20} = \frac{5}{20} = \frac{1}{4}$

Elementary Counting Techniques

Practice (p. 113)

1. 120
2. 336
3. 35
4. 24
5. 48
6. 15
7. 120
8. 625
9. 504
10. 5,040

ACT-Type Problems (p. 114)

1. **C** This is a combination problem.
$\frac{n!}{(n-r)!(r)!} = \frac{8!}{(8-2)!(2)!} = \frac{8!}{6! \times 2!}$

$= \frac{8 \times 7 \times \cancel{6} \times \cancel{5} \times \cancel{4} \times \cancel{3} \times \cancel{2} \times \cancel{1}}{\cancel{6} \times \cancel{5} \times \cancel{4} \times \cancel{3} \times \cancel{2} \times \cancel{1} \times 2 \times 1} = \frac{56}{2} = 28$

It is easier to use a calculator.

2. **J** This is a permutation problem.
The answer is 4!, or 24. There are 24 different ways to distribute the gifts.

3. **E** This is a product problem. There are 26 letters and 10 digits.

$26 \times 26 \times 26 \times 10 \times 10 \times 10 = 17{,}576{,}000.$
Use a calculator.

4. **K** This is a permutation problem. Use a calculator.

$$\frac{n!}{(n-r)!} = \frac{6!}{(6-5)!} = \frac{6 \times 5 \times 4 \times 3 \times 2 \times 1}{1} = 720$$

There are 720 ways for the reporters to be assigned to stories.

5. **D** This is a product problem. There are 10 digits, 0–9. Only 9 of them are eligible for the first slot, while the other two slots can be filled by any of the ten digits. $9 \times 10 \times 10 = 900$. There are 900 possible area codes whose first digit is not zero.

Writing Linear Expressions and Equations

Practice (p. 116)

1. B = amount at the beginning of the year
W = amount withdrawn during the year
(You may choose other letters for the variables.)
$B - W$

2. $\dfrac{Q + R + S + T + U}{5}$

3. $8r + 3(1.5r) = 8r + 4.5r = 12.5r$

4. $P + (I\% \times P)$

5. $w = \$3{,}155 - \$2{,}855$

6. $a = \$4 + (12 - 9)(\$0.75)$ or $a = \$4 + 3(\$0.75)$

7. Let A = amount of acid left
Let e = number of experiments
$A = 100 - 4e$

8. Let P = final payment
$P = 8{,}000f - 0.25(8{,}000f)$
$P = 8{,}000f - 2{,}000f$
$P = 6{,}000f$

9. $K + 3 = M - 29$
$K + 32 = M$

10. Let P = amount paid
$P = 20 + 0.08(20) + 0.15(20)$

ACT-Type Problems (p. 116)

1. **D** There are 50 copies of the virus on each infected computer. Multiply 50 by the number of infected computers.
$V = 50n$

2. **G** Since the account grows by \$250,000 in 2 years, in regular yearly amounts $\left(\dfrac{250{,}000}{2}\right)$, the amount in the account after x years is $\left(\dfrac{250{,}000}{2}\right)x$. So after x years there would be $500{,}000 + \left(\dfrac{250{,}000}{2}\right)x$ dollars in the account.

3. **D** The 30 chocolate bars are being divided into 12 groups. Divide 30 by 12 to find the number of chocolate bars each child receives.
$c = 30 \div 12$

4. **F** Joan works h hours and is paid \$5 per hour. Multiply $\$5 \times h$ to find her base salary. Her commission is 8% of the value of the bicycles sold. Convert 8% to a decimal and multiply by the value of the bicycles, v: $0.08 \times v$. Add Joan's base salary to her commission to find her total pay.
$5h + 0.08v$

5. **E** To find the average of 5 tests, find their sum and divide by 5. Having an average of 82 on the first four tests is like having an 82 four times.

$$\frac{82 + 82 + 82 + 82 + T}{5} = 85$$

Solving Linear Equations

Practice (p. 120)

1. $w = 16$
2. $x = -21$
3. $y = 5$
4. $z = -3.5$
5. $y = \dfrac{1}{5}$
6. $y = 10$
7. $v = 9$
8. $t = -9$
9. $m = -1.6$
10. $n = 1$
11. $w = 3$
12. $s = -22$
13. $d = 3$
14. $g = \dfrac{1}{2}$
15. $k = -2\dfrac{1}{2}$
16. $x = \dfrac{c - y}{2}$
17. $k = \dfrac{m - p}{3}$
18. $x = y - z$
19. $k = \dfrac{2y - r}{3}$
20. $x = 24$
21. $y = 31\dfrac{1}{2}$
22. $k = 256$
23. $a = \dfrac{5}{11}$
24. $b = 4\dfrac{3}{7}$
25. $t = \dfrac{2}{15}$
26. \$112.50
27. 70 home runs
28. 9
29. 7.5
30. 5 miles

ACT-Type Problems (p. 121)

1. **D**

Write the equation.	$5y - 3y + 13$	$= 26$
Combine like terms.	$2y + 13$	$= 26$
Subtract 13.	$2y$	$= 13$
Divide by 2.	y	$= \dfrac{13}{2} = 6\dfrac{1}{2}$
		$= 6.5$

2. K Solve for r.

Write the equation.	$4r - 5 = 17$
Add 5.	$4r = 22$
Divide by 4.	$r = 5.5$

Solve for s.

Write the equation.	$12 - 2s = 2$
Subtract 12.	$-2s = -10$
Divide by -2.	$s = 5$
Multiply r and s.	$rs = 5.5(5) = 27.5$

3. C m = Mike's age = 28 e = Emily's age = ?

Write the equation. $e = \frac{1}{2}m + 7$

Substitute 28 for m. $e = \frac{1}{2}(28) + 7 = 14 + 7 = 21$

Emily is 21 years old.

4. F Write the equation. $\frac{2}{3}h - 5 = h + 7$

Subtract 7. $\frac{2}{3}h - 12 = h$

Subtract $\frac{2}{3}h$. $-12 = \frac{1}{3}h$

Multiply by 3. $-36 = h$

5. C Write the equation.
$$\$2,225c - \$2,000 = P$$

Substitute \$49,175 for P.
$$\$2,225c - \$2,000 = \$49,175$$

Add \$2,000. $\$2,225c = \$51,175$

Divide by \$2,225. $c = 23$ computers

Cumulative ACT Practice
Pre-Algebra II (p. 122)

Correct your Pre-Algebra II Cumulative ACT Practice answers and work through the answer explanations. If you missed an item or guessed at the answer, circle the number of the item below and review the pages indicated.

1. B Topic: Ratio and Proportion (pp. 94–97)

Since you are making x dollars for every hour of work, multiply the number of hours worked, 8, by the amount paid per hour, x.

2. F Topic: Percent (pp. 86–89)

$60 \times 0.15 = 9$

3. E Topic: Ratio and Proportion (pp. 94–97)

1 centimeter = 6 inches

Set up a proportion. $\dfrac{\text{inches} \to}{\text{centimeters} \to}\dfrac{6}{1} = \dfrac{x}{6}$

Cross multiply and solve. $6 \times 6 = x$
$36 = x$

36 inches = 3 feet

The door will be 3 feet wide.

4. K Topic: Elementary Counting Techniques (pp. 110–114)

It matters which players fill the positions. This is a permutation problem. Use your calculator if it computes permutations.

Use $\dfrac{n!}{(n-r)!}$. $(n = 6, r = 3)$ $\dfrac{6!}{(6-3)!} = \dfrac{6!}{3!}$

$= \dfrac{6 \times 5 \times 4 \times \cancel{3 \times 2 \times 1}}{\cancel{3 \times 2 \times 1}} = 6 \times 5 \times 4 = 120$

There are 120 ways to fill the three positions.

5. B Topic: Statistics—Mean, Median, and Mode (pp. 98–101)

To find the mean, find the sum of the scores and divide by the number of scores. Multiply 7 (the number of scores) by 80 (the mean) to get 560.

The sum of the scores in choice B is 560.

6. J Topic: Probability (pp. 107–110)

If a pair of dice is thrown, there are 36 possible outcomes as shown in the table below.

Second Die

First Die \ Second Die	1	2	3	4	5	6
1	2	3	4	5	6	7
2	3	4	5	6	7	8
3	4	5	6	7	8	9
4	5	6	7	8	9	10
5	6	7	8	9	10	11
6	7	8	9	10	11	12

A probability of $\frac{1}{9}$ means the sum occurs 4 times out of the 36 possibilities. From the table, you can see that 5 is the only sum that occurs 4 times.

7. C Topic: Writing Linear Expressions and Equations (pp. 114–117)

10% of I can be written as $0.10 \cdot I$.

The amount Erin has placed into the account after M months is $0.10 \cdot I \cdot M$.

8. J Topic: Data Collection, Representation, and Interpretation (pp. 101–106)

A \$900 a year increase in take-home pay means an average monthly increase of \$75. (\$900 ÷ 12 months = \$75 per month) Jan's food budget is 18% of her monthly take-home pay. Multiply:

18% of \$75 = $0.18 \times \$75 = \$13.50$

9. D Topic: Ratio and Proportion (pp. 94–97)

Set up a proportion. Jordan $\to \dfrac{3}{18} = \dfrac{1}{A} \gets$ Andy

Cross multiply and solve. $3A = 18$
$A = 6$

If Jordan can run 18 miles, Andy can run 6 miles.

10. **H** Topics: Writing Linear Expressions and Equations (pp. 114–117); Solving Linear Equations (pp. 117–121)

Write an equation for the total cost of beads.

Total = $0.05R + 0.15B + 0.25G$

Substitute the number of beads and multiply.

Total = $0.05(4) + 0.15(5) + 0.25(3) = \$1.70$

Pre-Algebra Subtest (p. 123)

The answers and explanations for the Pre-Algebra Subtest items are shown with the topic each item tests. If you missed an item or guessed at the answer, circle the number of the item below and review the pages indicated.

1. **C** Topic: Factors, Divisibility, and Primes (pp. 58–61)

To find the factors of 40, start from the ends and work in. Find the factors in pairs. 1, 2, 4, 5, 8, 10, 20, 40

2. **G** Topics: Writing Linear Expressions and Equations (pp. 114–117); Solving Linear Equations (pp. 117–121)

y is the number of $1,300 type y computers.

x is the number of $1,500 type x computers.

The company sells 3 times as many x computers as y computers, so $x = 3y$.

The company sold a total of 10,000 computers, so $x + y = 10{,}000$ or $3y + y = 10{,}000$.

Solve the equation for y.

$$4y = 10{,}000$$
$$\frac{4y}{4} = \frac{10{,}000}{4}$$
$$y = 2{,}500$$

Since $x = 3y$, $x = 3(2{,}500) = 7{,}500$.

So, the company sold 2,500 type y computers and 7,500 type x computers.

The company made $7{,}500 \times \$1{,}500 = \$11{,}250{,}000$ from selling type x computers.

The company made $2{,}500 \times \$1{,}300 = \$3{,}250{,}000$ from selling type y computers.

The difference is $\$11{,}250{,}000 - \$3{,}250{,}000 = \$8{,}000{,}000$.

Write $8,000,000 as $\$8 \times 10^6$.

3. **D** Topic: Percent (pp. 86–89)

Remember: percent × base = percentage

Write an equation. $12\% \times ? = 6$

Divide. $? = 6 \div 0.12 = 50$

4. **G** Topic: Ratio and Proportion (pp. 94–97)

Change 1 foot to 12 inches.

Write a proportion.

$$\frac{\text{distance on map} \;\rightarrow}{\text{actual distance} \;\rightarrow} \frac{12 \text{ inches}}{x \text{ miles}} = \frac{1 \text{ inch}}{300 \text{ miles}}$$

Cross multiply. $x = 12 \times 300 = 3{,}600$

5. **B** Topic: Elementary Counting Techniques (pp. 110–114)

Who occupies a particular space is not important. This is a combination problem.

Use $\dfrac{n!}{n!(n-r)!}$. $(n = 7, r = 5)$

$$\frac{7!}{5!\,2!} = \frac{7 \times 6}{2 \times 1} = \frac{42}{2} = 21$$

6. **H** Topic: Statistics—Mean, Median, and Mode (pp. 98–101)

Find the sum of the scores:

$93 + 89 + 95 + 78 + 78 + 80 + 92 + 87 + 82 = 774$.

Divide by the number of scores: $774 \div 9 = 86$.

7. **E** Topic: Probability (pp. 107–110)

The probability of an event is written as a fraction:

$$\frac{\text{Total number of ways to obtain a 7}}{\text{Total number of possible outcomes}}$$

There are six ways to roll a 7 with a pair of dice: 1, 6; 6, 1; 2, 5; 5, 2; 3, 4; 4, 3.

There are 36 possible outcomes when a pair of dice is rolled.

$$\frac{6}{36} = \frac{1}{6}$$

8. **G** Topic: Fractions (pp. 64–67)

Rewrite the fractions with a common denominator.

$$\frac{1}{2} = \frac{20}{40} \quad \frac{2}{8} = \frac{10}{40} \quad \frac{3}{5} = \frac{24}{40} \quad \frac{7}{20} = \frac{14}{40} \quad \frac{11}{40}$$

Once all the fractions have the same denominator, use the numerators to write them in order from smallest to largest.

$$\frac{2}{8}, \frac{11}{40}, \frac{7}{20}, \frac{1}{2}, \frac{3}{5}$$

The fractions can also be written as decimals to compare.

$$\frac{1}{2} = 0.5 \quad \frac{2}{8} = 0.25 \quad \frac{3}{5} = 0.6 \quad \frac{7}{20} = 0.35 \quad \frac{11}{40} = 0.275$$

9. **D** Topic: Square Roots, Exponents, and Scientific Notation (pp. 61–64)

Add the weights: $50 + 50 + 25 + 25 + 45 = 195$.

Change 195 to scientific notation: $195 = 1.95 \times 10^2$.

10. **J** Topic: Ratio and Proportion (pp. 94–97)

Write a proportion. $\dfrac{\text{number of tables}}{\text{number of people}} \to \dfrac{4}{28} = \dfrac{x}{224}$

Cross multiply. $\quad 28x = 4 \times 224$

$$28x = 896$$

$$x = 896 \div 28 = 32$$

11. **B** Topic: Order of Operations (pp. 81–83)

Use the order of operations to simplify the numerator and the denominator. Remember: parentheses, exponents, multiplication and division, addition and subtraction.

$$\dfrac{3 \times \dfrac{1}{4} + 5 + 0.2 - 3.7 + 2 - \dfrac{1}{3}}{8 - 2.1 + \dfrac{2}{5} \times 8} = \dfrac{\dfrac{3}{4} + 5 + 0.2 - 3.7 + 2 - \dfrac{1}{3}}{8 - 2.1 + \dfrac{16}{5}}$$

$$\approx \dfrac{3.92}{9.1} \approx 3.92 \div 9.1$$

$$\approx 0.430\ldots \approx 0.43$$

12. **J** Topic: Elementary Counting Techniques (pp. 110–114)

The cars are going to just one tollbooth, so the other tollbooths do not matter. The question asks about the order of all the cars. This is a permutation problem.

Use $n!$ $(n = 4)$ $\quad n! = 4! = 4 \times 3 \times 2 \times 1 = 24$

There are 24 ways for the cars to line up.

13. **D** Topic: Data Collection, Representation, and Interpretation (pp. 101–106)

Track the graphs day by day for Item A and Item C to find the correct date, November 23. The Item A sales level is 14,000 while the Item C sales level is 28,000.

14. **H** Topic: Data Collection, Representation, and Interpretation (pp. 101–106)

Inspection shows that the mode is 110. That attendance appears twice. There are 17 games, so the median attendance is 9 from the top or the bottom. That attendance is 145. The difference between the mode and the median is $145 - 110 = 35$.

Chapter 8 Elementary Algebra
Evaluating Formulas and Expressions

Practice (p. 127)

1. 12 cm²

2. 0.49 cm²

3. 1.8 cm²

4. 3.51 cm²

5. 2.925 cm²

6. $\dfrac{1}{16}$ cm²

7. Circumference: $2\pi(2) = 4\pi$ cm or 12.56 cm
Area: $\pi(2)^2 = 4\pi$ cm² or 12.56 cm²

8. Circumference: 6π cm or 18.84 cm
(Remember: If $d = 6$, $r = 3$.)
Area: 9π cm² or 28.26 cm²

9. Circumference: 1.8π cm or 5.652 cm
Area: 0.81π cm² or 2.5434 cm²

10. 0.729 cm³

11. 36 cm³

12. 36π cm³ or 113.04 cm³

13. $d = 120$ miles

14. $r = 3$ feet per second

15. $t = 6$ hours

16. $1,687.50

17. 13π cm or 40.82 cm

18. 58 mph

19. 31.25 in.²

20. $4,500\pi$ mm³ or 14,130 mm³

21. $1,107

22. $2,720

ACT-Type Problems (p. 128)

1. **E** The face of a cube is a square. Use the formula for the area of a square to find the length of a side.

$$A = s^2$$
$$36 \text{ m}^2 = s^2$$
$$6 \text{ m} = s$$

Find the volume of the cube.

$$V = s^3$$
$$V = 6^3 \text{ m}^3$$
$$V = 216 \text{ m}^3$$

2. **K** Find the length of the side of a square with an area of 30.25 in.².

$$A = s^2$$
$$30.25 \text{ in.}^2 = s^2$$
$$\sqrt{30.25 \text{ in.}^2} = \sqrt{s^2}$$
$$5.5 \text{ in.} = s$$

Multiply by 4. $\quad 4 \times 5.5 = 22$

The sum of the sides of the square is 22 inches.

Find the length of the edge of a cube with a volume of 343 in.³

$$V = s^3$$
$$343 \text{ in.}^3 = s^3$$
$$\sqrt[3]{343 \text{ in.}^3} = \sqrt[3]{s^3}$$
$$7 \text{ in.} = s$$

Multiply by 12. $12 \times 7 = 84$

The sum of the edges of the cube is 84 inches.

$$22 \text{ in.} + 84 \text{ in.} = 106 \text{ in.}$$

3. **D** Find the radius of the circle.

$$C = 2\pi r$$
$$24\pi = 2\pi r$$
$$12 = r$$

Find the area of the circle.

$$A = \pi r^2$$
$$A = \pi (12)^2 \text{ cm}^2$$
$$A = \pi (144) \text{ cm}^2$$
$$A = 144\pi \text{ cm}^2$$

4. **F** Find the width of the rectangle.

$$A = lw$$
$$30 = 6w$$
$$5 = w$$

Add the length and the width.

6 in. + 5 in. = 11 in.

5. **B** Find the trip distance.

$$d = rt$$
$$d = 6(65)$$
$$d = 390$$

Find out how long it took Eric to travel 390 miles at a speed of 60 mph.

$$d = rt$$
$$390 = 60t$$
$$t = 6.5 \text{ hours}$$

Find the difference between John's time and Eric's time.

$$6.5 \text{ hours} - 6 \text{ hours} = 0.5 \text{ hour} = \frac{1}{2} \text{ hour}$$

Exponents and Radicals

Practice (p. 131)

1. 7^{11} or 1,977,326,743

2. 9^3 or 729

3. $2^3 \times 3^4$ or 648

4. 1

5. 3^7 or 2,187

6. $\frac{1}{125}$ or 5^{-3}

7. 1

8. 12^{10}

9. $\frac{1}{2}$

10. 6^5 or 7,776

11. 5

12. $12\sqrt{6}$

13. $2\sqrt[3]{7}$

14. $4\sqrt[3]{9}$

15. $9\sqrt{2}$

16. $\frac{7\sqrt{11}}{11}$

17. $\sqrt{2}$

18. $\sqrt{19}$

19. $\frac{13\sqrt[3]{1,369}}{74}$

20. $\frac{13\sqrt{2}}{4}$

ACT-Type Problems (p. 131)

1. **B** $13^{-2} \times 13^2 = 13^{(-2+2)} = 13^0 = 1$

2. **K** $\dfrac{7}{\sqrt{45}} = \dfrac{7}{\sqrt{45}} \times \dfrac{\sqrt{45}}{\sqrt{45}} = \dfrac{7\sqrt{9 \times 5}}{45} = \dfrac{7 \times 3 \times \sqrt{5}}{45}$

$$= \dfrac{7\sqrt{5}}{15}$$

3. **E** $5^4 \div 25^2 = 5^4 \div (5^2)^2 = 5^4 \div 5^4 = 5^{(4-4)} = 5^0 = 1$

4. **H** $\dfrac{4^7}{\sqrt{16}} = \dfrac{4^7}{4} = 4^6 = (2^2)^6 = 2^{12}$

5. **E** Write the reciprocal of $\dfrac{\sqrt{72}}{6^5}$. $\dfrac{6^5}{\sqrt{72}}$

Simplify. $\dfrac{6^5}{\sqrt{36\sqrt{2}}} = \dfrac{6^5}{6\sqrt{2}} = \dfrac{6^4\sqrt{2}}{\sqrt{2}\sqrt{2}} = 6^4\dfrac{\sqrt{2}}{2}$

$6^{12} \div 6^3 \times \dfrac{\sqrt{2}}{2} = 6^9\dfrac{\sqrt{2}}{2}$, which is not equal to $6^4\dfrac{\sqrt{2}}{2}$.

Operations with Radicals

Practice (p. 134)

1. $23\sqrt{6}$

2. $46\sqrt[3]{7}$

3. $91\sqrt{3}$

4. $-52\sqrt{2}$

5. $24\sqrt{3}$

6. $10\sqrt[3]{12}$

7. $2\sqrt{14}$

8. $3\sqrt{13}$

9. $4\sqrt{15}$

10. $\dfrac{\sqrt{13}}{13}$

11. $2\sqrt{2}$

12. $\dfrac{1}{3}$

13. $9\sqrt{2}$

14. $3\sqrt{13}$

15. $4\sqrt{3}$

16. $\dfrac{1}{12}$

17. $\dfrac{5\sqrt{2}}{2}$

18. 4

19. 0

20. $-4\sqrt{15}$

ACT-Type Problems (p. 134)

1. **C** Multiply the radicals and simplify.

$$\sqrt{22} \times \sqrt{14} = \sqrt{22 \times 14} = \sqrt{308}$$
$$= \sqrt{4 \times 77} = 2\sqrt{77}$$

2. **G** Simplify the radicals. Then combine similar radicals.

$$\sqrt{54} - \sqrt{96} = \sqrt{9 \times 6} - \sqrt{16 \times 6}$$
$$= 3\sqrt{6} - 4\sqrt{6} = -\sqrt{6}$$

3. **D** Simplify the radicals. Then add similar radicals.

$$3\sqrt{125} + 6\sqrt{80} = 3\sqrt{25 \times 5} + 6\sqrt{16 \times 5}$$
$$= 3 \times 5\sqrt{5} + 6 \times 4\sqrt{5} = 15\sqrt{5} + 24\sqrt{5}$$
$$= 39\sqrt{5}$$

4. **F** Simplify radicals where possible. Then combine similar radicals and multiply.

$$\sqrt{8}\left(\sqrt{13} - \sqrt{117}\right)$$
$$= \sqrt{4 \times 2}\left(\sqrt{13} - \sqrt{9 \times 13}\right)$$
$$= 2\sqrt{2}\left(\sqrt{13} - 3\sqrt{13}\right) = 2\sqrt{2}\left(-2\sqrt{13}\right)$$
$$= -4\sqrt{26}$$

5. D Simplify. $\dfrac{\sqrt{45}}{\sqrt{10}} = \dfrac{\sqrt{9}\sqrt{5}}{\sqrt{2}\sqrt{5}} = \dfrac{3}{\sqrt{2}} = \dfrac{3\sqrt{2}}{\sqrt{2}\sqrt{2}} = \dfrac{3\sqrt{2}}{2}$

Multiply numerator and denominator.
$3\sqrt{2} \times 2 = 6\sqrt{2}$

Polynomials

Practice (p. 139)

1. $3x^2 + 3x^2y + 10y$
2. $7x^2y^2 + 6x^2y + 24x$
3. $4x^2y - 6y^2x + 16x^2 - 8xy + 17x + 13$
4. $3x^2 + 6y^2 - 3y$
5. $-7x^2 + 11y^2 + 3xy - 7x + 3y$
6. $10x^5y - 2x^2 - 7y^2 - 20xy$
7. $5y^4 + 4y^3 + 10x^2y^2 + 2y - 19x$
8. $-14x^2y^4 - 4xy^3 - 18y^2 + 5x^2 - y$
9. $11x^5 + 4x^3y + 3xy^3 - 6x^2y + 6xy^2 + x^2 - 4y^2$
10. $6x^5 + 13x^4 - 4x^3 - 2x^2y^3 + 3xy^2 + 2x^2y^2$
11. $-x^5 - 6y^3 + 15x^2 + 17y^2 - 6x - 6y$
12. $-18x^5y - xy^5 + 15x^4 - 14y^3 - 1$
13. $-4x^4 - 20x^3y^2 + 15x^2y^3 + 13xy - 12x - 19y$
14. $-15x^9 + 15x^8 + 24x^4 + 3x^3y - 4x^2y + 5xy^3$
15. $-6x^3 + 9x^2 + 3y^5 + 3y^3 - xy^2 - 3x + 6$
16. $-20x^5 + 15x^3 - 20x^2y^2 + 5x^2$
17. $24x^3 + 16x^3y + 32x^2y - 48x^2$
18. $12x^2 + 30x + 12$
19. $14x^2 - 33x - 5$
20. $27x^2 - 60x - 32$
21. $x^2 + 5x - 4$
22. $x^2 + 9x - 3$
23. $x^3 + 2x^2 + 6x + 1$
24. $x^2 + 8x + 6$
25. $x^2 - 2x + 8$

ACT-Type Problems (p. 140)

1. **C** Change the signs of terms in the polynomial being subtracted.

$3x^5 - 2x^4y + 9x^2y^2 + 2xy - 5x^4y^2 + 7x^5 + 18x^2y^2 - 2xy$

Rearrange terms.

$-5x^4y^2 + 3x^5 + 7x^5 - 2x^4y + 9x^2y^2 + 18x^2y^2 + 2xy - 2xy$

Combine similar terms.

$-5x^4y^2 + 10x^5 - 2x^4y + 27x^2y^2$

2. **G** Multiply each term of the polynomial by the monomial.

$5x^2y(2x^3 - 6x^2y^2 + 4xy^2) = 10x^5y - 30x^4y^3 + 20x^3y^3$

3. **E** Use the FOIL method to multiply the two binomials.

$(2x^2 + 5)(x^2 - 7) = 2x^4 - 14x^2 + 5x^2 - 35$

Combine similar terms. $2x^4 - 9x^2 - 35$

4. **G**

$$
\begin{array}{r}
x^2 + 3x - 7 \\
2x + 1 \overline{)\ 2x^3 + 7x^2 - 11x - 7} \\
\underline{-(2x^3 + \ x^2)} \\
6x^2 - 11x \\
\underline{-(6x^2 + \ 3x)} \\
-14x - 7 \\
\underline{-(-14x - 7)} \\
0
\end{array}
$$

5. **C**

Write as a polynomial.
$2y^5x - 7y^3x^2 + 13y^2x - 6x + 10y^5x + 3y^2x^3$
$- 21y^2x + 5x$

Rearrange terms.
$2y^5x + 10y^5x - 7y^3x^2 + 3y^2x^3 + 13y^2x - 21y^2x$
$- 6x + 5x$

Combine similar terms.
$12y^5x - 7y^3x^2 + 3y^2x^3 - 8y^2x - x$

Factoring Polynomials

Practice (p. 143)

1. $2x^4(3 + 4y^2)$
2. $4y^3(x^2 - 3y + 4xy)$
3. $2x(6x^2 + 5x - 10)$
4. $(5w + 0.7z)(5w - 0.7z)$
5. $5(3k^2 + y^2)$
6. $3z^2(5x^2 + 3)$
7. $5xy(x^2y + 2 - 3xy - y)$
8. $4x^2y(2x^2y^4 - 4 + 3xy^2)$
9. $(x - 5)(x + 3)$
10. $12xyz(6x^3yz^3 + 5)$
11. $-14yz(y + 1)$
12. $4(x - 2y)(x + 2y)$
13. $(3x - 7)(2x + 5)$
14. $x^2z(7xz - 13z + 27)$
15. $(4x - 3y)(16x^2 + 12xy + 9y^2)$
16. $4x^2z^2(4x^2z - x - 2 + 5x^3z)$

17. $(2x - 9)(x + 6)$

18. $3(2x - 5)(2x - 1)$

19. $(5x + 7y)(25x^2 - 35xy + 49y^2)$

20. $(x + 2)(x - 1)$

ACT-Type Problems (p. 144)

1. **A** $144x^2y^2 - 169z^2$ is the difference of two squares.
    $$144x^2y^2 - 169z^2 = (12xy - 13z)(12xy + 13z)$$

2. **G** Group to show the parts to be factored.
    $$(6x^2 + 27x) - (14x + 63)$$
    Factor out $3x$ from the first part and 7 from the second part.
    $$3x(2x + 9) - 7(2x + 9)$$
    Factor out $(2x + 9)$ because it is the common factor in both terms.
    $$(2x + 9)(3x - 7)$$

3. **C** Choose the greatest common factor of the coefficient.   4
    Choose the smallest exponent for each variable. x: x^2
                                                                              y: y^2
    Factor out $4x^2y^2$. $4x^2y^2(6x - 3 + 4xy - 2x^3y^2)$

4. **H** Combine similar terms.
    $$(y - 2)(2y + 3) + (y - 2)$$
    Factor out $(y - 2)$.
    $$(y - 2)(2y + 3 + 1) = (y - 2)(2y + 4)$$
    Factor out 2. $2(y - 2)(y + 2)$

5. **D** $96x^2 - 16$ is the difference of two squares.
    $$(3x - 4)(3x + 4)$$
    Add the factors.
    $$(3x - 4) + (3x + 4) = 3x - 4 + 3x + 4 = 6x$$

Quadratic Equations

Practice (p. 148)

1. $x = \{-3, -1\}$

2. $x = \left\{\frac{3}{2}, 1\right\}$

3. $x = \{-3\}$

4. $x = \left\{-\frac{5}{3}, \frac{3}{2}\right\}$

5. $x = \left\{3\sqrt{2}, -3\sqrt{2}\right\}$

6. $x = \{0, -5\}$

7. $x = \left\{3, -\frac{7}{2}\right\}$

8. $x = \{2, -2\}$

9. $x = \left\{-\frac{2}{3}\right\}$

10. $x = \left\{5, -\frac{1}{4}\right\}$

11. $x = \{3, -2\}$

12. $x = \left\{\frac{6}{5}, -\frac{6}{5}\right\}$

13. $x = \{3, -3\}$

14. $x = \left\{\frac{7}{3}, -2\right\}$

15. $x = \left\{\frac{5}{2}\right\}$

16. $x = \{-2, -12\}$

17. $x = \left\{-\frac{1}{7}, \frac{1}{2}\right\}$

18. $x = \left\{-\frac{13}{2}, -3\right\}$

19. $x = \{-1\}$

20. $x = \left\{\frac{3}{5}, -\frac{7}{2}\right\}$

ACT-Type Problems (p. 148)

1. **E** $x^2 - 81 = 0$ is the difference of two squares.
    So $x^2 - 81 = (x - 9)(x + 9)$.
    $(x - 9)(x + 9) = 0$

    $x - 9 = 0$ $x + 9 = 0$
    Add 9. x $= 9$ Subtract 9. x $= -9$
    The solution set is $x = \{9, -9\}$.

2. **F** Rewrite the equation in quadratic form.
    $4x^2 - 24x = -36$ $\rightarrow$ $4x^2 - 24x + 36 = 0$
    $4x^2 - 24x + 36$ is a perfect square.
    So $4x^2 - 24x + 36 = (2x - 6)^2$.
    $(2x - 6)^2 = 0$ or $(2x - 6)(2x - 6) = 0$
    The factors are identical. Solve $2x - 6 = 0$ to find the solution to the quadratic equation.

    $2x - 6 = 0$
    Add 6. $2x$ $= 6$
    Divide by 2. x $= 3$
    The solution set is $x = \{3\}$.

3. **B** A quadratic equation has two solutions at most. Choice B has 3 solutions, so it cannot be the solution set for a quadratic equation.

4. **J** Find the solutions to the quadratic equation $(x - 3)(x + 5) = 0$.

    $x - 3 = 0$ $x + 5 =$ 0
    Add 3. x $= 3$ Subtract 5. x $= -5$
    Add the solutions. $3 + (-5) = -2$

5. **B** Rewrite the equation in quadratic form.
$$10x^2 = -21x + 10 \rightarrow 10x^2 + 21x - 10 = 0$$

Find the solutions to the quadratic equation
$10x^2 + 21x - 10 = 0$.

The first terms will either be
$(x\_\_\_)(10x\_\_\_)$ or $(2x\_\_\_)(5x\_\_\_)$.

The product of the outer terms is -10. The possible factors are:

$(x\_\_\_)(10x\_\_\_)$		or	$(2x\_\_\_)(5x\_\_\_)$	
-1	10		-1	10
1	-10		1	-10
10	-1		10	-1
-10	1		-10	1
-5	2		-5	2
5	-2		5	-2
2	-5		2	-5
-2	5		-2	5

The correct factored form is $(2x + 5)(5x - 2) = 0$.

Find the solutions to the quadratic equation.

	$2x + 5 = 0$		$5x - 2 = 0$
Subtract 5.	$2x = -5$	Add 2.	$5x = 2$
Divide by 2.	$x = -\dfrac{5}{2}$	Divide by 5.	$x = \dfrac{2}{5}$

Multiply the solutions. $-\dfrac{5}{2} \times \dfrac{2}{5} = -1$

Cumulative ACT Practice
Elementary Algebra (p. 149)

Correct your Elementary Algebra Cumulative ACT Practice answers and work through the answer explanations. If you missed an item or guessed at the answer, circle the number of the item below and review the pages indicated.

1. **C** Topic: Evaluating Formulas and Expressions (pp. 125–128)

Write an equation.	$E = I \cdot R$
Substitute for 12 for E and 2.5 for R.	$12 = I \cdot 2.5$
Divide by 2.5	$\dfrac{12}{2.5} = \dfrac{I \cdot 2.5}{2.5}$
	$4.8 = I$

2. **G** Topic: Quadratic Equations (pp. 144–148)

The solution set is $x = 2$ and $x = -4$, so the factored quadratic equation is $(x - 2)(x + 4) = 0$.

Multiply the factors to find the equation
$x^2 + 2x - 8 = 0$.

3. **A** Topic: Operations with Radicals (pp. 132–134)

$$\dfrac{30\sqrt{3}}{2\sqrt{5}}$$

Cancel.	$= \dfrac{15\sqrt{3}}{\sqrt{5}}$
Multiply numerator and denominator by denominator.	$= \dfrac{15\sqrt{15}}{\sqrt{25}}$
Simplify and cancel.	$= \dfrac{15\sqrt{15}}{5} = 3\sqrt{15}$

B and C cannot be simplified because the radicals are not similar.

D cannot be true because it has cubic roots.

E cannot be true because it simplifies to $\sqrt{15}$.

4. **G** Topic: Operations on Polynomials (pp. 135–140)

Align similar terms and add.

$$\begin{array}{r} a^2b^3 + 2ab - 3a \\ + \ 4a^2b^3 - 4ab \qquad + 3b \\ \hline 5a^2b^3 - 2ab - 3a + 3b \end{array}$$

F is not correct because the sum is
$2a^2b^2 + 4a^2b^3 - 8ab + 3a - 2b$.

H is not correct because the sum is
$5a^2b^3 - 2ab + 7a - 3b$.

J is not correct because the sum is
$5a^3b^2 + 6ab + 3a - 3b$.

K is not correct because the sum is
$-a^2b^2 + 4a^2b^3 - 2ab - 3a + 3b$.

5. **A** Topic: Operations on Polynomials (pp. 135–140)

Use long division to find the quotient.

$$\begin{array}{r} x^2 + 2x + 3 \\ 7x - 5\overline{)\ 7x^3 + 9x^2 + 11x - 15} \\ \underline{-(7x^3 - 5x^2)} \\ 14x^2 + 11x \\ \underline{-(14x^2 - 10x)} \\ 21x - 15 \\ \underline{-(21x - 15)} \\ 0 \end{array}$$

6. **J** Topic: Quadratic Equations (pp. 144–148); Factoring Polynomials (pp. 141–144)

	$x^2 - 5x + 6 = 0$	
Factor.	$(x - 2)(x - 3) = 0$	
	$x - 2 = 0$	$x - 3 = 0$
	$x = 2$	$x = 3$
Add the solutions.	$2 + 3 = 5$	

7. **B** Topic: Exponents and Radicals (pp. 129–131)

$2^5 \div \sqrt[3]{64}$

$= 32 \div \sqrt[3]{4^3}$

$= 32 \div 4$

$= 8$

8. F Topic: Evaluating Formulas and Expressions (pp. 125–128)

$$\frac{3^3 \div 3 - 6}{\left(\frac{1}{2}\right)^{-2} - 1}$$

Simplify numerator and denominator.

$$= \frac{3^2 - 6}{2^2 - 1}$$

$$= \frac{9 - 6}{4 - 1}$$

$$= \frac{3}{3}$$

$$= 1$$

9. D Topic: Operations with Radicals (pp. 132–134)

$5\sqrt{12} + 7\sqrt{108}$

$= 5\sqrt{4 \cdot 3} + 7\sqrt{36 \cdot 3}$

$= 5(2)\sqrt{3} + 7(6)\sqrt{3}$

$= 10\sqrt{3} + 42\sqrt{3}$

$= 52\sqrt{3}$

10. G Topic: Evaluating Formulas and Expressions (pp. 125–128)

$$\frac{(\sqrt[3]{8} - 2) + 3}{(-2)^{-1} + 1}$$

Simplify the numerator in this order:
(1) root, (2) parentheses, (3) addition.

$$= \frac{(2 - 2) + 3}{(-2)^{-1} + 1} = \frac{0 + 3}{(-2)^{-1} + 1}$$

$$= \frac{3}{(-2)^{-1} + 1}$$

Simplify the denominator in this order:
(1) power, (2) addition.

$$= \frac{3}{-\frac{1}{2} + 1} = \frac{3}{\frac{1}{2}}$$

Divide. $= 6$

Elementary Algebra Subtest (p. 150)

The answers and explanations for the Elementary Algebra Subtest items are shown with the topic each item tests. If you miss an item or guessed at the answer, circle the number of the item below and review the pages indicated.

1. B Topic: Evaluating Formulas and Expressions (pp. 125–128)

Substitute 2 for a, -4 for b, and 2 for c. Then evaluate.

$$\frac{-b \pm \sqrt{b^2 - 4ac}}{2a} = \frac{4 \pm \sqrt{16 - 4(2)(2)}}{4} = \frac{4 \pm \sqrt{16 - 16}}{4}$$

$$= \frac{4 \pm \sqrt{0}}{4} = \frac{4}{4} = 1$$

2. F Topic: Operations on Polynomials (pp. 135–140)

Use long division to find the quotient.

$$\begin{array}{r} x^2 + 2x + 5 \\ x + 2 \overline{) \ x^3 + 4x^2 + 9x + 10} \\ \underline{-(x^3 + 2x^2)} \\ 2x^2 + 9x \\ \underline{-(2x^2 + 4x)} \\ 5x + 10 \\ \underline{-(5x + 10)} \\ 0 \end{array}$$

3. A Topic: Quadratic Equations (pp. 144–148)

$$2x^2 + 2x - 12 = 0$$

Factor. $2(x + 3)(x - 2) = 0$

Solve. $x + 3 = 0 \qquad x - 2 = 0$

$$x = -3 \qquad x = 2$$

4. H Topic: Operations on Polynomials (pp. 135–140)

$$3x^2 + x\left(2x - \frac{3}{x}\right) + 7$$

Distribute x. $= 3x^2 + 2x^2 - 3 + 7$

Combine similar terms. $= 5x^2 + 4$

5. C Topic: Operations on Polynomials (pp. 135–140)

Align similar terms and add.

$$\begin{array}{r} x^3y + 2y^2x + 3x \\ + \ -4x^3y + 3y^2x - 2x \\ \hline -3x^3y + 5y^2x + x \end{array}$$

6. K Topic: Exponents and Radicals (pp. 129–131)

$$3^4 \div \frac{1}{\sqrt[4]{81}}$$

$$= 81 \times \sqrt[4]{81}$$

$$= 81 \times \sqrt[4]{3^4}$$

$$= 81 \times 3$$

$$= 243$$

7. C Topic: Quadratic Equations (pp. 144–148)

$$x^3 - 6x^2 + 5x = 0$$

Factor out x to create a quadratic expression. $x(x^2 - 6x + 5) = 0$

Factor the quadratic expression. $x(x - 5)(x - 1) = 0$

$$x = 0 \quad \text{or} \quad x - 5 = 0 \quad \text{or} \quad x - 1 = 0$$

$$x = 5 \qquad x = 1$$

The product of the solutions is $0 \cdot 5 \cdot 1 = 0$.

Note: We could have stopped when x was factored out. We knew then that one of the solutions was 0. Since the product of any number and 0 is 0, the product of the solutions would have to be 0.

8. H Topic: Factoring Polynomials (pp. 141–144)

Use FOIL to work backward from the answer choices. Choice H gives:

$$(3x - 7)(2x + 9)$$
$$= 6x^2 + 27x - 14x - 63$$
$$= 6x^2 + 13x - 63$$

9. A Topic: Evaluating Formulas and Expressions (pp. 125–128)

First convert 9,000 seconds to hours. 3,600 seconds = 1 hour, so 9,000 seconds $= \frac{9{,}000}{3{,}600}$ hours = 2.5 hours.

To find the speed, or rate, rewrite the distance equation in terms of r.

$$d = rt \qquad \frac{d}{t} = \frac{rt}{t} \qquad \frac{d}{t} = r$$

Substitute 140 miles for d and 2.5 hours for t.

$$\frac{140 \text{ miles}}{2.5 \text{ hours}} = r \quad 56 \text{ miles per hour} = r$$

10. F Topic: Factoring Polynomials (pp. 141–144)

Find the greatest common factor of the coefficients.

24, 6, 8, and 12 are all divisible by 2.

Find the smallest exponent of x.

The third and fourth terms contain x^1, or x.

Find the smallest exponent of y.

The fourth term contains y^1, or y.

The greatest common factor of the given expression is $2xy$. To factor it out, divide each term of the given expression by $2xy$.

$$24x^3y^2 \div 2xy = 12x^2y$$
$$6x^2y^3 \div 2xy = 3xy^2$$
$$-8xy^2 \div 2xy = -4y$$
$$12xy \div 2xy = 6$$

Putting the partial results together, we get $2xy(12x^2y + 3xy^2 - 4y + 6)$.

Cumulative ACT Practice Pre-Algebra/Elementary Algebra (p. 151)

Correct your Pre-Algebra/Elementary Algebra Cumulative ACT Practice answers and work through the answer explanations. If you missed an item or guessed at the answer, circle the number of the item below and review the pages indicated.

1. A Topic: Probability (pp. 107–110)

The probability that the marble will be blue $= \frac{\text{number of blue marbles}}{\text{total number of marbles}}$.

$$P(\text{Blue}) = \frac{2}{12} = \frac{1}{6}$$

2. F Topic: Evaluating Formulas and Expressions (pp. 125–128)

$$V = \pi r^2 h$$

Divide by πh. $\quad \frac{V}{\pi h} = \frac{\pi r^2 h}{\pi h}$

$$\frac{V}{\pi h} = r^2$$

Take the square root. $\quad \sqrt{\frac{V}{\pi h}} = \sqrt{r^2}$

$$\sqrt{\frac{V}{\pi h}} = r$$

3. E Topic: Quadratic Equations (pp. 144–148)

$$x^4 - 3x^3 - 18x^2 = 0$$
$$x^2(x^2 - 3x - 18) = 0$$

No more work needs to be done because we can see that one of the roots is $x = 0$. Therefore the product of the solutions will be 0.

4. J Topic: Probability (pp. 107–110)

There are 36 possible outcomes. Every third outcome is divisible by 3, so 12 out of the 36 outcomes are divisible by 3.

$$\frac{12}{36} = \frac{1}{3}$$

You could also make a table showing the outcomes.

Second Die

First Die

Result	*1*	*2*	*3*	*4*	*5*	*6*
1	2	3	4	5	6	7
2	3	4	5	6	7	8
3	4	5	6	7	8	9
4	5	6	7	8	9	10
5	6	7	8	9	10	11
6	7	8	9	10	11	12

Count to find that 12 of the 36 outcomes $\left(\frac{1}{3}\right)$ are divisible by 3.

5. B Topic: Percent (pp. 86–89)

$$\text{percent} = \frac{\text{part}}{\text{whole}}$$

Substitute 0.40 for percent, 60 for part, and x for whole. $\quad 0.40 = \frac{60}{x}$

Multiply by x. $\quad 0.40x = 60$

Divide by 0.4. $\quad x = 150$

6. **J** Topic: Operations on Polynomials (pp. 135–140)

Subtract means add the opposite.

$$
\begin{array}{r}
5x^2y + 2xy \quad\quad + y \\
-(3x^2y \quad\quad + 3x + 2y) \\
\hline
\end{array}
$$

$$
\begin{array}{r}
5x^2y + 2xy \quad\quad + y \\
-3x^2y \quad\quad - 3x - 2y \\
\hline
2x^2y + 2xy - 3x - y
\end{array}
$$

7. **C** Topic: Ratio and Proportion (pp. 94–97)

There are 3 feet in 1 yard.

So 100 yards is $100 \cdot 3$ feet $= 300$ feet.

8. **F** Topics: Writing Linear Expressions and Equations (pp. 114–117); Solving Linear Equations (pp. 117–121)

We know that twice as many double-scoop cones as single-scoop cones were sold. Since 300 double-scoop cones were sold, we know that 150 single-scoop cones were sold.

So $(300 \cdot \$2.50) + (150 \cdot \$1.50) = \$750 + \$225 = \$975$.

9. **A** Topic: Evaluating Formulas and Expressions (pp. 125–128)

$$(s^3 - t^3 + s)(t^2 + \sqrt{s})$$

Substitute 4 for s and -3 for t.
$$= (4^3 - (-3)^3 + 4)((-3)^2 + \sqrt{4})$$

Use the order of operations and simplify.
$$= (64 - (-27) + 4)(9 + 2)$$
$$= (64 + 27 + 4)(11)$$
$$= (95)(11) = 1{,}045$$

10. **J** Topic: Factors, Divisibility, and Primes (pp. 58–61)

Consider each answer in turn.

F. The factors are not primes.

G. The product of the factors is 16.

H. The product of the factors is 18.

J. All the factors are prime and the product is 24.

K. The product of the factors is 12.

So J is the correct answer. $24 = 2 \cdot 2 \cdot 2 \cdot 3$

11. **A** Topic: Evaluating Formulas and Expressions (pp. 125–128)

$$a = \sqrt{c^2 - b^2}$$

Substitute 13 for c and 5 for b.
$$= \sqrt{13^2 - 5^2}$$

Simplify and find the square root.
$$= \sqrt{169 - 25} = \sqrt{144} = 12$$

12. **H** Topic: Statistics—Mean, Median, and Mode (pp. 98–101)

Find the average, or mean.

$$\frac{40 + 37 + 42 + 36 + 35}{5} = 38 \text{ minutes}$$

13. **B** Topics: Writing Linear Expressions and Equations (pp. 114–117); Solving Linear Equations (pp. 117–121)

27.5 miles is 55 half-miles.

Write an expression. Multiply 55 times the cost per half-mile and add the initial fee.

$$55(\$0.15) + \$2.80 = \$11.05$$

14. **K** Topic: Quadratic Equations (pp. 144–148)

$$x^2 - 4x - 12 = 0$$

Factor the equation.
$$(x + 2)(x - 6) = 0$$
$$x + 2 = 0 \quad\quad x - 6 = 0$$
$$x = -2 \quad\quad x = 6$$

15. **E** Topic: Evaluating Formulas and Expressions (pp. 125–128)

Use the order of operations.

$$\frac{(5^2 - \sqrt{9} + (-2)^3) \cdot 2^{-2}}{\left(3^{-1} + \frac{2}{3}\right) \cdot 4^{-2}} = \frac{(25 - 3 + (-8)) \cdot \frac{1}{4}}{\left(\frac{1}{3} + \frac{2}{3}\right) \cdot \frac{1}{16}}$$

$$= \frac{14 \cdot \frac{1}{4}}{1 \cdot \frac{1}{16}} = \frac{\frac{7}{2}}{\frac{1}{16}} = \frac{7}{2} \cdot \frac{16}{1} = 56$$

16. **J** Topic: Operations on Polynomials (pp. 135–140)

Use long division.

$$
\begin{array}{r}
3x^2 - 2x + 5 \\
2x - 7 \overline{)6x^3 - 25x^2 + 24x - 35} \\
\underline{-(6x^3 - 21x^2)} \\
-4x^2 + 24x \\
\underline{-(-4x^2 + 14x)} \\
10x - 35 \\
\underline{-(10x - 35)} \\
0
\end{array}
$$

17. **C** Topic: Evaluating Formulas and Expressions (pp. 125–128)

$$A = \pi r^2$$

Divide by π.
$$\frac{A}{\pi} = \frac{\pi r^2}{\pi}$$
$$\frac{A}{\pi} = r^2$$

Take the square root.
$$\sqrt{\frac{A}{\pi}} = \sqrt{r^2}$$
$$\sqrt{\frac{A}{\pi}} = r$$

18. F Topic: Ratio and Proportion (pp. 94–97)

Jordan takes 39 shots.

Divide the number of shots by 3 to find the number of baskets.

$39 \div 3 = 13$

19. D Topic: Square Roots, Exponents, and Scientific Notation (pp. 61–64)

The exponent is positive.

Move the decimal point five places to the right.

$3.56 \times 10^5 = 356{,}000$

20. G Topic: Square Roots, Exponents, and Scientific Notation (pp. 61–64)

$$(5.73 \times 10^2) - (3.74 \times 10^3)$$
$$= (5.73 \times 10^2) - (37.4 \times 10^2)$$
$$= (5.73 - 37.4) \times 10^2 = -31.67 \times 10^2$$
$$= -3.167 \times 10^3$$

21. E Topic: Writing Linear Expressions and Equations (pp. 114–117)

Because a wash is $10, and a haircut is one and a half times that amount, a haircut will cost $15. The expression is $10x + 15y$.

22. F Topic: Evaluating Formulas and Expressions (pp. 125–128)

$$\frac{(rp)^2 - pq + 3}{(r - p + q)^2}$$

Substitute 4 for r, -1 for p, and 2 for q.
$$= \frac{(4 \cdot -1)^2 - (-1 \cdot 2) + 3}{(4 - (-1) + 2)^2}$$

Use the order of operations and evaluate.
$$= \frac{(-4)^2 + 2 + 3}{7^2} = \frac{16 + 2 + 3}{49}$$

$$= \frac{21}{49} = \frac{3}{7}$$

23. D Topic: Operations on Polynomials (pp. 135–140)

Align similar terms and add.

$$
\begin{array}{l}
r^2t + \quad t^2r - r^2 \qquad\quad + t \\
+\ 3r^2t - 2t^2r \qquad + 5t^2 - t \\
\hline
4r^2t - \quad t^2r - r^2 + 5t^2 \\
\end{array}
$$
$$= 4r^2t - t^2r + 5t^2 - r^2$$

24. H Topic: Evaluating Formulas and Expressions (pp. 125–128)

$$m = \frac{y_2 - y_1}{x_2 - x_1}$$

Substitute 2 for y_2, 3 for y_1, 7 for x_2, and 5 for x_1.
$$= \frac{2 - 3}{7 - 5}$$

Simplify.
$$= -\frac{1}{2}$$

25. D Topic: Evaluating Formulas and Expressions (pp. 125–128)

The formula for the area of a triangle is $A = \frac{1}{2}bh$.

$$A = \frac{1}{2}bh$$

Substitute 38 for A. $\qquad 38 = \frac{1}{2}bh$

Multiply by 2. $\qquad 2(38) = 2\left(\frac{1}{2}bh\right)$

$$76 = bh$$

The sum of the base and the height is 23 in all the answer choices.

Find the answer choice in which the product of the base and height is 76.

A. $11 \cdot 12 = 132$

B. $10 \cdot 13 = 130$

C. $8 \cdot 15 = 120$

D. $4 \cdot 19 = 76$

E. $3 \cdot 20 = 60$

D is the correct choice.

Pre-Algebra/Elementary Algebra Subtest (p. 154)

The answers and explanations for the Pre-Algebra/Elementary Algebra Subtest items are shown with the topic each item tests. If you missed an item or guessed at the answer, circle the number of the item below and review the pages indicated. You can use the chart at the end of these answer explanations to estimate your ACT Pre-Algebra/Elementary Algebra (EA) scale score.

1. B Topic: Probability (pp. 107–110)

P(not black) = P(blue or red)

$$= \frac{\text{number of blue} + \text{number of red}}{\text{number of socks}}$$
$$= \frac{5 + 3}{14} = \frac{8}{14} = \frac{4}{7}$$

2. H Topic: Evaluating Formulas and Expressions (pp. 125–128)

Writing r in terms of V means to solve $V = \frac{4}{3}\pi r^3$ for r.

$$V = \frac{4}{3}\pi r^3$$

Multiply by $\frac{3}{4}$.

$$\frac{3}{4}V = \frac{3}{4}\left(\frac{4}{3}\pi r^3\right)$$

$$\frac{3}{4}V = \pi r^3$$

Divide by π.

$$\frac{\frac{3}{4}V}{\pi} = \frac{\pi r^3}{\pi}$$

$$\frac{\frac{3}{4}V}{\pi} = r^3$$

Find the cube root.

$$\sqrt[3]{\frac{\frac{3}{4}V}{\pi}} = \sqrt[3]{r^3}$$

The equation is written in terms of r.

$$\sqrt[3]{\frac{\frac{3}{4}V}{\pi}} = r$$

3. D Topic: Quadratic Equations (pp. 144–148)

$$x^3 + 2x^2 - 15x = 0$$

Factor out x to create a quadratic expression. $x(x^2 + 2x - 15) = 0$

Factor the quadratic expression. $x(x + 5)(x - 3) = 0$

$$x = 0 \quad x + 5 = 0 \quad x - 3 = 0$$
$$x = -5 \qquad x = 3$$

Add to find the sum of the solutions.
$0 + (-5) + 3 = -2$

4. J Topic: Factors, Divisibility, and Primes (pp. 58–61)

Eliminate choices G and H. These choices contain factors that are not prime numbers.

$2 \times 2 \times 3 \times 3 \times 3 \times 5 = 540$

5. C Topic: Percent (pp. 86–89)

Write 14% as a decimal and multiply by 60.
$0.14 \cdot 60 = 8.4$

6. K Topic: Operations on Polynomials (pp. 135–140)

Align similar terms and subtract.

$$\begin{array}{r} 2ab^2 + 4a^2b - 3a^2b^2 \\ -\ (5ab^2 - a^2b + 6a^2b^2) \\ \hline -3ab^2 + 5a^2b - 9a^2b^2 \end{array}$$

7. B Topic: Ratio and Proportion (pp. 94–97)

Write a proportion.

$$\frac{12}{4.8} = \frac{x}{8}$$
$$4.8x = (12)(8)$$
$$4.8x = 96$$
$$x = 20$$

8. H Topics: Writing Linear Expressions and Equations (pp. 114–117); Quadratic Equations (pp. 144–148)

The two perpendicular sides are the length (l) and the width (w).

The product of the length and the width is $lw = 132$.

The sum of the length and the width is $l + w = 23$. So $l = 23 - w$.

Substitute $23 - w$ for l in the equation $lw = 132$.

$$(23 - w)w = 132$$

Distribute w. $23w - w^2 = 132$

Subtract 132. $23w - w^2 - 132 = 0$

Multiply by -1. $w^2 - 23w + 132 = 0$

Factor. $(w - 11)(w - 12) = 0$

$$w - 11 = 0 \text{ or } w - 12 = 0$$
$$w = 11 \text{ or } \qquad w = 12$$

The width must be the shorter side so $w = 11$ and $l = 12$.

9. C Topic: Evaluating Formulas and Expressions (pp. 125–128)

Substitute 2 for a, -4 for b, and 5 for c. Then evaluate.

$$\frac{b^3 - abc - b^a}{c^2 + bc} = \frac{(-4)^3 - 2(-4)(5) - (-4)^2}{5^2 + (-4)(5)}$$

$$= \frac{-64 + 40 - 16}{25 - 20} = \frac{-24 - 16}{5} = \frac{-40}{5} = -8$$

10. J Topic: Factors, Divisibility, and Primes (pp. 58–61)

Prime factorization means writing a number as the product of primes.

Use a factor tree to find the prime factorization.

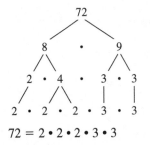

$72 = 2 \cdot 2 \cdot 2 \cdot 3 \cdot 3$

11. **E** Topic: Evaluating Formulas and Expressions (pp. 125–128)

Substitute -3 for a and 4 for b. Then evaluate.

$c = \sqrt{a^2 + b^2} = \sqrt{(-3)^2 + 4^2} = \sqrt{9 + 16}$

$\quad = \sqrt{25} = 5$

12. **H** Topic: Statistics—Mean, Median, and Mode (pp. 98–101)

To find the median, write the values in order.

$\quad$ 8, 9, 10, 11, 11, 13, 14, 15, 16, 17

There is an even number of values so average the two middle values.

$\quad \dfrac{11 + 13}{2} = 12$

13. **A** Topic: Writing Linear Expressions and Equations (pp. 114–117)

Multiply to find the total hourly earnings.
$\$7 \times 40$ hours $= \$280$

$100x$ represents the total earnings in dollars for selling x cars.

$100x + 280$ is the amount in dollars the salesman makes before taxes.

14. **K** Topic: Quadratic Equations (pp. 144–148)

Factor the quadratic equation. $\quad x^2 + 3x - 28 = (x - 4)(x + 7)$

Add the factors. $\qquad\qquad x - 4$

$\qquad\qquad\qquad\quad \underline{+ (x + 7)}$

$\qquad\qquad\qquad\qquad 2x + 3$

15. **E** Topic: Evaluating Formulas and Expressions (pp. 125–128)

Use the order of operations to evaluate.

$\dfrac{2 - 5 \cdot (4 - 7^2) + 5}{(3 - 1) \cdot 2 + 4} = \dfrac{2 - 5(-45) + 5}{2 \cdot 2 + 4} = \dfrac{2 + 225 + 5}{4 + 4}$

$= \dfrac{232}{8} = 29$

16. **J** Topic: Quadratic Equations (pp. 144–148)

$2x^2 + 12x + 18 = (2x + 6)(x + 3)$

Divide. $\dfrac{(2x + 6)}{x + 3} = \dfrac{2(x + 3)}{(x + 3)} = 2$

17. **A** Topic: Evaluating Formulas and Expressions (pp. 125–128)

Solve the equation for b. $\qquad a^2 + b^2 = c^2$

Subtract a^2. $\qquad a^2 + b^2 - a^2 = c^2 - a^2$

$\qquad\qquad\qquad\qquad\qquad b^2 = c^2 - a^2$

Take the square root $\qquad \sqrt{b^2} = \sqrt{c^2 - a^2}$

$\qquad\qquad\qquad\qquad\qquad b = \sqrt{c^2 - a^2}$

18. **K** Topic: Writing Linear Expressions and Equations (pp. 114–117)

Dog's age = Human age $\cdot$ 7

Dog's age = (Current year − Birth year) $\cdot$ 7

$\qquad = (2008 - 2000) \cdot 7 = 8 \cdot 7 = 56$

19. **C** Topic: Square Roots, Exponents, and Scientific Notation (pp. 61–64)

Move the decimal point eight places to the left and multiply by 10^8.

$1.28000000 \qquad 1.28 \times 10^8$

$\qquad$ 8 places left

20. **F** Topic: Square Roots, Exponents, and Scientific Notation (pp. 61–64)

Write each quantity to the same power of 10. $\quad (2.89 \times 10^1) + (5.4 \times 10^2) =$

$\qquad\qquad\qquad (2.89 \times 10^1) + (54 \times 10^1) =$

Add. $\quad (2.89 + 54) \times 10^1 = 56.89 \times 10^1$

Write in scientific notation. $\qquad 5.689 \times 10^2$

21. **B** Topic: Ratio and Proportion (pp. 94–97)

Chocolate: 1 gallon every 30 minutes (0.5 hour).

In 8 hours, $\dfrac{8}{0.5} = 16$ gallons.

Vanilla: 1 gallon every 0.25 hour.

In 8 hours, $\dfrac{0}{0.25} = 32$ gallons

Strawberry: 1 gallon every hour.

In 8 hours, $\dfrac{8}{1} = 8$ gallons.

Total gallons $= 16 + 32 + 8 = 56$ gallons

22. **K** Topic: Evaluating Formulas and Expressions (pp. 125–128)

Substitute 4 for s, 2 for t, and -3 for u. Then evaluate.

$\dfrac{st + ust + 4}{u^4} = \dfrac{4 \cdot 2 + (-3) \cdot 4 \cdot 2 + 4}{(-3)^4}$

$= \dfrac{8 - 24 + 4}{81} = \dfrac{-16 + 4}{81} = \dfrac{-12}{81} = \dfrac{-4}{27}$

23. **E** Topic: Operations on Polynomials (pp. 135–140)

Align similar terms and add.

$$\begin{array}{r} m^2n + 2mn \qquad\quad + 7n \\ + 3m^2n - 5mn + 2m \qquad\quad \\ \hline 4m^2n - 3mn + 2m + 7n \end{array}$$

24. **F** Topic: Quadratic Equations (pp. 144–148)

Factor the quadratic equation.
$$2x^2 - 5x - 12 = 0 \rightarrow (2x + 3)(x - 4) = 0$$

$$2x + 3 = 0 \qquad\qquad x - 4 = 0$$

$$\underline{-3 \qquad -3} \qquad \underline{+4 \qquad +4}$$

$$2x \qquad = -3 \qquad\qquad x \qquad = 4$$

$$x \qquad = -\frac{3}{2}$$

Multiply the solutions. $\left(-\frac{3}{2}\right)(4) = -\frac{12}{2} = -6$

Estimate Your ACT Pre-Algebra/Elementary Algebra (EA) Sub Score

The number correct on the Pre-Algebra/Elementary Algebra items leads to the EA mathematics scale score. Use the chart below to estimate your EA scale score. But remember, these items are not scattered throughout the test as they would be on the regular ACT, and you have just finished reviewing these concepts. All this means that the information given below is only an approximation, and that your actual EA score will probably be different, possibly quite different, from this score.

On national ACT administrations:

- The average number of EA items correct is about 11–12.
- About 10% of test takers get more than 20 correct.
- About 25% of test takers get more than 15 correct.
- About 25% of test takers get fewer than 8 correct.

Circle the number of items you answered correctly in the table below and locate the scale score. The highest scale score is 18 and the lowest is 1.

Approximate EA Scale Scores

Number Correct	Approximate Scale Score	Number Correct	Approximate Scale Score	Number Correct	Approximate Scale Score
24	18	16	12	6	6
22–23	17	14–15	11	4–5	5
21	16	12–13	10	3	4
20	15	10–11	9	2	3
19	14	9	8	1	2
17–18	13	7–8	7	0	1

Chapter 9 Intermediate Algebra
Solving Inequalities

Practice (p. 160)

1. $x > 4$ **2.** $x \le 35$ **3.** $y \ge -2\frac{1}{3}$ **4.** $y < -66$ **5.** $t \ge -0.75$ **6.** $k > 115.2$

7. $x \ge 2$ **8.** $t < -9$ **9.** $x < -\frac{1}{2}$ **10.** $x < -3$ **11.** $x \ge 9.5$ **12.** $x < 1$

13. $x \ge -2$ **14.** $x \ge 0.9$ **15.** $x \le -4.5$ **16.** $x < \frac{9}{14}$ **17.** $x < -15$ **18.** $x > -1$

19. $x \le 2\frac{1}{3}$ **20.** $x > 4$

ACT-Type Problems (p. 160)

1. **C**

Divide by 4.

$4x \geq -12$

$x \geq -3$

2. **K**

Add 7.

Divide by -3.

$-3x - 7 < 20$

$-3x < 27$

$x > -9$

3. **D**

Subtract 17.

Divide by -4.

$-4x + 17 \leq -3$

$-4x \leq -20$

$x \geq 5$

4. **F**

Add $5x$.

Add 12.

Divide by 3.

5 is not in the solution set.

$-5x + 3 > -2x - 12$

$3 > 3x - 12$

$15 > 3x$

$5 > x$

5. **D**

Add $4x$.

Subtract 11.

Divide by -5.

$-9x + 11 \leq -4x + 3$

$-5x + 11 \leq 3$

$-5x \leq -8$

$x \geq \dfrac{8}{5}$

$\dfrac{8}{5}$ is the smallest number in the solution set.

The reciprocal of $\dfrac{8}{5}$ is $\dfrac{5}{8}$.

Absolute Value Equations and Inequalities

Practice (p. 162)

1. $-\dfrac{10}{3} < x < 6$

2. $-3 \leq x \leq 15$

3. $x = 8$ or $x = -14$

4. $x \leq -\dfrac{1}{5}$ or $x \geq 5$

5. $x < -10$ or $x > 2$

6. $x = 5$ or $x = -3$

7. $x > \dfrac{2}{3}$ or $x < -\dfrac{7}{3}$

8. $x \geq 7$ or $x \leq 3$

9. $-\dfrac{4}{3} < x < 2$

10. $-5 \leq x \leq 1$

11. $2 < x < 4$

12. $x = 1$ or $x = -8$

13. $x \geq \dfrac{10}{7}$ or $x \leq -\dfrac{4}{7}$

14. $-\dfrac{8}{5} \leq x \leq 0$

15. $x > 3$ or $x < \dfrac{3}{2}$

16. $x = \dfrac{1}{3}$ or $x = -\dfrac{7}{3}$

17. $-\dfrac{1}{2} \leq x \leq \dfrac{7}{4}$

18. $-\dfrac{15}{2} < x < \dfrac{9}{2}$

19. $x \geq \dfrac{12}{5}$ or $x \leq -\dfrac{2}{5}$

20. $x > 13$ or $x < -15$

ACT-Type Problems (p. 162)

1. **D** You don't have time to solve every equation. Think the problem through. When the value is positive, $x = 16$ is a solution for A and D. Solve these equations for when the value is negative.

A. $-(x - 2) = 14$ $-x + 2 = 14$

$-x = 12$ $x = -12$

The other solution to A is not -10. We know that D is the answer without solving the equation. But here is the solution for D when the value is negative.

D. $-(x - 3) = 13$ $-x + 3 = 13$

$-x = 10$ $x = -10$

2. **F** Substitute values.

F. $|2(2) - 9| = |4 - 9| = |-5| = 5$

5 is not less than 5. The value 2 makes this inequality false. We do not have to substitute other values.

3. **C** $|7x - 5| \geq 9$

Case 1: $7x - 5 \geq 9$ $7x \geq 14$ $x \geq 2$

Case 2: $-(7x - 5) \geq 9$ $-7x + 5 \geq 9$

$-7x \geq 4$ $-x \geq \dfrac{4}{7}$ $x \leq -\dfrac{4}{7}$

$x \leq -\dfrac{4}{7}$ or $x \geq 2$

4. **F** Substitute 6 for x in the choices.

F. $2x = 12$ $x = 6$

Choices G, H, and K also have $x = 6$ as a solution. However, there is another solution for each of these choices because of the absolute value. So $x = 6$ is not the *complete* solution set for these choices.

5. **D** $|2x - 3| = 15$ when $2x - 3 = 15$ and when $2x - 3 = -15$.

Solve both equations.

$2x - 3 = 15$ and $2x - 3 = -15$

$2x = 18$ $2x = -12$

$x = 9$ $x = -6$

Find the product of the solutions. $9 \cdot -6 = -54$

Solving Systems of Linear Equations

Practice (p. 166)

1. $(0.75, 2)$

2. $\left(4\dfrac{1}{3}, 3\dfrac{1}{2}\right)$

3. $(1.25, -1)$

4. $(0.5, 1)$

5. $\left(-1\dfrac{2}{3}, 2\dfrac{2}{5}\right)$

6. $(2, 0)$

7. $(3.4, 0.55)$

8. $(-1, 2)$

9. $(14.5, -10)$

10. $(0.5, 0.875)$

11. $(-1.7, 2.8)$

12. $\left(-3\dfrac{2}{3}, -5\right)$

13. $(-10, -4)$

14. $(-0.25, -1)$

15. $(38.9, 19.4)$

16. $(-3, 0)$

17. $(-1, 2)$

18. $(-6, -13)$

19. 9 nickels and 6 dimes

20. 4.75 hours studying math and 3.25 hours studying science

21. 2 videos and 6 CDs

ACT-Type Problems (p. 166)

1. **A** Subtract one equation from the other.

$2x + 5y = 8$

$-2x - 4y = -7$

$y = 1$

Substitute 1 for y in one of the equations.

Solve for x.

$2x + 4(1) = 7$

$2x + 4 = 7$

$2x = 3$

$x = 1.5$

The solution is $x = 1.5$ and $y = 1$.

The solution as an ordered pair is $(1.5, 1)$.

2. G

Add the equations.

$$3x + 5y = 8$$
$$\underline{-3x + 5y = 8}$$
$$10y = 16$$

Solve for y.

$$y = 1.6$$

Substitute 1.6 for y in one
of the equations.

$$3x + 5(1.6) = 8$$

Solve for x.

$$3x + 8 = 8$$
$$3x = 0$$
$$x = 0$$

The solution is $x = 0$ and $y = 1.6$.
The solution as an ordered pair is (0,1.6).

3. C Multiply both sides of the second equation by 2.

$$2(5x + 5y = 6) \rightarrow 10x + 10y = 12$$

Subtract the equations.

$$7x + 10y = 12$$
$$\underline{-10x - 10y = -12}$$

Solve for x.

$$-3x = 0$$
$$x = 0$$

Substitute 0 for x in one
of the equations.

$$5(0) + 5y = 6$$

Solve for y.

$$5y = 6$$
$$y = 1.2$$

The solutions are $x = 0$ and $y = 1.2$.
The product of the solutions is $0 \times 1.2 = 0$.

4. J Write the two equations.

$$2x + 3y = 1$$
$$3x + 5y = 2$$

Multiply both sides of the first equation by 3.
$$3(2x + 3y = 1) \rightarrow 6x + 9y = 3$$

Multiply both sides of the second equation by 2.
$$2(3x + 5y = 2) \rightarrow 6x + 10y = 4$$

Subtract the equations.

$$6x + 9y = 3$$
$$\underline{-6x - 10y = -4}$$

Solve for y.

$$-y = -1$$
$$y = 1$$

Substitute 1 for y in one
of the equations.

$$2x + 3(1) = 1$$

Solve for x.

$$2x + 3 = 1$$
$$2x = -2$$
$$x = -1$$

The numbers are $x = -1$ and $y = 1$.
The solution as an ordered pair is $(-1,1)$.

5. B Write the two equations.

$$6x + 12y = 8$$
$$9x + 4y = 5$$

Multiply both sides of the second equation by 3.
$$3(9x + 4y = 5) \rightarrow 27x + 12y = 15$$

Subtract the equations.

$$6x + 12y = 8$$
$$\underline{-27x - 12y = -15}$$

Solve for x.

$$-21x = -7$$
$$x = \frac{1}{3}$$

Substitute $\frac{1}{3}$ for x in one of the equations.

$$9\left(\frac{1}{3}\right) + 4y = 5$$

Solve for y.

$$3 + 4y = 5$$
$$4y = 2$$
$$y = \frac{1}{2}$$

The solution is $x = \frac{1}{3}$ and $y = \frac{1}{2}$.

The solution as an ordered pair is $\left(\frac{1}{3} \text{ mile}, \frac{1}{2} \text{ mile}\right)$.

Rational and Radical Expressions

Practice (p. 169)

1. $\frac{1}{x^6} + x^{\left(\frac{2}{3}\right)}$

2. $x - 1$

3. 1

4. $\frac{1}{x^{\left(\frac{4}{3}\right)}}$

5. x^2

6. $x^{\left(\frac{2}{3}\right)} + x^{\left(\frac{4}{3}\right)}$

7. $1 + x^2$

8. $x^{\left(\frac{17}{4}\right)} + 1$

9. $x^6 + x$

10. $x + x^{\left(\frac{23}{30}\right)}$

11. Defined for all real values of x

12. Defined for all real values of x except $x = 0$

13. Defined for all real values of x except $x = \frac{1}{3}$

14. Defined for all real values of x except $x = -1$

15. Defined for all real values of x except $x = -3$

ACT-Type Problems (p. 170)

1. D

$$\frac{4x^3y^2 + 4x^2y^3}{x + y} = \frac{(4x^2y^2)(x + y)}{(x + y)} = 4x^2y^2$$

2. F

$$\frac{1}{\sqrt[3]{x}} \cdot \sqrt[3]{x} - \sqrt[2]{x} \cdot \sqrt{\frac{1}{x}}$$
$$= x^{\left(-\frac{1}{3}\right)} \cdot x^{\left(\frac{1}{3}\right)} - x^{\left(\frac{1}{2}\right)} \cdot x^{\left(-\frac{1}{2}\right)}$$
$$= x^{\left(-\frac{1}{3}+\frac{1}{3}\right)} - x^{\left(\frac{1}{2}+\left(-\frac{1}{2}\right)\right)}$$
$$= x^0 - x^0$$
$$= 1 - 1$$
$$= 0$$

3. A
The denominator $(2^{(3x-1)})$ can never equal zero. The expression is defined for all real numbers.

4. J
The numerator is defined when $x \geq 0$ since $\sqrt{x}$ is undefined for negative values. The denominator is defined everywhere except where it equals zero. Find the values of x for which the denominator equals zero.

$$x^2 + x - 6 = 0$$

Factor.

$$(x + 3)(x - 2) = 0$$
$$x + 3 = 0 \quad \text{or} \quad x - 2 = 0$$
$$x = -3 \qquad\qquad x = 2$$

The denominator equals zero when $x = -3$ or $x = 2$. The entire expression is defined for all non-negative values except 2.

5. B

$$\frac{-2x^2 + 2y^2}{x - y} = \frac{-2(x^2 - y^2)}{x - y} = \frac{-2(x - y)(x + y)}{(x - y)}$$
$$= -2x - 2y$$

Solving Quadratic Equations

Practice (p. 172)

1. $x = \dfrac{-b \pm \sqrt{b^2 - 4ac}}{2a}$

2. $a = -2, b = 8, c = 3$

3. -1 and 4

4. $\dfrac{2 + \sqrt{14}}{5}$ and $\dfrac{2 - \sqrt{14}}{5}$

5. -2 and $-\dfrac{1}{2}$ 6. -2 and 3

7. -4 and 1 8. $-\dfrac{1}{2}$ and 3

9. -8 and 7 10. -2 and 9

11. $-\dfrac{1}{4}$ and -2 12. -2 and -1

13. 2 and 3 14. $2\dfrac{1}{2}$ and $-1\dfrac{1}{2}$

15. -9 and $\dfrac{1}{3}$ 16. -7 and $\dfrac{3}{4}$

17. -1 and 2.5 18. 3 feet by 9 feet

19. 12 feet by 15 feet   20. 1.5 inches

ACT-Type Problems (p. 173)

1. **E**

Write the equation in standard form. $4x^2 + 2 = -14x$

Add $14x$. $4x^2 + 14x + 2 = 0$

Identify the values for a, b, and c. $a = 4, b = 14, c = 2$

Multiply a, b, and c. $4 \times 14 \times 2 = 112$

2. **H**

Write the equation in standard form. $8x^2 = -44x - 56$

Add $44x$. $8x^2 + 44x = -56$

Add 56. $8x^2 + 44x + 56 = 0$

Identify the values for a, b, and c. $a = 8, b = 44, c = 56$

Substitute the values of a, b, and c into the quadratic formula. $x = \dfrac{-b \pm \sqrt{b^2 - 4ac}}{2a}$

$= \dfrac{-44 \pm \sqrt{44^2 - 4(8)(56)}}{2(8)}$

$= \dfrac{-44 \pm \sqrt{1{,}936 - 1{,}792}}{16}$

$= \dfrac{-44 \pm \sqrt{144}}{16}$

$= \dfrac{-44 \pm 12}{16}$

$= \dfrac{-44 + 12}{16}$ or $\dfrac{-44 - 12}{16}$

The solutions are -3.5 and -2.

3. **C**

Identify the values for a, b, and c. $a = 2, b = 16, c = 24$

Substitute the values of a, b, and c into the quadratic formula. $x = \dfrac{-b \pm \sqrt{b^2 - 4ac}}{2a}$

$= \dfrac{-16 \pm \sqrt{16^2 - 4(2)(24)}}{2(2)}$

$= \dfrac{-16 \pm \sqrt{256 - 192}}{4}$

$= \dfrac{-16 \pm \sqrt{64}}{4}$

$= \dfrac{-16 \pm 8}{4}$

$= \dfrac{-16 + 8}{4}$ or $\dfrac{-16 - 8}{4}$

The solutions are -6 and -2.

4. **H**

Identify the values for a, b, and c. $a = 21, b = 0, c = 189$

Substitute the values of a, b, and c into the quadratic formula. $x = \dfrac{-b \pm \sqrt{b^2 - 4ac}}{2a}$

$= \dfrac{0 \pm \sqrt{0^2 - 4(21)(-189)}}{2(21)}$

$= \dfrac{\pm \sqrt{15{,}876}}{42}$

$= \dfrac{\pm 126}{42} = \dfrac{126}{42}$ or $\dfrac{-126}{42}$

The solutions are -3 and 3.

Add the solutions. $-3 + 3 = 0$

5. **A**

Write the equation in standard form. $-29x - 22 = -6x^2$

Add $6x^2$. $6x^2 - 29x - 22 = 0$

Identify the values for a, b, and c. $a = 6, b = -29, c = -22$

Substitute the values of a, b, and c into the quadratic formula. $x = \dfrac{-b \pm \sqrt{b^2 - 4ac}}{2a}$

$= \dfrac{-(-29) \pm \sqrt{(-29)^2 - 4(6)(-22)}}{2(6)}$

$= \dfrac{29 \pm \sqrt{841 + 528}}{12}$

$= \dfrac{29 \pm \sqrt{1{,}369}}{12}$

$= \dfrac{29 \pm 37}{12} = \dfrac{66}{12}$ or $\dfrac{-8}{12}$

The solutions are $-\dfrac{2}{3}$ and $\dfrac{11}{2}$.

Multiply the solutions. $-\dfrac{2}{3} \times \dfrac{11}{2} = -\dfrac{11}{3}$

Solving Quadratic Inequalities

Practice (p. 174)

1. $-4 < x < 4$
2. $-7 \leq x \leq 0$
3. $x < -5$ or $x > 5$
4. $x \leq 0$ or $x \geq 4$
5. $-4 \leq x \leq \frac{2}{3}$
6. $x \leq -\frac{3}{2}$ or $x \geq \frac{3}{2}$
7. $x \leq \frac{1}{2}$ or $x \geq 5$
8. $x < -2$ or $x > 10$
9. $3 < x < 9$
10. $-5 < x < 3$
11. $x < -4$ or $x > 4$
12. $x > \frac{3}{2}$ or $x < -1$
13. No solution. $(x - 1)(x - 1)$ cannot be less than 0 because the two factors always have the same sign.
14. $-2 \leq x \leq 2$
15. $1 \leq x \leq 5$
16. $-\sqrt{2} < x < \sqrt{2}$
17. $x < -4$ or $x > 4$
18. $x \geq 0$ or $x \leq -\frac{5}{4}$
19. $-9 \leq x \leq 0$
20. x is any real number.

ACT-Type Problems (p. 175)

1. **B**
$x^2 - 8x + 12 < 0$
$(x - 2)(x - 6) < 0$
$x - 2 < 0$ and $x - 6 > 0$ OR
$x - 2 > 0$ and $x - 6 < 0$
$x < 2$ and $x > 6$ $x > 2$ and $x < 6$
Impossible. $2 < x < 6$

2. **F**
$-\sqrt{3} < x < \sqrt{3}$ means $x > -\sqrt{3}$ and $x < \sqrt{3}$.
Working backwards, we find the product
$x + \sqrt{3} > 0$ and $x - \sqrt{3} < 0$.
Since one factor is positive and one is negative, their product is negative:
$(x + \sqrt{3})(x - \sqrt{3}) < 0$
Use FOIL to multiply. $x^2 - \sqrt{3}x + \sqrt{3}x - 3 < 0$
Simplify. $x^2 - 3 < 0$

3. **D**
$x^2 + 3x - 4 > 0$
$(x - 1)(x + 4) > 0$
$x - 1 > 0$ and $x + 4 > 0$ OR
$x - 1 < 0$ and $x + 4 < 0$
$x > 1$ and $x > -4$ $x < 1$ and $x < -4$
$x > 1$ OR $x < -4$

4. **H**
$x^2 - 3x \leq 0$
$x(x - 3) \leq 0$
$x \leq 0$ and $x - 3 \geq 0$ OR $x \geq 0$ and $x - 3 \leq 0$
$x \leq 0$ and $x \geq 3$ $x \geq 0$ and $x \leq 3$
Impossible. $0 \leq x \leq 3$

5. **E**
$-3 \leq x \leq 8$ means $x \geq -3$ and $x \leq 8$.
Working backwards, we find the factors
$x + 3 \geq 0$ and $x - 8 \leq 0$.
Since one factor is positive or zero and one is negative or zero, their product is negative or zero:
$(x + 3)(x - 8) \leq 0$
Use FOIL to multiply. $x^2 - 8x + 3x - 24 \leq 0$
Simplify. $x^2 - 5x - 24 \leq 0$

Complex Numbers

Practice (p. 177)

1. -18
2. $-i$
3. $7i$
4. $5i\sqrt{2}$
5. $4i\sqrt{3}$
6. $12 + 5i$
7. $27 + 3i$
8. $-16 - i$
9. $-4 + 12i$
10. $42 - 7i$
11. $5 + 16i$
12. $19 + 5i$
13. $-8 + 16i$
14. $-20.5 + 9i$
15. $\frac{3}{4} + \frac{7}{8}i$
16. $-57 + 41i$
17. $78 + 50i$
18. $-61 - 23i$
19. $-22 - 21i$
20. $-\frac{3}{8} + \frac{25}{48}i$
21. $-4i$
22. $-\frac{1}{4} - \frac{3}{4}i$
23. $1.4 - 1.2i$
24. $-\frac{1}{2} + \frac{1}{2}i$

ACT-Type Problems (p. 177)

1. **A** $-3(6) - (-3)(\sqrt{-25}) = -18 - (-3)(5i)$
$= -18 + 15i$

2. **J** $(5 - 2i) + (-3 + 7i) = (5 + (-3)) + (-2 + 7)i$
$= 2 + 5i$

3. **E**
$x^2 + 4 = 0$ $a = 1, b = 0, c = 4$
$x = \dfrac{-0 \pm \sqrt{0^2 - 4(1)(4)}}{2(1)}$
$x = \dfrac{\pm\sqrt{-16}}{2}$
$x = \dfrac{\pm 4i}{2} = \pm 2i$

4. **G** The roots are $x = 3i$ and $x = -3i$ so
$x - 3i = 0$ and $x + 3i = 0$.
Use FOIL. $(x - 3i)(x + 3i) = 0$
Combine like terms and $x^2 + 3ix - 3ix - 9i^2 = 0$
simplify. $x^2 - 9(-1) = 0$
                   $x^2 + 9 = 0$

5. **D**
$x^2 - 6x + 10 = 0$ $a = 1, b = -6, c = 10$
$x = \dfrac{-(-6) \pm \sqrt{(-6)^2 - 4(1)(10)}}{2(1)}$
$x = \dfrac{6 \pm \sqrt{36 - 40}}{2}$
$x = \dfrac{6 \pm \sqrt{-4}}{2} = \dfrac{6 \pm 2i}{2} = 3 \pm i$

Patterns, Sequences, and Modeling

Practice (p. 180)

1. 105, 170, 275
2. 11, 13, 15
3. 42, 31, 20
4. 76, 123, 199
5. 28, 33, 38
6. 216, 343, 512
7. 62.5, 31.25, 15.625
8. 1,458; 4,374; 13,122
9. $-1, -9, -17$
10. 23, 29, 31
11. 12, 14, 16
12. 64, 128, 256
13. 96, 192, 384
14. 36, 49, 64
15. $\dfrac{1}{64}, \dfrac{1}{128}, \dfrac{1}{256}$
16. $-5, -9, -13$
17. 13, 17, 19
18. 27, 81, 243
19. 63, 127, 255
20. $\dfrac{1}{8}, 9, \dfrac{1}{10}$

ACT-Type Problems (p. 180)

1. **B** This is an arithmetic sequence whose common difference is 9. Therefore $111 + 9 = 120$.

2. **G** You can start with 3, and just keep multiplying by 2 until you get the first five terms.

   You can also use the formula for the terms in a geometric sequence.
   $a_n = a_1 r^{(n-1)} = 3(2)^{(n-1)}$
   $a_1 = 3(2)^{(1-1)} = 3 \qquad a_2 = 3(2)^{(2-1)} = 6$
   $a_3 = 3(2)^{(3-1)} = 12 \qquad a_4 = 3(2)^{(4-1)} = 24$
   $a_5 = 3(2)^{(5-1)} = 48$

3. **D** This is a geometric expression with a common factor of 8. You can just multiply until you reach the seventh term.

   $2{,}048 \times 8 = 16{,}384$
   $16{,}384 \times 8 = 131{,}072$
   $131{,}072 \times 8 = 1{,}048{,}576$

   Or you can use the formula
   $a_n = a_1 r^{(n-1)}$
   $a_n = 4(8)^{(7-1)} = 4(8)^{(6)} = 4 \times 262{,}144 = 1{,}048{,}576$

4. **H** Each term in the sequence is n^2. Therefore, the ninth term is $9^2 = 81$.

5. **B** In the sequence 1, 2, 4, 8, 16, ... , each new number is formed by doubling the prior number. Therefore the next three numbers in the sequence will be 32, 64, 128.

Matrices

Practice (p. 183)

1. 3×3
2. 2×4
3. 3×2
4. 1×4
5. 4×3
6. $\begin{bmatrix} 15 & -6 & 17 & 11 \\ -6 & 13 & 2 & 10 \end{bmatrix}$

7. $\begin{bmatrix} 8 & 19 & -16 \\ 9 & -4 & -6 \\ 10 & -1 & 11 \end{bmatrix}$
8. $\begin{bmatrix} 3 & 23 & 0 \\ 19 & 2 & -8 \\ 21 & 10 & 1 \end{bmatrix}$

9. $\begin{bmatrix} 8 & 19 & -16 \\ 9 & -4 & -6 \\ 10 & -1 & 11 \end{bmatrix}$
10. Matrix D and Matrix H do not have the same dimension so they cannot be added.

11. $\begin{bmatrix} -7 & -2 & 5 & -8 \end{bmatrix}$
12. $\begin{bmatrix} 18 & 36 & -24 \\ 0 & 6 & -12 \\ -4 & 16 & 18 \end{bmatrix}$

13. $\begin{bmatrix} 70 & 21 & 35 & 63 \\ -21 & 42 & 35 & 14 \end{bmatrix}$
14. $\begin{bmatrix} -6 & 16 \\ -12 & -24 \\ -10 & -20 \end{bmatrix}$

15. $\begin{bmatrix} -0.5 & 1.5 & 1 \\ 4.5 & -0.5 & -4 \\ -2 & 5.5 & -3.5 \\ 6 & 1.5 & 2.5 \end{bmatrix}$
16. $\begin{bmatrix} 5 & 12 & -7 & 7 \\ 0 & -1 & 8 & -6 \end{bmatrix}$

17. $\begin{bmatrix} -10 & -11 & 8 \\ 9 & -10 & 6 \\ 14 & -17 & -7 \end{bmatrix}$
18. $\begin{bmatrix} -15 & -13 & 24 \\ 19 & -4 & 4 \\ 25 & -6 & -17 \end{bmatrix}$

19. $\begin{bmatrix} 10 & 11 & -8 \\ -9 & 10 & -6 \\ -14 & 17 & 7 \end{bmatrix}$
20. The matrices cannot be subtracted because their dimensions are not the same.

ACT-Type Problems (p. 184)

1. **A** The dimension of a matrix is the number of rows followed by the number of columns. This matrix has one row and three columns, so the dimension is 1×3.

2. **K** In order to be added, two matrices must have the same dimension. These two matrices cannot be added because their dimensions are different.

3. **C** To multiply a matrix by a scalar, multiply the scalar by each element in the matrix.
   $-3\begin{bmatrix} 4 & -7 & -\frac{1}{2} \end{bmatrix} = \begin{bmatrix} -3(4) & -3(-7) & -3\left(-\frac{1}{2}\right) \end{bmatrix}$
   $= \begin{bmatrix} -12 & 21 & \frac{3}{2} \end{bmatrix}$

4. **J** To add two matrices, add their corresponding elements.
   $\begin{bmatrix} 2 & 3 & 4 \end{bmatrix} + \begin{bmatrix} -2 & -3 & -4 \end{bmatrix}$
   $= \begin{bmatrix} (2 + (-2)) & 3 + (-3) & 4 + (-4) \end{bmatrix} = \begin{bmatrix} 0 & 0 & 0 \end{bmatrix}$

5. **B** Subtract corresponding elements.
   $\begin{bmatrix} 1 & 3 \\ 5 & 7 \\ 9 & 11 \end{bmatrix} - \begin{bmatrix} -1 & -3 \\ -5 & -7 \\ -9 & -11 \end{bmatrix} = \begin{bmatrix} 1-(-1) & 3-(-3) \\ 5-(-5) & 7-(-7) \\ 9-(-9) & 11-(-11) \end{bmatrix} = \begin{bmatrix} 2 & 6 \\ 10 & 14 \\ 18 & 22 \end{bmatrix}$

6. J Multiply the two matrices.

$$\begin{bmatrix} 24 & 38 & 13 \\ 32 & 52 & 11 \end{bmatrix} \times \begin{bmatrix} \$200 \\ \$100 \\ \$300 \end{bmatrix} = \begin{bmatrix} 24(\$200) + 38(\$100) + 13(\$300) \\ 32(\$200) + 52(\$100) + 11(\$300) \end{bmatrix}$$

$$= \begin{bmatrix} \$12,500 \\ \$14,900 \end{bmatrix}$$

$\$12,500 + \$14,900 = \$27,400$

7. A

$$[450 \quad 700 \quad 900] \times \begin{bmatrix} 28 & 81 & 41 \\ 30 & 36 & 32 \\ 51 & 25 & 29 \end{bmatrix} = [79,500 \quad 84,150 \quad 66,950]$$

8. F You don't have to use matrices. You can just think the problem through.

$(120 + 125 + 90)(\$22.50) + (75 + 60 + 35)(\$29)$
$+ (35 + 25 + 40)(\$36.50)$
$= 335(\$22.50) + 170(\$29) + 100(\$36.50)$
$= \$16,117.50$

Cumulative ACT Practice
Intermediate Algebra (p. 186)

Correct your Intermediate Algebra Cumulative ACT Practice answers and work through the answer explanations. If you missed at item or guessed at the answer, circle the number of the item below and review the pages indicated.

1. B Topic: Absolute Value Equations and Inequalities (pp. 161–163)

$|9x - 5| \le 4$

$9x - 5 \le 4$	OR	$-9x + 5 \le 4$
$9x \le 9$		$-9x \le -1$
$x \le 1$		$x \ge \dfrac{1}{9}$

So $\dfrac{1}{9} \le x \le 1$. The only answer choice that is in the solution set is $\dfrac{1}{8}$.

2. H Topic: Solving Systems of Linear Equations (pp. 163–167)

$5x + 3y = 2$
$6x + 4y = 9$

Multiply the top equation by 4.

$4(5x + 3y = 2) \rightarrow 20x + 12y = 8$

Multiply the bottom equation by -3.

$-3(6x + 4y = 9) \rightarrow \underline{-18x - 12y = -27}$

Add the equations. $\qquad\qquad 2x = -19$

Solve for x. $\qquad\qquad\qquad x = -9.5$

Substitute -9.5 for y in one of the equations. $\quad 5(-9.5) + 3y = 2$

Solve for y. $\qquad\qquad\quad -47.5 + 3y = 2$
$\qquad\qquad\qquad\qquad\qquad 3y = 49.5$
$\qquad\qquad\qquad\qquad\qquad\ y = 16.5$

Add -9.5 and 16.5. $\qquad -9.5 + 16.5 = 7$

3. C Topic: Matrices (pp. 180–185)

$$-4\begin{bmatrix} 4 & -2 \\ 7 & 3 \end{bmatrix} + \begin{bmatrix} -3 & 2 \\ 2 & 5 \end{bmatrix}$$

Multiply each term in the first matrix by the scalar 4.

$$= \begin{bmatrix} -16 & 8 \\ -28 & -12 \end{bmatrix} + \begin{bmatrix} -3 & 2 \\ 2 & 5 \end{bmatrix}$$

Add corresponding terms of the matrices.

$$= \begin{bmatrix} -19 & 10 \\ -26 & -7 \end{bmatrix}$$

4. J Topics: Complex Numbers (pp. 175–177); Solving Quadratic Equations (pp. 170–173)

Identify the values for a, b, and c. $\qquad a = 1, b = -2, c = 5$

Substitute the values of a, b, and c into the quadratic formula.

$x = \dfrac{-b \pm \sqrt{b^2 - 4ac}}{2a}$

$= \dfrac{-(-2) \pm \sqrt{(-2)^2 - 4(1)(5)}}{2(1)}$

$= \dfrac{2 \pm \sqrt{4 - 20}}{2}$

$= \dfrac{2 \pm \sqrt{-16}}{2}$

$= \dfrac{2 \pm 4i}{2} = 1 \pm 2i$

The solutions of the quadratic equation are $1 + 2i$ and $1 - 2i$.

5. E Topic: Absolute Value Equations and Inequalities (pp. 161–163)

$|x + 4| \ge 10$ when $x + 4 \ge 10$ or $-(x + 4) \ge 10$.

Solve both inequalities.

$x + 4 \ge 10$	$-(x + 4) \ge 10$
$x \ge 6$	$x + 4 \le -10$
	$x \le -14$

The solution set is $x \ge 6$ and $x \le -14$. So, 13 is part of the solution set.

6. K Topic: Solving Quadratic Inequalities (pp. 173–175)

$2x^2 + 6x - 20 \le 0$

$(2x - 4)(x + 5) \le 0$

$(2x - 4) \le 0$ and $(x + 5) \ge 0$	OR	$(2x - 4) \ge 0$ and $(x + 5) \le 0$
$x \le 2$ and $x \ge -5$		$x \ge 2$ and $x \le -5$
$-5 \le x \le 2$		Impossible.

The only answer choice not contained in the solution set is 4.

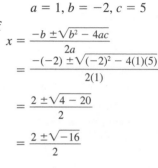

7. A Topic: Patterns, Sequences, and Modeling (pp. 178–180)

The sequence that is formed is 3, 5, 7, …

Each term can be described by the formula $2n + 1$ where n is the number of days.

For 10 days, $2(10) + 1 = 21$ roses.

8. J Topic: Solving Quadratic Equations (pp. 170–173)

The maximum number of roots of a function is the largest exponent value in the function. So, 5 is the maximum number of roots that might be found.

9. C Topic: Rational and Radical Expressions (pp. 167–170)

$$\left(\sqrt[4]{x}\right)^8 - \frac{1}{x^{-1}} - 1 + \frac{3}{x^{-2}} = \left(x^{\frac{1}{4}}\right)^8 - x - 1 + 3x^2$$
$$= x^2 - x - 1 + 3x^2$$
$$= 4x^2 - x - 1$$

10. K Topic: Solving Systems of Linear Equations (pp. 163–167)

To find when the two functions are equal, set the two functions equal to one another. Then solve for x.

$$k(x) = r(x)$$

Substitute. $3{,}000x + 3{,}800 = 3{,}500x + 3{,}300$

Solve for x. $500 = 500x$

$$1 = x$$

Since x is the number of years after 1990, add 1 to 1990 and get 1991.

Intermediate Algebra Subtest (p. 187)

The answers and explanations for the Intermediate Algebra Subtest items are shown with the topic each item tests. If you missed an item or guessed at the answer, circle the number of the item below and review the pages indicated.

1. D Topic: Absolute Value Equations and Inequalities (pp. 161–163)

$$|-3x + 3| > 6$$

$$\begin{array}{lll} -3x + 3 > 6 & \text{OR} & 3x - 3 > 6 \\ -3x > 3 & & 3x > 9 \\ x < -1 & & x > 3 \end{array}$$

The only answer choice not contained in the solution set is 3.

2. G Topic: Solving Systems of Linear Equations (pp. 163–167)

Multiply the top equation by 2.

$$2(2x + 5y = 16) \rightarrow 4x + 10y = 32$$

Subtract the equations. $\underline{-(4x + 7y = 23)}$

Solve for y. $3y = 9$

$$y = 3$$

Substitute 3 for x in one of the equations. $2x + 5(3) = 16$

Solve for x. $2x + 15 = 16$

$$2x = 1$$
$$x = 0.5$$

Multiply 3 and 0.5. $3 \times 0.5 = 1.5$

3. D Topic: Matrices (pp. 180–185)

Multiply each term in the first matrix by 2 and each term in the second matrix by 1.

$$2\begin{bmatrix} 4 & 3 \\ 2 & 1 \end{bmatrix} + -1\begin{bmatrix} 1 & 3 \\ 5 & 6 \end{bmatrix}$$

$$= \begin{bmatrix} 8 & 6 \\ 4 & 2 \end{bmatrix} + \begin{bmatrix} -1 & -3 \\ -5 & -6 \end{bmatrix}$$

Add corresponding terms of the matrices.

$$= \begin{bmatrix} 7 & 3 \\ -1 & -4 \end{bmatrix}$$

4. H Topics: Complex Numbers (pp. 175–177); Solving Quadratic Equations (pp. 170–173)

Write the equation in standard form.

$$2x^2 - 4x = -4 \rightarrow 2x^2 - 4x + 4 = 0$$

Identify the values for a, b, and c. $a = 2, b = -4, c = 4$

Substitute the values of a, b, and c into the quadratic formula. $x = \dfrac{-b \pm \sqrt{b^2 - 4ac}}{2a}$

$$= \dfrac{-(-4) \pm \sqrt{(-4)^2 - 4(2)(4)}}{2(2)}$$

$$= \dfrac{4 \pm \sqrt{16 - 32}}{4}$$

$$= \dfrac{4 \pm \sqrt{-16}}{4}$$

$$= \dfrac{4 \pm 4i}{4} = 1 \pm i$$

The solutions of the quadratic equation are $1 + i$ and $1 - i$.

5. E Topic: Absolute Value Equations and Inequalities (pp. 161–163)

$|x - 4| \leq 9$ means $x - 4 \leq 9$ or $x - 4 \geq -9$.

Solve both inequalities.

$$\begin{array}{ll} x - 4 \leq 9 & x - 4 \geq -9 \\ x \leq 13 & x \geq -5 \end{array}$$

The solution set is $x \leq 13$ and $x \geq -5$ or $-5 \leq x \leq 13$. So, 14 is not in the solution set.

6. K Topic: Solving Quadratic Inequalities (pp. 173–175)

$$P(x) = 20x^2 - 1{,}000$$

Substitute 1,000 for $P(x)$. $1{,}000 \leq 20x^2 - 1{,}000$

Since the profit must be *at least* $1,000, use $\leq$.

Solve for x. $0 \leq 20x^2 - 2{,}000$

Divide by 20. $0 \leq x^2 - 100$

Factor. $0 \leq (x + 10)(x - 10)$

Solve.

$$\begin{array}{ll} x + 10 \geq 0 \text{ and } x - 10 \geq 0 & \text{or } x + 10 \leq 0 \text{ and } x - 10 \leq 0 \\ x \geq -10 \text{ and } x \geq 10 & x \leq -10 \text{ and } x \leq 10 \\ x \geq 10 & x \leq -10 \end{array}$$

Reject. Can't have negative customers.

The boutique needs at least 10 customers each month.

7. **C** Topic: Patterns, Sequences, and Modeling
(pp. 178–180)

Look at the pattern. Every week there are twice as many plus 5 more ducks than the week before.

Weeks	0	1	2	3
Ducks	5	15	35	75

Continue the pattern. Week 6 is the first time they will see more than 315 ducks.

Weeks	0	1	2	3	4	5	6
Ducks	5	15	35	75	155	315	635

8. **G** Topic: Solving Quadratic Equations (pp. 170–173)

A polynomial cannot have more roots than the value of its largest exponent, so the maximum number of roots is 4.

9. **D** Topic: Rational and Radical Expressions (pp. 167–170)

$$x^{\frac{1}{2}} \cdot \sqrt[3]{x^2} + \frac{1}{\sqrt[4]{x^{-2}}}$$

Write radicals with fractional exponents.

$$= x^{\frac{1}{2}} \cdot x^{\frac{2}{3}} + \frac{1}{(x^{-2})^{\frac{1}{4}}}$$

Write fractional exponents to be added with common denominators.

$$= x^{\frac{3}{6}} \cdot x^{\frac{4}{6}} + \frac{1}{(x^{-2})^{\frac{1}{4}}}$$

Simplify.

$$= x^{\frac{7}{6}} + \frac{1}{x^{-\frac{1}{2}}} = x^{\frac{7}{6}} + x^{\frac{1}{2}}$$

Chapter 10 Coordinate Geometry
Graphing Inequalities on a Number Line

Practice (p. 190)

1.

2.

3.

4.

5.

6.

7.

8.

9.

10.

11.

12.

13.

14.

15.

16.

17.

18.

19.

20.

ACT-Type Problems (p. 190)

1. **B** The graph has an open circle on −6, which means −6 is not included. The arrow goes on forever to the right, representing numbers greater than −6. Therefore, the inequality is $x > -6$.

2. **J** The graph has a closed circle at −5, which means −5 is included, and has an open circle at 1, which means 1 is excluded. Only the section between the two numbers is shaded, so the answer is $-5 \le x < 1$.

3. **A** The inequality $-3 < x < 4$ represents the section of the number line between −3 and 4 but not including −3 and 4. This is shown by the graph in A.

4. **J**

Rewrite the inequality	$3x - 5 \le 5x + 7$
Subtract 5x.	$-2x - 5 \le 7$
Add 5.	$-2x \le 12$
Divide by −2.	$x \ge -6$

$x \ge -6$ is represented by the graph shown in J.

5. **E** List all the integers shown on the graph. Remember that the open circle at 4 means 4 is excluded.

$-6, -5, -4, -3, -2, -1, 0, 1, 2, 3$

Then find the sum.

$-6 + (-5) + (-4) + (-3) + (-2) + (-1) + 0 + 1 + 2 + 3 = -15$

Graphing Equations on the Coordinate Plane

Practice (p. 197)

1.

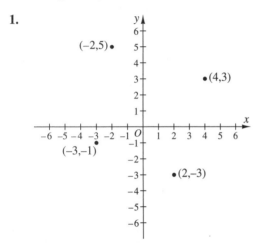

2. $(0,1)$ is on the y-axis. $(2,0)$ is on the x-axis.

3.

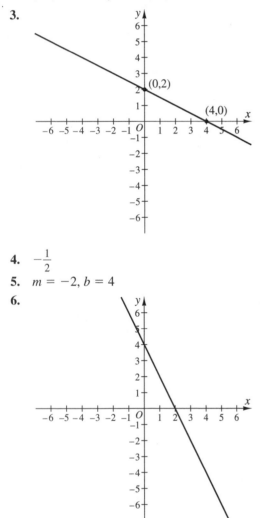

4. $-\dfrac{1}{2}$

5. $m = -2, b = 4$

6.

7. zero 8. positive 9. negative
10. undefined 11. false 12. true
13. true 14. true 15. false
16. true 17. false 18. $\dfrac{3}{2}$
19. $\dfrac{2}{3}$ 20. neither

ACT-Type Problems (p. 198)

1. **C** Rewrite the equation in slope-intercept form.
$$6x - 9y = 3$$
Subtract $6x$. $\quad -9y = -6x + 3$

Divide by -9. $\quad y = \dfrac{2}{3}x - \dfrac{1}{3}$

The equation is now in the form $y = mx + b$ where m is the slope and b is the y-intercept, so $m = \dfrac{2}{3}$ and $b = -\dfrac{1}{3}$.

Add the two values.

$m + b = \dfrac{2}{3} + \left(-\dfrac{1}{3}\right) = \dfrac{1}{3}$

2. **F** To find the slope of the line, find two points on the line. The points $(0,4)$ and $(2,0)$ are on the line. Then use the formula $m = \dfrac{y_2 - y_1}{x_2 - x_1}$ to find the slope.

$m = \dfrac{y_2 - y_1}{x_2 - x_1} = \dfrac{0 - 4}{2 - 0} = -\dfrac{4}{2} = -2$

The graph crosses the y-axis at 4, so that is the y-intercept.

Choice F is the equation of a line with slope -2 and y-intercept 4.

3. **B** Rewrite the equation in slope-intercept form.
$$12x - 3y = 17$$
Subtract $12x$. $\quad -3y = -12x + 17$

Divide by -3. $\quad y = 4x - \dfrac{17}{3}$

The equation is now in slope-intercept form, $y = mx + b$, so the slope, m, is 4.

Parallel lines have the same slope, so the slope of the parallel line is also 4.

4. **J** There is an infinite number of solutions when two equations represent the same line. You want to find the equation that represents the same line as $y = 5x - 7$.

Write each choice in slope-intercept form until you find one that results in $y = 5x - 7$.

Here is what happens when you write choice J in slope-intercept form.
$$4y - 20x = -28$$
Add $20x$. $\quad 4y = 20x - 28$

Divide by 4. $\quad y = 5x - 7$

This is the same equation as the one given. So the lines are the same, producing an infinite number of solutions.

5. E Rewrite the equation in slope-intercept form.

$$3y - 4x = 21$$

Add $4x$.

$$3y = 4x + 21$$

Divide by 3.

$$y = \frac{4}{3}x + 7$$

Find the slope of a line perpendicular to the given line.

The slope, m, of the line is $\frac{4}{3}$. Since the product of the slopes of perpendicular lines is -1, the slope of a perpendicular line is $-\frac{3}{4}$.

Write the equation for the line.

$y = mx + b$ is the slope-intercept equation of a line. In this case, $m = -\frac{3}{4}$ and the y-intercept, b, is -3.

$y = -\frac{3}{4}x - 3$ is the equation of the line perpendicular to the given line with a y-intercept of -3.

Distance and Midpoint Formulas

Practice (p. 201)

1. $d = \sqrt{(x_2 - x_1)^2 + (y_2 - y_1)^2}$

2. $d = \sqrt{13}$

3. $d = \sqrt{205}$

4. $\left(\frac{x_1 + x_2}{2}, \frac{y_1 + y_2}{2}\right)$

5. $(5,3)$

6. $\left(5\frac{1}{2}, 3\frac{1}{2}\right)$

7. $\sqrt{41}$

8. $\sqrt{145}$

9. $2\sqrt{10}$

10. 5

11. $2\sqrt{37}$

12. $2\sqrt{13}$

13. $\sqrt{65}$

14. $(3,0.5)$

15. $(-0.5,-2)$

16. $(7,6)$

17. $(-1.5,2)$

18. $(3,6)$

19. $(4,6)$

20. $(-5,5.5)$

ACT-Type Problems (p. 202)

1. B Use the distance formula.

$$d = \sqrt{(x_2 - x_1)^2 + (y_2 - y_1)^2}$$
$$= \sqrt{(5 - (-4))^2 + (-3 - 7)^2}$$
$$= \sqrt{9^2 + (-10)^2}$$
$$= \sqrt{181}$$

2. K Use the midpoint formula.

$$\left(\frac{x_1 + x_2}{2}, \frac{y_1 + y_2}{2}\right) = \left(\frac{2 + 3}{2}, \frac{6 + 8}{2}\right) = \left(\frac{5}{2}, 7\right)$$

The midpoint is $(2.5,7)$.

3. A Use the midpoint formula.

$$\left(\frac{x_1 + x_2}{2}, \frac{y_1 + y_2}{2}\right) = \left(\frac{-1 + 3}{2}, \frac{5 + 8}{2}\right) = \left(\frac{2}{2}, \frac{13}{2}\right)$$

The midpoint is $(1,6.5)$, so the y-coordinate of the midpoint is 6.5.

Find the length of the line segment.

$$d = \sqrt{(x_2 - x_1)^2 + (y_2 - y_1)^2}$$
$$= \sqrt{(3 - (-1))^2 + (8 - 5)^2}$$
$$= \sqrt{4^2 + 3^2} = \sqrt{16 + 9} = \sqrt{25} = 5$$

Add the length of the line segment and the y-coordinate of the midpoint.

$$5 + 6.5 = 11.5$$

4. H Use the midpoint formula, $\left(\frac{x_1 + x_2}{2}, \frac{y_1 + y_2}{2}\right)$.

Substitute known values, $x_1 = 7$ and $y_1 = 4$.

$$\left(\frac{7 + x_2}{2}, \frac{4 + y_2}{2}\right) = (3,4)$$

Solve for x_2 and y_2.

$$\frac{7 + x_2}{2} = 3 \qquad \frac{4 + y_2}{2} = 4$$

Multiply by 2. $\qquad 7 + x_2 = 6 \qquad 4 + y_2 = 8$

Subtract. $\qquad\qquad\quad x_2 = -1 \qquad\quad y_2 = 4$

$(-1,4)$ is the other endpoint.

5. D Identify the points.

$\overline{AB} \quad (x_1,y_1) \rightarrow (-4,3) \qquad \overline{AC} \quad (x_1,y_1) \rightarrow (-4,3)$
$\quad\quad\quad (x_2,y_2) \rightarrow (5,6) \qquad\qquad\quad (x_2,y_2) \rightarrow (2,-2)$

$\overline{BC} \quad (x_1,y_1) \rightarrow (5,6)$
$\quad\quad\quad (x_2,y_2) \rightarrow (2,-2)$

Use the distance formula.

Find the length of $\overline{AB}$.

$$AB = \sqrt{(x_2 - x_1)^2 + (y_2 - y_1)^2}$$
$$= \sqrt{(5 - (-4))^2 + (6 - 3)^2} = \sqrt{9^2 + 3^2}$$
$$= \sqrt{81 + 9} = \sqrt{90} = 3\sqrt{10} \approx 9.49$$

Find the length of $\overline{AC}$.

$$AC = \sqrt{(x_2 - x_1)^2 + (y_2 - y_1)^2}$$
$$= \sqrt{(2 - (-4))^2 + (-2 - 3)^2}$$
$$= \sqrt{6^2 + (-5)^2} = \sqrt{36 + 25} = \sqrt{61} \approx 7.81$$

Find the length of $\overline{BC}$.

$$BC = \sqrt{(x_2 - x_1)^2 + (y_2 - y_1)^2}$$
$$= \sqrt{(2 - 5)^2 + (-2 - 6)^2}$$
$$= \sqrt{(-3)^2 + (-8)^2} = \sqrt{9 + 64} = \sqrt{73} \approx 8.54$$

So, $AC < BC < AB$.

Graphing Systems of Inequalities on the Coordinate Plane

Practice (p. 206)

1. dotted
2. solid
3. solid
4. dotted
5. The part of the graph that is common to both inequalities.

6.

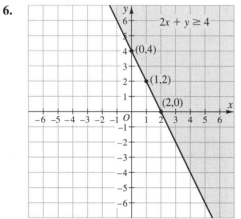

$2x + y \geq 4$

(0,4)
(1,2)
(2,0)

7.

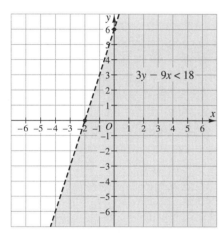

$3y - 9x < 18$

8.

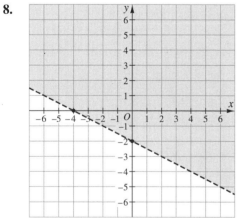

9.

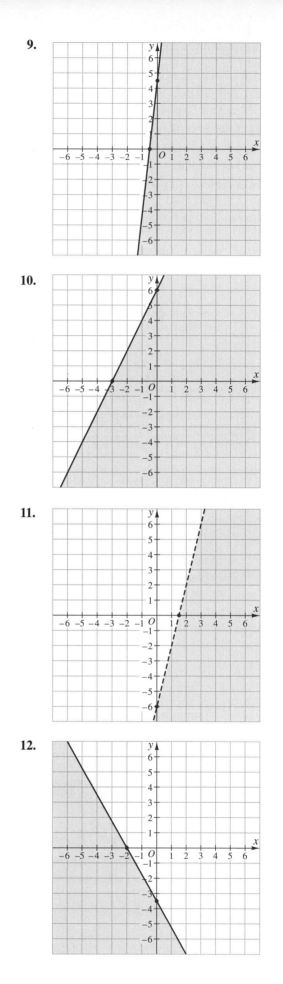

10.

11.

12.

13.

14.

15.

16.

17.

18.

19.

20.

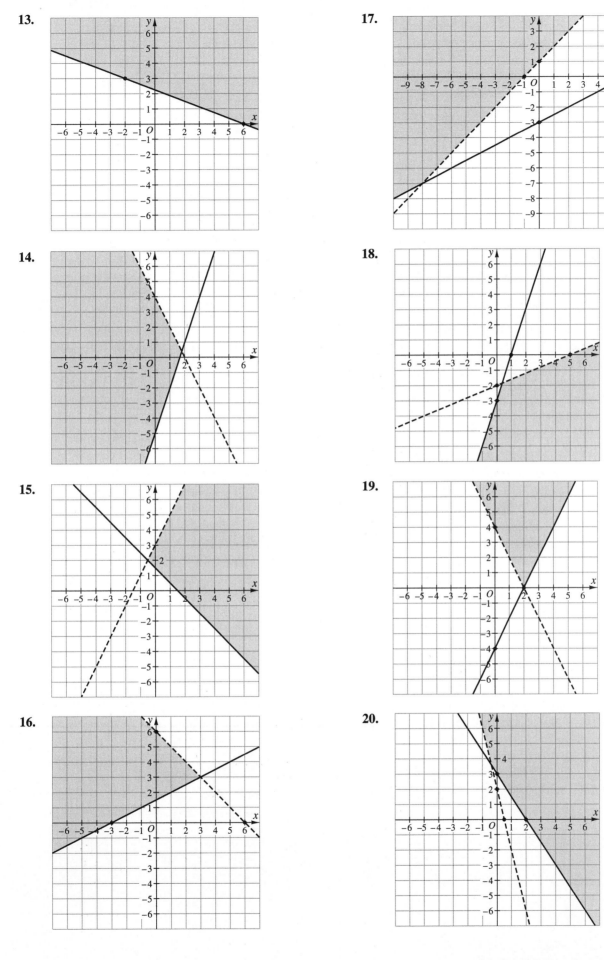

1. **D** Substitute coordinates in the equation.

   All the coordinates produce true statements, except the coordinates in choice D.

   Here is what happens when you substitute coordinates (2,13) for x and y.

   $$2y - 6x \leq 13$$
   Substitute. $\qquad 2(13) - 6(2) \leq 13$
   $$26 - 12 \leq 13$$
   $$14 \leq 13$$

   This is not a true statement, so (2,13) is not a solution of the inequality.

2. **H** Only the inequalities in choices F, G, and H have a solid boundary line.

   The x- and y-intercepts of the graph are (4,0) and (0,−3).

   These points satisfy the equation in H.

3. **A** Here is what would happen if you sketched the graph of the system in choice A. Use a solid line to graph the equation $4y + 5x = 20$ because the inequality is $\geq$. Find a point on one side of boundary line that makes the inequality true and shade that side. Use a dotted line to graph the equation $2y - 4x = -4$ because the inequality is $<$. Find a point on one side of boundary line that makes the inequality true and shade that side. The portion of the coordinate plane in which the two shaded regions intersect along with the boundary lines match the graph in A.

4. **F** Use a solid line to graph the equation $-2x + 5y = 10$ because the inequality is $\leq$.

   Find a point on one side of the boundary line that makes the inequality true and shade that side. The graph matches choice F.

5. **B** Use a solid line to graph the equation $8x - 6y = -24$ because the inequality is $\geq$. Find a point on one side of boundary line that makes the inequality true and shade that side.

   Use a dotted line to graph the equation $-14x - 7y < -28$ because the inequality is $<$. Find a point on one side of boundary line that makes the inequality true and shade that side.

   The portion of the coordinate plane in which the two shaded regions intersect, along with the boundary lines, match the graph in B.

Graphing Conic Sections

Practice (p. 214)

1. Center = (−2,3); radius = 4
2. $(x + 2)^2 = 4$
3. $(x - 5)^2 + (y + 7)^2 = 10^2$

4.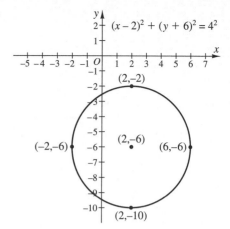

5. (2,−3)
6. vertical
7. $(2, -3 + \sqrt{5})$ and $(2, -3 - \sqrt{5})$
8. (4,2), (−6,2), (−1,6), and (−1,−2)
9. (−4,2) and (2,2)

10.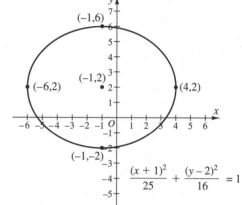

11. vertical
12. horizontal
13. horizontal
14. vertical
15. $y - 4 = 6(x - 3)^2$
16. (3,4)
17. (1,3) and (1,−9)
18. $(-3 - \sqrt{24}, 2)$ and $(-3 + \sqrt{24}, 2)$

19.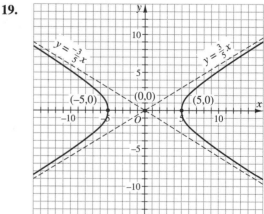

20.

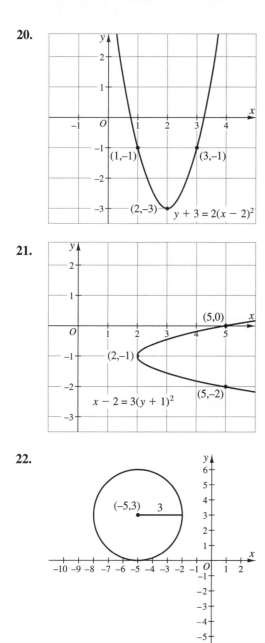

21.

22.

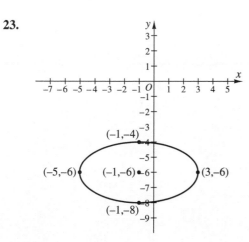

23.

24.

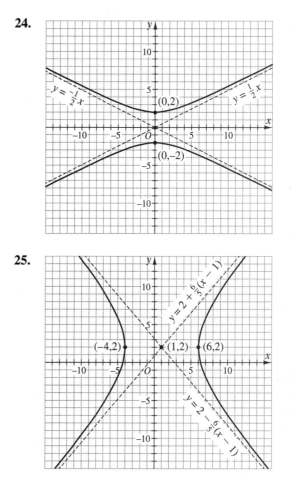

25.

ACT-Type Problems (p. 215)

1. **D** The standard form for the equation of a circle is $(x - h)^2 + (y - k)^2 = r^2$.

   Write the equation $x^2 - 6x + y^2 + 2y + 9 = 0$ in standard form.

   Add the square of half the y-coefficient to each side. $x^2 - 6x + y^2 + 2y + 9 + 1 = 0 + 1$

   Rearrange. $x^2 - 6x + 9 + y^2 + 2y + 1 = 1$

   Group. $(x^2 - 6x + 9) + (y^2 + 2y + 1) = 1$

   Write the trinomial squares. $(x - 3)^2 + (y + 1)^2 = 1$

   The center of this circle is at $(3, -1)$.

2. F Determine whether the major axis is horizontal or vertical.

The standard form of the equation for an ellipse is $\frac{(x-h)^2}{a^2} + \frac{(y-k)^2}{b^2} = 1$ where (h,k) is the center of the ellipse. If $a > b$ then the major axis is horizontal. If $b > a$ the major axis is vertical.

Write the equation in standard form.

$$\frac{(x-4)^2}{49} + \frac{(y+7)^2}{36} = 1 \rightarrow \frac{(x-4)^2}{7^2} + \frac{(y-(-7))^2}{6^2} = 1$$

So, $a = 7$ and $b = 6$. Since $a > b$ the major axis is horizontal.

Determine the center of the ellipse and the equation of the major axis.

The center of the ellipse is at $(4,-7)$. The equation for the major axis is $y = -7$.

3. C The standard form of the equation for a parabola is $y - k = a(x + h)^2$.

Rewrite the equation in standard form.

$$y = x^2 - 8x + 10$$

Subtract 10. $\quad y - 10 = x^2 - 8x$

Add the square of the x-coefficient. $\quad y - 10 + 16 = x^2 - 8x + 16$

Write the perfect square. $\quad y + 6 = (x - 4)^2$

In the standard form of the equation for a parabola $y - k = a(x - h)^2$ the vertex is at (h,k). The vertex is at $(4,-6)$.

Add the x-coordinate of the vertex and the y-coordinate of the vertex.

$$-6 + 4 = -2$$

4. H The standard form of the equation for an ellipse is $\frac{(x-h)^2}{a^2} + \frac{(y-k)^2}{b^2} = 1$.

This is what happens when you write the equation in H in standard form.

$$4x^2 + 9y^2 = 36$$

Divide by 36. $\quad \frac{x^2}{9} + \frac{y^2}{4} = 1$

So $a = 3$ and $b = 2$. This is the equation of an ellipse.

5. D Rewrite the equation $4y^2 + 8y - 9x^2 + 18x = 41$ in standard form.

$$4y^2 + 8y - 9x^2 + 18x = 41$$

Factor out 4 from the y variables and -9 from x variables.

$$4(y^2 + 2y) - 9(x^2 - 2x) = 41$$

Add the square of half the y-coefficient.

$$4(y^2 + 2y + 1) - 9(x^2 - 2x) = 41 + 4$$

Add the square of half the x-coefficient.

$$4(y^2 + 2y + 1) - 9(x^2 - 2x + 1) = 45 - 9$$
$$4(y^2 + 2y + 1) - 9(x^2 - 2x + 1) = 36$$

Write the perfect square.

$$4(y+1)^2 - 9(x-1)^2 = 36$$

Divide both sides by 36. $\quad \frac{4(y+1)^2}{36} - \frac{9(x-1)^2}{36} = \frac{36}{36}$

Simplify. $\quad \frac{(y+1)^2}{9} - \frac{(x-1)^2}{4} = 1$

The standard equation of a hyperbola whose transverse axis is vertical is $\frac{(y-k)^2}{a^2} - \frac{(x-h)^2}{b^2} = 1$, where (h,k) is the center. The center of this hyperbola is at $(1,-1)$. The graph in D is correct.

Cumulative ACT Practice
Coordinate Geometry (p. 216)

Correct your Coordinate Geometry Cumulative ACT Practice answers and work through the answer explanations. If you missed an item or guessed at the answer, circle the number of the item below and review the pages indicated.

1. B Topic: Graphing Equations on the Coordinate Plane (pp. 191–199)

The slope-intercept form of an equation is $y = mx + b$ where m is the slope and b is the y-intercept. The y-intercept of this line is 4. So we know that $y = mx + 4$. We need to find the slope.

$$m = \frac{y_2 - y_1}{x_2 - x_1} = \frac{4 - 0}{0 - 4} = \frac{4}{-4} = -1$$

The equation of the line is $y = -x + 4$.

2. K Topic: Graphing Equations on the Coordinate Plane (pp. 191–199)

The product of the slopes of perpendicular lines is -1. Since the slope of the given line is 2, the slope of a line perpendicular to it will be $-\frac{1}{2}$. The y-intercept of the new line is -3. So the equation for the new line is $y = -\frac{1}{2}x - 3$.

3. B Topic: Graphing Conic Sections (pp. 208–215)

The general equation of a circle is $(x - h)^2 + (y - k)^2 = r^2$, where (h,k) is the center of the circle and r is the radius. Thus, the equation of a circle with a center of $(1,2)$ and a radius of 3 is $(x - 1)^2 + (y - 2)^2 = 9$.

4. G Topic: Distance and Midpoint Formulas (pp. 199–202)

Use the distance formula, $d = \sqrt{(x_2 - x_1)^2 + (y_2 - y_1)^2}$. Substitute $(3,5)$ and $(-2,10)$ and simplify.

$$d = \sqrt{(x_2 - x_1)^2 + (y_2 - y_1)^2}$$
$$= \sqrt{(-2 - 3)^2 + (10 - 5)^2}$$
$$= \sqrt{(-5)^2 + 5^2} = \sqrt{50} = \sqrt{25 \cdot 2} = 5\sqrt{2}$$

5. **E** Topic: Graphing Equations on the Coordinate Plane (pp. 191–199)

Write the equation in slope-intercept form, $y = mx + b$, where m is the slope.

$y - 5 = 3(x - 2) \rightarrow y - 5 = 3x - 6 \rightarrow y = 3x - 1$

The slope is 3.

You can also use the original equation to find the slope since it is in point-slope form, $y - y_1 = m(x - x_1)$, where m is the slope and in this case is 3.

6. **J** Topic: Graphing Systems of Inequalities on the Coordinate Plane (pp. 202–208)

The equation of the boundary line is $y = \frac{3}{2}x - 3$. To determine whether $\leq$ or $\geq$ should be used, try a point on the coordinate plane in the shaded region and see which inequality forms a true statement. For example, choose (3,0) because it is easy to work with a point that has 0 as a coordinate.

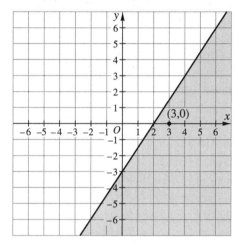

Write the inequality.	$y \,[\leq \text{ or } \geq]\, \frac{3}{2}x - 3$
Substitute.	$0 \,[\leq \text{ or } \geq]\, \frac{3}{2}(3) - 3$
Evaluate.	$0 \,[\leq \text{ or } \geq]\, 4.5 - 3$
	$0 \,[\leq \text{ or } \geq]\, 1.5$
	$0 \leq 1.5$

The inequality $\leq$ makes the statement true, so $y \leq \frac{3}{2}x - 3$.

7. **C** Topic: Graphing Conic Sections (pp. 208–215)

Equations in the form $\frac{x^2}{4} - \frac{y^2}{9} = 1$ are hyperbolas.

8. **H** Topic: Graphing Equations on the Coordinate Plane (pp. 191–199)

Write the equation in slope-intercept form.

$2y + 6x = 10 \rightarrow 2y = -6x + 10 \rightarrow y = -3x + 5$

The slope of the line is -3. The slope of a parallel line is also -3.

Find the y-intercept of a line that has a slope of -3 and passes through $(-1, 2)$.

Substitute -1 for x and 2 for y. Then solve for b.

$y = -3x + b \rightarrow 2 = -3(-1) + b \rightarrow$
$2 = 3 + b \rightarrow -1 = b$

So, the equation of the line is $y = -3x - 1$.

9. **B** Topic: Distance and Midpoint Formulas (pp. 199–202)

Use the midpoint formula, $\left(\frac{x_1 + x_2}{2}, \frac{y_1 + y_2}{2} \right)$.

Substitute values in the formula and simplify.

$\left(\frac{x_1 + x_2}{2}, \frac{y_1 + y_2}{2} \right) = \left(\frac{5 + (-3)}{2}, \frac{-2 + 7}{2} \right) = \left(\frac{2}{2}, \frac{5}{2} \right) = (1, 2.5)$

10. **H** Topic: Distance and Midpoint Formulas (pp. 199–202)

Use the distance formula,

$d = \sqrt{(x_2 - x_1)^2 + (y_2 - y_1)^2}$. Substitute (9,0) and (4,10) and simplify.

$d = \sqrt{(x_2 - x_1)^2 + (y_2 - y_1)^2}$

$= \sqrt{(4 - 9)^2 + (10 - 0)^2}$

$= \sqrt{(-5)^2 + 10^2} = \sqrt{125} = \sqrt{25 \cdot 5} = 5\sqrt{5}$

Coordinate Geometry Subtest (p. 217)

The answers and explanations for the Coordinate Geometry Subtest items are shown with the topic each item tests. If you missed an item or guessed at the answer, circle the number of the item below and review the pages indicated.

1. **C** Topic: Graphing Equations on the Coordinate Plane (pp. 191–199)

The slope-intercept form of a linear equation is $y = mx + b$, where m is the slope. The slope of the line is the change in y-values divided by the change in x-values from one point to another. For this line, the y-value moves -3 while the x-value moves 5.

So, the slope of the line is $-\frac{3}{5}$. The line intercepts the y-axis at 3, so the y-intercept is 3.

The equation shown in the graph is $y = -\frac{3}{5}x + 3$.

2. **K** Topic: Graphing Equations on the Coordinate Plane (pp. 191–199)

Parallel lines have the same slope. So, all lines parallel to the line with equation $y = 3x + 8$ have a slope of 3. The y-intercept of the given line is -7.

Substitute $m = 3$ and $b = -7$ into the equation $y = mx + b$ to get $y = 3x - 7$.

3. B Topic: Graphing Conic Sections (pp. 208–215)

The general equation for a circle is $(x - h)^2 + (y - k)^2 = r^2$, where h is the distance the x-value of the center is shifted from the origin, k is the distance the y-value of the center is shifted from the origin, and r is the length of the radius.

For the circle shown in the graph, $h = 3$, $k = -2$, and $r = 4$.

So, the equation of this circle is
$(x - 3)^2 + (y - (-2))^2 = 4^2$ or
$(x - 3)^2 + (y + 2)^2 = 16$.

4. G Topic: Distance and Midpoint Formulas (pp. 199–202)

Use the distance formula to find the distance between each pair of points.

$AB = \sqrt{(-1 - 4)^2 + (-2 - 2)^2}$
$\quad = \sqrt{(-5)^2 + (-4)^2} = \sqrt{25 + 16} = \sqrt{41}$

$AC = \sqrt{(-1 - (-2))^2 + (-2 - 4)^2}$
$\quad = \sqrt{1^2 + (-6)^2} = \sqrt{1 + 36} = \sqrt{37}$

$BC = \sqrt{(4 - (-2))^2 + (2 - 4)^2}$
$\quad = \sqrt{6^2 + (-2)^2} = \sqrt{36 + 4} = \sqrt{40}$

The length of the longest side is $\sqrt{41}$.

5. C Topic: Graphing Equations on the Coordinate Plane (pp. 191–199)

Write the equation in slope-intercept form to find the slope.

$$3y + 2x - 9 = 0$$
$$3y + 2x = 9$$
$$3y = -2x + 9$$
$$y = -\frac{2}{3}x = 3$$

The slope is $-\frac{2}{3}$.

6. F Topic: Graphing Systems of Inequalities on the Coordinate Plane (pp. 202–208)

Solve the inequality for y.
$$-2y + 4x \leq 8$$
$$-2y \leq -4x + 8$$
$$y \geq 2x - 4$$
The graph in F matches this inequality.

7. B Topic: Graphing Conic Sections (pp. 208–215)

The general equation for an ellipse is $\frac{x^2}{a^2} + \frac{y^2}{b^2} = 1$.

So, $\frac{x^2}{9} + \frac{y^2}{16} = 1$ is the equation of an ellipse.

8. J Topic: Graphing Equations on the Coordinate Plane (pp. 191–199)

Write the equation in slope-intercept form.
$4y - 2x = 12 \rightarrow 4y = 2x + 12 \rightarrow y = \frac{1}{2}x + 3$
The product of the slopes of perpendicular lines is -1, so the slope of a perpendicular line is -2.

Find the y-intercept of a line that has a slope of -2 and passes through $(2,3)$.

Substitute 2 in for x and 3 for y. Then solve for b.
$y = -2x + b \rightarrow 3 = -2(2) + b \rightarrow$
$\quad 3 = -4 + b \rightarrow 7 = b$

So, the equation of the line is $y = -2x + 7$.

9. C Topic: Distance and Midpoint Formulas (pp. 199–202)

Use the midpoint formula, $\left(\frac{x_1 + x_2}{2}, \frac{y_1 + y_2}{2}\right)$.

Substitute values in the formula and simplify.

$\left(\frac{x_1 + x_2}{2}, \frac{y_1 + y_2}{2}\right) = \left(\frac{-2 + 5}{2}, \frac{4 + (-7)}{2}\right) = \left(\frac{3}{2}, -\frac{3}{2}\right)$
$= (1.5, -1.5)$

Cumulative ACT Practice Intermediate Algebra/Coordinate Geometry (p. 219)

Correct your Intermediate Algebra/Coordinate Geometry Cumulative ACT Practice answers and work through the answer explanations. If you missed an item or guessed at the answer, circle the number of the item below and review the pages indicated.

1. A Topic: Rational and Radical Expressions (pp. 167–170)

The function $f(x) = \frac{1}{x^2 - 4}$ is undefined when $x^2 - 4 = 0$.

Since $x^2 - 4 = 0$ when $x^2 = 4$, the two values of x that will make the function undefined are 2 and -2.

Find the product of these two values. $2(-2) = -4$

2. K Topic: Solving Systems of Linear Equations (pp. 163–167)

Let $c =$ the number of cats sold and $d =$ the number of dogs sold.

Each cat costs $50 and each dog costs three times as much or $150.

Write a system of equations.

The total number of pets sold was 25.	$c + d =$	25
The total amount collected was $2,650.	$50c + 150d =$	2,650

Multiply the first equation by -150 to eliminate d and solve for c.

$$-150c - 150d = -3,750$$
$$\underline{50c + 150d = \quad 2,650}$$
$$-100c \qquad\quad = -1,100$$
$$c \qquad\quad = \quad 11$$

3. C Topic: Distance and Midpoint Formulas (pp. 199–202)

Solve for y to write the equation in slope-intercept form, $y = mx + b$, where b is the y-intercept.

$3y - x = 6 \rightarrow 3y = x + 6 \rightarrow y = \frac{1}{3}x + 2$
The y-intercept is 2.

4. F Topic: Absolute Value Equations and Inequalities (pp. 161–163)

The absolute value of the difference between the coordinates of two points on a number line is equal to the distance between the points.

Therefore, $|x - 5| = d$.

5. E Topic: Graphing Equations on the Coordinate Plane (pp. 191–199)

Use the slope formula $m = \frac{y_2 - y_1}{x_2 - x_1}$. Substitute $(4, -1)$ for (x_1, y_1) and $(3, 2)$ for (x_2, y_2).

$$m = \frac{y_2 - y_1}{x_2 - x_1} = \frac{2 - (-1)}{3 - 4} = \frac{3}{-1} = -3$$

6. H Topics: Solving Quadratic Equations (pp. 170–173); Complex Numbers (pp. 175–177)

The general form of a quadratic equation is $ax^2 + bx + c = 0$.

In this equation, $a = 1$, $b = -6$, and $c = 10$.

Use the quadratic equation, $x = \frac{-b \pm \sqrt{b^2 - 4ac}}{2a}$, to solve for x.

$$x = \frac{-b \pm \sqrt{b^2 - 4ac}}{2a} = \frac{6 \pm \sqrt{6^2 - 4 \cdot 1 \cdot 10}}{2 \cdot 1}$$

$$= \frac{6 \pm \sqrt{36 - 40}}{2} = \frac{6 \pm \sqrt{-4}}{2} = \frac{6 \pm 2i}{2} = 3 \pm i$$

7. B Topic: Graphing Equations on the Coordinate Plane (pp. 191–199)

The slope of a vertical line is undefined because the change in the x-values is 0. Since parallel lines have the same slopes and the general equation of a vertical line is $x = a$, $x = 2$ is the correct equation.

8. J Topic: Graphing Equations on the Coordinate Plane (pp. 191–199)

Evaluate the function $f(x) = x^3$ for $x = -4$.

$f(-4) = (-4)^3 = -64$; therefore the point is $(-4, -64)$.

9. C Topic: Complex Numbers (pp. 175–177)

$(2i + 5) + (-3 - 5i) = 2i + (-5i) + 5 + (-3)$
$= -3i + 2$

10. J Topic: Graphing Conic Sections (pp. 208–215)

Since the largest denominator is in the term containing y, the major axis will be parallel to the y-axis. The equation must be for a vertical line, whose general form is $x = a$.

The quantity $(x - 3)$ tells us the center of the ellipse is shifted three units to the right of the origin. Therefore, the x-coordinate of the center is 3. The major axis of an ellipse passes through the center of the ellipse. The equation for the major axis must be $x = 3$.

11. B Topic: Matrices (pp. 180–185)

Multiply each term in the matrix by the scalar. Then add corresponding terms.

$$-2\begin{bmatrix} 1 & -2 \\ 3 & 5 \end{bmatrix} + 4\begin{bmatrix} 2 & -3 \\ 6 & 4 \end{bmatrix}$$

$$= \begin{bmatrix} -2 & 4 \\ -6 & -10 \end{bmatrix} + \begin{bmatrix} 8 & -12 \\ 24 & 16 \end{bmatrix} = \begin{bmatrix} 6 & -8 \\ 18 & 6 \end{bmatrix}$$

12. J Topic: Graphing Systems of Inequalities on the Coordinate Plane (pp. 202–208)

Find the equations for the boundary lines. Choose points on the x- and y-axes.

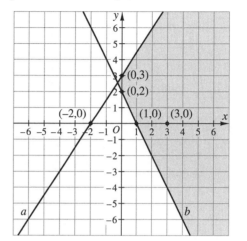

Boundary line a contains points $(-2, 0)$ and $(0, 3)$.

Use the slope formula.

$$m = \frac{3 - 0}{0 - (-2)} = \frac{3}{2}$$

The y-intercept is 3. So the equation of the boundary line is $y = \frac{3}{2}x + 3$.

Boundary line b contains points $(1, 0)$ and $(0, 2)$.

Use the slope formula.

$$m = \frac{2 - 0}{0 - 1} = \frac{2}{-1} = -2$$

The y-intercept is 2. So the equation of the boundary line is $y = -2x + 2$.

These are equations for the boundary lines. Now find the inequalities for the shaded region. Because both boundary lines are solid, the inequalities will be $\leq$ or $\geq$. Pick a point in the shaded region. Choose $(3, 0)$, because it is easy to work with a point for which one or both of the coordinates are 0.

Write the inequality. $y\ [\geq \text{ or } \leq]\ \frac{3}{2}x + 3$

Substitute. $0\ [\geq \text{ or } \leq]\ \frac{3}{2}(3) + 3$

Simplify. $0\ [\geq \text{ or } \leq]\ 4.5 + 3 \rightarrow 0 \leq 7.5$

So, $y \leq \frac{3}{2}x + 3$.

Write the inequality. $y\ [\geq \text{ or } \leq]\ -2x + 2$

Substitute. $0\ [\geq \text{ or } \leq]\ -2(3) + 3$

Simplify. $0\ [\geq \text{ or } \leq]\ -6 + 3 \rightarrow 0 \geq -3$

So, $y \geq -2x + 2$.

13. **C** Topic: Patterns, Sequences, and Modeling (pp. 178–180)

$2 \cdot 4 = 8$

$8 \cdot 4 = 32$

$32 \cdot 4 = 128$

$128 \cdot 4 = 512$

The sequence is increasing by a factor of 4.

14. **F** Topic: Graphing Conic Sections (pp. 208–215)

The high point of a negative quadratic equation is the vertex of the graph. Given the general form of a quadratic equation $f(x) = ax^2 + bx + c$, the x-value of the vertex is $-\dfrac{b}{2a}$. This equation is the general quadratic equation using t instead of x.

$p(t) = -t^2 + 12t + 32$, so $a = -1$ and $b = 12$.

$-\dfrac{b}{2a} = -\dfrac{12}{2(-1)} = \dfrac{-12}{-2} = 6$, so 6 seconds is the time at which the rocket reaches its highest point.

15. **B** Topic: Graphing Conic Sections (pp. 208–215)

The general equation of a circle is $(x - h)^2 + (y - k)^2 = r^2$. The radius is r and the center is (h,k), where h is the horizontal shift from the origin, and k is the vertical shift from the origin. Since the circle was shifted down 6 units and to the right 3 units the center is $(3,-6)$. The radius is 8.

Substitute $h = 3$, $k = -6$, and $r = 8$ into the equation of a circle and simplify.

$(x - 3)^2 + (y - (-6))^2 = 8^2 \rightarrow (x - 3)^2 + (y + 6)^2 = 64$

Intermediate Algebra/Coordinate Geometry Subtest (p. 221)

The answers and explanations for the Intermediate Algebra/Coordinate Geometry Subtest items are shown with the topic each item tests. If you missed an item or guessed at the answer, circle the number of the item below and review the pages indicated. You can use the chart at the end of these answer explanations to estimate your ACT Intermediate Algebra/Coordinate Geometry (AG) scale score.

1. **B** Topic: Rational and Radical Expressions (pp. 167–170)

Consider the expression $\sqrt{9 - b^2}$.

The square root of a negative number is not a real number.

So for the expression to be a real number, $9 - b^2$ cannot be less than 0.

The smallest and largest values of b will occur when $9 - b^2 = 0$.

Solve this equation.

$9 - b^2 = 0$

$b^2 = 9$

$b = \pm 3$

The smallest value for b is -3 and the largest value for b is 3.

The sum of the smallest and largest values is $3 + (-3) = 0$.

2. **H** Topic: Solving Systems of Linear Equations (pp. 163–167)

Write and solve a system of equations.

Let x = adult tickets which cost $6.50 and y = children's tickets which cost $3.00.

Write a system of equations.

The total number of tickets sold was 105. $x + y = 105$

The total amount collected was $567. $6.5x + 3y = 567$

Multiply the top equation by -3.

$-3x - 3y = -315$

$\underline{6.5x + 3y = 567}$

Solve. $3.5x \quad = 252$

$x \quad = 72$

Substitute $x = 72$ into one of the original equations to find y.

$x + y = 105 \rightarrow 72 + y = 105 \rightarrow y = 33$

72 is 39 more than 33.

3. **A** Topic: Graphing Conic Sections (pp. 208–215)

Write the ordered pairs for the points.

$(1,1) \quad (1,-1) \quad (4,2) \quad (4,-2)$

Each x-value equals the square of the y-value.

So, the graph is of the equation $x = y^2$.

4. **J** Topic: Graphing Equations on the Coordinate Plane (pp. 191–199)

Solve for y to write the equation in slope-intercept form.

$2y - 5x + 2 = 0 \rightarrow 2y - 5x = -2 \rightarrow$
$2y = 5x - 2 \rightarrow y = \dfrac{5}{2}x - 1$

5. **A** Topic: Absolute Value Equations and Inequalities (pp. 161–163)

The distance between any two points on a number line is equal to the absolute value of the difference between the coordinate points. Therefore, $d = |x - (-3)|$ or $d = |x + 3|$.

6. **H** Topic: Graphing Equations on the Coordinate Plane (pp. 191–199)

Use the slope formula $m = \dfrac{y_2 - y_1}{x_2 - x_1}$. Substitute $(5,2)$ for (x_1,y_1) and $(-3,4)$ for (x_2,y_2).

$m = \dfrac{y_2 - y_1}{x_2 - x_1} = \dfrac{4 - 2}{-3 - 5} = \dfrac{2}{-8} = -\dfrac{1}{4}$

7. **E** Topics: Solving Quadratic Equations (pp. 170–173); Complex Numbers (pp. 175–177)

The determinant of a quadratic equation is $b^2 - 4ac$.

The roots are real when the determinant is greater than or equal to 0.

Find the determinant for $x^2 - 2x + 4 = 0$ by substituting $a = 1$, $b = -2$, and $c = 4$.

$b^2 - 4ac = (-2)^2 - 4(1)(4) = 4 - 16 = -12$

There are no real roots. The answer is 0.

8. **F** Topic: Graphing Equations on the Coordinate Plane (pp. 191–199)

A line with a slope of 0 is horizontal.

A line perpendicular to it must be vertical.

The equation of a vertical line is in the form $x = a$.

So $x = 5$ is perpendicular to a line with a slope of 0.

9. **C** Topic: Graphing Equations on the Coordinate Plane (pp. 191–199)

Start with point $(6, -4)$. Subtract $-4 - (4)$ to get $y = -8$.

The x-coordinate changes 3 units when the y-coordinate changes -4 units.

Add $6 + 3$ to get $x = 9$.

$(9, -8)$ is the point on the graph that has a y-coordinate of -8.

10. **J** Topic: Complex Numbers (pp. 175–177)

Multiply numerator and denominator by $2i + 4$ to remove the i from the denominator.

$$\frac{5i + 3}{2i - 4} = \frac{(5i + 3)(2i + 4)}{(2i - 4)(2i + 4)}$$

$$= \frac{10i^2 + 20i + 6i + 12}{4i^2 + 8i - 8i - 16}$$

$i^2 = -1$

$$= \frac{10(-1) + 20i + 6i + 12}{4(-1) - 16}$$

$$= \frac{-10 + 26i + 12}{-20} = \frac{2 + 26i}{-20} = \frac{1 + 13i}{-10}$$

11. **C** Topic: Solving Systems of Linear Equations (pp. 163–167)

Write an equation to show that $R(x) = S(x)$. Then solve for x.

$$6500x + 8000 = 7000x + 6000$$
$$6500x = 7000x - 2000$$
$$-500x = -2000$$
$$x = 4$$

The number of computers manufactured will be the same 4 years after the start of production.

Add to find the year. $1980 + 4 = 1984$

12. **G** Topic: Graphing Conic Sections (pp. 208–215)

The general equation for a hyperbola is

$$\frac{(x - h)^2}{a^2} - \frac{(y - k)^2}{b^2} = 1.$$

Write the given equation in general form.

$$\frac{(x - 2)^2}{4} - \frac{(y + 1)^2}{9} = 1 \rightarrow \frac{(x - 2)^2}{2^2} - \frac{(y - (-1))^2}{3^2} = 1$$

The term $\frac{(x - 2)^2}{4}$ is positive so the transverse axis is parallel to the x-axis.

All lines parallel to the x-axis have an equation in the form $y = b$ where b is the y-intercept.

In the quantity $(y - (-1))$, $k = -1$, so the equation is $y = -1$.

13. **A** Topic: Matrices (pp. 180–185)

Multiply each term in the matrix by the scalar. Then subtract corresponding terms.

$$3\begin{bmatrix} 4 & -2 \\ -7 & 6 \end{bmatrix} - 5\begin{bmatrix} -1 & -3 \\ 5 & 2 \end{bmatrix} = \begin{bmatrix} 12 & -6 \\ -21 & 18 \end{bmatrix} - \begin{bmatrix} -5 & -15 \\ 25 & 10 \end{bmatrix}$$

$$= \begin{bmatrix} 17 & 9 \\ -46 & 8 \end{bmatrix}$$

14. **H** Topic: Graphing Systems of Inequalities on the Coordinate Plane (pp. 202–208)

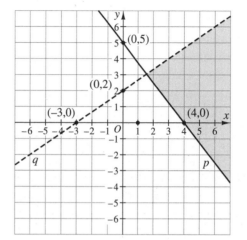

Label one boundary line p and the other boundary line q.

Find the slope of line p. Two points on the line are $(4,0)$ and $(0,5)$.

$$m = \frac{5 - 0}{0 - 4} = -\frac{5}{4}$$

Boundary line p is solid. So the inequality must be $\leq$ or $\geq$.

The answer must be choice G or choice H because these are the only choices with a slope of $-\frac{5}{4}$ and $\leq$ or $\geq$.

Find the slope of boundary line q. Two points on the line are $(0,2)$ and $(-3,0)$.

$$m = \frac{0 - 2}{-3 - 0} = \frac{-2}{-3} = \frac{2}{3}$$

So, H is the correct choice.

15. **A** Topic: Patterns, Sequences, and Modeling (pp. 178–180)

In the sequence 120, 96, 76.8, 61.44, 49.152, terms are found by multiplying the previous term by $\frac{4}{5}$.

You can verify this by dividing any term by the term before it. This means the terms are decreasing by the factor $\frac{4}{5}$.

16. F Topic: Distance and Midpoint Formulas (pp. 199–202)

Find each point.

For one point, $x = 1$ and $y = \sqrt{1} = 1$. So the point is $(1,1)$.

For the other point, $x = 9$ and $y = \sqrt{9} = 3$. So the point is $(9,3)$.

Use the distance formula to find the distance between the two points.

$$d = \sqrt{(x_2 - x_1)^2 + (y_2 - y_1)^2}$$
$$= \sqrt{(9 - 1)^2 + (3 - 1)^2} = \sqrt{8^2 + 2^2}$$
$$= \sqrt{64 + 4} = \sqrt{68} = 2\sqrt{17}$$

17. D Topic: Solving Quadratic Inequalities (pp. 173–175)

$P(x) = x^2 + 30x$

Make a table to reflect the answer choices.

A. Actual production: 140 in 4 hours.
$P(x) = x^2 + 30x = (4)^2 + (30)(4)$
$\quad = 16 + 120 = 136$
Production is greater than or equal to $P(x)$. There is a profit.

B. Actual production: 64 in 2 hours.
$P(x) = x^2 + 30x = (2)^2 + (30)(2)$
$\quad = 4 + 60 = 64$
Production is greater than or equal to $P(x)$. There is a profit.

D. Actual production: 250 in 7 hours
$P(x) = x^2 + 30x = (7)^2 + (30)(7)$
$\quad = 49 + 210 = 259$
Production is less than $P(x)$. There is no profit.
This is the correct answer.

Seven hours after the start of the day the company should have produced at least $P(7) = 7^2 + 30(7) = 259$ bicycles. Since they only produced 250 bicycles they made no profit.

18. F Topic: Graphing Conic Sections (pp. 208–215)

The general equation of a circle is $(x - h)^2 + (y - k)^2 = r^2$, where h is the horizontal shift, k is the vertical shift, and r is the radius.

Substitute -5 for h, 3 for k, and 5 for r. Then write the equation in standard form.

$$(x - (-5))^2 + (y - 3)^2 = 5^2 \rightarrow (x + 5)^2 + (y - 3)^2 = 25$$

Estimate Your ACT Intermediate Algebra/Coordinate Geometry (AG) Subscore

The number correct on the Intermediate Algebra/Coordinate Geometry items leads to the AG mathematics scale score. Use the chart below to estimate your AG scale score. But remember these items are not scattered throughout the test as they would be on a real ACT, and you have just finished reviewing these concepts. All this means that the information given below is only an approximation, and that your actual AG score will probably be different, possibly quite different, from this score.

On national ACT administrations:

- The average number of AG items correct is about 6.
- About 10% of test takers get more than 12 or 13 correct.
- About 25% of test takers get more than 9 correct.
- About 25% of test takers get fewer than 4 correct.

Circle the number of items you answered correctly in the table below and locate the scale score. The highest scale score is 18 and the lowest is 1.

Approximate AG Scale Scores

Number Correct	Approximate Scale Score	Number Correct	Approximate Scale Score
18	18	7–8	10
17	17	6	9
16	16	4–5	8
15	15	3	6–7
13–14	14	2	4–5
12	13	1	2–3
10–11	12	0	1
9	11		

Chapter 11 Plane Geometry
Basic Elements of Plane Geometry

Practice (p. 226)

1. **a.** 2 points

   b.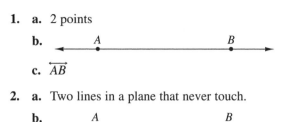

   c. $\overleftrightarrow{AB}$

2. **a.** Two lines in a plane that never touch.

   b.

   c. $\overleftrightarrow{AB} \parallel \overleftrightarrow{CD}$

3. **a.** Two lines that intersect and form a right angle.

   b.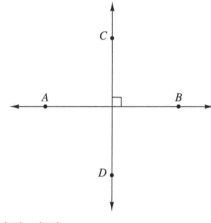

   c. $\overleftrightarrow{AB} \perp \overleftrightarrow{CD}$

4. **a.** 3 noncollinear points

   b.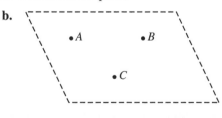

5. **a.** A portion of a line with two endpoints.

   b.

   c. $\overline{AB}$

6. **a.** A portion of a line with one endpoint and extending infinitely in one direction.

   b.

   c. $\overrightarrow{AB}$

7. Here are two possibilities.

   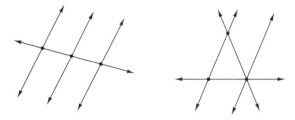

8. Parallel lines are lines in the same plane that never intersect. Skew lines lie in different planes and never intersect.

9.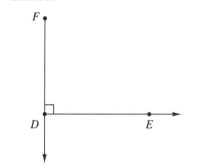

10. An infinite number

ACT-Type Problems (p. 226)

1. **E** The overbar is the symbol for "segment." The symbol $\perp$ means "perpendicular."

2. **G** Draw the figure.

   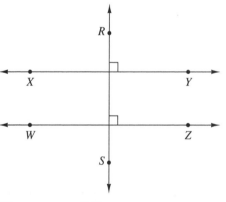

   $\overleftrightarrow{XY}$ is parallel to $\overleftrightarrow{WZ}$.

3. **E** The notation $\overrightarrow{KL}$ means a ray that has endpoint K and passes through L. A ray is a portion of a line beginning at a specific point and extending infinitely in one direction.

4. **H** The opposite edges of a piece of paper never intersect. They are parallel.

5. **B** Two letters written together without an overbar represent the length of a segment.

Angles

Practice (p. 230)

1. **a.** An angle consists of two rays with a common endpoint called the vertex.

   b.

   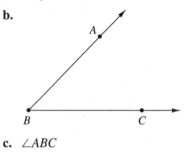

   c. ∠ABC

2. **a.** A right angle

   b.

   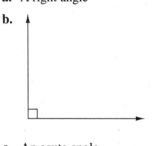

3. **a.** An acute angle

   b.

   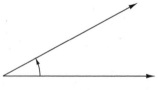

4. **a.** An obtuse angle

   b.

   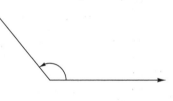

5. **a.** A reflex angle

   b.

   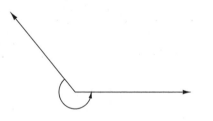

6. **a.** A straight angle

   b.

7. **a.** Complementary angles are two angles that have a sum of 90°.

   b.

   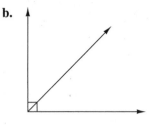

8. **a.** Supplementary angles are two angles that have a sum of 180°.

   b.

   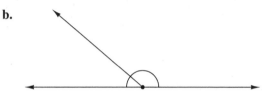

9. Two angles are congruent if they have the same angle measure.

10. 78°

11. 144°

12. 90°

13. **a.** ∠1 and ∠8, ∠1 and ∠2, ∠2 and ∠7, ∠7 and ∠8, ∠3 and ∠4, ∠3 and ∠6, ∠4 and ∠5, ∠5 and ∠6

    b. Supplementary

14. **a.** ∠1 and ∠7, ∠2 and ∠8, ∠3 and ∠5, ∠4 and ∠6

    b. Congruent

15. **a.** ∠8 and ∠6, ∠7 and ∠5, ∠1 and ∠3, ∠2 and ∠4

    b. Congruent

16. **a.** ∠2 and ∠6, ∠7 and ∠3

    b. Congruent

17. **a.** ∠8 and ∠4, ∠1 and ∠5

    b. Congruent

18. **a.** ∠7 and ∠6, ∠2 and ∠3

    b. Supplementary

19. **a.** ∠1 and ∠4, ∠8 and ∠5

    b. Supplementary

20. 105°

21. 105°

22. 65°

23. 70°

24. 110°

25. **a.** 55°; **b.** 125°; **c.** 125°; **d.** 55°; **e.** 125°; **f.** 125°; **g.** 55°

26. 120°; 60°

1. **A** Same-side exterior angles are congruent only when parallel lines intersect a perpendicular transversal.

2. **J** Find m∠6.

   ∠3 and ∠6 are alternate interior angles.

   Since $\overleftrightarrow{AB} \parallel \overleftrightarrow{CD}$, ∠3 and ∠6 are congruent.

	m∠3 = 50°
∠3 and ∠6 are congruent.	m∠3 = m∠6 = 50°
∠6 and ∠8 are supplementary.	m∠6 + m∠8 = 180°
Substitute 50° for m∠6.	50° + m∠8 = 180°
Subtract 50°.	m∠8 = 130°

3. **E** Find m∠7.

   We know that m∠8 = 120° and that ∠7 and ∠8 are supplementary, therefore

	m∠7 + m∠8 = 180°
Substitute 120° for m∠8.	m∠7 + 120° = 180°
Subtract 120°.	m∠7 = 60°

   Find m∠1.

   m∠8 = 120°

   ∠1 and ∠8 are congruent alternate exterior angles.

   ∠1 ≅ ∠8 so m∠1 = m∠8 = 120°.

   m∠1 = 120°

   Multiply m∠7 by m∠1.

   60° × 120° = 7,200°

4. **H** ∠4 and ∠6 are same-side interior angles.

	m∠4 + m∠6 = 180°
∠4 and ∠6 are supplementary.	
Substitute 60° for m∠6.	m∠4 + 60° = 180°
Subtract 60°.	m∠4 = 120°

5. **B** ∠14 is supplementary to ∠13. Other supplementary angles can be identified as vertical angles, corresponding angles, same-side interior angles, alternate interior angles, and alternate exterior angles to ∠14.

Quadrilaterals

Practice (p. 234)

1. If it is possible to draw a segment that passes outside the quadrilateral, the figure is concave. If all segments are contained in the interior of the figure, the quadrilateral is convex.

2. Trapezoid

3. Parallelogram (that is not a rectangle), trapezoid, rhombus

4. Square

5. Rectangle

6. Rhombus

7. Rectangle

8. In both figures, all sides are the same length, the diagonals are perpendicular, and opposite sides are parallel.

9. Trapezoid

10. Parallelogram

ACT-Type Problems (p. 234)

1. **D** A parallelogram is a quadrilateral that has opposite sides that are parallel and congruent. A square meets these criteria.

2. **H** A rectangle has congruent diagonals. None of the other choices is necessarily true.

3. **E** A square has the properties of both a rectangle and a rhombus. A rectangle has four congruent angles; a rhombus has perpendicular diagonals.

4. **G** The angles of a rhombus have the same properties as the angles in a parallelogram. Consecutive angles are supplementary.

5. **D** Since all sides of a square or of a rhombus are the same length, the perimeter of each is 4*s*. A trapezoid could have a perimeter equal to 4 times the length of one of its sides. For example, only one side could be 4 and the perimeter could still be 16.

General Properties of Triangles

Practice (p. 236)

1. **a.** Scalene
   b. Isosceles
   c. Equilateral

2. **a.** Acute
   b. Right
   c. Obtuse

3. 60°

4. 50°, 80°

5. 65° each

6. 10°

7. m∠*a* = 40°

8. The measurements are not realistic because 2 + 3 is not greater than 5. The sum of the lengths of any two sides of a triangle is always greater than the length of the third side.

9. m∠*a* = 45°

10. No. 8 + 3 = 11. The sum of the lengths of any two sides of a triangle must be more than the length of the third side.

11. Yes. The sum of any two sides is greater than the third side.

1. **B** Find the measure of $\angle B$.

   Since $\triangle ABC$ is an isosceles triangle, $m\angle A = m\angle C = 70°$. Therefore $m\angle A = 70°$.

   The sum of the measures of the angles of a triangle is 180°.

$$m\angle A + m\angle B + m\angle C = 180°$$

Substitute.　$70° + m\angle B + 70° = 180°$

Solve for $m\angle B$.　$140° + m\angle B = 180°$

$$m\angle B = 40°$$

   Find $m\angle C + m\angle B$.

   $m\angle C + m\angle B = 70° + 40° = 110° \neq 100°$

2. **J** $\triangle ABC$ is an isosceles triangle. The base angle A measures 40°, so the other base angle, C, also measures 40°. The sum of the measures of the angles of a triangle is 180°. So the measure of angle B is equal to $180° - (40° + 40°) = 180° - 80° = 100°$.

3. **A** The sum of the lengths of any two sides of a triangle is greater than the length of the third side. In choice A, $3 + 5 = 8$, which is less than the third side, 9.

4. **J** 45°, 45°, and 90° are the angle measures of some right triangles, but they cannot be the angle measures of any obtuse triangle. (An obtuse triangle contains one angle whose measure is greater than 90°.) The other statements are true.

5. **C** Of the choices given, only an equilateral triangle always has at least two angles of equal measure. (In fact, it has three angles of equal measure.) The other types of triangles given could have three different angle measures.

Right Triangles

Practice (p. 241)

1. 45°, 45°, and 90°　　2. $h = 10$

3. $a = 6$　　　　　　　4. $b = 4$ and $h = 4\sqrt{2}$

5. $b = 8$ and $a = 8\sqrt{3}$　6. $h = \sqrt{61}$

7. $b = 5$　　　　　　　8. $a = 3$

9. $b = 3$ and $h = 3\sqrt{2}$　10. $b = 6\sqrt{3}$ and $h = 12$

1. **C** You may notice that the lengths of the sides are a 5-12-13 Pythagorean Triple. Therefore $h = 13$.

   You can also use the Pythagorean Theorem.

$$(\text{leg}_1)^2 + (\text{leg}_2)^2 = h^2$$
$$5^2 + 12^2 = h^2$$
$$25 + 144 = h^2$$
$$169 = h^2$$
$$\sqrt{169} = \sqrt{h^2}$$
$$13 = h$$

2. **F** Find the length of the hypotenuse.

   $\triangle ABC$ is a 30–60–90 triangle. According to the diagram, the side opposite the 30° angle is 6.

   The hypotenuse of a 30–60–90 triangle is twice the length of the leg opposite the 30° angle.

   $h = 2 \times 6 = 12$

   Find the length of the other side.

   The length of the leg opposite the 60° angle in a 30–60–90 triangle is $\sqrt{3}$ times the length of the other leg.

   $b = 6\sqrt{3}$

   Find the product of h and b.

   $h \times b = 12 \times 6\sqrt{3} = 72\sqrt{3}$

3. **D** Find the length of the other leg.

   The legs of an isosceles right triangle are the same length. Therefore $a = 10$.

   Find the length of the hypotenuse.

   The length of the hypotenuse of a right isosceles triangle is leg $\times \sqrt{2}$. Therefore $h = 10\sqrt{2}$.

   Find $a + h$.

   $a + h = 10 + 10\sqrt{2}$

4. **G** Use the Pythagorean Theorem to solve this problem. Look at the figure below.

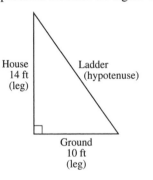

House 14 ft (leg)　Ladder (hypotenuse)

Ground 10 ft (leg)

$$(\text{leg})^2 + (\text{leg})^2 = (\text{hyp})^2$$
$$10^2 + 14^2 = \text{hyp}^2$$
$$100 + 196 = \text{hyp}^2$$
$$296 = \text{hyp}^2$$
$$\sqrt{296} = \text{hyp} \approx 17.2, \text{ the length of the ladder}$$
rounded to the nearest tenth.

5. **B** In a 30–60–90 triangle, the hypotenuse (side AC) is twice the length of the side opposite the 30° angle (side AB). $AC = 8$, so $AB = 4$.

   In a 30–60–90 triangle, the side opposite the 60° angle (side BC) is $\sqrt{3}$ times as long as the shorter leg (side AB). So $BC = 4\sqrt{3}$.

   Find the sum of AB and BC: $4 + 4\sqrt{3}$

Similar Triangles

Practice (p. 246)

1. Corresponding angles of similar triangles are congruent.

2. The ratios of corresponding sides of similar triangles are proportional.

3. The triangles are similar because corresponding angles are congruent.

4. $z = 81.2$
5. $x = 46.4$
6. $m\angle C = 100°$
7. $m\angle X = 30°$
8. $m\angle Z = 100°$
9. $m\angle Y = 50°$
10. $m\angle R = 77°$
11. $m\angle E = 77°$
12. $m\angle D = 62°$
13. $m\angle F = 41°$
14. $EF = 8$
15. $DF = 9$
16. $m\angle L = 75°$
17. $m\angle T = 75°$
18. $m\angle U = 75°$
19. $m\angle V = 30°$
20. $KM = 24$
21. $LM = 24$
22. $AB = 25$, $FD = 56$; $\triangle ABC$: $15 + 35 + 25 = 75$; $\triangle DEF$: $24 + 40 + 56 = 120$; $120 - 75 = 45$. The perimeter of $\triangle DEF$ is 45 units longer than the perimeter of $\triangle ABC$.

23.

	Perimeter	Area
$\triangle PQR$	108 units	486 square units
$\triangle XYR$	60 units	150 square units

ACT-Type Problems (p. 248)

1. **C** Corresponding sides of similar triangles are proportional, therefore

$$\frac{DE}{AB} = \frac{DF}{AC}.$$

Substitute known values.

$$\frac{1.5}{1} = \frac{DF}{2}$$

Multiply by 2. $\quad 2 \times 1.5 = DF = 3$

2. **G** The ordering of the letters in the names of the triangles shows that $\overline{RS}$ corresponds to $\overline{YZ}$.

3. **C** $m\angle Q = 27°$ because $\triangle ABC$ and $\triangle QRS$ are similar.

Find the measure of $\angle R$.

$m\angle Q + m\angle S + m\angle R = 180°$, because these angles are interior angles of a triangle.

Substitute these values.

$m\angle Q = 27°$ and $m\angle S = 35°$.

$\quad 27° + 35° + m\angle R = 180°$

Solve. $\qquad\qquad 62° + m\angle R = 180°$

Subtract 62°. $\qquad\quad m\angle R = 118°$

Find the measure of $\angle B$.

$\triangle ABC$ is similar to $\triangle QRS$. $\angle B$ and $\angle R$ are corresponding angles.

Therefore $m\angle B = m\angle R = 118°$.

Add $m\angle B + m\angle Q$.

$118° + 27° = 145°$

4. **J** $\triangle DEF$ is similar to $\triangle XYZ$, therefore $\frac{YZ}{EF} = \frac{XY}{DE}$.

Substitute known values.

$$\frac{YZ}{5} = \frac{1}{4}$$

Multiply by 5. $\quad YZ = \left(\frac{1}{4}\right) \times 5 = \frac{5}{4} = 1.25 \neq 1.2$

5. **B** $\triangle ABE$ is similar to $\triangle CDE$. Corresponding sides are proportional.

$$\frac{AB}{CD} = \frac{AE}{CE} \qquad\qquad \frac{AB}{CD} = \frac{BE}{BD}$$

$$\frac{3}{2} = \frac{8}{CE} \qquad\qquad\quad \frac{3}{2} = \frac{5}{DE}$$

$3CE = 16 \qquad\qquad\quad 3DE = 10$

$CE = 5\frac{1}{3} \qquad\qquad\quad DE = 3\frac{1}{3}$

$AC = 8 - 5\frac{1}{3} \qquad\quad BD = 5 - 3\frac{1}{3}$

$AC = 2\frac{2}{3} \qquad\qquad\quad BD = 1\frac{2}{3}$

Add AC and BD.

$2\frac{2}{3} + 1\frac{2}{3} = 4\frac{1}{3}$

Concept of Proof and Proof Techniques

Practice (p. 252)

1.

Statement	Reason
1. $\angle 2 \cong \angle 6$	1. Given
2. $\angle 2 \cong \angle 3$	2. Vertical angles are congruent.
3. $\angle 3 \cong \angle 6$	3. If two angles are congruent to the same angle, then they are congruent to each other.

2. Yes. SAS congruence theorem.

3.

Statement	Reason
1. $m\angle BCA = 70°$	1. Given
2. $m\angle BCA + m\angle A + m\angle B = 180°$	2. The sum of the measures of the angles of a triangle is equal to 180°.
3. $\angle BCA + m\angle BCD = 180°$	3. They are supplementary angles.
4. $m\angle BCA + m\angle A + m\angle B = m\angle BCA + m\angle BCD$	4. Substitution
5. $m\angle A + m\angle B = m\angle BCD$	5. Subtract $m\angle BCA$ from both sides.

4. $\overline{BC} \cong \overline{EF}$ and $\angle B \cong \angle E$ or $\overline{AC} \cong \overline{DF}$ and $\angle A \cong \angle D$

5. $\triangle ABC \cong \triangle FED$. The order of the vertices is important when naming a triangle.

6. Not necessarily. Angle Angle Angle (AAA) does not establish congruence. For example, corresponding angles are congruent in the triangles below. The triangles are similar, but not congruent.

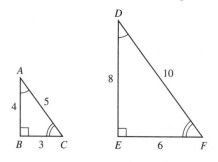

7. Yes. Since $\angle ACB \cong \angle ECD$ (vertical angles), we can show that $\angle A \cong \angle E$. We know from the figure that $\overline{AB} \cong \overline{ED}$ and $\angle B \cong \angle D$. So $\triangle ABC \cong \triangle EDC$ by Angle Side Angle (ASA).

8.

Statement	Reason
1. $\overline{AB} \cong \overline{DE}$	1. Given
2. $\angle B \cong \angle E$	2. Given
3. $\overline{BC} \cong \overline{EF}$	3. Given
4. $\triangle ABC \cong \triangle DEF$	4. SAS Congruence
5. $\angle A \cong \angle D$	5. Corresponding parts of congruent triangles are congruent.

9. $\angle A \cong \angle E$. You know $\angle B \cong \angle D$ (given), and $\angle ACB \cong \angle ECD$ (vertical angles). Knowing that another pair of angles is congruent would not help. You need to establish that two corresponding sides are congruent.

10. No. The angles of the triangles are not listed in corresponding order. $\triangle ABC \cong \triangle FED$.

ACT-Type Problems (p. 254)

1. **B** To be able to complete the proof as shown, we must know that $m\angle CBA = m\angle CDE$.

2. **H** Since we are given that $\triangle ABC \cong \triangle EDF$ we do not know that $\overline{CA}$ is congruent to $\overline{FD}$. We do know that $\overline{CA}$ is congruent to $\overline{FE}$.

3. **D** To be able to prove that $EF = GH$ we would need to know more. For example, if we knew $\triangle EBF \cong \triangle HCG$, we would be able to prove choice D.

4. **H** $\overline{AB} \cong \overline{EF}$ could not be proved because these segments are not corresponding parts of congruent triangles. Note also that with the given information we could not prove that these sides are *not* congruent.

5. **E** Corresponding angles of congruent triangles have the same measure. Use this relationship to determine that $m\angle A$ is 40°.

The sum of the measures of the angles in a triangle is 180°. Use this relationship to determine that $m\angle C$ is 25°.

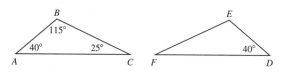

Circles

Practice (p. 257)

1. A diameter is a line segment that passes through the center of the circle and has its endpoints on the circle.

2. A radius of a circle is a line segment with the center for one endpoint and the other endpoint lying on the circle.

3. Center: C; radius: $\overline{CD}$ (or $\overline{CA}$ or $\overline{CB}$); diameter: $\overline{AB}$

4. A chord is a line segment whose endpoints lie on the circle.

5. A secant is a line containing a chord.

6. A tangent is a line that touches a circle at exactly one point.

7. Chord: $\overline{CD}$ (or $\overline{XB}$); secant: $\overleftrightarrow{BX}$ (or $\overleftrightarrow{CD}$); tangent: $\overleftrightarrow{EF}$

8. A central angle is an angle formed by two radii.

9. An inscribed angle is an angle formed by two chords with the vertex lying on the circle.

10. Central angle: $\angle STU$; inscribed angle: $\angle PQR$

11. The inscribed angle intercepts $\overparen{PR}$ and the central angle intercepts $\overparen{SU}$.

12. $m\overparen{PQ} = 50°$ 13. $m\overparen{RS} = 60°$

14. $m\overparen{AB} = 40°$ 15. $m\overparen{CD} = 136°$

16. $m\angle\theta = 120°$ 17. $m\angle\alpha = 90°$

18. $m\angle\theta = 72°$ 19. $m\angle\alpha = 35°$

20. $m\angle\beta = 115°$

21. a. $m\overparen{AD} = 120°$; $\frac{120°}{360°} = \frac{1}{3}$; $\frac{1}{3} \times 6$ units $= 2$ units. Distance from A to D is 2 units.

    b. $m\overparen{DF} = 80°$; $\frac{80°}{360°} = \frac{2}{9}$; $\frac{2}{9} \times 6$ units $= 1\frac{1}{3}$ units. Distance from D to F is $1\frac{1}{3}$ units.

    c. $m\overparen{BFC} = 155°$; $\frac{155°}{360°} = \frac{31}{72}$; $\frac{31}{72} \times 6$ units $= \frac{31}{12} = 2\frac{7}{12}$ units. Distance from B to C is $2\frac{7}{12}$ units.

1. **D** A chord is a line segment whose endpoints are on the circle.

2. **G** An inscribed angle is an angle formed by two chords whose vertex is on the circle.

3. **D** The measure of a central angle is equal to the measure of the arc that it intercepts. Therefore, the measure of arc *RS* is 84°.

4. **K** Find m∠θ.

   ∠θ is an inscribed angle.

   Therefore m∠θ = $\left(\frac{1}{2}\right)$ (measure of arc *AB*).

   m∠θ = $\left(\frac{1}{2}\right)$ 180°

   m∠θ = 90°

   Find m∠α.

   ∠α is a central angle, therefore m∠α = measure of arc *AB*.

   m∠α = 180°

   Find m∠θ + m∠α.

   m∠θ + m∠α = 90° + 180° = 270°

5. **B** Find the measures of the angles to see if △*ABC* is a special triangle.

   ∠*A* is an inscribed angle that intercepts arc *BC*, therefore:

   m∠*A* = $\left(\frac{1}{2}\right)$ × measure of arc *BC*.

   m∠*A* = $\left(\frac{1}{2}\right)$ × 90°

   m∠*A* = 45°

   The problem gives the measure of arc *AC* as twice the measure of arc *BC*, therefore:

   m$\overset{\frown}{AC}$ = 2 × m$\overset{\frown}{BC}$ = 2 × 90° = 180°

   ∠*B* is an inscribed angle that intercepts arc *AC*, therefore:

   m∠*B* = $\left(\frac{1}{2}\right)$ × m$\overset{\frown}{AC}$

   m∠*B* = $\left(\frac{1}{2}\right)$ × 180°

   m∠*B* = 90°

   The sum of the angles of a triangle is 180°, therefore:

   m∠*A* + m∠*B* + m∠*C* = 180°

   Substitute. 45° + 90° + m∠*C* = 180°

   135° + m∠*C* = 180°

   Subtract 135°. m∠*C* = 45°

   △*ABC* is an isosceles right triangle.

   Find the length of *AC*.

   AC is the hypotenuse of △*ABC*. The figure shows that *BC* is equal to 2. In an isosceles right triangle the relationship between the hypotenuse and a leg is:

   hypotenuse = leg × $\sqrt{2}$

   hypotenuse = $2\sqrt{2}$

   Since *AC* is the hypotenuse, *AC* = $2\sqrt{2}$.

Transformations in the Plane

Practice (p. 262)

1.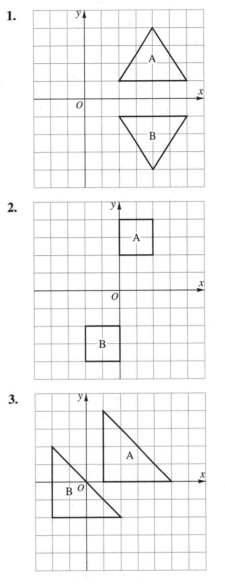

2.

3.

4. Rotate triangle A +90° (90° counterclockwise) about the point (3,1).

5. Reflect (flip) triangle A across the line *y* = 1.

6.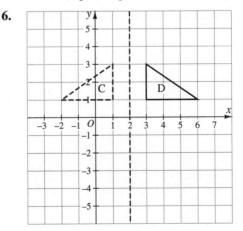

7.

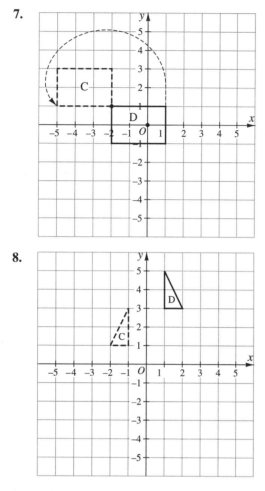

8.

9. Reflection (flip) across the line $y = 1$

10. 90° rotation about the point $(-3.5, 2.5)$

ACT-Type Problems (p. 265)

1. **C** The base of A became the altitude of B. It must have rotated 90°.

2. **J** In J, the circle moves up 4 units.

3. **B** The left side of A lies on $x = 1$.

4. **J** The result of the transformation in choice J is shown as A' below.

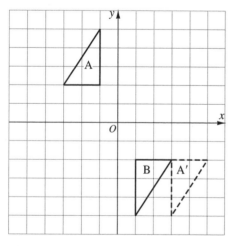

388 Mathematics

---

5. **B** Rectangle A is 2 units from the y-axis while rectangle B is 1 unit from the y-axis. Rectangle A can never be reflected across the x-axis or the y-axis to create rectangle B.

Geometric Formulas

Practice (p. 269)

1. $A = 32$ square units
2. $h = 5$ cm
3. $A = 18$ square units
4. $b = 8$ in.
5. $w = 5$ units
6. $A = 63$ square units
7. $A = 36$ square units
8. $s = 11$ units
9. $A = 25\pi$ cm^2 ≈ 78.5 cm^2 and $C = 10\pi$ cm ≈ 31.4 cm
10. $r = 4$ m
11. $h = 3$ units
12. $A = 36$ square inches
13. $A = (15)(20) + \frac{1}{2}\left(\left(\frac{15}{2}\right)^2 \pi\right)$

    $A = 300 + \frac{1}{2}(56.25\pi)$

    $A = 300 + 28.125\pi ≈ 388.3125$ square feet
14. $b = 4$ meters
15. $h = 8$ meters
16. $r = 6$ units
17. The width of the rectangle is twice the radius of the circle. The base and height of the triangle are equal to the width of the rectangle.

    Area of the rectangle $= 10 \times 6 = 60$ square units
    Area of the triangle $= \frac{1}{2} \times 6 \times 6 = 18$ square units
    Area of the semicircle $= \frac{1}{2} \times \pi \times 3^2 = 4.5\pi$
    Total area $= 60 + 18 + 4.5\pi = 78 + 4.5\pi$ square units
18. 240 square units
19. 20π square units
20. 16 units

ACT-Type Problems (p. 271)

1. **D** $h = 5$ $b_1 = 6$ $b_2 = 11$

    Formula. $A = \frac{h}{2}(b_1 + b_2)$

    Substitute. $A = \frac{5}{2}(6 + 11) = 2.5(17) = 42.5$

2. **G** Find the radius.

    $A = 49\pi$ and $A = \pi r^2$

    $49\pi = \pi r^2$

    Divide by π. $49 = r^2$

    Find the square root. $\sqrt{r^2} = \sqrt{49}$

    $r = 7$

    Find the circumference.

    Formula. $C = 2\pi r$
    Substitute. $C = 2 \times \pi \times 7$
                   $C = 14\pi$

3. **D** Sketch the figure.

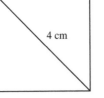

Two sides of the square form the legs of a 45–45–90 triangle whose hypotenuse is the diagonal of the square. Find the length of the legs using the Pythagorean Theorem.

$$s^2 + s^2 = 4^2$$
$$2s^2 = 16$$
$$s^2 = 8$$
$$s = \sqrt{8}$$

The area of the square equals the square of one side.

$$A = s^2$$
$$A = \left(\sqrt{8}\right)^2$$
$$A = 8 \text{ cm}^2$$

4. **G** To find the area of the shaded region, find the area of the square and subtract the area of the circle.

The side of the square is equal to the diameter of the circle. Since the radius is 5, the diameter is 10. The area of the square is the square of its side.

$$A = 10^2 = 100$$

Now find the area of the circle.

$$A = \pi r^2 = \pi(5)^2 = 25\pi$$

Subtracting the area of the circle from the area of the square, we get:

Area of the shaded region $= 100 - 25\pi$.

5. **C** To find the area of the right triangle, we need the lengths of its legs. In an isosceles right triangle, the hypotenuse is equal to the length of a leg times $\sqrt{2}$. Since the hypotenuse is $8\sqrt{2}$, the legs must measure 8. Now find the area of the triangle.

$$A_T = \frac{1}{2}bh$$
$$A_T = \frac{1}{2} \cdot 8 \cdot 8$$
$$A_T = 32$$

Next find the area of the semicircle. The radius is half the diameter, or $\frac{1}{2} \cdot 8\sqrt{2}$. The semicircle is half of a circle.

$$A_C = \frac{1}{2}\pi r^2$$
$$A_C = \frac{1}{2}\pi\left(4\sqrt{2}\right)^2$$
$$A_C = \frac{1}{2}\pi(32)$$
$$A_C = 16\pi$$

Add the area of the semicircle to the area of the triangle to find the area of the figure.

$$A_T + A_C = 32 + 16\pi$$

Geometry in Three Dimensions

Practice (p. 275)

1. 314 cm²
2. 9 cm
3. 12 cm
4. 141.3 cm³
5. 2 cm
6. $V = 64$ cubic units
7. $s = 6$ units
8. $h = 5$ cm
9. $V = 56$ cubic units
10. $V = 36\pi \approx 113.0$ cubic units
11. $r = 12$ cm
12. $V = 81\pi$ in.³
13. $h = 12$ units
14. $d = 8$ cm
15. **a.** $V = 60$ cubic units
    b. $V = 30$ cubic units
16. $A = 150\pi \approx 471$ square units
17. $V = \frac{19{,}652\pi}{3} \approx 20{,}569.1$ cubic units
18. $r = 16$
19. 1,000 cubic inches
20. 216 cm²

ACT-Type Problems (p. 277)

1. **C**
   $$SA = 2(11 \cdot 5 + 11 \cdot 4 + 5 \cdot 4)$$
   $$= 2(55 + 44 + 20) = 238 \text{ cm}^2$$

2. **K** The surface area of a square pyramid is the sum of the areas of its faces. Each side of the square base is 6 cm, so the area of the square base is 36 cm².
   132 cm² = 36 cm² + 4 • area of each triangular face, because the triangle faces are congruent.
   132 cm² = 36 cm² + 4 • area of each triangle
   −36 cm² −36 cm²
   96 cm² = 4 • area of each triangle
   96 cm² ÷ 4 = (4 • area of each triangle) ÷ 4
   24 cm² = area of each triangle

3. **C** $V = \frac{4}{3} \cdot \pi \cdot r^3$

$3{,}052.08 = \frac{4}{3} \cdot \pi \cdot r^3$

$2{,}289.06 = 3.14 \cdot r^3$ Multiply both sides by $\frac{3}{4}$.

$729 = r^3$ Divide both sides by 3.14.

$\sqrt[3]{729} = \sqrt[3]{r^3}$ Take the cube root of both sides.

$9 = r$

4. **K** The base of the cylinder has an area of 25π and the height of the cylinder is 16.

Formula: $V = \pi r^2 h$

Substitute. $V = 25\pi \times 16 = 400\pi$

5. **D**

Find the radius (r). $V = 7{,}776\pi$

$V = \frac{4}{3} \times \pi r^3$

$7{,}776\pi = \frac{4}{3} \times \pi r^3$

Divide by π. $7{,}776 = \frac{4}{3} r^3$

Multiply by 3. $23{,}328 = 4 \times r^3$

Divide by 4. $5{,}832 = r^3$

Find the cube root. $\sqrt[3]{5{,}832} = \sqrt[3]{r}$

$18 = r$

Find the area of the circle with radius (r) = 18.

$A = \pi r^2$

$A = \pi (18)^2$

$A = 324\pi$

Cumulative ACT Practice
Plane Geometry (p. 278)

Correct your Plane Geometry Cumulative ACT Practice answers and work through the answer explanations. If you missed an item or guessed at the answer, circle the number of the item below and review the pages indicated.

1. **A** Topic: Angles (pp. 227–231)

Vertical angles have equal measures.

So the measure of $\angle \alpha = 30°$.

2. **H** Topic: Circles (pp. 255–260)

The area of a circle is πr^2.

The area of this circle is $49\pi = (7^2)\pi$. So the radius of the circle, r, is 7.

Use the circumference formula. $C = 2\pi r$

Substitute and evaluate. $C = 2\pi(7) = 14\pi$

3. **E** Topic: Similar Triangles (pp. 244–248)

The corresponding sides of similar triangles are proportional. $\overline{LN}$ is $\frac{1}{3}$ the length of $\overline{QS}$. So the lengths of the sides of $\triangle LMN$ are $\frac{1}{3}$ the lengths of the corresponding sides of $\triangle QRS$. $QR = 15$ and $\overline{LM}$ corresponds to $\overline{QR}$. So $LM = \left(\frac{1}{3}\right) QR = 5$.

4. **H** Topic: Concept of Proof and Proof Techniques (pp. 249–254)

The figure shows that the only answer choice that could be used to prove the two triangles congruent is the Angle Side Angle postulate.

5. **D** Topic: General Properties of Triangles (pp. 235–238)

In an isosceles triangle the base angles are congruent. $\angle G$ and $\angle E$ are base angles.

So $m\angle G = m\angle E = 50°$.

The sum of the measure of the angles in a triangle is 180°.

$m\angle G + m\angle E + m\angle F = 180°$

$50° + 50° + m\angle F = 180°$

$100° + m\angle F = 180°$

$m\angle F = 80°$

6. **G** Topic: Transformations in the Plane (pp. 260–267)

If you rotate $\overline{AB}$ 180° about the point $(1, -1)$ it will be placed directly on top of $\overline{CD}$.

7. **D** Topic: Geometry in Three Dimensions (pp. 271–277)

Write the formula for the volume of a sphere. $V = \frac{4}{3}\pi r^3$

Substitute. $288\pi = \frac{4}{3}\pi r^3$

Solve for r. $288 = \frac{4}{3}r^3$

$864 = 4r^3$

$216 = r^3$

$6 = r$

8. **F** Topic: Angles (pp. 227–231)

When two parallel lines are cut by a transversal the corresponding angles are congruent. $\angle 4$ and $\angle 8$ are corresponding angles. So $m\angle 4 = 40° = m\angle 8$.

9. **C** Topic: Angles (pp. 227–231)

$\angle ABC \cong \angle 1$ because they are vertical angles.

So $m\angle ABC = 30°$ because $m\angle 1 = 30°$.

We know that $m\angle BAC = 70°$ and that $m\angle BAC + m\angle ABC + m\angle ACB = 180°$ because they are interior angles of a triangle.

So $70° + 30° + m\angle ACB = 180°$

$100° + m\angle ACB = 180°$

$m\angle ACB = 80°$

$\angle ACB \cong \angle 2$ because they are vertical angles. So $m\angle 2 = 80°$.

10. **H** Topic: Right Triangles (pp. 238–243)

A 30–60–90 triangle has special properties. Let the length of the side opposite the 30° angle equal x. The length of the side opposite the 90° angle will be $2x$ and the length of the side opposite the 60° angle will be $x\sqrt{3}$. The length of the side opposite the 30° angle in this triangle is 3, so the length of the side opposite the 90° angle will be $2(3) = 6$. The length of the side opposite the 60° angle is $3\sqrt{3}$. Therefore the sum of the two is $6 + 3\sqrt{3}$.

11. **B** Topic: Concept of Proof and Proof Techniques (pp. 249–254)

We know that corresponding sides of $\triangle CBA$ and $\triangle FDE$ are the same length.

$\triangle CBA \cong \triangle FDE$ by the SSS postulate.

12. **J** Topic: Geometry in Three Dimensions (pp. 271–277)

Write the formula for the volume of a right circular cylinder. $\quad V = \pi r^2 h$

Substitute values. $\quad V = \pi \cdot 5^2 \cdot 5$

Evaluate. $\quad V = 125\pi$

13. **D** Topic: Right Triangles (pp. 238–243)

We know that $\angle D \cong \angle F$ because $\triangle DEF$ is isosceles. $\angle DCE \cong \angle FCE = 90°$ because the base and altitude are perpendicular. So, $\triangle DCE \cong \angle FCE$ by the AAS postulate. Since $\angle DEC \cong \angle FEC$, we know they each take up one half of $\angle DEF$, or 60°. We now have two 30–60–90 triangles as shown in the diagram below.

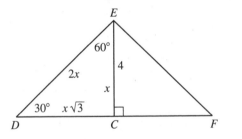

In particular, $\triangle DCE$ is a 30–60–90 triangle. A 30–60–90 triangle has special properties. The length of the side opposite the 30° angle is x. The length of the side opposite the 90° angle is $2x$ and the length of the side opposite the 60° angle is $x\sqrt{3}$. $\overline{EC}$ is opposite the 30° angle and its length is 4, so $x = 4$. So $\overline{DC}$ is $x\sqrt{3}$ or $4\sqrt{3}$. Since $\overline{DC} \cong \overline{FC}$, we double $4\sqrt{3}$ to get the length of the entire base, $\overline{DF}$. Therefore $\overline{DF} = 8\sqrt{3}$.

Use the area formula with $b = 8\sqrt{3}$ and $h = 4$. $\quad A = \frac{1}{2}bh$

$$A = \frac{1}{2} \cdot 8\sqrt{3} \cdot 4$$

$$A = 16\sqrt{3}$$

14. **K** Topic: Transformations in the Plane (pp. 260–267)

A horizontal slide left 3 units plus a horizontal flip across the line $x = -2$ will not place $\triangle ABC$ on $\triangle DEF$.

15. **C** Topics: Right Triangles (pp. 238–243); Geometric Formulas (pp. 267–271)

The right triangle is a 45–45–90 triangle. In this type of right triangle, the lengths of the legs are equal. So base = height = 6 feet.

Use the area formula. $\quad A = \frac{1}{2}bh$

$$A = \frac{1}{2} \cdot 6 \text{ ft} \cdot 6 \text{ ft} = 18 \text{ ft}^2$$

Plane Geometry Subtest (p. 280)

The answers and explanations for the Plane Geometry Subtest items are shown with the topic each item tests. If you missed an item or guessed at the answer, circle the number of the item below and review the pages indicated.

1. **D** Topic: Angles (pp. 227–231)

Linear pairs are supplementary; the sum of their measures is 180°.

We know that $\theta = 2\alpha$ and $\theta + \alpha = 180°$.

Substitute 2α for θ. $\quad 2\alpha + \alpha = 180°$

Divide both sides of the equation by 3. $\quad \dfrac{3\alpha}{3} = \dfrac{180°}{3}$

$$\alpha = 60°$$

Substitute 60° for α. $\quad \theta = 2(60) = 120°$

2. **K** Topic: Geometric Formulas (pp. 267–271)

Circumference: $C = 2\pi r$

Substitute 24π for C and solve for r.

$24\pi = 2\pi r \rightarrow \dfrac{24\pi}{2\pi} = \dfrac{2\pi r}{2\pi} \rightarrow 12 = r$

Area: $A = \pi r^2$

Substitute 12 for r and solve for A.

$A = \pi(12)^2$

$A = 144\pi$

3. **E** Topic: Similar Triangles (pp. 244–248)

The triangles are similar so the lengths of their sides are proportional. The sides in $\triangle DEF$ are twice as long as the sides in $\triangle ABC$.

So $EF = 18$ and $DF = 22$.

The length of the longest side is 22.

4. **K** Topic: Concept of Proof and Proof Techniques (pp. 249–254)

From the diagram we can see that $\overline{AB} \cong \overline{DE}$, $\angle B \cong \angle E$, and $\overline{BC} \cong \overline{EF}$. The two triangles are congruent by Side Angle Side.

5. **B** Topic: Quadrilaterals (pp. 232–235)

Since l_1 is parallel to $\overline{CD}$, a parallelogram is formed. Opposite angles in a parallelogram have equal measures, therefore m$\angle D$ = 50°.

Since it is an isosceles trapezoid, m$\angle D$ = m$\angle A$, so m$\angle A$ = 50°.

Finally we know $\triangle AEB$ is a right triangle because $\overline{BE} \perp \overline{AD}$.

So 50° + 90° + m$\angle ABE$ = 180°.

m$\angle ABE$ = 40°

6. **J** Topic: Transformations in the Plane (pp. 260–267)

A horizontal flip across the line $x = -2$ places $\triangle ABC$ on top of $\triangle DEF$.

7. **E** Topic: Geometry in Three Dimensions (pp. 271–277)

Volume of a right circular cylinder is $\pi r^2 h$, where πr^2 represents the area of the base and h is the height of the cylinder.

Substitute and calculate.

$\pi r^2 h$ = (49π cm²)(14 cm) = 686π cm³

8. **F** Topic: Angles (pp. 227–231)

The measure of $\angle 1$ is 115°, so m$\angle 5$ = 115° because they are alternate exterior angles. Alternate exterior angles have the same measure.

9. **D** Topic: Angles (pp. 227–231)

m$\angle ACB$ = 58°, because it is supplementary to the 122° angle.

m$\angle ABC$ = 73°, because it is an alternate interior angle to the 73° angle.

m$\angle ACB$ + m$\angle ABC$ + m$\angle BAC$ = 180° because they are the angles of a triangle.

Substitute.   58° + 73° + m$\angle BAC$ = 180°

Evaluate.   m$\angle BAC$ = 180° − 131°

                              m$\angle BAC$ = 49°

10. **G** Topic: Right Triangles (pp. 238–243); Concept of Proof and Proof Techniques (pp. 249–254)

$\triangle ABC$ is an isosceles right triangle. $AC = 3\sqrt{2}$ because the hypotenuse is $\sqrt{2}$ times the leg.

Since $\triangle ABC$ is isosceles, $AB = BC$. By SAS, then, we know that $\triangle ADB \cong \triangle CDB$. By this congruency, m$\angle ADB$ = m$\angle CDB$ = 90° and m$\angle ABD$ = m$\angle CBD$ = 45°. This means that $\triangle BDC$ is also an isosceles right triangle with hypotenuse of length 3.

The length of leg $DC = \frac{3\sqrt{2}}{2}$ because DC is half of AC. So the length of BD, the other leg of the triangle, is also equal to $\frac{3\sqrt{2}}{2}$ or $\frac{3}{2}\sqrt{2}$.

11. **A** Topics: Angles (pp. 227–231); Concept of Proof and Proof Techniques (pp. 249–254)

m$\angle BCA$ = m$\angle DCE$ because they are vertical angles.

Since BD and AE bisect one another, $\overline{AC} \cong \overline{EC}$ and $\overline{BC} \cong \overline{DC}$.

Therefore, by the SAS postulate, $\triangle ACB \cong \triangle ECD$.

12. **H** Topics: Geometric Formulas (pp. 267–271); Geometry in Three Dimensions (pp. 271–277)

The circle inscribed in the square has a diameter of 18 cm so the radius is 9 cm.

Volume of a sphere. $V = \frac{4}{3}\pi r^3$

Substitute and solve. $V = \frac{4}{3}\pi r(9)^3$

                              $V = 972\pi$ cm³

13. **E** Topics: General Properties of Triangles (pp. 235–238); Geometric Formulas (pp. 267–271)

Show that the three shaded triangles are congruent.

Look at $\triangle SRC$. $\overline{SC}$ and $\overline{RC}$ are 4 units long because S and R are midpoints of the larger segments which are 8 units long. So $\triangle SRC$ is an isosceles triangle and $\angle R$ and $\angle S$ are congruent because they are opposite congruent sides.

$\angle C$ = 60° because $\triangle ABC$ is an equilateral triangle. So m$\angle R$ + m$\angle S$ = 120°. $\angle R$ and $\angle S$ each measure 60° because they are congruent. This means $\triangle SRC$ is an equilateral triangle because each angle measures 60°. Therefore, each side is 4 inches long.

The same conditions apply to $\triangle QBR$ and $\triangle AQS$, so these three triangles are congruent. We will find the area of one of these triangles and multiply by 3 to find the total shaded area.

Find the height of $\triangle SRC$.

To find the area of each triangle, first find the height of one of them. Use the Pythagorean Theorem to find the height of $\triangle SRC$.

The altitude of an equilateral triangle bisects the base. Half the base is 2.

Find the height of $\triangle SRC$. $4^2 = (2)^2 + (\text{height})^2$

                              $16 - 4 = (\text{height})^2$

                              $\sqrt{12}$ = height

                              $2\sqrt{3}$ = height

Find the area of $\triangle SRC$.

$A = \frac{1}{2}bh$

$A = \frac{1}{2}(4)(2\sqrt{3}) = 4\sqrt{3}$ inches²

Find the total shaded area.

The total shaded area consists of 3 of these triangles, so the total area is

$(3)(4\sqrt{3})$ inches² = $12\sqrt{3}$ inches²

14. **F** Topic: Transformations in the Plane (pp. 260–267)

As shown in the diagram, a horizontal slide -4 units and then a flip across the line $y = 2$ will place $\triangle ABC$ on top of $\triangle DEF$.

Chapter 12 Trigonometry
Right Triangle Trigonometry

Practice (p. 286)

1. $\sin \theta = \dfrac{4}{5}$

2. $\cos \theta = \dfrac{3}{5}$

3. $\tan \theta = \dfrac{4}{3}$

4. $\csc \theta = \dfrac{5}{4}$

5. $\sec \theta = \dfrac{5}{3}$

6. $\cot \theta = \dfrac{3}{4}$

7. $\sin \theta = \dfrac{5}{13}$

8. $\cos \theta = \dfrac{15}{17}$

9. $\csc \theta = \dfrac{13}{12}$

10. $\sin \theta = \dfrac{\sqrt{5}}{3}$

ACT-Type Problems (p. 287)

1. **C**

hypotenuse $= 15$

opposite $= 9$

adjacent $= 12$

$\sec \theta = \dfrac{\text{hypotenuse}}{\text{adjacent}} = \dfrac{15}{12}$

2. **G** Side a is opposite the 40° angle. The side adjacent to this angle measures 6 units.

Formula. $\tan \theta = \dfrac{\text{opposite}}{\text{adjacent}}$

$\tan 40° = \dfrac{a}{6}$

Multiply by 6. $6 \times \tan 40° = a$

$6 \times 0.8391 \approx a$

$5.03 \approx a$

3. **E** $\csc \theta$ is equal to the inverse of $\sin \theta$. So if $\sin \theta = \dfrac{7}{25}$ then $\csc \theta = \dfrac{25}{7}$.

4. **H** Think of the pole as the side opposite the angle and the rope as the hypotenuse. Therefore,

$\theta = 50°$

hypotenuse $= 40$ feet

opposite $= ?$

Formula. $\sin \theta = \dfrac{\text{opposite}}{\text{hypotenuse}}$

$\sin 50° = \dfrac{\text{opposite}}{40 \text{ feet}}$

Multiply by 40. $40 \text{ feet} \times \sin 50° = \text{opposite}$

$30.64 \text{ feet} \approx \text{opposite}$

The pole is approximately 30.64 feet tall.

5. **D**

$\theta = 30°$

opposite $= 20{,}000$ feet

adjacent $= ?$

Formula. $\tan \theta = \dfrac{\text{opposite}}{\text{adjacent}}$

$\tan 30° = \dfrac{20{,}000 \text{ feet}}{\text{adjacent}}$

Multiply by adj. $\text{adj} \times \tan 30° = 20{,}000 \text{ feet}$

Divide by $\tan 30°$. $\text{adj} = \dfrac{20{,}000 \text{ feet}}{\tan 30°} \approx 34{,}662 \text{ feet}$

The plane will have flown a ground distance of 34,662 feet.

Trigonometric Identities

Practice (p. 290)

1. $\sin \theta = \pm\frac{1}{2}$
2. $\cos \theta = \pm\frac{\sqrt{2}}{2}$
3. $\csc^2 \theta$
4. $\cos 2\theta$
5. $\cos \theta$
6. $\cos^2 2\theta$
7. $\sec^2 \theta$
8. π
9. $\tan^2 \frac{\theta}{2}$
10. 1

ACT-Type Problems (p. 291)

1. **C** $1 + \tan^2 \theta = \sec^2 \theta$, so $\dfrac{1 + \tan^2 \theta}{\sec \theta} = \sec \theta$

   Therefore, $\dfrac{1}{\dfrac{1 + \tan^2 \theta}{\sec \theta}} \cdot 2\sin \theta = \dfrac{1}{\sec \theta} \cdot 2\sin \theta$

   $= \cos \theta \cdot 2\sin \theta = 2\sin \theta \cos \theta = \sin 2\theta$

2. **F** $\sin 2\theta = 2\sin \theta \cos \theta$, so $\dfrac{\sin 2\theta}{2 \cos \theta} = \sin \theta$

   Therefore, $\dfrac{\sin 2\theta}{2 \cos \theta} \cdot \sin \theta + \cos^2 \theta$

   $= \sin \theta \cdot \sin \theta + \cos^2 \theta = \sin^2 \theta + \cos^2 \theta = 1$

3. **B** $\cos^2 \theta - \sin^2 \theta = \cos 2\theta$

   but $\cos^2 \theta + \sin^2 \theta = 1$

4. **J** $1 - \dfrac{\sin^2 \theta + \cos^2 \theta}{\csc^2 \theta} = 1 - \dfrac{1}{\csc^2 \theta} = 1 - \sin^2 \theta$

   $= \cos^2 \theta$

5. **A** $(2\cos^2 \theta - 1) - (1 - 2\sin^2 \theta)$

   $= \cos 2\theta - \cos 2\theta = 0$

   or

   $(2\cos^2 \theta - 1) - (1 - 2\sin^2 \theta)$

   $= 2\cos^2 \theta - 1 - 1 + 2\sin^2 \theta$

   $= 2(\cos^2 \theta + \sin^2 \theta) - 2 = 2(1) - 2 = 0$

Unit-Circle Trigonometry

Practice (p. 296)

1. $55°$
2. $45°$
3. $65°$
4. $24°$

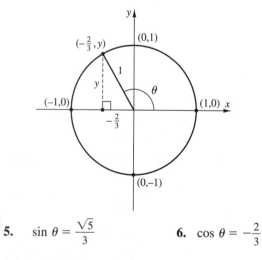

5. $\sin \theta = \dfrac{\sqrt{5}}{3}$
6. $\cos \theta = -\dfrac{2}{3}$

7. $\tan \theta = \dfrac{-\sqrt{5}}{2}$
8. $\csc \theta = \dfrac{3\sqrt{5}}{5}$
9. $\sec \theta = -\dfrac{3}{2}$
10. $\cot \theta = \dfrac{-2\sqrt{5}}{5}$

ACT-Type Problems (p. 296)

1. **E** 245° is 65° from 180°, therefore the reference angle is 65°.

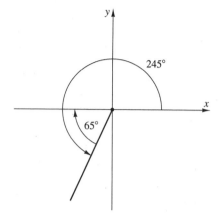

2. **G** 396° is 36° from 360°. Therefore the reference angle is 36°.

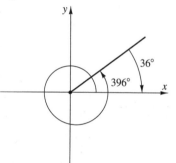

3. **E** Find the reference angle. 300° is 60° from 360°. The reference angle is 60°.

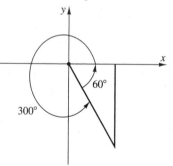

Find sin 60°.

$\sin \theta = \dfrac{\text{opposite}}{\text{hypotenuse}}$

Using the properties of the 30–60–90 triangle we can see that $\sin 60° = \dfrac{\sqrt{3}}{2}$.

Find sin 300°. A 300° angle is in the fourth quadrant and sine is negative in the fourth quadrant. Therefore, $\sin 300° = -\dfrac{\sqrt{3}}{2}$.

4. **J** Find the reference angle. 225° is 45° from 180°, therefore the reference angle is 45°.

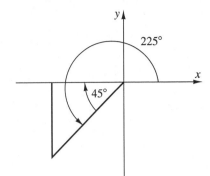

Find cos 45°.

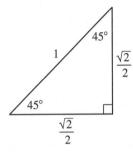

$$\cos \theta = \frac{\text{adjacent}}{\text{hypotenuse}}$$

This is a 45–45–90 triangle so we know that $\cos 45° = \frac{\sqrt{2}}{2}$. Find cos 225°.

A 225° angle is in the third quadrant and the cosine is negative in the third quadrant. So $\cos 225° = -\frac{\sqrt{2}}{2}$.

5. **E** Find the length of the side adjacent to θ. Notice that the triangle is a 3–4–5 right triangle, so the side adjacent to θ is 4.

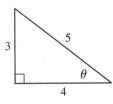

Find tan θ.

$$\tan \theta = \frac{\text{opposite}}{\text{adjacent}} = \frac{3}{4}$$

θ is in the third quadrant because $\pi < \theta < \frac{3\pi}{2}$.

The tangent is positive in the third quadrant.

$$\tan \theta = \frac{3}{4}$$

Graphs of Trigonometric Functions

Practice (p. 300)

1. Amplitude = 2

2. Period = 4π

3. $y = 2\cos \frac{1}{2}x$

4. Amplitude is undefined for cosecant.

5. Period = 2π

6.

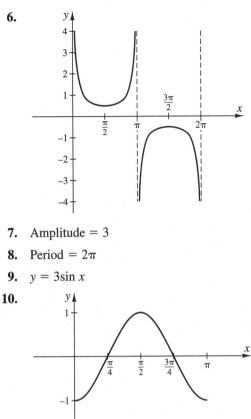

7. Amplitude = 3

8. Period = 2π

9. $y = 3\sin x$

10.

ACT-Type Problems (p. 301)

1. **C** The graph has an amplitude of $\frac{1}{2}$ and a period of $\frac{1}{2}(2\pi)$, or π.

2. **H** The amplitude is 3 and the period is π. The product is $3 \times \pi = 3\pi$.

3. **B** The period of the graph is the period of $y = \cot x$ divided by π, or $\pi \div \pi = 1$.

4. **H** The graph is an inverted sine curve with an amplitude of $\frac{1}{2}$ and a period of $\frac{\pi}{2}$.

5. **E** The period is $\frac{\pi}{2}$. The graph has the same form as the graph of $y = \tan x$.

Cumulative ACT Practice Trigonometry (p. 303)

Correct your Trigonometry Cumulative ACT Practice answers and work through the answer explanations. If you missed an item or guessed at the answer, circle the number of the item below and review the pages indicated.

1. **A** Topic: Right Triangle Trigonometry (pp. 283–287)

$$\tan \theta = \frac{\text{opposite}}{\text{adjacent}} = \frac{4}{3}$$

2. H Topic: Graphs of Trigonometric Functions (pp. 297–301)

The graph is one period of a sine curve.

3. B Topic: Right Triangle Trigonometry (pp. 283–287)

Follow these steps.

$$\cos \theta = \frac{\text{adjacent}}{\text{hypotenuse}}$$

$$\cos 37° = \frac{x}{140 \text{ feet}}$$

$(140 \text{ ft}) \times (\cos 37°) = x$

$x = (140 \text{ ft}) \times (\cos 37°)$

4. K Topic: Trigonometric Identities (pp. 288–291)

$(\cos^2 \theta + \sin^2 \theta) \times \frac{\sin \theta}{\cos \theta} = 1 \times \tan \theta = \tan \theta$

5. B Topic: Right Triangle Trigonometry (pp. 283–287)

This is a right triangle since the wall is perpendicular to the ground. Use the sine ratio to solve the problem.

$$\sin 30° = \frac{t}{15}$$

$$\frac{1}{2} = \frac{t}{15}$$

$2t = 15$

$t = 7.5$

The ladder meets the wall 7.5 feet above the ground.

Trigonometry Subtest (p. 304)

The answers and explanations for the Trigonometry Subtest items are shown with the topic each item tests. If you missed an item or guessed at the answer, circle the number of the item below and review the pages indicated.

1. B Topic: Right Triangle Trigonometry (pp. 283–287)

$$\sin \theta = \frac{\text{opposite}}{\text{hypotenuse}}$$

$$\sin \theta = \frac{4}{5}$$

2. F Topic: Graphs of Trigonometric Functions (pp. 297–301)

This is one period of the graph of $\cos \theta$.

3. E Topic: Trigonometric Identities (pp. 288–291)

Write the equation. $\cos \theta \times \sin^2 \theta + \cos^3 \theta = 1$
Factor out $\cos \theta$. $(\cos \theta)(\sin^2 \theta + \cos^2 \theta) = 1$
Substitute.
$\sin^2 \theta + \cos^2 \theta = 1$. $(\cos \theta)(1) = 1$
  $\cos \theta = 1$

$\cos \theta = 1$ when $\theta = 0°$. $\theta = 0°$

4. H Topic: Trigonometric Identities (pp. 288–291)

Write the expression. $(\sec^2 \theta - 1)(\csc^2 \theta)$

Substitute.
$\sec^2 \theta - 1 = \tan^2 \theta$. $(\tan^2 \theta)(\csc^2 \theta)$

Substitute.

$$\tan^2 \theta = \frac{\sin^2 \theta}{\cos^2 \theta}. \quad \csc^2 \theta = \frac{1}{\sin^2 \theta}$$

$$\left(\frac{\sin^2 \theta}{\cos^2 \theta}\right)\left(\frac{1}{\sin^2 \theta}\right)$$

Cancel $(\sin^2 \theta)$. $\left(\frac{\cancel{\sin^2 \theta}}{\cos^2 \theta}\right)\left(\frac{1}{\cancel{\sin^2 \theta}}\right)$

$$\frac{1}{\cos^2 \theta} = \sec^2 \theta$$

5. B Topic: Right Triangle Trigonometry (pp. 283–287)

$$\sin \theta = \frac{\text{opposite}}{\text{hypotenuse}}$$

$$\sin \theta = \frac{5,000}{10,000}$$

$$\sin \theta = \frac{1}{2}$$

$\theta = 30°$

6. G Topic: Right Triangle Trigonometry (pp. 283–287)

$$\sec \theta = \frac{\text{hypotenuse}}{\text{adjacent}} \qquad \sec \theta = \sqrt{2}$$

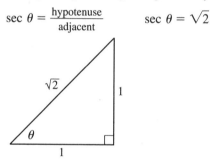

The easiest way to represent this relationship is with a right triangle. The 1–1–$\sqrt{2}$ relationship identifies this as a 45–45–90 triangle and the measure of angle θ is 45°.

7. C Topic: Right Triangle Trigonometry (pp. 283–287)

$$\tan 46° = \frac{\text{opposite}}{\text{adjacent}}$$

$$\tan 46° = \frac{\text{opposite}}{5 \text{ feet}}$$

$5 \text{ feet} \times \tan 46° = \text{opposite}$

8. K Topic: Right Triangle Trigonometry (pp. 283–287)

We know that $\cot \theta = \frac{\text{adjacent}}{\text{opposite}}$. We also know that $\cot \theta = x^2$. An easy way to represent this relationship is to use a right triangle with the side opposite θ equal to 1 and the adjacent side equal to x^2.

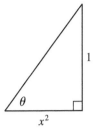

Use the Pythagorean Theorem to find the hypotenuse of the triangle.

$a^2 + b^2 = c^2$

$(1)^2 + (x^2)^2 = c^2$

$1 + x^4 = c^2$

$\sqrt{x^4 + 1} = c$

$\sec \theta = \frac{\text{hypotenuse}}{\text{adjacent}} = \frac{\sqrt{x^4 + 1}}{x^2}$

Cumulative ACT Practice
Plane Geometry/Trigonometry (p. 305)

Correct your Plane Geometry/Trigonometry Cumulative ACT Practice answers and work through the answer explanations. If you missed an item or guessed at the answer, circle the number of the item below and review the pages indicated.

1. C Topic: Geometric Formulas (pp. 267–271)

Since $\triangle ABC$ is an isosceles triangle and the base angles are $\angle A$ and $\angle C$, then the opposite sides, $\overline{AB}$ and $\overline{BC}$, are equal. Since $BD = \frac{4}{5}(AB)$ and $BC = 10$ inches, $BD = \frac{4}{5}(10 \text{ inches}) = 8$ inches. Using the length of the height, $BD = 8$ inches, and the length of the base, $AC = 12$ inches, we can find the area of the triangle.

$A = \frac{1}{2}bh = \frac{1}{2} \times 12 \text{ in.} \times 8 \text{ in.} = \frac{1}{2} \times 96 \text{ in.}^2 = 48 \text{ in.}^2$

2. J Topic: General Properties of Triangles (pp. 235–238)

The sum of the measures of the angles of a triangle is 180°.

$27° + 92° + m\angle FGE = 180°$

$119° = m\angle FGE = 180°$

$m\angle FGE = 61°$

3. B Topic: Right Triangles (pp. 238–243)

Apply the Pythagorean Theorem to each set of sides and notice that for choices A, C, D, and E, $a^2 + b^2 = c^2$. For choice B, however, we have:

$a^2 + b^2 \overset{?}{=} c^2$

$4^2 + 5^2 \overset{?}{=} 6^2$

$16 + 25 \overset{?}{=} 36$

$41 \neq 36$

Since 41 does not equal 36, this cannot be a right triangle.

4. J Topic: Geometry in Three Dimensions (pp. 271–277)

Use the volume formula. $V = \frac{4}{3}\pi r^3$

Substitute. $V = \frac{4}{3}\pi \times (6 \text{ m})^3$

Solve. $V = \frac{4}{3}\pi \times 216 \text{ m}^3$

$V = 288\pi \text{ m}^3$

5. B Topic: Concept of Proof and Proof Techniques (pp. 249–254)

Given: $\angle A \cong \angle D$, $\angle C \cong \angle F$, $\overline{BC} \cong \overline{EF}$

If two pairs of angles are congruent the third pair must be congruent, so $\angle B \cong \angle E$.

The two triangles are congruent by Angle Side Angle.

6. H Topic: Transformations in the Plane (pp. 260–267)

When points are flipped across the line $y = x$, the x-coordinates become the y-coordinates and the y-coordinates become the x-coordinates. Therefore, since we start with the endpoints P (2,1) and Q (4,−3) the new endpoints are P' (1,2) and Q' (−3,4).

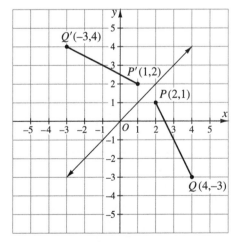

7. A Topic: Angles (pp. 227–231)

When two parallel lines are cut by a transversal, angles have special properties.

$\angle 5 \cong \angle 8$ because they are vertical angles

$\angle 4 \cong \angle 8$ because they are corresponding angles

$\angle 1 \cong \angle 8$ because they are alternate exterior angles

8. **K** Topic: Similar Triangles (pp. 244–248)

When two triangles are similar the areas are proportional by the square of the factor. We can see from the diagram that the lengths of the sides of $\triangle DEF$ are one-third the lengths of the sides of $\triangle ABC$. So we know that the area of $\triangle DEF$ is one-ninth the area of $\triangle ABC$.

Area of $\triangle DEF = \dfrac{1}{9}$ area of $\triangle ABC$

Area of $\triangle DEF = \dfrac{1}{9}(36) = 4$

9. **C** Topic: Geometric Formulas (pp. 267–271)

$\overline{AC}$ is three times as long as $\overline{BC}$, and $BC = 1$ inch, so $AC = 3$ inches. We know $AB = AC - BC$ so $AB = 3$ in. $-$ 1 in. $=$ 2 in. Since $ACFD$ is a rectangle, $AF = CD = 2$ in. So $AF = 2$ in. We can now find the area of $ABEF$.

$A = AB \times AF = 2$ in. $\times$ 2 in. $= 4$ in.²

10. **H** Topic: Circles (pp. 255–260)

Since $\angle ABC$ is a right angle it cuts off exactly half of the circumference.

So the arc length of $\overset{\frown}{AC} = \dfrac{1}{2} \times 30$ in. $= 15$ in.

11. **A** Topic: Geometry in Three Dimensions (pp. 271–277)

Use the formula for the volume of a sphere.	$V = \dfrac{4}{3}\pi r^3$
Substitute.	$V = \dfrac{4}{3}\pi \times (3)^3$
Solve.	$V = \dfrac{4}{3}\pi \times 27 = 36\pi$
Use the formula for the volume of a cylinder.	$V = \pi r^2 h$
Substitute the volume of the sphere for V. Substitute 4 for h.	$36\pi = \pi r^2(4)$
Solve for r.	$9 = r^2$
	$3 = r$

12. **G** Topic: Transformations in the Plane (pp. 260–267)

Rotating $\triangle ABC$ about the origin 270° would place it in the 4th quadrant, and not on top of $\triangle DEF$.

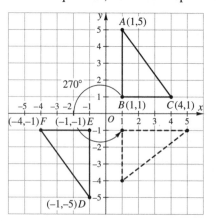

13. **C** Topic: Graphs of Trigonometric Functions (pp. 297–301)

$y = \cos\dfrac{\pi}{2} = 0$

14. **H** Topic: Unit-Circle Trigonometry (pp. 292–296)

$\tan\theta = \dfrac{\text{opposite}}{\text{adjacent}} = \dfrac{3}{4}$

By using the Pythagorean Theorem we can find the length of the hypotenuse.

$c = \sqrt{a^2 + b^2} = \sqrt{\text{opp}^2 + \text{adj}^2} = \sqrt{3^2 + 4^2}$

$\quad = \sqrt{9 + 16} = \sqrt{25} = 5$

$\sin\theta = \dfrac{\text{opposite}}{\text{hypotenuse}} = \dfrac{3}{5}$

However, since we are in the third quadrant the sine is negative. Therefore, the correct answer is $-\dfrac{3}{5}$.

15. **E** Topic: Trigonometric Identities (pp. 288–291)

$\dfrac{1 - \cos^2\theta}{(\sin^2\theta)(\cot^2\theta)} = \dfrac{\sin^2\theta}{(\sin^2\theta)(\cot^2\theta)} = \dfrac{1}{\cot^2\theta} = \tan^2\theta$

16. **G** Topic: Right Triangle Trigonometry (pp. 283–287)

$\sin 30° = \dfrac{x}{15}$

$15 \times \sin 30° = x$

$15 \times 0.5 = x$

$7.5 = x$

Plane Geometry/Trigonometry Subtest (p. 308)

The answers and explanations for the Plane Geometry/Trigonometry Subtest items are shown with the topic each item tests. If you missed an item or guessed at the answer, circle the number of the item below and review the pages indicated. You can use the chart at the end of these answer explanations to estimate your ACT Plane Geometry/Trigonometry (GT) scale score.

1. **B** Topic: Quadrilaterals (pp. 232–235)

The diagonals of a parallelogram bisect one another. Therefore if $AE = 9$, $AD = 18$. Since $BC = \dfrac{2}{3}AD$ then $BC = \dfrac{2}{3}(18) = 12$. Since $BC = 12$, BE must equal 6.

2. **G** Topics: General Properties of Triangles (pp. 235–238); Angles (pp. 227–231)

Since there are 180° in a triangle, $70° + 60° + \angle BCA = 180°$. Then m$\angle BCA = 50°$. It is also given that $\vec{BE} \parallel \vec{AD}$. So m$\angle CBE$ is also 50° because it is alternate interior to $\angle BCA$.

3. D Topic: Circles (pp. 255–260)

A circle measures 360°.

We know that $\overline{AB}$ is a diameter, so m$\overset{\frown}{AB}$ is 180° because it is half of the circle.

Since $\angle ABC$ is an inscribed angle, its measure is half the measure of the arc it intercepts. So, m$\overset{\frown}{AC}$ is $30° \times 2 = 60°$.

$$m\overset{\frown}{BC} = 360° - m\overset{\frown}{AB} - m\overset{\frown}{AC}$$
$$m\overset{\frown}{BC} = 360° - 180° - 60°$$
$$m\overset{\frown}{BC} = 120°$$

4. G Topic: Geometry in Three Dimensions (pp. 271–277)

Volume of a cone $= \frac{1}{3}\pi r^2 h$ where πr^2 is the area of the base.

Substitute 32π m³ for the volume.	$32\pi \text{ m}^3 = \frac{1}{3}\pi r^2 h$
Substitute 16π m² for the area of the base.	$32\pi \text{ m}^3 = \frac{1}{3} \cdot 16\pi h \text{ m}^2$
Multiply both sides of the equation by 3.	$96\pi \text{ m}^3 = 16\pi h \text{ m}^2$
Divide both sides by 16π m².	$\dfrac{96\pi \text{ m}^3}{16\pi \text{ m}^2} = \dfrac{16\pi h \text{ m}^2}{16\pi \text{ m}^2}$
	$6 \text{ m} = h$

5. E Topic: Concept of Proof and Proof Techniques (pp. 249–254)

Answers A, B, C, and D all give information that could be used to prove that the triangles are congruent.

A. by Side Angle Side

B. by Side Angle Side

C. by Side Side Side

D. by Angle Side Angle

Choice E states that all the angles are congruent. This information would be useful only to prove that the triangles are similar.

6. H Topic: Transformations in the Plane (pp. 260–267)

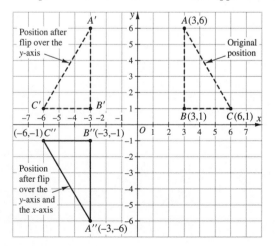

The dotted triangle ($A'B'C'$) shows triangle ABC flipped over the y-axis. Triangle $A''B''C''$ shows triangle $A'B'C'$ flipped over the x-axis. The vertices of triangle $A''B''C''$ are: A'' $(-3,-6)$, B'' $(-3,-1)$, and C'' $(-6,-1)$. You might also notice that these two flips place the entire triangle in the third quadrant where all x- and y-values are negative.

7. C Topic: Right Triangles (pp. 238–243)

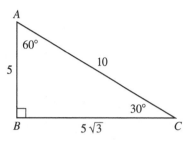

In a 30–60–90 triangle such as this one, the lengths of the sides always have these relationships:

    The side opposite the 30° angle is x.

    The side opposite the 60° angle is $x\sqrt{3}$.

    The side opposite the 90° angle is $2x$.

The side opposite 60° is $5\sqrt{3}$, so $x = 5$.

The side opposite 30° is x, or 5.

The side opposite 90° is $2x$, or 10.

Perimeter $= 10 + 5 + 5\sqrt{3} = 15 + 5\sqrt{3}$

8. F Topic: Angles (pp. 227–231)

Supplementary angles are angles whose sum equals 180°.

$\angle 2$, $\angle 4$, $\angle 5$, $\angle 7$, $\angle 10$, $\angle 12$, $\angle 13$, and $\angle 15$ are all pairs congruent through some combination of linear pairs, vertical angles, or corresponding angles, and are supplementary to $\angle 11$.

9. D Topics: General Properties of Triangles (pp. 235–238); Concept of Proof and Proof Techniques (pp. 249–254)

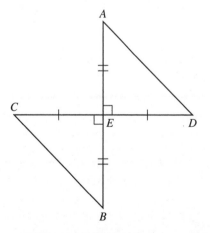

$\angle CEB \cong \angle DEA$ because they are right angles and vertical angles.

$\overline{CE} \cong \overline{DE}$ and $\overline{AE} \cong \overline{BE}$ because $\overline{AB}$ and $\overline{CD}$ bisect each other.

$\triangle AED \cong \triangle BEC$ by Side Angle Side.

10. F Topics: General Properties of Triangles (pp. 235–238); Geometric Formulas (pp. 267–271)

$\triangle ABC$ is similar to $\triangle DEF$, so $\overline{EF} = 3$ cm. Using the Pythagorean Theorem or the 3–4–5 ratio, $\overline{DE} = 4$ cm.

Use the formula for the area of a triangle.

Area $= \frac{1}{2}bh$

$A = \frac{1}{2}(3)(4) = 6$ cm²

11. D Topic: Geometric Formulas (pp. 267–271)

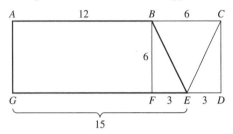

Find the lengths of line segments.

$BC = 6$. $\overline{AB}$ is twice as long as $\overline{BC}$, so $AB = 12$.

$BC = 6$ so $FD = BF = 6$ because they are sides of the same square.

E is the midpoint of $\overline{FD}$, so $FE = 3$.

$GF = 12$ because $\overline{GF}$ is the side opposite $\overline{AB}$ in the rectangle $ABFG$.

Since $GF + FE = GE$, $GE = 12 + 3 = 15$

Use the formula to find the area of the trapezoid.

Area $= \frac{1}{2}(b_1 + b_2)h$.

From information above:

$b_1 = AB = 12 \qquad b_2 = GE = 15 \qquad h = BF = 6$

By substitution: $\qquad A = \frac{1}{2}(12 + 15)6$

$A = 81$ inches²

12. K Topics: General Properties of Triangles (pp. 235–238); Circles (pp. 255–260); Concept of Proof and Proof Techniques (pp. 249–254)

List the things you know about the figure.

$\triangle ABC$ is isosceles because $\overline{CA}$ and $\overline{CB}$ are radii of the same circle.

$\overline{AD} = \overline{DB}$ because point D is the midpoint of $\overline{AB}$.

Test the answer choices.

Choice F: True. The segment joining the vertex of an isosceles triangle to the midpoint of the base is the altitude of the triangle. It is perpendicular to the base.

Choice G: True. The corresponding sides have equal lengths (Side Side Side).

Choice H: True. $\overline{CA}$ and $\overline{CB}$ are radii of the same circle.

Choice J: True. The angles are corresponding angles of congruent triangles.

Choice K: False. It is impossible for any two angles of a triangle to be supplementary because the sum of all three angle measures is 180°.

13. D Topic: Geometry in Three Dimensions (pp. 271–277)

Volume of a cylinder $= \pi r^2 h$

Volume of a sphere $= \frac{4}{3}\pi r^3$

Write an equation with the volume of the cylinder equal to the volume of the sphere.

$\pi r^2 h = \frac{4}{3}\pi r^3$

Divide both sides by π. $\qquad r^2 h = \frac{4}{3}r^3$

Substitute 12 for the height of the cylinder. $\qquad 12r^2 = \frac{4}{3}r^3$

Divide each side by r^2. $\qquad 12 = \frac{4}{3}r$

Multiply each side by $\frac{3}{4}$. $\qquad \frac{3}{4}(12) = \frac{3}{4}\left(\frac{4}{3}r\right)$

$9 = r$

The radius is 9 meters.

14. G Topic: Transformations in the Plane (pp. 260–267)

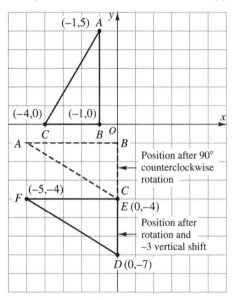

A 90° counterclockwise rotation about the origin and then a vertical shift −3 units (down) would place $\triangle ABC$ on top of $\triangle FED$.

15. B Topic: Graphs of Trigonometric Functions (pp. 297–301)

Since sec θ is the inverse of cos θ, sec $\theta = \frac{1}{\cos \pi}$.

Wherever cos $\theta = 0$, sec θ is undefined. The graph below shows cos $\frac{\pi}{2} - 0$. So when $\theta = \frac{\pi}{2}$, sec θ is undefined.

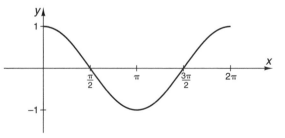

16. K Topic: Unit-Circle Trigonometry (pp. 292–296)

Find the third side of the triangle. Notice that this is a 5–12–13 triangle.

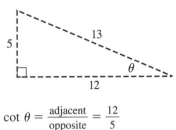

$\cot \theta = \dfrac{\text{adjacent}}{\text{opposite}} = \dfrac{12}{5}$

Since $\sin \theta > 0$, θ is located in the first or second quadrant. In the first quadrant, $\cot \theta = \dfrac{12}{5}$. In the second quadrant, $\cot \theta = \dfrac{12}{5}$.

17. A Topic: Trigonometric Identities (pp. 288–291)

$$\dfrac{1 + \cot^2 \theta}{1 + \tan^2 \theta}$$

$$= \dfrac{\csc^2 \theta}{\sec^2 \theta}$$

$$= \dfrac{\dfrac{1}{\sin^2 \theta}}{\dfrac{1}{\cos^2 \theta}}$$

$$= \dfrac{1}{\sin^2 \theta} \cdot \dfrac{\cos^2 \theta}{1}$$

$$= \dfrac{\cos^2 \theta}{\sin^2 \theta}$$

$$= \cot^2 \theta$$

18. H Topic: Right Triangle Trigonometry (pp. 283–287)

Find $\sin 70°$.	$\sin 70° \approx 0.9397$
Write an equation.	$\sin 70° = \dfrac{\text{height}}{30 \text{ feet}}$
Substitute.	$(30 \text{ feet})(0.9397) \approx h$
Solve.	$28.191 \approx h$

To the nearest hundredth, $h \approx 28.19$.

Estimate Your ACT Plane Geometry/Trigonometry (GT) Subscore

The number correct on the Plane Geometry/Trigonometry items leads to the GT mathematics scale score. Use the chart below to estimate your GT scale score. But remember these items are not scattered throughout the test as they would be on a real ACT, and you have just finished reviewing these concepts. All this means is that the information given below is only an approximation, and that your actual GT score will probably be different, possibly quite different, from this score.

On national ACT administrations:

- The average number of GT items correct is about 8.
- About 10% of test takers get more than 14 correct.
- About 25% of test takers get more than 9 or 10 correct.
- About 25% of test takers get fewer than 5 correct.

Circle the number of items you answered correctly in the table below and locate the scale score. The highest scale score is 18 and the lowest is 1.

Approximate GT Scale Scores

Number Correct	Approximate Scale Score	Number Correct	Approximate Scale Score
18	18	7	9
17	16	5–6	8
16	15	4	7
14–15	14	3	5–6
13	13	2	3–4
12	12	1	2
10–11	11	0	1
8–9	10		

Section III ■

Model Mathematics Tests

Chapter 14 ▪

Model Mathematics **ACT II**
With Answers Explained

This Mathematics Model ACT II is just like a real ACT. Take this test after you take the Diagnostic Mathematics ACT.

Take this model test under simulated test conditions. Allow 60 minutes to answer the 60 Mathematics items. Use a pencil to mark the answer sheet on page 407, and answer the questions in the Test 2 (Mathematics) section.

Use the answer key on page 417 to mark the answer sheet. Review the answer explanations on pages 417–424. You may decide to retake this test later. There are additional answer sheets for this purpose following page 605.

The test scoring chart on page 464 shows you how to convert the number correct to ACT scale scores. Other charts on pages 468–469 show you how to find the Pre-Algebra/Elementary Algebra, Intermediate Algebra/Coordinate Geometry, and Plane Geometry/Trigonometry subscores.

DO NOT leave any answers blank. There is no penalty for guessing on the ACT. Remember that the test is yours. You may mark up, write on, or draw on the test.

When you are ready, note the time and begin.

ANSWER SHEET

The ACT answer sheet looks something like this one. Use a No. 2 pencil
to completely fill the circle corresponding to the correct answer.
If you erase, erase completely; incomplete erasures may be read as answers.

TEST 1—English

1 Ⓐ Ⓑ Ⓒ Ⓓ	11 Ⓐ Ⓑ Ⓒ Ⓓ	21 Ⓐ Ⓑ Ⓒ Ⓓ	31 Ⓐ Ⓑ Ⓒ Ⓓ	41 Ⓐ Ⓑ Ⓒ Ⓓ	51 Ⓐ Ⓑ Ⓒ Ⓓ	61 Ⓐ Ⓑ Ⓒ Ⓓ	71 Ⓐ Ⓑ Ⓒ Ⓓ
2 Ⓕ Ⓖ Ⓗ Ⓙ	12 Ⓕ Ⓖ Ⓗ Ⓙ	22 Ⓕ Ⓖ Ⓗ Ⓙ	32 Ⓕ Ⓖ Ⓗ Ⓙ	42 Ⓕ Ⓖ Ⓗ Ⓙ	52 Ⓕ Ⓖ Ⓗ Ⓙ	62 Ⓕ Ⓖ Ⓗ Ⓙ	72 Ⓕ Ⓖ Ⓗ Ⓙ
3 Ⓐ Ⓑ Ⓒ Ⓓ	13 Ⓐ Ⓑ Ⓒ Ⓓ	23 Ⓐ Ⓑ Ⓒ Ⓓ	33 Ⓐ Ⓑ Ⓒ Ⓓ	43 Ⓐ Ⓑ Ⓒ Ⓓ	53 Ⓐ Ⓑ Ⓒ Ⓓ	63 Ⓐ Ⓑ Ⓒ Ⓓ	73 Ⓐ Ⓑ Ⓒ Ⓓ
4 Ⓕ Ⓖ Ⓗ Ⓙ	14 Ⓕ Ⓖ Ⓗ Ⓙ	24 Ⓕ Ⓖ Ⓗ Ⓙ	34 Ⓕ Ⓖ Ⓗ Ⓙ	44 Ⓕ Ⓖ Ⓗ Ⓙ	54 Ⓕ Ⓖ Ⓗ Ⓙ	64 Ⓕ Ⓖ Ⓗ Ⓙ	74 Ⓕ Ⓖ Ⓗ Ⓙ
5 Ⓐ Ⓑ Ⓒ Ⓓ	15 Ⓐ Ⓑ Ⓒ Ⓓ	25 Ⓐ Ⓑ Ⓒ Ⓓ	35 Ⓐ Ⓑ Ⓒ Ⓓ	45 Ⓐ Ⓑ Ⓒ Ⓓ	55 Ⓐ Ⓑ Ⓒ Ⓓ	65 Ⓐ Ⓑ Ⓒ Ⓓ	75 Ⓐ Ⓑ Ⓒ Ⓓ
6 Ⓕ Ⓖ Ⓗ Ⓙ	16 Ⓕ Ⓖ Ⓗ Ⓙ	26 Ⓕ Ⓖ Ⓗ Ⓙ	36 Ⓕ Ⓖ Ⓗ Ⓙ	46 Ⓕ Ⓖ Ⓗ Ⓙ	56 Ⓕ Ⓖ Ⓗ Ⓙ	66 Ⓕ Ⓖ Ⓗ Ⓙ	
7 Ⓐ Ⓑ Ⓒ Ⓓ	17 Ⓐ Ⓑ Ⓒ Ⓓ	27 Ⓐ Ⓑ Ⓒ Ⓓ	37 Ⓐ Ⓑ Ⓒ Ⓓ	47 Ⓐ Ⓑ Ⓒ Ⓓ	57 Ⓐ Ⓑ Ⓒ Ⓓ	67 Ⓐ Ⓑ Ⓒ Ⓓ	
8 Ⓕ Ⓖ Ⓗ Ⓙ	18 Ⓕ Ⓖ Ⓗ Ⓙ	28 Ⓕ Ⓖ Ⓗ Ⓙ	38 Ⓕ Ⓖ Ⓗ Ⓙ	48 Ⓕ Ⓖ Ⓗ Ⓙ	58 Ⓕ Ⓖ Ⓗ Ⓙ	68 Ⓕ Ⓖ Ⓗ Ⓙ	
9 Ⓐ Ⓑ Ⓒ Ⓓ	19 Ⓐ Ⓑ Ⓒ Ⓓ	29 Ⓐ Ⓑ Ⓒ Ⓓ	39 Ⓐ Ⓑ Ⓒ Ⓓ	49 Ⓐ Ⓑ Ⓒ Ⓓ	59 Ⓐ Ⓑ Ⓒ Ⓓ	69 Ⓐ Ⓑ Ⓒ Ⓓ	
10 Ⓕ Ⓖ Ⓗ Ⓙ	20 Ⓕ Ⓖ Ⓗ Ⓙ	30 Ⓕ Ⓖ Ⓗ Ⓙ	40 Ⓕ Ⓖ Ⓗ Ⓙ	50 Ⓕ Ⓖ Ⓗ Ⓙ	60 Ⓕ Ⓖ Ⓗ Ⓙ	70 Ⓕ Ⓖ Ⓗ Ⓙ	

TEST 2—Mathematics

1 Ⓐ Ⓑ Ⓒ Ⓓ Ⓔ	9 Ⓐ Ⓑ Ⓒ Ⓓ Ⓔ	17 Ⓐ Ⓑ Ⓒ Ⓓ Ⓔ	25 Ⓐ Ⓑ Ⓒ Ⓓ Ⓔ	33 Ⓐ Ⓑ Ⓒ Ⓓ Ⓔ	41 Ⓐ Ⓑ Ⓒ Ⓓ Ⓔ	49 Ⓐ Ⓑ Ⓒ Ⓓ Ⓔ	57 Ⓐ Ⓑ Ⓒ Ⓓ Ⓔ
2 Ⓕ Ⓖ Ⓗ Ⓙ Ⓚ	10 Ⓕ Ⓖ Ⓗ Ⓙ Ⓚ	18 Ⓕ Ⓖ Ⓗ Ⓙ Ⓚ	26 Ⓕ Ⓖ Ⓗ Ⓙ Ⓚ	34 Ⓕ Ⓖ Ⓗ Ⓙ Ⓚ	42 Ⓕ Ⓖ Ⓗ Ⓙ Ⓚ	50 Ⓕ Ⓖ Ⓗ Ⓙ Ⓚ	58 Ⓕ Ⓖ Ⓗ Ⓙ Ⓚ
3 Ⓐ Ⓑ Ⓒ Ⓓ Ⓔ	11 Ⓐ Ⓑ Ⓒ Ⓓ Ⓔ	19 Ⓐ Ⓑ Ⓒ Ⓓ Ⓔ	27 Ⓐ Ⓑ Ⓒ Ⓓ Ⓔ	35 Ⓐ Ⓑ Ⓒ Ⓓ Ⓔ	43 Ⓐ Ⓑ Ⓒ Ⓓ Ⓔ	51 Ⓐ Ⓑ Ⓒ Ⓓ Ⓔ	59 Ⓐ Ⓑ Ⓒ Ⓓ Ⓔ
4 Ⓕ Ⓖ Ⓗ Ⓙ Ⓚ	12 Ⓕ Ⓖ Ⓗ Ⓙ Ⓚ	20 Ⓕ Ⓖ Ⓗ Ⓙ Ⓚ	28 Ⓕ Ⓖ Ⓗ Ⓙ Ⓚ	36 Ⓕ Ⓖ Ⓗ Ⓙ Ⓚ	44 Ⓕ Ⓖ Ⓗ Ⓙ Ⓚ	52 Ⓕ Ⓖ Ⓗ Ⓙ Ⓚ	60 Ⓕ Ⓖ Ⓗ Ⓙ Ⓚ
5 Ⓐ Ⓑ Ⓒ Ⓓ Ⓔ	13 Ⓐ Ⓑ Ⓒ Ⓓ Ⓔ	21 Ⓐ Ⓑ Ⓒ Ⓓ Ⓔ	29 Ⓐ Ⓑ Ⓒ Ⓓ Ⓔ	37 Ⓐ Ⓑ Ⓒ Ⓓ Ⓔ	45 Ⓐ Ⓑ Ⓒ Ⓓ Ⓔ	53 Ⓐ Ⓑ Ⓒ Ⓓ Ⓔ	
6 Ⓕ Ⓖ Ⓗ Ⓙ Ⓚ	14 Ⓕ Ⓖ Ⓗ Ⓙ Ⓚ	22 Ⓕ Ⓖ Ⓗ Ⓙ Ⓚ	30 Ⓕ Ⓖ Ⓗ Ⓙ Ⓚ	38 Ⓕ Ⓖ Ⓗ Ⓙ Ⓚ	46 Ⓕ Ⓖ Ⓗ Ⓙ Ⓚ	54 Ⓕ Ⓖ Ⓗ Ⓙ Ⓚ	
7 Ⓐ Ⓑ Ⓒ Ⓓ Ⓔ	15 Ⓐ Ⓑ Ⓒ Ⓓ Ⓔ	23 Ⓐ Ⓑ Ⓒ Ⓓ Ⓔ	31 Ⓐ Ⓑ Ⓒ Ⓓ Ⓔ	39 Ⓐ Ⓑ Ⓒ Ⓓ Ⓔ	47 Ⓐ Ⓑ Ⓒ Ⓓ Ⓔ	55 Ⓐ Ⓑ Ⓒ Ⓓ Ⓔ	
8 Ⓕ Ⓖ Ⓗ Ⓙ Ⓚ	16 Ⓕ Ⓖ Ⓗ Ⓙ Ⓚ	24 Ⓕ Ⓖ Ⓗ Ⓙ Ⓚ	32 Ⓕ Ⓖ Ⓗ Ⓙ Ⓚ	40 Ⓕ Ⓖ Ⓗ Ⓙ Ⓚ	48 Ⓕ Ⓖ Ⓗ Ⓙ Ⓚ	56 Ⓕ Ⓖ Ⓗ Ⓙ Ⓚ	

TEST 3—Reading

1 Ⓐ Ⓑ Ⓒ Ⓓ	6 Ⓕ Ⓖ Ⓗ Ⓙ	11 Ⓐ Ⓑ Ⓒ Ⓓ	16 Ⓕ Ⓖ Ⓗ Ⓙ	21 Ⓐ Ⓑ Ⓒ Ⓓ	26 Ⓕ Ⓖ Ⓗ Ⓙ	31 Ⓐ Ⓑ Ⓒ Ⓓ	36 Ⓕ Ⓖ Ⓗ Ⓙ
2 Ⓕ Ⓖ Ⓗ Ⓙ	7 Ⓐ Ⓑ Ⓒ Ⓓ	12 Ⓕ Ⓖ Ⓗ Ⓙ	17 Ⓐ Ⓑ Ⓒ Ⓓ	22 Ⓕ Ⓖ Ⓗ Ⓙ	27 Ⓐ Ⓑ Ⓒ Ⓓ	32 Ⓕ Ⓖ Ⓗ Ⓙ	37 Ⓐ Ⓑ Ⓒ Ⓓ
3 Ⓐ Ⓑ Ⓒ Ⓓ	8 Ⓕ Ⓖ Ⓗ Ⓙ	13 Ⓐ Ⓑ Ⓒ Ⓓ	18 Ⓕ Ⓖ Ⓗ Ⓙ	23 Ⓐ Ⓑ Ⓒ Ⓓ	28 Ⓕ Ⓖ Ⓗ Ⓙ	33 Ⓐ Ⓑ Ⓒ Ⓓ	38 Ⓕ Ⓖ Ⓗ Ⓙ
4 Ⓕ Ⓖ Ⓗ Ⓙ	9 Ⓐ Ⓑ Ⓒ Ⓓ	14 Ⓕ Ⓖ Ⓗ Ⓙ	19 Ⓐ Ⓑ Ⓒ Ⓓ	24 Ⓕ Ⓖ Ⓗ Ⓙ	29 Ⓐ Ⓑ Ⓒ Ⓓ	34 Ⓕ Ⓖ Ⓗ Ⓙ	39 Ⓐ Ⓑ Ⓒ Ⓓ
5 Ⓐ Ⓑ Ⓒ Ⓓ	10 Ⓕ Ⓖ Ⓗ Ⓙ	15 Ⓐ Ⓑ Ⓒ Ⓓ	20 Ⓕ Ⓖ Ⓗ Ⓙ	25 Ⓐ Ⓑ Ⓒ Ⓓ	30 Ⓕ Ⓖ Ⓗ Ⓙ	35 Ⓐ Ⓑ Ⓒ Ⓓ	40 Ⓕ Ⓖ Ⓗ Ⓙ

TEST 4—Science Reasoning

1 Ⓐ Ⓑ Ⓒ Ⓓ	6 Ⓕ Ⓖ Ⓗ Ⓙ	11 Ⓐ Ⓑ Ⓒ Ⓓ	16 Ⓕ Ⓖ Ⓗ Ⓙ	21 Ⓐ Ⓑ Ⓒ Ⓓ	26 Ⓕ Ⓖ Ⓗ Ⓙ	31 Ⓐ Ⓑ Ⓒ Ⓓ	36 Ⓕ Ⓖ Ⓗ Ⓙ
2 Ⓕ Ⓖ Ⓗ Ⓙ	7 Ⓐ Ⓑ Ⓒ Ⓓ	12 Ⓕ Ⓖ Ⓗ Ⓙ	17 Ⓐ Ⓑ Ⓒ Ⓓ	22 Ⓕ Ⓖ Ⓗ Ⓙ	27 Ⓐ Ⓑ Ⓒ Ⓓ	32 Ⓕ Ⓖ Ⓗ Ⓙ	37 Ⓐ Ⓑ Ⓒ Ⓓ
3 Ⓐ Ⓑ Ⓒ Ⓓ	8 Ⓕ Ⓖ Ⓗ Ⓙ	13 Ⓐ Ⓑ Ⓒ Ⓓ	18 Ⓕ Ⓖ Ⓗ Ⓙ	23 Ⓐ Ⓑ Ⓒ Ⓓ	28 Ⓕ Ⓖ Ⓗ Ⓙ	33 Ⓐ Ⓑ Ⓒ Ⓓ	38 Ⓕ Ⓖ Ⓗ Ⓙ
4 Ⓕ Ⓖ Ⓗ Ⓙ	9 Ⓐ Ⓑ Ⓒ Ⓓ	14 Ⓕ Ⓖ Ⓗ Ⓙ	19 Ⓐ Ⓑ Ⓒ Ⓓ	24 Ⓕ Ⓖ Ⓗ Ⓙ	29 Ⓐ Ⓑ Ⓒ Ⓓ	34 Ⓕ Ⓖ Ⓗ Ⓙ	39 Ⓐ Ⓑ Ⓒ Ⓓ
5 Ⓐ Ⓑ Ⓒ Ⓓ	10 Ⓕ Ⓖ Ⓗ Ⓙ	15 Ⓐ Ⓑ Ⓒ Ⓓ	20 Ⓕ Ⓖ Ⓗ Ⓙ	25 Ⓐ Ⓑ Ⓒ Ⓓ	30 Ⓕ Ⓖ Ⓗ Ⓙ	35 Ⓐ Ⓑ Ⓒ Ⓓ	40 Ⓕ Ⓖ Ⓗ Ⓙ

Model Mathematics **ACT II**

60 Questions—60 Minutes

INSTRUCTIONS: Find the solution to each problem, choose the correct answer choice, then darken the appropriate oval on your answer sheet. Do not spend too much time on any one problem. Answer as many problems as you can easily and then work on the remaining problems within the time limit for this test. Check pages 417–424 for answers and explanations.

Unless the problem indicates otherwise:

- figures are NOT necessarily drawn to scale
- geometric figures are plane figures
- a *line* is a straight line
- an *average* is the arithmetic mean

You may use a calculator.

---

1. $\triangle ABC$ is an isosceles triangle with the measure of $\angle C = 110°$. What is the measure of $\angle B$?

- **A.** 35°
- **B.** 45°
- **C.** 70°
- **D.** 90°
- **E.** 120°

2. Mary has saved $160 of the $250 she needs for rent. She takes home $6 an hour from her part-time job. Mary works the same number of hours each day for 3 days. How many hours does Mary have to work each day to take home enough money to pay the balance of her rent?

- **F.** 3 hours
- **G.** 4 hours
- **H.** 5 hours
- **J.** 6 hours
- **K.** 7 hours

3. Basil earns $6 an hour when he works up to 8 hours in a day. Basil earns $9 an hour for each hour over 8 hours a day that he works. If Basil earned $84 on Tuesday, how many hours did he work that day?

- **A.** 4
- **B.** 6
- **C.** 8
- **D.** 12
- **E.** 16

4. In the equation $d = rt$, where d = distance, r = rate, and t = time. If you travel a distance of 260 miles in 4 hours, at what rate are you traveling?

- **F.** 45 mph
- **G.** 60 mph
- **H.** 65 mph
- **J.** 70 mph
- **K.** 85 mph

5. In the figure below, lines p and ℓ are parallel. Line t is a transversal that crosses ℓ and p but is not perpendicular to them. Which pair of angles is not supplementary?

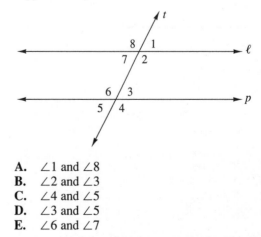

- **A.** $\angle 1$ and $\angle 8$
- **B.** $\angle 2$ and $\angle 3$
- **C.** $\angle 4$ and $\angle 5$
- **D.** $\angle 3$ and $\angle 5$
- **E.** $\angle 6$ and $\angle 7$

GO ON TO THE NEXT PAGE.

6. In scientific notation $15,000 \times 2,500 = ?$

    F. 3.75×10^{-7}
    G. 6×10^{0}
    H. 1.75×10^{4}
    J. 4.75×10^{6}
    K. 3.75×10^{7}

7. If $0.04 \times n = 3.626$, then $n = ?$

    A. 0.09065
    B. 9.065
    C. 90.65
    D. 906.5
    E. 9,065

8. When $x \neq 3$, $\dfrac{x^2 - 9}{x - 3} = ?$

    F. $(x - 3)(x + 3)$
    G. $(x + 3)$
    H. $(x + 9)$
    J. $(x + 12)$
    K. $(x + 15)$

9. What is the value of the expression $3y^2 - 2y + 3$ for $y = 3$?

    A. 18
    B. 21
    C. 24
    D. 26
    E. 36

10. A sweater is on sale for $38.25, after a 15% reduction from the original price. What was the original price of the sweater?

    F. $32.50
    G. $40.00
    H. $44.25
    J. $45.00
    K. $50.00

11. For all x, $(x - 4)(x^2 + 4x + 16) = ?$

    A. $8x^2 + 4x + 24$
    B. $x^3 - 64$
    C. $(x + 8)^2$
    D. $(x + 4)^3$
    E. $4(x + 4)^2$

12. If a car travels 75 miles in 2 hours, how far will the car travel at the same rate in 5.5 hours?

    F. 150 miles
    G. 200 miles
    H. 206.25 miles
    J. 400 miles
    K. 412.5 miles

13. In the diagram below, line segment $\overline{AB}$ is 8 units long and line segment $\overline{CD}$ is 6 units long. The two segments are perpendicular bisectors of each other. Another line segment, $\overline{DB}$, joins the endpoints of segments $\overline{AB}$ and $\overline{CD}$. How long is line segment $\overline{DB}$?

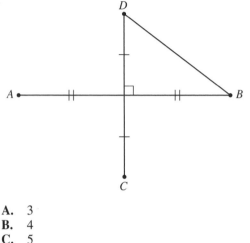

    A. 3
    B. 4
    C. 5
    D. 6
    E. 7

14. A floor in the shape of a triangle has an area of 24 square feet. If the base is 2 feet less than the height, which equation would be used to find the height of the triangle?

    F. $48 = (h + 2)h$
    G. $24 = (h + 2)h$
    H. $48 = h^2$
    J. $48 = (h - 2)h$
    K. $48 = (h - 2)h + 2h$

15. In Joan's bank account, which pays simple interest, $2,500 would grow at a fixed rate to $3,125 in 5 years. If she were to place $5,000 in this account, how much interest would she earn after 7 years?

    A. $1,750
    B. $6,250
    C. $7,000
    D. $17,500
    E. $28,000

16. A window company produces a rectangular window that is 12 feet high and 8 feet wide. They also produce a square window that has the same area as the rectangular window. How long, in feet, is each side of the square window?

    F. 8
    G. $\sqrt{6}$
    H. $4\sqrt{6}$
    J. 4
    K. 16

GO ON TO THE NEXT PAGE.

17. A video store has 20 copies of a movie that each rent for $3 a day. If 75% of these copies are rented on one day, and 90% of these copies are rented the next day, how much money was collected from renting this movie on both days?

    A. $45
    B. $54
    C. $85
    D. $99
    E. $100

18. In the figure below, line ℓ and line m are parallel. Line t is a transversal that crosses both lines ℓ and m. Given that $\angle 1$ is 35°, what is the measure of $\triangle BAD$?

    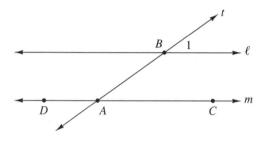

    F. 35°
    G. 55°
    H. 135°
    J. 145°
    K. 215°

19. What is the value of x in the following system of equations?

    $2x + 3y = 4$
    $9x + 7.5y = 18$

    A. 1
    B. 2
    C. 3
    D. 4
    E. 5

20. $\left(\dfrac{3}{5} + \dfrac{3}{15}\right) \times 5 - \left(\dfrac{3}{4} \div \dfrac{1}{4}\right) = ?$

    F. 0
    G. $\dfrac{1}{4}$
    H. $\dfrac{1}{2}$
    J. 1
    K. 3

21. Which of the following expressions is equivalent to $-(3x + 5) - 2x + 2x(2x + 3)$?

    A. $-4x^2 + x + 5$
    B. $15x - 5$
    C. $4x^2 - 5x$
    D. $-4x^2 + 11x - 2$
    E. $4x^2 + x - 5$

22. Simplify the following expression: $\dfrac{c^2 + 6c + 9}{2c + 6}$. $(c \neq -3)$

    F. $(c + 3)$
    G. $(c - 3)$
    H. $\dfrac{(c + 3)}{2}$
    J. $\dfrac{(c - 3)}{2}$
    K. $\dfrac{(c + 3)^2}{2}$

23. If $x = 7$ in the equation $3y + 4 - 2x = 0$, what is the value of y?

    A. $\dfrac{1}{3}$
    B. $\dfrac{2}{3}$
    C. 3
    D. $3\dfrac{1}{3}$
    E. 6

24. In $\triangle ABC$, segment $\overline{BD}$ is perpendicular to segment $\overline{AC}$. If the measure of $\angle DAB$ is 25°, and the measure of $\angle ABC$ is 120°, what is the measure of $\angle DBC$?

    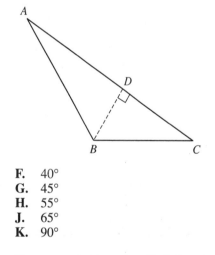

    F. 40°
    G. 45°
    H. 55°
    J. 65°
    K. 90°

25. Six runners begin a race. If all the runners finish and there are no ties, in how many different orders can the runners cross the finish line?

    A. 8,776
    B. 720
    C. 36
    D. 30
    E. 20

GO ON TO THE NEXT PAGE.

26. For all positive values m, n, and p, with $m > n > p$, which of the following statements is always true?

F. $m + n < p$
G. $2p + m > n$
H. $p + n < m$
J. $2n - m > p$
K. $2m + n < p$

27. In the right triangle shown, cosine $\theta = \frac{6}{10}$. How long is leg b?

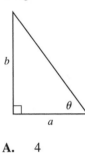

A. 4
B. 8
C. 9
D. 10
E. 12

28. Find the value(s) of y in the following equation:

$$\frac{2y + 3}{\frac{y}{3} + 2} = \frac{3y}{6 + y}.$$

F. $\{-3\}$
G. $\{3, 6\}$
H. $\{3\}$
J. $\{-6\}$
K. $\{-3, -6\}$

29. In the figure shown, lines ℓ and m are parallel, and line t is a transversal that crosses both lines ℓ and m. If the measure of $\angle ABC$ is $30°$, what is the measure of $\angle FCH$?

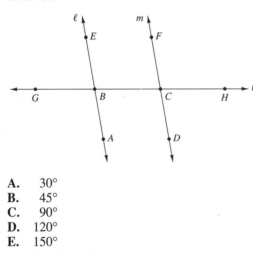

A. 30°
B. 45°
C. 90°
D. 120°
E. 150°

30. $2|-8| + 3|-6| - |-2| = ?$

F. -32
G. -30
H. 8
J. 30
K. 32

31. If $x - 6 = -x^2$, find the possible value(s) of x.

A. $\{-2, -3\}$
B. $\{2, 3\}$
C. $\{2, -3\}$
D. $\{4, 6\}$
E. $\{-4, -6\}$

32. A carpenter cuts a rectangular piece of plywood in half diagonally to make two congruent triangles. If the width of the rectangle is 4 feet, and the cut along the diagonal is $4\sqrt{5}$ feet, what is the length of the rectangle?

F. 6 feet
G. $5\sqrt{4}$ feet
H. 5 feet
J. 8 feet
K. $3\sqrt{5}$ feet

33. What is the radius of a circle whose equation is $(x - 2)^2 + (y + 1)^2 = 54$?

A. $3\sqrt{5}$
B. 2
C. $3\sqrt{6}$
D. 8
E. $2\sqrt{5}$

34. In slope-intercept form, what is the equation of a line having a slope of -2 and a y-intercept of 5?

F. $y = -2x + 5$
G. $y = 5x - 2$
H. $2x + y + 5 = 0$
J. $5x - 2y + 1 = 0$
K. $y = 3x + 5$

35. The scores on a math test are shown in the chart below. What is the average score on this test?

Score	No. of Students
90	4
85	6
80	2
75	1

A. 90
B. 85
C. 82.5
D. 80
E. 75

GO ON TO THE NEXT PAGE.

36. In the two circles shown below, the radius of the larger circle is three times the radius of the smaller circle. If the area of the larger circle is 60.84π, what is the area of the smaller circle?

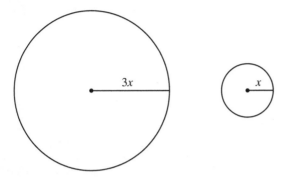

- **F.** 51.84π
- **G.** 40.96π
- **H.** 31.36π
- **J.** 20.28π
- **K.** 6.76π

37. What is the equation of a line that is perpendicular to and has the same y-intercept as the line $y = \frac{2}{3}x + 3$?

- **A.** $y = \frac{2}{3}x + 3$
- **B.** $y = -\frac{2}{3}x - 3$
- **C.** $y = -\frac{3}{2}x - 3$
- **D.** $y = -\frac{3}{2}x + 3$
- **E.** $y = \frac{3}{2}x - 3$

38. For which real values of x is $\frac{3xy - 2}{|x| - 2}$ defined?

- **F.** All real values
- **G.** All real values except 2
- **H.** All real values except 2 and -2
- **J.** All real values except $\frac{1}{2}$
- **K.** All real values except 0

39. What is the slope of the line passing through points $(2,4)$ and $(4,5)$?

- **A.** $\frac{1}{2}$
- **B.** $\frac{1}{3}$
- **C.** $\frac{2}{3}$
- **D.** $-\frac{1}{2}$
- **E.** $\frac{3}{2}$

40. The formula for calculating voltage of a circuit is $E = IR$, where E = volts, I = amperes, and R = ohms. If a circuit contains 4 amperes and 2.25 ohms, what is the voltage of the circuit?

- **F.** 1.8 volts
- **G.** 4 volts
- **H.** 6 volts
- **J.** 9 volts
- **K.** 12 volts

41. What figure will be created by connecting each of the following points with a line segment?

$$P\,(-2,-2) \qquad Q\,(2,4) \qquad R\,(6,10)$$

- **A.** A line segment
- **B.** An equilateral triangle
- **C.** An isosceles triangle
- **D.** A circle
- **E.** A scalene triangle

42. $\frac{x^8 + x^6 - x^4}{x^4 + 4x^4} = ?$, where $x \neq 0$.

- **F.** $\frac{x^4 + x^2 + x}{5}$
- **G.** $\frac{x^2 + x^2 + 1}{x + 4}$
- **H.** $\frac{x + 2}{x}$
- **J.** $\frac{x^4 + x^2 - 1}{x^4}$
- **K.** $\frac{x^4 + x^2 - 1}{5}$

43. Jack is twice Beth's age, and Alice, who is 14, is four years older than Beth. How old is Jack?

- **A.** 20
- **B.** 24
- **C.** 28
- **D.** 32
- **E.** 36

44. What is the product of all the solutions to the equation $2x^2 - 7x + 6 = 0$?

- **F.** $1\frac{1}{2}$
- **G.** 3
- **H.** $3\frac{1}{2}$
- **J.** 5
- **K.** $5\frac{1}{2}$

GO ON TO THE NEXT PAGE.

45. What is the value of $(\cos 30°)(\sec \frac{\pi}{3}) +$
$(\csc 60°)(\sin \frac{\pi}{6})$?

   A. $\sqrt{3} + \frac{1}{\sqrt{3}}$

   B. $\frac{3}{2} + \sqrt{3}$

   C. $\frac{2}{3} + \sqrt{3}$

   D. $\sqrt{\frac{3}{3}} + \frac{1}{3}$

   E. $\frac{1}{3} + \sqrt{3}$

46. A pizzeria sells 12 cheese pizzas and 9 mushroom
pizzas for \$210.75. If the mushroom pizzas cost
\$10.75 each, how much does each cheese pizza cost?

   F.   \$8.75
   G.   \$9.25
   H.   \$9.50
   J.   \$10.00
   K.   \$10.25

47. Which of the following number lines shows the
solution to the inequality $(x - 4)(-x - 3) \le 0$?

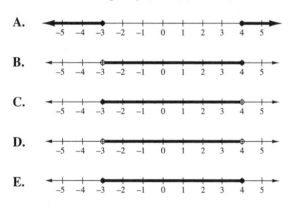

48. In $\triangle XYZ$ in the diagram below, $XY = 1$, $YZ = 1$, and
$XZ = \sqrt{2}$. What is the measure of $\angle X$?

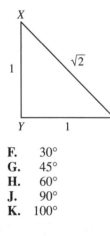

   F.   30°
   G.   45°
   H.   60°
   J.   90°
   K.   100°

49. If $\tan \theta = \frac{1}{2}$, and the measure of $\angle \theta$ is between
0° and 180°, what is $\sec \theta$?

   A. $\sqrt{5}$

   B. $\frac{4}{8}$

   C. 2

   D. $\frac{\sqrt{5}}{5}$

   E. $\frac{\sqrt{5}}{2}$

50. Simplify: $\frac{(a^2 - b^2)}{2(a + b) + (a + b)} = ?$ $(a > 0, b > 0)$

   F. $\frac{a + b}{3}$

   G. $\frac{a + b}{a - b}$

   H. $\frac{a - b}{3}$

   J. $\frac{b - a}{3}$

   K. $\frac{a - b}{2(a + b)}$

51. Triangle ABC below is an equilateral triangle and
side $\overline{AB}$ is 6 units long. $\overline{BD}$ is a perpendicular
bisector of $\overline{AC}$. What is the height of triangle ABC?

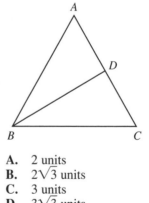

   A.   2 units
   B. $2\sqrt{3}$ units
   C.   3 units
   D. $3\sqrt{3}$ units
   E.   4 units

52. A line in the standard (x,y) coordinate plane contains
the point (5,3) and has a slope of $-\frac{2}{3}$. Which of the
following points lies on this line?

   F.   (2,5)
   G.   (−6,3)
   H.   (3,4)
   J.   (−2,−5)
   K.   (3,5)

GO ON TO THE NEXT PAGE.

53. Jonathan places the bottom of a ladder 9 feet from a wall, as shown in the diagram below. The ladder forms a 50° angle with the ground and touches the wall 15 feet above the ground. Using another ladder, Jonathan again forms a 50° angle with the ground. This ladder touches the wall 9 feet above the ground. How far away from the wall is the bottom of the second ladder?

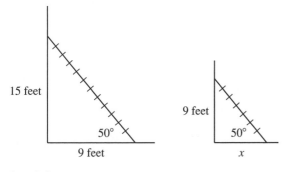

15 feet

50°

9 feet

9 feet

50°

x

A. 3 feet
B. 4.5 feet
C. 5.4 feet
D. 6 feet
E. 6.2 feet

54. Which of the following is the solution to

$$\frac{|6 - x|}{5} < 4?$$

F. $-26 < x < -14$
G. $14 < x < -26$
H. $-26 < x < 14$
J. $26 < x < -14$
K. $-14 < x < 26$

55. In the figure below, the lengths of $\overline{AD}$, $\overline{DC}$, and $\overline{BC}$ are given in units. What is the area of $\triangle ABC$ in square units?

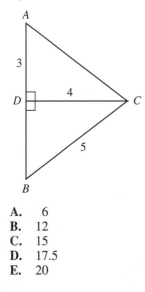

A

3

D

4

C

5

B

A. 6
B. 12
C. 15
D. 17.5
E. 20

56. Which of the following represents $(7 + 3i)(6 - 4i) + \sqrt{-25}$ in standard form?

F. $42 + 23i$
G. $54 - 5i$
H. $84 + 18i$
J. $5i^2 + 18i + 72$
K. $12i^2 - 5i + 42$

57. In the figure shown, the length of the minor arc $\overset{\frown}{AB}$ is $\frac{8\pi}{9}$. The measure of $\angle ACB$ is 40° and point C is the center of the circle. What is the radius of this circle?

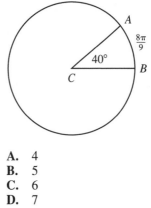

A

$\frac{8\pi}{9}$

40°

B

C

A. 4
B. 5
C. 6
D. 7
E. 8

58. Which of the following statements is true for the slopes m_1 and m_2 of lines ℓ_1 and ℓ_2 in the diagram below?

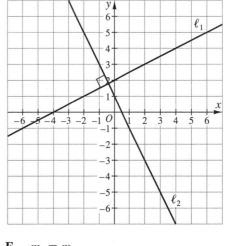

F. $m_1 = m_2$
G. $m_1 \times m_2 = 0$
H. $m_1 \times m_2 = -1$
J. $m_1 = \frac{1}{m_2}$
K. $m_1 = -1m_2$

GO ON TO THE NEXT PAGE.

59. In the right triangle *XYZ* shown below, the measure of *XY* is 6 and the measure of *YZ* is 8. What is the cosine of ∠*Z*?

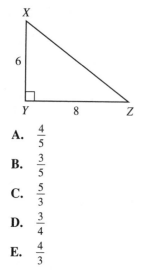

A. $\dfrac{4}{5}$

B. $\dfrac{3}{5}$

C. $\dfrac{5}{3}$

D. $\dfrac{3}{4}$

E. $\dfrac{4}{3}$

60. The line $x = 4$ is graphed on a coordinate plane. Which equation represents a line perpendicular to the line $x = 4$?

F. $x = 2$

G. $y = 4x$

H. $y = 2$

J. $y = 2x - 4$

K. $x - 2y + 6 = 0$

END OF TEST 2

Model Mathematics ACT II

1.	A	16.	H	31.	C	46.	H
2.	H	17.	D	32.	J	47.	A
3.	D	18.	J	33.	C	48.	G
4.	H	19.	B	34.	F	49.	E
5.	D	20.	J	35.	B	50.	H
6.	K	21.	E	36.	K	51.	D
7.	C	22.	H	37.	D	52.	F
8.	G	23.	D	38.	H	53.	C
9.	C	24.	H	39.	A	54.	K
10.	J	25.	B	40.	J	55.	B
11.	B	26.	G	41.	A	56.	G
12.	H	27.	B	42.	K	57.	A
13.	C	28.	K	43.	A	58.	H
14.	J	29.	E	44.	G	59.	A
15.	A	30.	K	45.	A	60.	H

ACT II MATHEMATICS Answers Explained

1. **A.** Two of the angles in an isosceles triangle have the same angle measure. Angle *B* cannot have a measure of 110° because the total for the angle measures for the triangle would be more than 180°. So angles *A* and *B* must have the same angle measure.

$180° - 110° = 70°$ (the total measure of angles *A* and *B*)

$70° \div 2 = 35°$ (the measure of angle *B*)

2. **H.** Think: Rent − amount saved = amount needed

$250 − $160 = $90

Mary earns the same amount of money on each of three days because she works the same number of hours. Divide $90 by 3.

$90 \div 3 = $30 (amount needed each day)

Mary earns $6 an hour. Divide $30 by $6 to find the number of hours she needs to work to earn $30 a day.

$30 \div $6/hour = 5 hours

Mary needs to work 5 hours each day.

3. **D.** If Basil works 8 hours, he earns

($6)(8) = $48 (regular earnings).

Since Basil earned $84, we know he worked overtime. Find the amount of overtime pay.

$84 − $48 = $36   (overtime earnings)

Find the number of overtime hours.

$36   ÷ $9/hour   = 4 hours
(overtime earnings)   (overtime pay)

Basil worked 4 overtime hours and 8 regular hours.

8 + 4 = 12

Basil worked 12 hours on Tuesday.

4. **H.** distance = rate × time

$d = rt$

$260 = 4r$

$r = 65$ mph

5. D. The sum of the measures of supplementary angles is 180°. Answers A and C cannot be correct because $\angle 1$ and $\angle 8$, and $\angle 4$ and $\angle 5$ are linear pairs, which are supplementary. Answers B and E cannot be correct because $\angle 2$ and $\angle 3$, and $\angle 6$ and $\angle 7$ are same-side interior angles, which are supplementary. By elimination, the answer must be D. $\angle 3$ and $\angle 5$ are vertical angles, which means that they are congruent. Line t is not perpendicular to l or to p, so these angles do not measure 90°. Therefore, the total of the measures cannot be 180° and they are not supplementary.

6. K. Write each of the numbers in scientific notation.
$15{,}000 = 1.5 \times 10^4 \qquad 2{,}500 = 2.5 \times 10^3$
Multiply. $(1.5 \times 10^4)(2.5 \times 10^3) =$
$\qquad (1.5 \times 2.5)(10^4 \times 10^3) = 3.75 \times 10^7$

7. C. Write the equation. $\quad 0.04 \times n = 3.626$
Solve for n. $\qquad n = 3.626 \div 0.04 = 90.65$

8. G. Factor the numerator.
$x^2 - 9 = (x + 3)(x - 3)$
Substitute $(x + 3)(x - 3)$ for $x^2 - 9$ and simplify.
$\dfrac{x^2 - 9}{x - 3} = \dfrac{(x + 3)\cancel{(x - 3)}}{\cancel{(x - 3)}} = (x + 3)$

9. C. Substitute 3 for y.
$3y^2 - 2y + 3$
$= 3(3)^2 - 2(3) + 3$
$= 3(9) - 6 + 3$
$= 27 - 6 + 3$
$= 24$

10. J. Since the sweater was reduced by 15% of the original price, the sweater now costs 85% of the original price $(0.85p)$. Write an equation to find the original price (p).
$0.85p = \$38.25$
$p = \dfrac{\$38.25}{0.85}$
$\quad = \$45.00$
The original price of the sweater was $45.

11. B. These are the factors for the difference of cubes $(x^3 - y^3)$ where $y = 4$.

12. H. The number of miles the car travels in 1 hour:
$75 \div 2 = 37.5$
The number of miles the car travels in 5.5 hours:
$37.5 \times 5.5 = 206.25$
Alternative solution:
Set up a proportion. $\quad \dfrac{\text{miles} \rightarrow}{\text{hours} \rightarrow} \dfrac{75}{2} = \dfrac{x}{5.5}$
Cross multiply and solve. $\quad 2x = (75)(5.5)$
$\qquad\qquad\qquad\qquad\quad 2x = 412.5$
$\qquad\qquad\qquad\qquad\quad\ x = 206.25$

13. C.

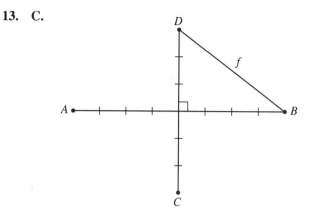

Because line segments $\overline{AB}$ and $\overline{CD}$ are perpendicular bisectors, half of each line segment forms a right triangle with segment $\overline{DB}$. The lengths of the legs of the right triangle are 3 units and 4 units. This must be a 3–4–5 right triangle. Therefore, line segment $\overline{DB}$ has a length of 5.
Alternative solution:
Use the Pythagorean Theorem.
$c^2 = a^2 + b^2$
$f^2 = 3^2 + 4^2$
$f^2 = 9 + 16$
$f^2 = 25$
$f = \pm\sqrt{25}$
$f = \pm 5$
The length of a line segment must be positive, so $f = 5$.

14. J. The formula for the area of a triangle is
$A = \dfrac{1}{2}bh$.
Substitute the given values in the formula. The area is 24 square feet. The base is 2 feet less than the height, or $h - 2$.
$24 = \dfrac{1}{2}(h - 2)h$
Multiply both sides of the equation by 2.
$48 = (h - 2)h$

15. A. First use $I = PRT$ (T is time in years) to find the rate at which this account earns money.
$I = 3{,}125 - 2{,}500 = 625$
$I = P \times R \times T$
$625 = (2{,}500)(R)(5)$
$625 = 12{,}500R$
$R = 0.05$ or 5%
Now find the amount of interest $5,000 would earn in 7 years at a rate of 5%. Substitute $5,000 for P, 0.05 for R, and 7 for T.
$I = PRT$
$I = (5{,}000)(0.05)(7)$
$I = 1{,}750$
Joan would earn $1,750 in interest.

16. H. To find the area of the rectangular window multiply $l \times w$.

$12 \times 8 = 96$

The formula for the area of a square is $A = s^2$.

The square has the same area as the rectangle. Write

$s^2 = 96$

$s = \sqrt{96}$

$s = \sqrt{16 \times 6}$

$s = 4\sqrt{6}$ feet

The length of a side of the square is $4\sqrt{6}$ feet.

17. D. The number of movies rented the first day is $0.75 \times 20 = 15$.

The number of movies rented the next day is $0.90 \times 20 = 18$.

The total number of rentals for both days is $18 + 15 = 33$.

Therefore, the video store collected $33 \times \$3 = \$99$ total for the two days.

18. J.

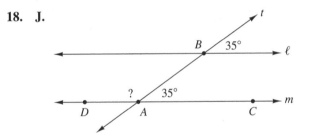

Line t is a transversal that crosses parallel lines. $\angle 1$ and $\angle BAC$ are corresponding angles and therefore are congruent. Since the measure of $\angle 1$ is $35°$, the measure of $\angle BAC$ is also $35°$. $\angle BAC$ and $\angle BAD$ are supplementary, so the measure of $\angle BAD + 35° = 180°$. The measure of $\angle BAD$ is $145°$.

19. B. Solve simultaneous equations.

$2x + 3y = 4$ Multiply by 5. $10x + 15y = 20$
$9x + 7.5y = 18$ Multiply by -2. $\underline{-18x - 15y = -36}$
                    Add. $-8x = -16$
                                       $x = 2$

20. J. Use the order of operations.

$\left(\dfrac{3}{5} + \dfrac{3}{15}\right) \times 5 - \left(\dfrac{3}{4} \div \dfrac{1}{4}\right)$

$= \left(\dfrac{9}{15} + \dfrac{3}{15}\right) \times 5 - \left(\dfrac{3}{4} \div \dfrac{1}{4}\right)$ Rewrite with common denominators.

$= \left(\dfrac{12}{15}\right) \times 5 - \left(\dfrac{3}{4} \times \dfrac{4}{1}\right)$ Work within the parentheses first.

$= \left(\dfrac{4}{5}\right) \times 5 - (3)$ Multiply, then subtract.

$= 4 - 3 = 1$

21. E. Simplify the expression.

$-(3x + 5) - 2x + 2x(2x + 3)$ Use the distributive property.

$= -3x - 5 - 2x + 4x^2 + 6x$

$= 4x^2 + 6x - 3x - 2x - 5$ Combine like terms.

$= 4x^2 + x - 5$

22. H.

$\dfrac{c^2 + 6c + 9}{2c + 6}$

$= \dfrac{(c + 3)(c + 3)}{2(c + 3)}$ Factor the numerator and denominator.

$= \dfrac{(c + 3)}{2}$ Simplify.

23. D. Substitute 7 for x. Simplify the equation and solve for y.

$3y + 4 - 2x = 0$
$3y + 4 - 2(7) = 0$
$3y + 4 - 14 = 0$
$3y - 10 = 0$
$3y = 10$
$y = 3\dfrac{1}{3}$

24. H.

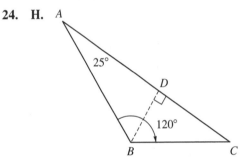

Since $\overline{BD}$ is perpendicular to $\overline{AC}$, the measure of $\angle BDA$ is $90°$. We know that the measure of $\angle DAB$ is $25°$. The sum of the angle measures in a triangle is $180°$, so $25° + 90° +$ the measure of $\angle ABD$ is $180°$. The measure of $\angle ABD$ is $65°$. We know that the measure of $\angle ABC$ is $120°$, so the measure of $\angle DBC + 65°$ is $120°$. Therefore, the measure of $\angle DBC$ is $55°$.

25. B. Use factorial to find the number of permutations.

$n! = 6! = 6 \times 5 \times 4 \times 3 \times 2 \times 1 = 720$

There are 720 different orders in which the runners can finish the race.

26. G. The important words in this problem are *is always true*. That means that if you think of even one example in which an answer choice is not true, then that choice is eliminated. Test the answer choices by substituting values for m, n, and p. Let $m = 5$, $n = 4$, and $p = 3$.

Choice F. $m + n < p$ $5 + 4 < 3$ is false.
Choice G. $2p + m > n$ $2(3) + 5 > 4$
                                  $11 > 4$ is true.

G is always true. Since $m > n$, any positive number added to m will also be greater than n.

27. B.

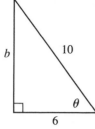

Since cosine $\theta = \dfrac{6}{10}$, we know that the length of leg a is 6 and the length of the hypotenuse is 10. Use the Pythagorean Theorem.

$a^2 + b^2 = c^2$

$(6)^2 + b^2 = 10^2$

$36 + b^2 = 100$

$b^2 = 100 - 36$

$b^2 = 64$

$b = \pm\sqrt{64}$

$b = \pm 8$

The length of a side must be positive, so $b = 8$.

28. K. The fractions form a proportion. Cross multiply and solve the resulting quadratic equation to find the value of y.

$\dfrac{2y + 3}{\frac{y}{3} + 2} = \dfrac{3y}{6 + y}$

$(2y + 3)(6 + y) = \left(\dfrac{y}{3} + 2\right)(3y)$ Cross multiply.

$12y + 2y^2 + 18 + 3y = 6y + y^2$ Multiply.

$2y^2 + 15y + 18 = 6y + y^2$ Combine like terms.

$y^2 + 9y + 18 = 0$ Write in quadratic form.

$(y + 3)(y + 6) = 0$ Factor to solve.

$y = -3 \quad y = -6$

$y = \{-3, -6\}$

29. E. $\angle ABC$ and $\angle FCB$ are congruent alternate interior angles. So, the measure of $\angle FCB$ is also 30°. $\angle FCB$ and $\angle FCH$ are supplementary; the sum of their measures is 180°.

$30° + \angle FCH = 180°$

$\angle FCH = 150°$

30. K. Solve. Remember that the absolute value of a number is always positive.

$2|-8| + 3|-6| - |-2|$

$= (2)(8) + (3)(6) - 2$

$= 16 + 18 - 2$

$= 32$

31. C.

$x - 6 = -x^2$

$x^2 + x - 6 = 0$

$(x - 2)(x + 3) = 0$ Factor.

$x = 2 \qquad x = -3$

$x = \{2, -3\}$

32. J.

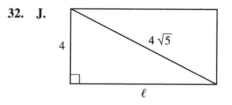

Use the Pythagorean Theorem.

$a^2 + b^2 = c^2$

$4^2 + \ell^2 = (4\sqrt{5})^2$, where ℓ is the length of the rectangle

$16 + \ell^2 = 80$

$\ell^2 = 64$

$\ell = \pm\sqrt{64}$

$\ell = \pm 8$

The length must be a positive number, so the length is 8.

33. C. The standard equation of a circle is $(x - h)^2 + (y - k)^2 = r^2$. To find the answer to this question, solve for r.

$r^2 = 54$

$r = \pm\sqrt{54} = \pm\sqrt{9 \times 6}$

$r = \pm 3\sqrt{6}$

The radius of a circle must be positive, so $r = 3\sqrt{6}$.

34. F. The slope-intercept equation of a line is $y = mx + b$, where x and y show the location on the coordinate plane, m is the slope, and b is the y-intercept. Substitute the given values for m and the y-intercept. The slope-intercept equation of this line is $y = -2x + 5$.

35. B. Multiply each score by its frequency and add.

$90 \times 4 = \quad 360$

$85 \times 6 = \quad 510$

$80 \times 2 = \quad 160$

$75 \times 1 = \quad \underline{75}$

$ 1,105$

To find the average score, divide the sum of the scores by the number of scores.

$1,105 \div 13 = 85$

36. K. First find the radius of the larger circle. The formula for the area of a circle is $A = \pi r^2$. The area of the larger circle is 60.84π, so we can say

$$A = \pi r^2$$
$$60.84\pi = \pi r^2$$
$$60.84 = r^2$$
$$\pm\sqrt{60.84} = r$$
$$r = \pm 7.8$$

A radius has to be positive, so the radius of the larger circle is 7.8. Divide by 3 to find the radius of the smaller circle.

$$7.8 \div 3 = 2.6$$

Find the area of the smaller circle.

$$A = \pi r^2$$
$$A = (2.6)^2\pi$$
$$A = 6.76\pi$$

Alternative solution:

Radius of the larger circle $= r$

Radius of the smaller circle $= \dfrac{r}{3}$

Area of the smaller circle $\left(\dfrac{r}{3}\right)^2 \pi = \dfrac{1}{9}r^2\pi$

The area of the smaller circle is $\dfrac{1}{9}$ the area of the

larger circle or $\dfrac{60.84\pi}{9} = 6.76\pi$.

37. D. The product of the slopes of perpendicular lines is -1. In other words, the slope of one line is the negative reciprocal of the other. Since the slope of $y = \dfrac{2}{3}x + 3$ is $\dfrac{2}{3}$, the slope of the line perpendicular to it is $-\dfrac{3}{2}$. The y-intercept is the same for both lines, 3. So the equation of the perpendicular line is $y = -\dfrac{3}{2}x + 3$.

38. H. $\dfrac{3xy - 2}{|x| - 2}$ is defined except when $|x| - 2 = 0$.

$|x| - 2 = 0$ when $x = 2$ or when $x = -2$
($|2| = 2$ and $|-2| = 2$)
The expression is defined for all real values of x except 2 and -2.

39. A. The slope is the change in y divided by the change in x.

Use the equation $m = \dfrac{y_2 - y_1}{x_2 - x_1}$ to find the slope of the line.

$$m = \dfrac{y_2 - y_1}{x_2 - x_1}$$
$$m = \dfrac{5 - 4}{4 - 2}$$
$$m = \dfrac{1}{2}$$

40. J. Use $E = IR$. Substitute $I = 4$ and $R = 2.25$.

$E = IR$
$E = 4 \times 2.25$
$E = 9$

The voltage in the circuit is 9 volts.

41. A.

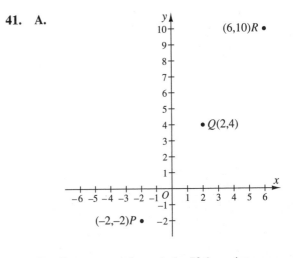

The figure cannot be a circle. If the points are on the same line, the figure is a line segment. If the points are not on the same line, the figure is a triangle and we will have to figure out what kind of triangle it is.

Equations for the same line have the same slope. From P to Q: up 6, over 4. From Q to R: up 6 over 4. The slopes are the same, so all three points lie on the same line. Connecting the points forms a line segment.

Alternative solution:

You can also use the slope formula to see if the points are on the same line. Check the slope for the segments $\overline{PQ}$ and $\overline{QR}$. (You can also check the slopes for $\overline{PQ}$ and $\overline{PR}$ or $\overline{QR}$ and $\overline{PR}$.)

$$m = \dfrac{y_2 - y_1}{x_2 - x_1}$$

Slope of $\overline{PQ} = \dfrac{4 - (-2)}{2 - (-2)} = \dfrac{6}{4} = \dfrac{3}{2}$

Slope of $\overline{QR} = \dfrac{10 - 4}{6 - 2} = \dfrac{6}{4} = \dfrac{3}{2}$

The slopes are the same, so all three points lie on the same line. Connecting the points forms a line segment.

You could also *carefully* sketch the three points to find the answer. Remember, though, that a slight sketching error could lead to the wrong answer.

42. K. Simplify. $\dfrac{x^8 + x^6 - x^4}{x^4 + 4x^4} = \dfrac{x^8 + x^6 - x^4}{5x^4}$

$$= \dfrac{x^4(x^4 + x^2 - 1)}{5x^4}$$

$$= \dfrac{x^4 + x^2 - 1}{5}$$

43. A. Write equations to fit the descriptions in the problem. Let J = Jack's age, A = Alice's age, and B = Beth's age.

$J = 2B$ Jack is twice Beth's age.

$A = 14$ Alice is 14.

$A = B + 4$ Alice is 4 years older than Beth.

Since both the second and third equations are equal to A, we can set them equal to each other and solve for B.

$B + 4 = 14$

$B = 10$

To find Jack's age, substitute 10 for B in the equation for Jack's age.

$J = 2B$

$J = 2(10) = 20$

Jack is 20 years old.

44. G. Solve the quadratic equation by factoring.

$2x^2 - 7x + 6 = 0$

$(2x - 3)(x - 2) = 0$

$2x - 3 = 0$ $x - 2 = 0$

$2x = 3$ $x = 2$

$x = \dfrac{3}{2}$

Multiply the two solutions.

$\dfrac{3}{2} \cdot 2 = 3$

45. A.

$(\cos 30°)(\sec \tfrac{\pi}{3}) + (\csc 60°)(\sin \tfrac{\pi}{6})$

$= \left(\dfrac{\sqrt{3}}{2} \times 2\right) + \left(\dfrac{2}{\sqrt{3}} \times \dfrac{1}{2}\right)$

$= \dfrac{2\sqrt{3}}{2} + \dfrac{2}{2\sqrt{3}}$

$= \dfrac{\sqrt{3}}{1} + \dfrac{1}{\sqrt{3}}$

$= \sqrt{3} + \dfrac{1}{\sqrt{3}}$

46. H. Write an equation to find the cost of a cheese pizza. Use x to represent the cost of a cheese pizza.

$12x + 9(\$10.75) = \$210.75$

$12x + \$96.75 \quad\; = \$210.75$

$12x \qquad\qquad\;\; = \$114$

$x \qquad\qquad\quad\; = \$9.50$

47. A. Solve the inequality $(x - 4)(-x - 3) \leq 0$.

In order for the product to be less than or equal to zero, either one factor must equal zero or one factor must be positive and one negative. (Remember, a positive times a negative is a negative.)

$(x - 4)(-x - 3) \leq 0$ when

$x - 4 \leq 0$ and $-x - 3 \geq 0$ OR $x - 4 \geq 0$ and $-x - 3 \leq 0$

$x \leq 4$ and $x \leq -3$ OR $x \geq 4$ and $x \geq -3$

       $x \leq -3$ OR $x \geq 4$

x can be less than or equal to -3 or x can be greater than or equal to 4. This inequality matches choice A.

48. G. You may recognize that $\triangle XYZ$ is a 45–45–90 triangle. Since $\overline{XZ}$ is the longest side, the measure of $\angle Y$ is 90°. The measures of $\angle X$ and $\angle Z$ must then be 45°.

Alternative solution:

Use the Pythagorean Theorem, $a^2 + b^2 = c^2$, to determine that $\triangle XYZ$ is a right triangle.

Since $\overline{XY}$ and $\overline{YZ}$ are the same length, the measures of $\angle X$ and $\angle Z$ will be equal.

$90° + 2(m\angle X) = 180°$

$m\angle X = 45°$

49. E. Recall that

$\tan \theta = \dfrac{\text{opposite}}{\text{adjacent}} = \dfrac{1}{2}$

$\sec \theta = \dfrac{\text{hypotenuse}}{\text{adjacent}} = \dfrac{c}{2}$

Use the Pythagorean Theorem to find the length of the hypotenuse. We know that the legs have lengths 1 and 2.

$c^2 = a^2 + b^2$

$c^2 = 1^2 + 2^2$

$c^2 = 5$

$c = \pm\sqrt{5}$

You know that θ is between 0° and 180° and the tangent is positive. This means θ is between 0° and 90°, so the secant is positive. Therefore, $c = +\sqrt{5}$.

$\text{secant } \theta = \dfrac{\text{hypotenuse}}{\text{adjacent}} = \dfrac{\sqrt{5}}{2}.$

Alternative solution:

Use a trigonometric identity and substitute.

$\sec^2 \theta = 1 + \tan^2 \theta$

The tangent of θ is $\dfrac{1}{2}$.

$\sec^2 \theta = 1 + \left(\dfrac{1}{2}\right)^2$

$\sec^2 \theta = 1 + \dfrac{1}{4} = \dfrac{5}{4}$

$\sec \theta = \sqrt{\dfrac{5}{4}} = \dfrac{\sqrt{5}}{2}$

50. H. Factor and simplify.

$\dfrac{(a^2 - b^2)}{2(a + b) + (a + b)}$

$= \dfrac{(a + b)(a - b)}{3(a + b)}$

$= \dfrac{a - b}{3}$

51. D.

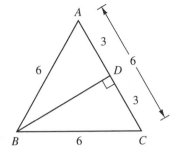

Each side of the triangle has a length of 6 units. Since $\overline{BD}$ is a perpendicular bisector of $\overline{AC}$, the length of $\overline{DC}$ is 3 units and $\angle BDC$ is 90°. We have formed a new triangle BDC, which is a right triangle. Use the Pythagorean Theorem to find the height of the triangle.

$$(DB)^2 + (DC)^2 = (BC)^2$$
$$(DB)^2 + 3^2 = 6^2$$
$$(DB)^2 = 6^2 - 3^2$$
$$(DB)^2 = 36 - 9$$
$$(DB)^2 = 27$$
$$DB = \sqrt{27}$$
$$DB = 3\sqrt{3}$$

52. F. The slope is the change in y divided by the change in x. Try each of the answer choices until you find a slope of $-\dfrac{2}{3}$.

Choice F:

$$m = \frac{y_2 - y_1}{x_2 - x_1}$$
$$m = \frac{3 - 5}{5 - 2} = -\frac{2}{3}$$

A line passing through (5,3) and (2,5) has a slope of $-\dfrac{2}{3}$.

53. C. Since the wall and the ground form a 90° angle in both figures, and both ladders form a 50° angle with the ground, the angles formed by both ladders at the wall each measure 40° (90° + 50° + 40° = 180°), so the two triangles are similar. In similar triangles, the ratio of corresponding sides is proportional. Write a proportion to represent the height and distance from the wall of each triangle and solve for x.

$$\frac{15}{9} = \frac{9}{x}$$

$15x = 81$ Cross multiply.

$x = 5.4$ Solve.

54. K. Solve. $\dfrac{|6 - x|}{5} < 4$

$$|6 - x| < 20$$

$6 - x < 20$ and $6 - x > -20$
$ -x < 14$ and $-x > -26$
$ x > -14$ and $x < 26$
$$-14 < x < 26$$

55. B. The two smaller triangles are congruent because they are both 3–4–5 right triangles. Find the area of one of the smaller triangles.

$$A = \frac{1}{2}bh$$
$$A = \frac{1}{2} \times 3 \times 4$$
$$A = 6$$

Since there are two triangles, $6 \times 2 = 12$.

Alternative solution:

Add the bases of the smaller triangles and find the area of the larger triangle.

$$AD + DB = 3 + 3 = 6$$
$$A = \frac{1}{2}bh$$
$$A = \frac{1}{2} \times 6 \times 4$$
$$A = 12$$

56. G. Remember: $i^2 = -1$

$$(7 + 3i)(6 - 4i) + \sqrt{-25}$$
$$= 42 - 28i + 18i - 12i^2 + \sqrt{-25}$$
$$= 42 - 28i + 18i - 12i^2 + 5i$$
$$= 42 - 10i - 12(-1) + 5i$$
$$= 42 - 10i + 12 + 5i$$
$$= 54 - 5i$$

57. A. The circumference of a circle is $2\pi r$. Since 40° is one-ninth of 360°, the length of arc $\widehat{AB}$ is $\dfrac{1}{9}$ of the circumference, or $\dfrac{1}{9}(2\pi r)$. We know that the length of arc $\widehat{AB}$ is $\dfrac{8\pi}{9}$. Substitute to find the measure of the radius.

$$AB = \frac{1}{9}(2\pi r)$$
$$\frac{8\pi}{9} = \frac{1}{9} \times 2\pi r$$
$$\frac{8\pi}{9} = \frac{2\pi r}{9}$$
$$\frac{8\pi}{2\pi} = r$$
$$4 = r$$

58. H. The product of the slopes of perpendicular lines is -1. Since the lines are perpendicular, $m_1 \times m_2 = -1$.

59. A. Use the Pythagorean Theorem to find the length of $\overline{XZ}$.

$$a^2 + b^2 = c^2$$
$$6^2 + 8^2 = (XZ)^2$$
$$36 + 64 = (XZ)^2$$
$$100 = (XZ)^2$$
$$10 = XZ$$

$$\cos Z = \frac{\text{adjacent}}{\text{hypotenuse}} = \frac{8}{10} = \frac{4}{5}$$

You might also recognize this as a 6–8–10 right triangle.

60. H. Look at the equation $x = 4$. It means that for any y, x will always be 4. This is a vertical line 4 units to the right of the y-axis. A line perpendicular to a vertical line must be a horizontal line. The only horizontal line given is $y = 2$. This equation means that for any x, y will always be 2.

Chapter 15 ▪

Model Mathematics **ACT III**
With Answers Explained

This Mathematics Model ACT III is just like a real ACT. Take this test after you take Model ACT II.

Take this model test under simulated test conditions. Allow 60 minutes to answer the 60 Mathematics items. Use a pencil to mark the answer sheet on page 427, and answer the questions in the Test 2 (Mathematics) section.

Use the answer key on page 438 to mark the answer sheet. Review the answer explanations on pages 438–444. You may decide to retake this test later. There are additional answer sheets for this purpose following page 605.

The test scoring chart on page 464 shows you how to convert the number correct to ACT scale scores. Other charts on pages 470–471 show you how to find the Pre-Algebra/Elementary Algebra, Intermediate Algebra/Coordinate Geometry, and Plane Geometry/Trigonometry subscores.

DO NOT leave any answers blank. There is no penalty for guessing on the ACT. Remember that the test is yours. You may mark up, write on, or draw on the test.

When you are ready, note the time and begin.

ANSWER SHEET

The ACT answer sheet looks something like this one. Use a No. 2 pencil to completely fill the circle corresponding to the correct answer.
If you erase, erase completely; incomplete erasures may be read as answers.

TEST 1—English

1 Ⓐ Ⓑ Ⓒ Ⓓ	11 Ⓐ Ⓑ Ⓒ Ⓓ	21 Ⓐ Ⓑ Ⓒ Ⓓ	31 Ⓐ Ⓑ Ⓒ Ⓓ	41 Ⓐ Ⓑ Ⓒ Ⓓ	51 Ⓐ Ⓑ Ⓒ Ⓓ	61 Ⓐ Ⓑ Ⓒ Ⓓ	71 Ⓐ Ⓑ Ⓒ Ⓓ
2 Ⓕ Ⓖ Ⓗ Ⓙ	12 Ⓕ Ⓖ Ⓗ Ⓙ	22 Ⓕ Ⓖ Ⓗ Ⓙ	32 Ⓕ Ⓖ Ⓗ Ⓙ	42 Ⓕ Ⓖ Ⓗ Ⓙ	52 Ⓕ Ⓖ Ⓗ Ⓙ	62 Ⓕ Ⓖ Ⓗ Ⓙ	72 Ⓕ Ⓖ Ⓗ Ⓙ
3 Ⓐ Ⓑ Ⓒ Ⓓ	13 Ⓐ Ⓑ Ⓒ Ⓓ	23 Ⓐ Ⓑ Ⓒ Ⓓ	33 Ⓐ Ⓑ Ⓒ Ⓓ	43 Ⓐ Ⓑ Ⓒ Ⓓ	53 Ⓐ Ⓑ Ⓒ Ⓓ	63 Ⓐ Ⓑ Ⓒ Ⓓ	73 Ⓐ Ⓑ Ⓒ Ⓓ
4 Ⓕ Ⓖ Ⓗ Ⓙ	14 Ⓕ Ⓖ Ⓗ Ⓙ	24 Ⓕ Ⓖ Ⓗ Ⓙ	34 Ⓕ Ⓖ Ⓗ Ⓙ	44 Ⓕ Ⓖ Ⓗ Ⓙ	54 Ⓕ Ⓖ Ⓗ Ⓙ	64 Ⓕ Ⓖ Ⓗ Ⓙ	74 Ⓕ Ⓖ Ⓗ Ⓙ
5 Ⓐ Ⓑ Ⓒ Ⓓ	15 Ⓐ Ⓑ Ⓒ Ⓓ	25 Ⓐ Ⓑ Ⓒ Ⓓ	35 Ⓐ Ⓑ Ⓒ Ⓓ	45 Ⓐ Ⓑ Ⓒ Ⓓ	55 Ⓐ Ⓑ Ⓒ Ⓓ	65 Ⓐ Ⓑ Ⓒ Ⓓ	75 Ⓐ Ⓑ Ⓒ Ⓓ
6 Ⓕ Ⓖ Ⓗ Ⓙ	16 Ⓕ Ⓖ Ⓗ Ⓙ	26 Ⓕ Ⓖ Ⓗ Ⓙ	36 Ⓕ Ⓖ Ⓗ Ⓙ	46 Ⓕ Ⓖ Ⓗ Ⓙ	56 Ⓕ Ⓖ Ⓗ Ⓙ	66 Ⓕ Ⓖ Ⓗ Ⓙ	
7 Ⓐ Ⓑ Ⓒ Ⓓ	17 Ⓐ Ⓑ Ⓒ Ⓓ	27 Ⓐ Ⓑ Ⓒ Ⓓ	37 Ⓐ Ⓑ Ⓒ Ⓓ	47 Ⓐ Ⓑ Ⓒ Ⓓ	57 Ⓐ Ⓑ Ⓒ Ⓓ	67 Ⓐ Ⓑ Ⓒ Ⓓ	
8 Ⓕ Ⓖ Ⓗ Ⓙ	18 Ⓕ Ⓖ Ⓗ Ⓙ	28 Ⓕ Ⓖ Ⓗ Ⓙ	38 Ⓕ Ⓖ Ⓗ Ⓙ	48 Ⓕ Ⓖ Ⓗ Ⓙ	58 Ⓕ Ⓖ Ⓗ Ⓙ	68 Ⓕ Ⓖ Ⓗ Ⓙ	
9 Ⓐ Ⓑ Ⓒ Ⓓ	19 Ⓐ Ⓑ Ⓒ Ⓓ	29 Ⓐ Ⓑ Ⓒ Ⓓ	39 Ⓐ Ⓑ Ⓒ Ⓓ	49 Ⓐ Ⓑ Ⓒ Ⓓ	59 Ⓐ Ⓑ Ⓒ Ⓓ	69 Ⓐ Ⓑ Ⓒ Ⓓ	
10 Ⓕ Ⓖ Ⓗ Ⓙ	20 Ⓕ Ⓖ Ⓗ Ⓙ	30 Ⓕ Ⓖ Ⓗ Ⓙ	40 Ⓕ Ⓖ Ⓗ Ⓙ	50 Ⓕ Ⓖ Ⓗ Ⓙ	60 Ⓕ Ⓖ Ⓗ Ⓙ	70 Ⓕ Ⓖ Ⓗ Ⓙ	

TEST 2—Mathematics

1 Ⓐ Ⓑ Ⓒ Ⓓ Ⓔ	9 Ⓐ Ⓑ Ⓒ Ⓓ Ⓔ	17 Ⓐ Ⓑ Ⓒ Ⓓ Ⓔ	25 Ⓐ Ⓑ Ⓒ Ⓓ Ⓔ	33 Ⓐ Ⓑ Ⓒ Ⓓ Ⓔ	41 Ⓐ Ⓑ Ⓒ Ⓓ Ⓔ	49 Ⓐ Ⓑ Ⓒ Ⓓ Ⓔ	57 Ⓐ Ⓑ Ⓒ Ⓓ Ⓔ
2 Ⓕ Ⓖ Ⓗ Ⓙ Ⓚ	10 Ⓕ Ⓖ Ⓗ Ⓙ Ⓚ	18 Ⓕ Ⓖ Ⓗ Ⓙ Ⓚ	26 Ⓕ Ⓖ Ⓗ Ⓙ Ⓚ	34 Ⓕ Ⓖ Ⓗ Ⓙ Ⓚ	42 Ⓕ Ⓖ Ⓗ Ⓙ Ⓚ	50 Ⓕ Ⓖ Ⓗ Ⓙ Ⓚ	58 Ⓕ Ⓖ Ⓗ Ⓙ Ⓚ
3 Ⓐ Ⓑ Ⓒ Ⓓ Ⓔ	11 Ⓐ Ⓑ Ⓒ Ⓓ Ⓔ	19 Ⓐ Ⓑ Ⓒ Ⓓ Ⓔ	27 Ⓐ Ⓑ Ⓒ Ⓓ Ⓔ	35 Ⓐ Ⓑ Ⓒ Ⓓ Ⓔ	43 Ⓐ Ⓑ Ⓒ Ⓓ Ⓔ	51 Ⓐ Ⓑ Ⓒ Ⓓ Ⓔ	59 Ⓐ Ⓑ Ⓒ Ⓓ Ⓔ
4 Ⓕ Ⓖ Ⓗ Ⓙ Ⓚ	12 Ⓕ Ⓖ Ⓗ Ⓙ Ⓚ	20 Ⓕ Ⓖ Ⓗ Ⓙ Ⓚ	28 Ⓕ Ⓖ Ⓗ Ⓙ Ⓚ	36 Ⓕ Ⓖ Ⓗ Ⓙ Ⓚ	44 Ⓕ Ⓖ Ⓗ Ⓙ Ⓚ	52 Ⓕ Ⓖ Ⓗ Ⓙ Ⓚ	60 Ⓕ Ⓖ Ⓗ Ⓙ Ⓚ
5 Ⓐ Ⓑ Ⓒ Ⓓ Ⓔ	13 Ⓐ Ⓑ Ⓒ Ⓓ Ⓔ	21 Ⓐ Ⓑ Ⓒ Ⓓ Ⓔ	29 Ⓐ Ⓑ Ⓒ Ⓓ Ⓔ	37 Ⓐ Ⓑ Ⓒ Ⓓ Ⓔ	45 Ⓐ Ⓑ Ⓒ Ⓓ Ⓔ	53 Ⓐ Ⓑ Ⓒ Ⓓ Ⓔ	
6 Ⓕ Ⓖ Ⓗ Ⓙ Ⓚ	14 Ⓕ Ⓖ Ⓗ Ⓙ Ⓚ	22 Ⓕ Ⓖ Ⓗ Ⓙ Ⓚ	30 Ⓕ Ⓖ Ⓗ Ⓙ Ⓚ	38 Ⓕ Ⓖ Ⓗ Ⓙ Ⓚ	46 Ⓕ Ⓖ Ⓗ Ⓙ Ⓚ	54 Ⓕ Ⓖ Ⓗ Ⓙ Ⓚ	
7 Ⓐ Ⓑ Ⓒ Ⓓ Ⓔ	15 Ⓐ Ⓑ Ⓒ Ⓓ Ⓔ	23 Ⓐ Ⓑ Ⓒ Ⓓ Ⓔ	31 Ⓐ Ⓑ Ⓒ Ⓓ Ⓔ	39 Ⓐ Ⓑ Ⓒ Ⓓ Ⓔ	47 Ⓐ Ⓑ Ⓒ Ⓓ Ⓔ	55 Ⓐ Ⓑ Ⓒ Ⓓ Ⓔ	
8 Ⓕ Ⓖ Ⓗ Ⓙ Ⓚ	16 Ⓕ Ⓖ Ⓗ Ⓙ Ⓚ	24 Ⓕ Ⓖ Ⓗ Ⓙ Ⓚ	32 Ⓕ Ⓖ Ⓗ Ⓙ Ⓚ	40 Ⓕ Ⓖ Ⓗ Ⓙ Ⓚ	48 Ⓕ Ⓖ Ⓗ Ⓙ Ⓚ	56 Ⓕ Ⓖ Ⓗ Ⓙ Ⓚ	

TEST 3—Reading

1 Ⓐ Ⓑ Ⓒ Ⓓ	6 Ⓕ Ⓖ Ⓗ Ⓙ	11 Ⓐ Ⓑ Ⓒ Ⓓ	16 Ⓕ Ⓖ Ⓗ Ⓙ	21 Ⓐ Ⓑ Ⓒ Ⓓ	26 Ⓕ Ⓖ Ⓗ Ⓙ	31 Ⓐ Ⓑ Ⓒ Ⓓ	36 Ⓕ Ⓖ Ⓗ Ⓙ
2 Ⓕ Ⓖ Ⓗ Ⓙ	7 Ⓐ Ⓑ Ⓒ Ⓓ	12 Ⓕ Ⓖ Ⓗ Ⓙ	17 Ⓐ Ⓑ Ⓒ Ⓓ	22 Ⓕ Ⓖ Ⓗ Ⓙ	27 Ⓐ Ⓑ Ⓒ Ⓓ	32 Ⓕ Ⓖ Ⓗ Ⓙ	37 Ⓐ Ⓑ Ⓒ Ⓓ
3 Ⓐ Ⓑ Ⓒ Ⓓ	8 Ⓕ Ⓖ Ⓗ Ⓙ	13 Ⓐ Ⓑ Ⓒ Ⓓ	18 Ⓕ Ⓖ Ⓗ Ⓙ	23 Ⓐ Ⓑ Ⓒ Ⓓ	28 Ⓕ Ⓖ Ⓗ Ⓙ	33 Ⓐ Ⓑ Ⓒ Ⓓ	38 Ⓕ Ⓖ Ⓗ Ⓙ
4 Ⓕ Ⓖ Ⓗ Ⓙ	9 Ⓐ Ⓑ Ⓒ Ⓓ	14 Ⓕ Ⓖ Ⓗ Ⓙ	19 Ⓐ Ⓑ Ⓒ Ⓓ	24 Ⓕ Ⓖ Ⓗ Ⓙ	29 Ⓐ Ⓑ Ⓒ Ⓓ	34 Ⓕ Ⓖ Ⓗ Ⓙ	39 Ⓐ Ⓑ Ⓒ Ⓓ
5 Ⓐ Ⓑ Ⓒ Ⓓ	10 Ⓕ Ⓖ Ⓗ Ⓙ	15 Ⓐ Ⓑ Ⓒ Ⓓ	20 Ⓕ Ⓖ Ⓗ Ⓙ	25 Ⓐ Ⓑ Ⓒ Ⓓ	30 Ⓕ Ⓖ Ⓗ Ⓙ	35 Ⓐ Ⓑ Ⓒ Ⓓ	40 Ⓕ Ⓖ Ⓗ Ⓙ

TEST 4—Science Reasoning

1 Ⓐ Ⓑ Ⓒ Ⓓ	6 Ⓕ Ⓖ Ⓗ Ⓙ	11 Ⓐ Ⓑ Ⓒ Ⓓ	16 Ⓕ Ⓖ Ⓗ Ⓙ	21 Ⓐ Ⓑ Ⓒ Ⓓ	26 Ⓕ Ⓖ Ⓗ Ⓙ	31 Ⓐ Ⓑ Ⓒ Ⓓ	36 Ⓕ Ⓖ Ⓗ Ⓙ
2 Ⓕ Ⓖ Ⓗ Ⓙ	7 Ⓐ Ⓑ Ⓒ Ⓓ	12 Ⓕ Ⓖ Ⓗ Ⓙ	17 Ⓐ Ⓑ Ⓒ Ⓓ	22 Ⓕ Ⓖ Ⓗ Ⓙ	27 Ⓐ Ⓑ Ⓒ Ⓓ	32 Ⓕ Ⓖ Ⓗ Ⓙ	37 Ⓐ Ⓑ Ⓒ Ⓓ
3 Ⓐ Ⓑ Ⓒ Ⓓ	8 Ⓕ Ⓖ Ⓗ Ⓙ	13 Ⓐ Ⓑ Ⓒ Ⓓ	18 Ⓕ Ⓖ Ⓗ Ⓙ	23 Ⓐ Ⓑ Ⓒ Ⓓ	28 Ⓕ Ⓖ Ⓗ Ⓙ	33 Ⓐ Ⓑ Ⓒ Ⓓ	38 Ⓕ Ⓖ Ⓗ Ⓙ
4 Ⓕ Ⓖ Ⓗ Ⓙ	9 Ⓐ Ⓑ Ⓒ Ⓓ	14 Ⓕ Ⓖ Ⓗ Ⓙ	19 Ⓐ Ⓑ Ⓒ Ⓓ	24 Ⓕ Ⓖ Ⓗ Ⓙ	29 Ⓐ Ⓑ Ⓒ Ⓓ	34 Ⓕ Ⓖ Ⓗ Ⓙ	39 Ⓐ Ⓑ Ⓒ Ⓓ
5 Ⓐ Ⓑ Ⓒ Ⓓ	10 Ⓕ Ⓖ Ⓗ Ⓙ	15 Ⓐ Ⓑ Ⓒ Ⓓ	20 Ⓕ Ⓖ Ⓗ Ⓙ	25 Ⓐ Ⓑ Ⓒ Ⓓ	30 Ⓕ Ⓖ Ⓗ Ⓙ	35 Ⓐ Ⓑ Ⓒ Ⓓ	40 Ⓕ Ⓖ Ⓗ Ⓙ

Model Mathematics ACT III

60 Questions—60 Minutes

INSTRUCTIONS: Find the solution to each problem, choose the correct answer choice, then darken the appropriate oval on your answer sheet. Do not spend too much time on any one problem. Answer as many problems as you can easily and then work on the remaining problems within the time limit for this test. Check pages 438–444 for answers and explanations.

Unless the problem indicates otherwise:

- figures are NOT necessarily drawn to scale
- geometric figures are plane figures
- a *line* is a straight line
- an *average* is the arithmetic mean

You may use a calculator.

---

1. The Smith children receive an allowance beginning at age 8. Each child receives an allowance equal to $0.50 times his or her age. How much more allowance will a 15-year-old receive than a 9-year-old?

   A. $ 2.00
   B. $ 3.00
   C. $ 4.50
   D. $ 7.50
   E. $15.00

2. What is the value of $\left(\frac{2}{5} + \frac{1}{2}\right) + \left(\frac{3}{5} - \frac{2}{3}\right)$ in simplest form?

   F. $\frac{1}{15}$

   G. $\frac{2}{3}$

   H. $\frac{3}{4}$

   J. $\frac{5}{6}$

   K. $\frac{9}{10}$

3. Mark polled 25 households and found there was an average of 2.5 children per household. However, when looking back at the calculations, Mark realized that he divided by 24, instead of 25 to find the average. What is the actual average number of children in the 25 households surveyed?

   A. 2
   B. 2.2
   C. 2.4
   D. 2.5
   E. 3.5

GO ON TO THE NEXT PAGE.

4. In the figure below, points *A* and *B* lie on the circle and *C* is the center of the circle. If the measure of ∠*ACB* is 80°, what is the measure of ∠*CAB*?

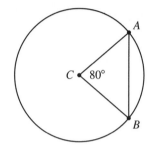

- **F.** 30°
- **G.** 40°
- **H.** 50°
- **J.** 60°
- **K.** 90°

5. $\sqrt{32 - 7} = ?$

- **A.** 3
- **B.** 4
- **C.** 5
- **D.** 6
- **E.** 7

6. A diver takes three measurements straight down from the surface of the ocean as shown in the following figure. The distance from the surface to point *y* is 16 feet. The distance from point *x* to point *z* is 23 feet. If the distance from the surface to point *z* is 33 feet, what is the distance from point *x* to point *y*?

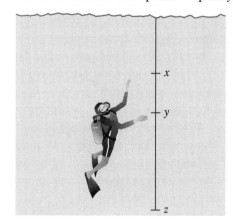

- **F.** 6 feet
- **G.** 16 feet
- **H.** 17 feet
- **J.** 23 feet
- **K.** 39 feet

7. In the figure below, ℓ and *m* are parallel lines cut by a transversal *t*. What is the difference between the value of *a* and the value of *b*?

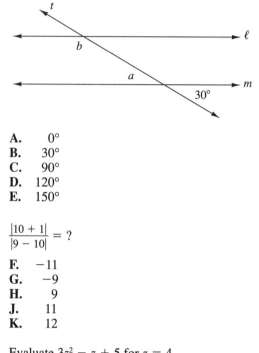

- **A.** 0°
- **B.** 30°
- **C.** 90°
- **D.** 120°
- **E.** 150°

8. $\dfrac{|10 + 1|}{|9 - 10|} = ?$

- **F.** −11
- **G.** −9
- **H.** 9
- **J.** 11
- **K.** 12

9. Evaluate $3z^2 - z + 5$ for $z = 4$.

- **A.** 49
- **B.** 53
- **C.** 67
- **D.** 122
- **E.** 145

10. In the figure below, lines *c* and *d* are parallel. They are both cut by two lines, *a* and *b*, which intersect each other. The measure of ∠*x* is 90° and the measure of ∠*y* is 40°. What is the measure of ∠*z*?

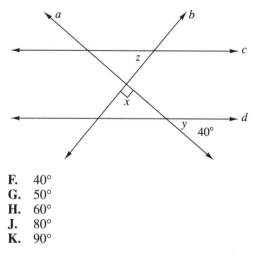

- **F.** 40°
- **G.** 50°
- **H.** 60°
- **J.** 80°
- **K.** 90°

GO ON TO THE NEXT PAGE.

11. Joshua earns $7 per hour of work. If his wages are decreased by D dollars per hour, which of the following equations represents the amount of money Joshua will earn with the new wage if he works 20 hours?

    A. $20(\$7.00 - D)$
    B. $20(D - \$7.00)$
    C. $20(\$7.00 + D)$
    D. $\$7.00 - 20D$
    E. $20(\$7.00) - D$

12. Solve for y: $5y + 3d = 8d + 6$.

    F. $y = 5d + \dfrac{6}{5}y$

    G. $d = \dfrac{5y + 6}{5}$

    H. $y = \dfrac{5d + 6}{5}$

    J. $d = \dfrac{5d + 6}{5}$

    K. $y = \dfrac{-5d + 6}{5}$

13. Jennifer and Dave each drove their own cars on a trip. Jennifer drove 25 miles farther than Dave. Dave drove 150 miles, and Jennifer drove $1t5$ miles, where t represents the digit in the tens place. What is the value of t?

    A.   2
    B.   4
    C.   5
    D.   7
    E.   9

14. Joel jogs 3.5 miles every $\dfrac{1}{2}$ hour. If he continues at this rate, how many miles will Joel jog in 5 hours?

    F.   25 miles
    G.   30 miles
    H.   35 miles
    J.   40 miles
    K.   45 miles

15. Triangle ABC and triangle ADE shown below are congruent right triangles. Both point B and point D lie on line XY. The measure of $\angle ACB$ is 70° and the measure of $\angle BAD$ is 80°. What is the measure of $\angle CBX$?

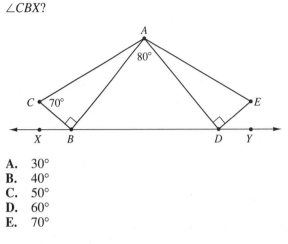

    A.   30°
    B.   40°
    C.   50°
    D.   60°
    E.   70°

16. Which of the following figures is a rhombus?

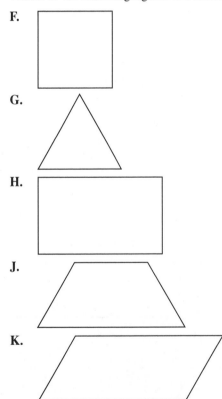

    F.

    G.

    H.

    J.

    K.

GO ON TO THE NEXT PAGE.

17. Which of the following expressions contains only numbers that are both multiples of 4 and factors of 32?

    A. {4, 8, 24, 32}
    B. {4, 20, 16, 28}
    C. {4, 8, 12, 16}
    D. {4, 8, 16, 32}
    E. {4, 16, 24, 32}

18. For all z, $(12z^7 + 5z^4 - 3z^2 + 8z) + (-6z^7 - 2z^4 + z^2 + 4) = ?$

    F. $6z^7 + 3z^4 - 2z^2 + 8z + 4$
    G. $18z^7 + 7z^4 + 4z^2 + 4z + 8$
    H. $-6z^7 - 3z^4 + 2z^2 - 8z - 4$
    J. $3z^7 - 6z^4 - z^2 + 4z - 2$
    K. $12z^7 - 6z^4 - 4z^2 + 16z + 8$

19. What is the value of x if $x^n = 3^{-n}$?

    A. $\dfrac{1}{3^n}$
    B. $\dfrac{1}{3}$
    C. 3^n
    D. $3^{\frac{1}{n}}$
    E. $\sqrt[n]{3}$

20. When $x \neq 0$, which of the following is a simplified version of $\dfrac{x-2}{2x} + \dfrac{4x+1}{3x}$?

    F. $\dfrac{11x}{6x} - \dfrac{4x}{6}$
    G. $\dfrac{5x-1}{5x}$
    H. $\dfrac{11x-4}{6x}$
    J. $\dfrac{5x-4}{6x}$
    K. $\dfrac{11}{6} - \dfrac{2}{3}x$

21. What is the area, in square units, of the figure below?

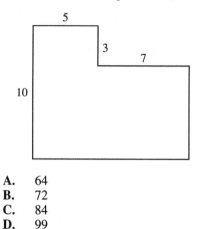

    A. 64
    B. 72
    C. 84
    D. 99
    E. 120

22. When C number of chores are divided equally among 11 workers, there are 4 chores left over. How many chores will be left if there are $(C + 9)$ chores?

    F. 2
    G. 3
    H. 4
    J. 5
    K. 6

23. For which of the following pairs of numbers is the average $3t - 2$?

    A. $5t + 1, 3t - 2$
    B. $4t + 2, 2t - 6$
    C. $6t - 4, 4t + 3$
    D. $5t + 2, 2t - 1$
    E. $7t + 5, 3t - 4$

24. For $x \neq -4$, which of the following is equivalent to $\dfrac{3x^3 + 11x^2 + 16}{x + 4}$?

    F. $3x^2 + x + 4$
    G. $3x^2 - x + 4$
    H. $3x^2 - x - 4$
    J. $3x^2 + x - 4$
    K. $3x^2 + x^2 + 4$

25. What is the value of s in $(s + 5y)(2y - 3) = 10y^2 - 3s + 9y$?

    A. 8
    B. 10
    C. 12
    D. 14
    E. 16

GO ON TO THE NEXT PAGE.

26. Solve for t: $|3t - 4| = 8$.

F. $4\frac{1}{2}$

G. $-\frac{1}{2}$

H. $\frac{3}{4}, \frac{4}{5}$

J. $4, -\frac{4}{3}$

K. $-\frac{2}{3}$

27. In a right triangle, if $\sin \theta = \frac{5}{13}$, what is the value of $\csc \theta$?

A. $\frac{5}{12}$

B. $\frac{12}{13}$

C. $\frac{13}{12}$

D. $\frac{12}{5}$

E. $\frac{13}{5}$

28. What is the y-intercept of the line with the equation $10y + 2x = 4x + 5y + 7$?

F. $\frac{2}{7}$

G. $\frac{2}{5}$

H. $\frac{7}{5}$

J. 5

K. 7

29. An 18-foot flagpole casts a shadow 24 feet long. How long is the distance between the top of the flagpole and the end of the shadow?

A. 28 feet
B. 30 feet
C. 32 feet
D. 38 feet
E. 42 feet

30. Simplify. $\frac{2a^{-2}b^3}{4a^3c^{-4}} = ?$ $(a \neq 0, c \neq 0)$

F. $\frac{a^{-2}b^3}{2c^{-4}}$

G. $2a^{-4}b^3$

H. $\frac{b^3c^4}{a^5}$

J. $\frac{2b^3c^4}{a^5}$

K. $\frac{b^3c^4}{2a^5}$

31. What is the distance between the points $(6, -5)$ and $(2, 3)$ in a standard (x, y) coordinate plane?

A. $4\sqrt{3}$
B. 7
C. 8
D. $4\sqrt{5}$
E. 10

32. Factor $6x^4 + 11x^3 - 10x^2$ completely.

F. $x(2x - 3)(4x + 6)$
G. $x^2(3x - 2)(2x + 5)$
H. $x^2(6x^2 + 11x - 10)$
J. $(3x + 1)(5x + 4)$
K. $x^2(4x - 5)(3x + 2)$

33. Kim runs a mile on Sunday and a mile every day that week. Her time decreases each day as shown by the graph below. How much time can Kim expect to take to run a mile on Saturday of that week?

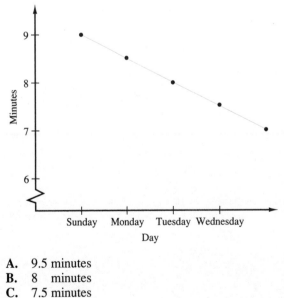

A. 9.5 minutes
B. 8 minutes
C. 7.5 minutes
D. 6.5 minutes
E. 6 minutes

GO ON TO THE NEXT PAGE.

34. The polygon below is a regular hexagon with the length of $\overline{AB}$ equal to 4 units. $\overline{FD} \perp \overline{CD}$. If the measure of $\angle FCD$ is 60°, how long is $\overline{CF}$?

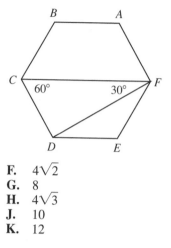

F. $4\sqrt{2}$
G. 8
H. $4\sqrt{3}$
J. 10
K. 12

35. In triangle LMN below, the measure of $\angle L$ is 45° and the measure of $\angle M$ is 90°. If the length of $\overline{LM}$ is 7 units, how long is $\overline{MN}$?

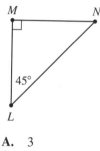

A. 3
B. 5
C. 6
D. 7
E. 9

36. What figure on the regular (x,y) coordinate plane is represented by the equation $(x + 2)^2 + (y - 3)^2 = 9$?

F. Ellipse
G. Circle
H. Parabola
J. Sphere
K. Parallel lines

37. If the lengths of the legs of a right triangle are 4 units and 6 units long, how long is the hypotenuse of the triangle?

A. $2\sqrt{13}$
B. $3\sqrt{5}$
C. 8
D. 10
E. $4\sqrt{15}$

38. In the figure below, points A, B, C, and D form a square. What is the area of the figure in square units?

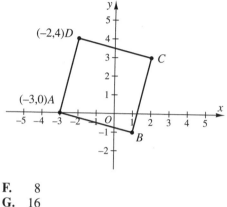

F. 8
G. 16
H. 17
J. 31
K. 36

39. The following set includes the solutions for which of the following equations?

$\{-3, 0, 1, 5, 7\}$

A. $x^2 + x - 2 = 0$
B. $x^2 - x - 12 = 0$
C. $x^2 - 4x - 21 = 0$
D. $x^2 - 5x - 36 = 0$
E. $x^2 - 5x - 66 = 0$

40. For which real values of x is $\dfrac{\sqrt[3]{x} - x}{4^{4x-8}}$ defined?

F. All real numbers
G. All real numbers except $x = 0$
H. All real numbers except $x = \dfrac{1}{4}$
J. All real numbers except $x = 2$
K. All real numbers except $x = 4$

41. Which of the following is an equation of a line having a slope equal to $\dfrac{2}{3}$ and y-intercept equal to 4?

A. $3y - 2x = 12$
B. $4y - 3x = 16$
C. $3y - 2x = 6$
D. $5y - 3x = 20$
E. $2y - 3x = 8$

GO ON TO THE NEXT PAGE.

42. What is the smallest number in the solution set to the inequality $-2x - 16 \leq 2x - 4$?

    F.   12
    G.   4
    H.   3
    J. -3
    K. -4

43. Two points that lie on a certain line are $(-1,4)$ and $(2,5)$. What is the slope of the line?

    A. $-\dfrac{3}{1}$
    B. -1
    C. $\dfrac{1}{3}$
    D. $\dfrac{1}{2}$
    E. $\dfrac{2}{3}$

44. In the figure below, $ABCD$ is a rectangle. If the measure of $\angle DEB$ is $135°$, what is the sine of $\angle ADE$?

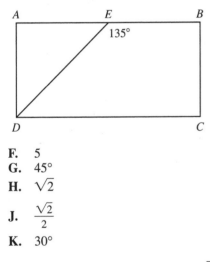

    F. 5
    G. $45°$
    H. $\sqrt{2}$
    J. $\dfrac{\sqrt{2}}{2}$
    K. $30°$

45. What are all the values of y so that $\sqrt{y} < y$?

    A. $y > 1$
    B. $y > 0$
    C. $y < 0$
    D. $-1 < y < 1$
    E. $y \geq 0$

46. Which of the following values for t will make the largest value of s in the equation $s = t^3 - t$?

    F. -4
    G. -2
    H.   0
    J.   2
    K.   4

47. In right triangle XYZ below, the measure of $\angle Y$ is $90°$. The length of $\overline{XY}$ is 9 and the length of $\overline{XZ}$ is $5\sqrt{10}$. How long is $\overline{ZY}$?

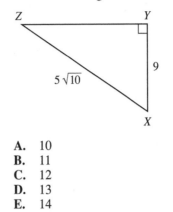

    A. 10
    B. 11
    C. 12
    D. 13
    E. 14

48. The solutions of $x^2 - 2xt - 8t^2 = 0$ when the equation is solved for x are

    F. t and $2t$
    G. $-2t$ and $4t$
    H. $2t$ and $-4t$
    J. $4t$ and $8t$
    K. $2t$ and $6t$

49. In a right triangle, one leg is 5 units long and the hypotenuse is 13 units long. What is the area of this triangle in square units?

    A. 30
    B. 32.5
    C. 60
    D. 65
    E. 78

50. A scientist observed a thermometer starting at zero degrees. The thermometer moved a certain number of degrees (F) by the first observation and another number of degrees (S) from the first to the second observation. The equations below describe the observations.

$$4F + 10S = 20$$

$$6F + 9S = 6$$

What is the temperature at the second observation?

    F. $5°$
    G. $4°$
    H. $-1°$
    J. $-4°$
    K. $-5°$

GO ON TO THE NEXT PAGE.

51. If a line passes through points (9,2) and (5,4), which of the following would be the slope of a line parallel to the given line?

A. $-\dfrac{2}{1}$

B. $-\dfrac{1}{2}$

C. $\dfrac{1}{2}$

D. 2

E. $\dfrac{3}{2}$

52. In the diagram below, what is cot θ?

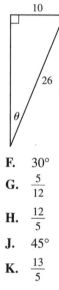

F. 30°

G. $\dfrac{5}{12}$

H. $\dfrac{12}{5}$

J. 45°

K. $\dfrac{13}{5}$

53. In the figure below, $\overline{AB} \perp \overline{DE}$. The measure of $\angle CAD$ is equal to the measure of $\angle CEB$. The length of $\overline{CA}$ is 6 units and the length of $\overline{CE}$ is 18 units. If the length of $\overline{AD}$ is 10 units, what is the length of $\overline{EB}$?

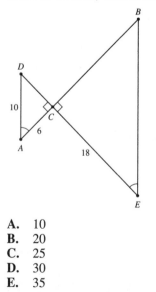

A. 10
B. 20
C. 25
D. 30
E. 35

54. In the diagram below, triangle *ABC* is similar to triangle *XYZ*. If the lengths are all measured in inches, how long is $\overline{XZ}$?

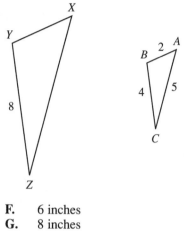

F. 6 inches
G. 8 inches
H. 10 inches
J. 12 inches
K. 14 inches

55. One period of the curve $y = 2 \cos \theta$ is shown below. The height of the curve at 2π is 2. The height at $\dfrac{\pi}{2}$ and $\dfrac{3\pi}{2}$ is 0 and the height at π is -2. What will the height of the curve be at $\dfrac{18\pi}{4}$?

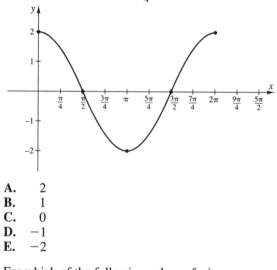

A. 2
B. 1
C. 0
D. -1
E. -2

56. For which of the following values of z is $\dfrac{5z - 5}{2z^2 + 7z - 15}$ undefined?

F. 5
G. 1
H. $-\dfrac{2}{3}$
J. $-\dfrac{3}{2}$
K. -5

GO ON TO THE NEXT PAGE.

57. A movie critic is tracking the attendance of three independent films: *Oyster Point, Quinn's Island, and Thalian* (O, I, and T, respectively). The matrices below show the attendance numbers for opening night at the Downtown, Main Square, and Country theaters, and the cost of the ticket at each theater. What was the total sales for all the movies at these theaters on opening night?

	O	I	T		Cost
Downtown	91	87	75		$8
Main square	68	53	47		$10
Country	47	35	22		$9

- **A.** $924
- **B.** $1,497
- **C.** $2,273
- **D.** $4,694
- **E.** $5,470

58. Which of the following is the equation for the graph below?

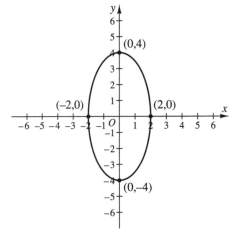

- **F.** $y = 2x^2$
- **G.** $(x + 2)^2 + (y + 4)^2 = 4$
- **H.** $\dfrac{x^2}{4} + \dfrac{y^2}{16} = 1$
- **J.** $\dfrac{x^2}{16} + \dfrac{y^2}{4} = 1$
- **K.** $x - 2y^2 = 0$

59. What is the solution set for $-3x^2 - 3x > -18$?

- **A.** $-2 < x < -3$
- **B.** $2 < x < 3$
- **C.** $-3 < x < -2$
- **D.** $-3 < x < 2$
- **E.** $-2 < x < 3$

60. $\angle ABC$ lies in the (x,y) coordinate plane. Line $\overleftrightarrow{BA}$ is represented by the equation $y = 3x + 5$, and line $\overleftrightarrow{BC}$ is represented by the equation $y = 6x + 8$. Where is the vertex of the angle located?

- **F.** (0,0)
- **G.** (1,1)
- **H.** (3,−1)
- **J.** (−1,2)
- **K.** (−2,3)

END OF TEST 2

✔ ANSWERS

1. B	16. F	31. D	46. K
2. J	17. D	32. G	47. D
3. C	18. F	33. E	48. G
4. H	19. B	34. G	49. A
5. C	20. H	35. D	50. H
6. F	21. D	36. G	51. B
7. D	22. F	37. A	52. H
8. J	23. B	38. H	53. D
9. A	24. G	39. C	54. H
10. G	25. C	40. F	55. C
11. A	26. J	41. A	56. K
12. H	27. E	42. J	57. D
13. D	28. H	43. C	58. H
14. H	29. B	44. J	59. D
15. B	30. K	45. A	60. J

ACT III MATHEMATICS Answers Explained

1. **B.** $15(\$0.50) - 9(\$0.50) = \$3.00$

2. **J.** $\left(\dfrac{2}{5} + \dfrac{1}{2}\right) + \left(\dfrac{3}{5} - \dfrac{2}{3}\right)$

$= \left(\dfrac{12}{30} + \dfrac{15}{30}\right) + \left(\dfrac{18}{30} - \dfrac{20}{30}\right)$

$= \dfrac{27}{30} - \dfrac{2}{30}$

$= \dfrac{25}{30} = \dfrac{5}{6}$

3. **C.** Let T equal the number of children that were counted in the households.

$\dfrac{T}{24} = 2.5$

$T = 60$

There are 60 children in the households.

Now divide 60 by the actual number of households polled.

$\dfrac{60}{25} = 2.4$

The actual average number of children in the 25 households is 2.4.

4. **H.** The sum of the measures of the angles of a triangle is 180°.

$80° + m\angle A + m\angle B = 180°$

$m\angle A + m\angle B = 100°$

Since $\overline{CA}$ and $\overline{CB}$ are radii, they are congruent. So $\triangle ABC$ is isosceles, and $\angle A$ and $\angle B$ are congruent. The measure of each angle is $100° \div 2$ or 50°.

5. **C.** $\sqrt{(32 - 7)} = \sqrt{25} = 5$

6. **F.** Label the diagram with the given distances.

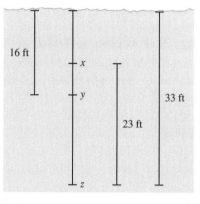

$16 \text{ feet} + 23 \text{ feet} - \text{distance } xy = 33 \text{ feet}$

$39 \text{ feet} - \text{distance } xy = 33 \text{ feet}$

$- \text{distance } xy = -6 \text{ feet}$

$\text{distance } xy = 6 \text{ feet}$

7. D. $m\angle a = 30°$ and $m\angle b = 150°$ because $\angle a$ and $\angle b$ are same-side interior angles and the sum of these angles is $180°$.

$m\angle a - m\angle b = 150° - 30° = 120°$

8. J. The absolute value of a number is always positive.

$\dfrac{|10+1|}{|9-10|} = \dfrac{|11|}{|-1|} = \dfrac{11}{1} = 11$

9. A. Substitute 4 for z in the expression and evaluate.

$3(4)^2 - 4 + 5 = 3(16) - 4 + 5 = 48 - 4 + 5 = 49$

10. G. $\angle z$ is an angle in a triangle. One angle of the triangle has a measure of $90°$ because it is the vertical angle of $\angle x$. Another angle of the triangle has a measure of $40°$ because it is a corresponding angle to $\angle y$.

$90° + 40° + m\angle z = 180°$
$130° + m\angle z = 180°$
$m\angle z = 50°$

11. A. \$7.00 decreased by D is written $(\$7.00 - D)$. Twenty hours of work is $20(\$7.00 - D)$.

12. H. Solve for y.

$5y + 3d = 8d + 6$
$5y = 5d + 6$
$y = \dfrac{5d+6}{5}$

13. D. Jennifer drove $150 + 25 = 175$ miles. $t = 7$

14. H. Write a proportion. $\dfrac{3.5 \text{ miles}}{0.5 \text{ hour}} = \dfrac{x \text{ miles}}{5 \text{ hours}}$

Cross multiply. $0.5x = 17.5$
Solve. $x = 35 \text{ miles}$

Alternative solution:

5 hours is 10 times $\dfrac{1}{2}$ hour, so $10 \times 3.5 = 35$.

15. B.

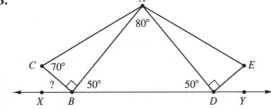

Since $\triangle ABC$ and $\triangle ADE$ are congruent, $\overline{AB}$ and $\overline{AD}$ are congruent. Thus, $\angle ABD$ and $\angle ADB$ are congruent because $\triangle ABD$ is isosceles.

$m\angle BAD = 80°$ so $m\angle ABD$ and $m\angle ADB$ are both $50°$.

$m\angle CBX + m\angle CBA + m\angle ABD = 180°$
$m\angle CBX + 90° + 50° = 180°$
$m\angle CBX = 180° - 90° - 50° = 40°$

16. F. A rhombus is a quadrilateral with opposite sides parallel and all four sides congruent. Every square is a rhombus.

17. D. Multiples of 4: $\{4, 8, 12, 16, 20, 24, 28, 32, 36,\dots\}$
Factors of 32: $\{1, 2, 4, 8, 16, 32\}$

18. F. $(12z^7 + 5z^4 - 3z^2 + 8z) + (-6z^7 - 2z^4 + z^2 + 4)$
$= 12z^7 - 6z^7 + 5z^4 - 2z^4 - 3z^2 + z^2 + 8z + 4$ Group like terms.
$= 6z^7 + 3z^4 - 2z^2 + 8z + 4$ Combine like terms.

19. B. We know that
$x^n = 3^{-n}$ and $3^{-n} = \dfrac{1}{3^n}$
so $x^n = \dfrac{1}{3^n} = \left(\dfrac{1}{3}\right)^n$
$x = \dfrac{1}{3}$

20. H. Simplify.

$\dfrac{x-2}{2x} + \dfrac{4x+1}{3x}$

$= \dfrac{3x-6}{6x} + \dfrac{8x+2}{6x}$ Rewrite the fractions with a common denominator.

$= \dfrac{3x - 6 + 8x + 2}{6x}$ Combine the fractions.

$= \dfrac{11x - 4}{6x}$

21. D.

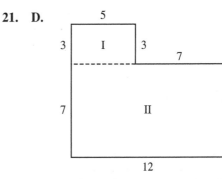

Split the figure into two rectangles. Find the area of each rectangle and add.
$A_\text{I} = 5 \times 3 = 15$
$A_\text{II} = 7 \times 12 = 84$
Total area $= 15 + 84 = 99$

22. F. When C is divided by 11, there is a remainder of 4.

$C = 11n + 4$
$C + 9 = (11n + 4) + 9$
$C + 9 = 11n + 13$
One more 11 can be taken from 13, so
$C + 9 = 11n + 2$
There are two chores left over.

23. B. Since the average is $3t - 2$, the sum of the number pair is twice $3t - 2$, or $6t - 4$. Find the sum for each answer choice.

Choice A: $(5t + 1) + (3t - 2)$
  $= 5t + 3t + 1 - 2 = 8t - 1$ No.

Choice B: $(4t + 2) + (2t - 6)$
  $= 4t + 2t + 2 - 6 = 6t - 4$ Yes.

Check your answer by finding the average.

$\dfrac{6t - 4}{2} = 3t - 2$

24. G. Factor the numerator and simplify.

$\dfrac{3x^3 + 11x^2 + 16}{x + 4}$

$= \dfrac{(3x^2 - x + 4)(x + 4)}{(x + 4)}$

$= 3x^2 - x + 4$

25. C. Use the FOIL method to multiply the binomials. It may help to rewrite the first factor.

$(5y + s)(2y - 3) = 10y^2 - 3s + 9y$

$10y^2 + 2sy - 15y - 3s = 10y^2 - 3s + 9y$
      Multiply the binomials.

$2sy - 15y = 9y$ Cancel the equal terms ($10y^2$ and $-3s$) on both sides of the equal sign.

$2sy = 24y$ Add $15y$ to both sides.

$s = 12$ Divide both sides by $2y$.

26. J. $|-8| = 8$ and $|8| = 8$. So $|3t - 4| = 8$ can be written as two equations. Solve each equation.

$3t - 4 = 8$ $-(3t - 4) = 8$
  $3t = 12$ $-3t + 4 = 8$
  $t = 4$ $-3t = 4$
        $t = -\dfrac{4}{3}$

27. E. $\csc\theta = \dfrac{1}{\sin\theta} = \dfrac{1}{\frac{5}{13}} = \dfrac{13}{5}$

28. H. Rewrite the equation in slope-intercept form, $y = mx + b$, where b is the y-intercept.

$10y + 2x = 4x + 5y + 7$

  $5y + 2x = 4x + 7$ Subtract $5y$ from both sides.

    $5y = 2x + 7$ Subtract $2x$ from both sides.

    $y = \dfrac{2}{5}x + \dfrac{7}{5}$ Divide all terms by 5.

        $\uparrow$

      y-intercept

29. B. Draw a picture.

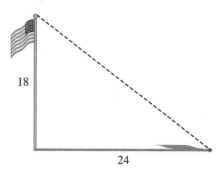

The figure forms a right triangle. Use the Pythagorean Theorem to find the distance.

$18^2 + 24^2 = c^2$

$324 + 576 = c^2$

  $900 = c^2$

  $\sqrt{900} = c$

    $30 = c$

The distance is 30 feet.

30. K. Note that two factors have negative exponents. Rewrite the factors with positive exponents, placing a^{-2} as a^2 in the denominator and c^{-4} as c^4 in the numerator. Then simplify.

$\dfrac{2a^{-2}b^3}{4a^3c^{-4}} = \dfrac{2b^3c^4}{4a^3a^2} = \dfrac{2b^3c^4}{4a^5} = \dfrac{b^3c^4}{2a^5}$

31. D. Use the distance formula.

$d = \sqrt{(x_2 - x_1)^2 + (y_2 - y_1)^2}$

$= \sqrt{(2 - 6)^2 + (3 + 5)^2}$

$= \sqrt{(-4)^2 + (8^2)}$

$= \sqrt{16 + 64}$

$= \sqrt{80}$

$= \sqrt{16 \times 5}$

$= 4\sqrt{5}$

32. G. First factor x^2 out of each term. Then factor the remaining quadratic expression.

$6x^4 + 11x^3 - 10x^2 = x^2(6x^2 + 11x - 10)$
      $= x^2(3x - 2)(2x + 5)$

33. E. The graph tells us that Kim's time decreases at a rate of 1 minute every 2 days or 0.5 minute a day. Her time was 9 minutes on Sunday and Saturday is 6 days away, so $9 - 0.5(6) = 6$ minutes.

Kim can expect to take 6 minutes to run a mile on Saturday.

34. G. Since $\overline{FD}$ is perpendicular to $\overline{CD}$, the measure of $\angle FDC$ is 90°. Triangle CDF is a 30–60–90 triangle. In this type of triangle, the hypotenuse is twice the length of the side opposite the 30° angle. Since this is a regular polygon, the length of $\overline{CD}$ is 4 units and the measurement of $\overline{CF}$ is twice as long, or 8 units.

35. D. Triangle LMN is a 45–45–90 triangle. In this type of triangle, the lengths of the legs are the same. So $\overline{MN}$ equals 7 units.

36. G. The equation is in the standard form for a circle.

37. A. Use the Pythagorean Theorem.
$$c^2 = a^2 + b^2$$
$$= 4^2 + 6^2$$
$$= 16 + 36$$
$$= 52$$
$$c = \sqrt{52}$$
$$= \sqrt{4 \times 13}$$
$$= 2\sqrt{13}$$

38. H. To find the length of one side, use the distance formula. Then square the length of the side to find the area.
$$d = \sqrt{(x_1 - x_2)^2 + (y_1 - y_2)^2}$$
$$= \sqrt{(-2 - (-3))^2 + (4 - 0)^2}$$
$$= \sqrt{(1)^2 + (4)^2}$$
$$= \sqrt{17}$$
$$d^2 = 17$$

39. C. Look for the equation whose solutions are in the set $\{-3, 0, 1, 5, 7\}$.
Begin with choice A.
$$x^2 + x - 2 = 0$$
$$(x + 2)(x - 1) = 0$$
$$x + 2 = 0 \text{ and } x - 1 = 0$$
$$x = -2 \text{ and } x = 1$$
-2 is not in the set.
Try choice B.
$$x^2 - x - 12 = 0$$
$$(x - 4)(x + 3) = 0$$
$$x - 4 = 0 \text{ and } x + 3 = 0$$
$$x = 4 \text{ and } x = -3$$
4 is not in the set.
Try choice C.
$$x^2 - 4x - 21 = 0$$
$$(x - 7)(x + 3) = 0$$
$$x - 7 = 0 \text{ and } x + 3 = 0$$
$$x = 7 \text{ and } x = -3$$
Yes! 7 and -3 are in the set.

40. F. The expression $\dfrac{\sqrt[3]{x} - x}{4^{4x-8}}$ is defined except if $4^{4x-8} = 0$. 4^{4x-8} is never equal to zero, so the expression is defined for all real numbers.

41. A. Rewrite the answer choices in slope-intercept form.
Begin with choice A.
$$3y - 2x = 12$$
$$3y = 2x + 12$$
$$y = \frac{2}{3}x + 4 \quad \text{Yes!}$$
$$\uparrow \qquad \uparrow$$
$$\text{slope} \quad \text{y-intercept}$$

42. J. Solve.
$$-2x - 16 \le 2x - 4$$
$$-4x \le 12$$
$$x \ge -3$$
The smallest number in the solution set is -3.

43. C. The slope of a line is
$$m = \frac{y_2 - y_1}{x_2 - x_1} = \frac{(5 - 4)}{2 - (-1)} = \frac{1}{3}$$

44. J.

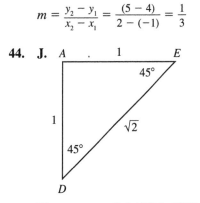

The measure of $\angle AED$ is $180° - 135°$, or $45°$. Since $ABCD$ is a rectangle, the measure of $\angle DAE$ is $90°$. Triangle ADE is a 45–45–90 triangle.
Let $AD = 1$. Then $AE = 1$ and $DE = \sqrt{2}$.
$$\sin \angle ADE = \frac{\text{opposite}}{\text{hypotenuse}}$$
$$= \frac{1}{\sqrt{2}} = \frac{1}{\sqrt{2}} \cdot \frac{\sqrt{2}}{\sqrt{2}} = \frac{\sqrt{2}}{2}$$

45. A. $\sqrt{y} < y$ when y is positive and greater than 1.

46. K. Look at the answer choices.
Choices F and G: Since t^3 is negative when t is negative, s for these choices would also be negative.
Check the other choices.
Choice H: t is 0 and s is 0.
Choice J: $s = 2^3 - 2 = 8 - 2 = 6$.
Choice K: $s = 4^3 - 4 = 64 - 4 = 60$. This is the largest value for s.

47. D. Use the Pythagorean Theorem.
$$(XY)^2 + (ZY)^2 = (XZ)^2$$
$$9^2 + (ZY)^2 = (5\sqrt{10})^2$$
$$81 + (ZY)^2 = 25 \times 10$$
$$81 + (ZY)^2 = 250$$
$$(ZY)^2 = 169$$
$$ZY = \sqrt{169}$$
$$ZY = 13$$

48. G. Factor.
$$x^2 - 2xt - 8t^2 = 0$$
$$(x + 2t)(x - 4t) = 0$$
$$x + 2t = 0 \quad \text{or} \quad x - 4t = 0$$
$$x = -2t \qquad\qquad x = 4t$$

49. A. The base of a right triangle is one leg and the height of the triangle is the other leg. To find the length of the other leg, use the Pythagorean Theorem.
$$a^2 + b^2 = c^2$$
$$5^2 + b^2 = 13^2$$
$$25 + b^2 = 169$$
$$b^2 = 144$$
$$b = \sqrt{144}$$
$$b = 12$$

Now find the area.
$$A = \frac{1}{2}bh$$
$$= \frac{1}{2}(12)(5)$$
$$= \frac{1}{2}(60)$$
$$= 30$$

50. H. Solve the two equations simultaneously.

$4F + 10S = 20 \;\rightarrow\; 12F + 30S = 60$ Multiply both sides by 3.

$6F + 9S = 6 \;\rightarrow\; \underline{12F + 18S = 12}$ Multiply both sides by 2.

$\qquad\qquad\qquad\qquad 12S = 48$ Subtract.

$\qquad\qquad\qquad\qquad\; S = 4$ Solve for S.

Substitute 4 for S in one of the equations and solve for F.
$$6F + 9(4) = 6$$
$$6F + 36 = 6$$
$$6F = -30$$
$$F = -5$$
The thermometer moved down 5° from 0, then up 4°. The temperature at the second observation is $0 - 5 + 4$, or $-1°$.

51. B. Find the slope of the line that passes through the given points.
$$m = \frac{4 - 2}{5 - 9} = \frac{2}{-4} = -\frac{1}{2}$$
Since parallel lines have the same slope, the slope of the line parallel to the given line is also $-\frac{1}{2}$.

52. H. We need to find the length of the other leg of the triangle. Use the Pythagorean Theorem.
$$a^2 + b^2 = c^2$$
$$10^2 + b^2 = 26^2$$
$$100 + b^2 = 676$$
$$b^2 = 576$$
$$b = \sqrt{576} = 24$$
$$\cot \theta = \frac{\text{adjacent}}{\text{opposite}} = \frac{24}{10} = \frac{12}{5}$$

53. D. $\triangle CAD$ is similar to $\triangle CEB$ by angle-angle-angle. Since the two triangles are similar, the ratio of corresponding sides is proportional. $\overline{CA}$ corresponds to $\overline{CE}$ and $\overline{AD}$ corresponds to $\overline{EB}$. Write a proportion to find the length of $\overline{EB}$.
$$\frac{CA}{CE} = \frac{AD}{EB}$$
$$\frac{6}{18} = \frac{10}{EB}$$
$$6(EB) = 180$$
$$EB = 30$$

54. H. Since $\triangle ABC$ is similar to $\triangle XYZ$, the ratio of corresponding sides is proportional. $\overline{YZ}$ corresponds to $\overline{BC}$ and $\overline{XZ}$ corresponds to $\overline{AC}$. Write a proportion to find the length of $\overline{XZ}$.
$$\frac{YZ}{BC} = \frac{XZ}{AC}$$
$$\frac{8}{4} = \frac{XZ}{5}$$
$$4(XZ) = 40$$
$$XZ = 10$$

55. C. Since the height of the curve is 0 at $\frac{\pi}{2}$ and $\frac{3\pi}{2}$, we can expect the height to be 0 at $\frac{5\pi}{2}$, $\frac{7\pi}{2}$, and $\frac{9\pi}{2}$, or $\frac{18\pi}{4}$.

56. K. The expression is undefined when the denominator is equal to 0. Solve $2z^2 + 7z - 15 = 0$

$(2z - 3)(z + 5) = 0$

$2z - 3 = 0 \qquad z + 5 = 0$

$2z = 3 \qquad\qquad z = -5$

$z = \dfrac{3}{2}$

Alternative solution:

The expression will be undefined when the denominator is 0. Try the answer choices.

Choice F: $z = 5$

$2z^2 + 7z - 15 = 2(5)^2 + 7(5) - 15$

$= 50 + 35 - 15$

$= 70 \quad$ No.

Choice G: $z = 1$

$2z^2 + 7z - 15 = 2(1)^2 + 7(1) - 15$

$= 2 + 7 - 15$

$= -6 \quad$ No.

Choice H: $z = -\dfrac{2}{3}$

$2z^2 + 7z - 15 = 2\left(-\dfrac{2}{3}\right)^2 + 7\left(-\dfrac{2}{3}\right) - 15$

$= 2\left(\dfrac{4}{9}\right) - \dfrac{14}{3} - 15$

$= \dfrac{8}{9} - \dfrac{42}{9} - 15$

$= -\dfrac{34}{9} - 15$

$= -18\dfrac{7}{9} \quad$ No.

Choice J: $z = -\dfrac{3}{2}$

$2z^2 + 7z - 15 = 2\left(-\dfrac{3}{2}\right)^2 + 7\left(-\dfrac{3}{2}\right) - 15$

$= \dfrac{18}{4} - \dfrac{21}{2} - 15$

$= \dfrac{18}{4} - \dfrac{42}{4} - 15$

$= -\dfrac{24}{4} - 15$

$= -6 - 15$

$= -21 \quad$ No.

Choice K: $z = -5$

$2z^2 + 7z - 15 = 2(-5)^2 + 7(-5) - 15$

$= 50 - 35 - 15$

$= 50 - 50$

$= 0 \quad$ Yes!

57. D. You don't have to use matrices. You can just think the problem through.

$(91 + 68 + 47)(\$8) + (87 + 53 + 35)(\$10)$

$+ (75 + 47 + 22)(\$9)$

$= 206(\$8) + 175(\$10) + 144(\$9) = \$4{,}694$

58. H. The graph is an ellipse. Look at the answer choices. F and K are equations of parabolas and G is the equation of a circle. H and J are equations of ellipses.

Look at Choice H. Test each point on the graph in the equation $\dfrac{x^2}{4} + \dfrac{y^2}{16} = 1$.

Try $(-2,0)$.

$\dfrac{(-2)^2}{4} + \dfrac{(0)^2}{16} = 1$

$\dfrac{4}{4} + 0 = 1$

$1 = 1 \quad$ True.

Try $(0,4)$.

$\dfrac{(0)^2}{4} + \dfrac{(4)^2}{16} = 1$

$0 + \dfrac{16}{16} = 1$

$1 = 1 \quad$ True.

Try $(2,0)$.

$\dfrac{(2)^2}{4} + \dfrac{(0)^2}{16} = 1$

$\dfrac{4}{4} + 0 = 1$

$1 = 1 \quad$ True.

Try $(0,-4)$.

$\dfrac{(0)^2}{4} + \dfrac{(-4)^2}{16} = 1$

$0 + \dfrac{16}{16} = 1$

$1 = 1 \quad$ True.

$\dfrac{x^2}{4} + \dfrac{y^2}{16} = 1$ is the equation of the ellipse shown.

59. D.

$-3x^2 - 3x > -18$

$x^2 + x < 6 \qquad$ Divide by -3.

$x^2 + x - 6 < 0 \qquad$ Write in standard quadratic form.

$(x + 3)(x - 2) < 0 \qquad$ Factor.

The inequality is less than 0. The product of the factors must be negative, so the factors must have different signs.

$x + 3 < 0$ and $x - 2 > 0$ or $x + 3 > 0$ and $x - 2 < 0$

$x < -3 \qquad x > 2 \qquad\qquad x > -3 \qquad x < 2$

$\qquad$ Impossible. $\qquad\qquad -3 < x < 2$

The solution set is $-3 < x < 2$.

60. **J.** Since $\angle ABC$ is created by the intersection of lines $\overrightarrow{BA}$ and $\overrightarrow{BC}$, we can find the location of the vertex by finding out where these two lines intersect.

$$y = 6x + 8 \qquad\qquad\qquad y = 6x + 8$$
$$y = 3x + 5 \text{ Multiply by } -2. \quad \underline{-2y = -6x - 10}$$
$$-y = -2$$
$$y = 2$$

Substitute y into both of the equations to find x. This allows you to check your work at the same time.

$$y = 3x + 5 \quad \text{and} \quad y = 6x + 8$$
$$2 = 3x + 5 \qquad\qquad 2 = 6x + 8$$
$$-3 = 3x \qquad\qquad\quad -6 = 6x$$
$$-1 = x \qquad\qquad\quad\; -1 = x$$

Both answers are the same, so you have done the problem correctly. The vertex is at point $(-1, 2)$.

Chapter 16 ▪

Model Mathematics **ACT IV**
With Answers Explained

This Mathematics Model ACT IV is just like a real ACT. Take this test after you take Model ACT III.

Take this model test under simulated test conditions. Allow 60 minutes to answer the 60 Mathematics items. Use a pencil to mark the answer sheet on page 447, and answer the questions in the Test 2 (Mathematics) section.

Use the answer key on page 458 to mark the answer sheet. Review the answer explanations on pages 458–463. You may decide to retake this test later. There are additional answer sheets for this purpose following page 605.

The test scoring chart on page 464 shows you how to convert the number correct to ACT scale scores. Other charts on pages 472–473 show you how to find the Pre-Algebra/Elementary Algebra, Intermediate Algebra/Coordinate Geometry, and Plane Geometry/Trigonometry subscores.

DO NOT leave any answers blank. There is no penalty for guessing on the ACT. Remember that the test is yours. You may mark up, write on, or draw on the test.

When you are ready, note the time and begin.

ANSWER SHEET

The ACT answer sheet looks something like this one. Use a No. 2 pencil
to completely fill the circle corresponding to the correct answer.
If you erase, erase completely; incomplete erasures may be read as answers.

TEST 1—English

1 Ⓐ Ⓑ Ⓒ Ⓓ	11 Ⓐ Ⓑ Ⓒ Ⓓ	21 Ⓐ Ⓑ Ⓒ Ⓓ	31 Ⓐ Ⓑ Ⓒ Ⓓ	41 Ⓐ Ⓑ Ⓒ Ⓓ	51 Ⓐ Ⓑ Ⓒ Ⓓ	61 Ⓐ Ⓑ Ⓒ Ⓓ	71 Ⓐ Ⓑ Ⓒ Ⓓ
2 Ⓕ Ⓖ Ⓗ Ⓙ	12 Ⓕ Ⓖ Ⓗ Ⓙ	22 Ⓕ Ⓖ Ⓗ Ⓙ	32 Ⓕ Ⓖ Ⓗ Ⓙ	42 Ⓕ Ⓖ Ⓗ Ⓙ	52 Ⓕ Ⓖ Ⓗ Ⓙ	62 Ⓕ Ⓖ Ⓗ Ⓙ	72 Ⓕ Ⓖ Ⓗ Ⓙ
3 Ⓐ Ⓑ Ⓒ Ⓓ	13 Ⓐ Ⓑ Ⓒ Ⓓ	23 Ⓐ Ⓑ Ⓒ Ⓓ	33 Ⓐ Ⓑ Ⓒ Ⓓ	43 Ⓐ Ⓑ Ⓒ Ⓓ	53 Ⓐ Ⓑ Ⓒ Ⓓ	63 Ⓐ Ⓑ Ⓒ Ⓓ	73 Ⓐ Ⓑ Ⓒ Ⓓ
4 Ⓕ Ⓖ Ⓗ Ⓙ	14 Ⓕ Ⓖ Ⓗ Ⓙ	24 Ⓕ Ⓖ Ⓗ Ⓙ	34 Ⓕ Ⓖ Ⓗ Ⓙ	44 Ⓕ Ⓖ Ⓗ Ⓙ	54 Ⓕ Ⓖ Ⓗ Ⓙ	64 Ⓕ Ⓖ Ⓗ Ⓙ	74 Ⓕ Ⓖ Ⓗ Ⓙ
5 Ⓐ Ⓑ Ⓒ Ⓓ	15 Ⓐ Ⓑ Ⓒ Ⓓ	25 Ⓐ Ⓑ Ⓒ Ⓓ	35 Ⓐ Ⓑ Ⓒ Ⓓ	45 Ⓐ Ⓑ Ⓒ Ⓓ	55 Ⓐ Ⓑ Ⓒ Ⓓ	65 Ⓐ Ⓑ Ⓒ Ⓓ	75 Ⓐ Ⓑ Ⓒ Ⓓ
6 Ⓕ Ⓖ Ⓗ Ⓙ	16 Ⓕ Ⓖ Ⓗ Ⓙ	26 Ⓕ Ⓖ Ⓗ Ⓙ	36 Ⓕ Ⓖ Ⓗ Ⓙ	46 Ⓕ Ⓖ Ⓗ Ⓙ	56 Ⓕ Ⓖ Ⓗ Ⓙ	66 Ⓕ Ⓖ Ⓗ Ⓙ	
7 Ⓐ Ⓑ Ⓒ Ⓓ	17 Ⓐ Ⓑ Ⓒ Ⓓ	27 Ⓐ Ⓑ Ⓒ Ⓓ	37 Ⓐ Ⓑ Ⓒ Ⓓ	47 Ⓐ Ⓑ Ⓒ Ⓓ	57 Ⓐ Ⓑ Ⓒ Ⓓ	67 Ⓐ Ⓑ Ⓒ Ⓓ	
8 Ⓕ Ⓖ Ⓗ Ⓙ	18 Ⓕ Ⓖ Ⓗ Ⓙ	28 Ⓕ Ⓖ Ⓗ Ⓙ	38 Ⓕ Ⓖ Ⓗ Ⓙ	48 Ⓕ Ⓖ Ⓗ Ⓙ	58 Ⓕ Ⓖ Ⓗ Ⓙ	68 Ⓕ Ⓖ Ⓗ Ⓙ	
9 Ⓐ Ⓑ Ⓒ Ⓓ	19 Ⓐ Ⓑ Ⓒ Ⓓ	29 Ⓐ Ⓑ Ⓒ Ⓓ	39 Ⓐ Ⓑ Ⓒ Ⓓ	49 Ⓐ Ⓑ Ⓒ Ⓓ	59 Ⓐ Ⓑ Ⓒ Ⓓ	69 Ⓐ Ⓑ Ⓒ Ⓓ	
10 Ⓕ Ⓖ Ⓗ Ⓙ	20 Ⓕ Ⓖ Ⓗ Ⓙ	30 Ⓕ Ⓖ Ⓗ Ⓙ	40 Ⓕ Ⓖ Ⓗ Ⓙ	50 Ⓕ Ⓖ Ⓗ Ⓙ	60 Ⓕ Ⓖ Ⓗ Ⓙ	70 Ⓕ Ⓖ Ⓗ Ⓙ	

TEST 2—Mathematics

1 Ⓐ Ⓑ Ⓒ Ⓓ Ⓔ	9 Ⓐ Ⓑ Ⓒ Ⓓ Ⓔ	17 Ⓐ Ⓑ Ⓒ Ⓓ Ⓔ	25 Ⓐ Ⓑ Ⓒ Ⓓ Ⓔ	33 Ⓐ Ⓑ Ⓒ Ⓓ Ⓔ	41 Ⓐ Ⓑ Ⓒ Ⓓ Ⓔ	49 Ⓐ Ⓑ Ⓒ Ⓓ Ⓔ	57 Ⓐ Ⓑ Ⓒ Ⓓ Ⓔ
2 Ⓕ Ⓖ Ⓗ Ⓙ Ⓚ	10 Ⓕ Ⓖ Ⓗ Ⓙ Ⓚ	18 Ⓕ Ⓖ Ⓗ Ⓙ Ⓚ	26 Ⓕ Ⓖ Ⓗ Ⓙ Ⓚ	34 Ⓕ Ⓖ Ⓗ Ⓙ Ⓚ	42 Ⓕ Ⓖ Ⓗ Ⓙ Ⓚ	50 Ⓕ Ⓖ Ⓗ Ⓙ Ⓚ	58 Ⓕ Ⓖ Ⓗ Ⓙ Ⓚ
3 Ⓐ Ⓑ Ⓒ Ⓓ Ⓔ	11 Ⓐ Ⓑ Ⓒ Ⓓ Ⓔ	19 Ⓐ Ⓑ Ⓒ Ⓓ Ⓔ	27 Ⓐ Ⓑ Ⓒ Ⓓ Ⓔ	35 Ⓐ Ⓑ Ⓒ Ⓓ Ⓔ	43 Ⓐ Ⓑ Ⓒ Ⓓ Ⓔ	51 Ⓐ Ⓑ Ⓒ Ⓓ Ⓔ	59 Ⓐ Ⓑ Ⓒ Ⓓ Ⓔ
4 Ⓕ Ⓖ Ⓗ Ⓙ Ⓚ	12 Ⓕ Ⓖ Ⓗ Ⓙ Ⓚ	20 Ⓕ Ⓖ Ⓗ Ⓙ Ⓚ	28 Ⓕ Ⓖ Ⓗ Ⓙ Ⓚ	36 Ⓕ Ⓖ Ⓗ Ⓙ Ⓚ	44 Ⓕ Ⓖ Ⓗ Ⓙ Ⓚ	52 Ⓕ Ⓖ Ⓗ Ⓙ Ⓚ	60 Ⓕ Ⓖ Ⓗ Ⓙ Ⓚ
5 Ⓐ Ⓑ Ⓒ Ⓓ Ⓔ	13 Ⓐ Ⓑ Ⓒ Ⓓ Ⓔ	21 Ⓐ Ⓑ Ⓒ Ⓓ Ⓔ	29 Ⓐ Ⓑ Ⓒ Ⓓ Ⓔ	37 Ⓐ Ⓑ Ⓒ Ⓓ Ⓔ	45 Ⓐ Ⓑ Ⓒ Ⓓ Ⓔ	53 Ⓐ Ⓑ Ⓒ Ⓓ Ⓔ	
6 Ⓕ Ⓖ Ⓗ Ⓙ Ⓚ	14 Ⓕ Ⓖ Ⓗ Ⓙ Ⓚ	22 Ⓕ Ⓖ Ⓗ Ⓙ Ⓚ	30 Ⓕ Ⓖ Ⓗ Ⓙ Ⓚ	38 Ⓕ Ⓖ Ⓗ Ⓙ Ⓚ	46 Ⓕ Ⓖ Ⓗ Ⓙ Ⓚ	54 Ⓕ Ⓖ Ⓗ Ⓙ Ⓚ	
7 Ⓐ Ⓑ Ⓒ Ⓓ Ⓔ	15 Ⓐ Ⓑ Ⓒ Ⓓ Ⓔ	23 Ⓐ Ⓑ Ⓒ Ⓓ Ⓔ	31 Ⓐ Ⓑ Ⓒ Ⓓ Ⓔ	39 Ⓐ Ⓑ Ⓒ Ⓓ Ⓔ	47 Ⓐ Ⓑ Ⓒ Ⓓ Ⓔ	55 Ⓐ Ⓑ Ⓒ Ⓓ Ⓔ	
8 Ⓕ Ⓖ Ⓗ Ⓙ Ⓚ	16 Ⓕ Ⓖ Ⓗ Ⓙ Ⓚ	24 Ⓕ Ⓖ Ⓗ Ⓙ Ⓚ	32 Ⓕ Ⓖ Ⓗ Ⓙ Ⓚ	40 Ⓕ Ⓖ Ⓗ Ⓙ Ⓚ	48 Ⓕ Ⓖ Ⓗ Ⓙ Ⓚ	56 Ⓕ Ⓖ Ⓗ Ⓙ Ⓚ	

TEST 3—Reading

1 Ⓐ Ⓑ Ⓒ Ⓓ	6 Ⓕ Ⓖ Ⓗ Ⓙ	11 Ⓐ Ⓑ Ⓒ Ⓓ	16 Ⓕ Ⓖ Ⓗ Ⓙ	21 Ⓐ Ⓑ Ⓒ Ⓓ	26 Ⓕ Ⓖ Ⓗ Ⓙ	31 Ⓐ Ⓑ Ⓒ Ⓓ	36 Ⓕ Ⓖ Ⓗ Ⓙ
2 Ⓕ Ⓖ Ⓗ Ⓙ	7 Ⓐ Ⓑ Ⓒ Ⓓ	12 Ⓕ Ⓖ Ⓗ Ⓙ	17 Ⓐ Ⓑ Ⓒ Ⓓ	22 Ⓕ Ⓖ Ⓗ Ⓙ	27 Ⓐ Ⓑ Ⓒ Ⓓ	32 Ⓕ Ⓖ Ⓗ Ⓙ	37 Ⓐ Ⓑ Ⓒ Ⓓ
3 Ⓐ Ⓑ Ⓒ Ⓓ	8 Ⓕ Ⓖ Ⓗ Ⓙ	13 Ⓐ Ⓑ Ⓒ Ⓓ	18 Ⓕ Ⓖ Ⓗ Ⓙ	23 Ⓐ Ⓑ Ⓒ Ⓓ	28 Ⓕ Ⓖ Ⓗ Ⓙ	33 Ⓐ Ⓑ Ⓒ Ⓓ	38 Ⓕ Ⓖ Ⓗ Ⓙ
4 Ⓕ Ⓖ Ⓗ Ⓙ	9 Ⓐ Ⓑ Ⓒ Ⓓ	14 Ⓕ Ⓖ Ⓗ Ⓙ	19 Ⓐ Ⓑ Ⓒ Ⓓ	24 Ⓕ Ⓖ Ⓗ Ⓙ	29 Ⓐ Ⓑ Ⓒ Ⓓ	34 Ⓕ Ⓖ Ⓗ Ⓙ	39 Ⓐ Ⓑ Ⓒ Ⓓ
5 Ⓐ Ⓑ Ⓒ Ⓓ	10 Ⓕ Ⓖ Ⓗ Ⓙ	15 Ⓐ Ⓑ Ⓒ Ⓓ	20 Ⓕ Ⓖ Ⓗ Ⓙ	25 Ⓐ Ⓑ Ⓒ Ⓓ	30 Ⓕ Ⓖ Ⓗ Ⓙ	35 Ⓐ Ⓑ Ⓒ Ⓓ	40 Ⓕ Ⓖ Ⓗ Ⓙ

TEST 4—Science Reasoning

1 Ⓐ Ⓑ Ⓒ Ⓓ	6 Ⓕ Ⓖ Ⓗ Ⓙ	11 Ⓐ Ⓑ Ⓒ Ⓓ	16 Ⓕ Ⓖ Ⓗ Ⓙ	21 Ⓐ Ⓑ Ⓒ Ⓓ	26 Ⓕ Ⓖ Ⓗ Ⓙ	31 Ⓐ Ⓑ Ⓒ Ⓓ	36 Ⓕ Ⓖ Ⓗ Ⓙ
2 Ⓕ Ⓖ Ⓗ Ⓙ	7 Ⓐ Ⓑ Ⓒ Ⓓ	12 Ⓕ Ⓖ Ⓗ Ⓙ	17 Ⓐ Ⓑ Ⓒ Ⓓ	22 Ⓕ Ⓖ Ⓗ Ⓙ	27 Ⓐ Ⓑ Ⓒ Ⓓ	32 Ⓕ Ⓖ Ⓗ Ⓙ	37 Ⓐ Ⓑ Ⓒ Ⓓ
3 Ⓐ Ⓑ Ⓒ Ⓓ	8 Ⓕ Ⓖ Ⓗ Ⓙ	13 Ⓐ Ⓑ Ⓒ Ⓓ	18 Ⓕ Ⓖ Ⓗ Ⓙ	23 Ⓐ Ⓑ Ⓒ Ⓓ	28 Ⓕ Ⓖ Ⓗ Ⓙ	33 Ⓐ Ⓑ Ⓒ Ⓓ	38 Ⓕ Ⓖ Ⓗ Ⓙ
4 Ⓕ Ⓖ Ⓗ Ⓙ	9 Ⓐ Ⓑ Ⓒ Ⓓ	14 Ⓕ Ⓖ Ⓗ Ⓙ	19 Ⓐ Ⓑ Ⓒ Ⓓ	24 Ⓕ Ⓖ Ⓗ Ⓙ	29 Ⓐ Ⓑ Ⓒ Ⓓ	34 Ⓕ Ⓖ Ⓗ Ⓙ	39 Ⓐ Ⓑ Ⓒ Ⓓ
5 Ⓐ Ⓑ Ⓒ Ⓓ	10 Ⓕ Ⓖ Ⓗ Ⓙ	15 Ⓐ Ⓑ Ⓒ Ⓓ	20 Ⓕ Ⓖ Ⓗ Ⓙ	25 Ⓐ Ⓑ Ⓒ Ⓓ	30 Ⓕ Ⓖ Ⓗ Ⓙ	35 Ⓐ Ⓑ Ⓒ Ⓓ	40 Ⓕ Ⓖ Ⓗ Ⓙ

Model Mathematics **ACT IV**

60 Questions—60 Minutes

INSTRUCTIONS: Find the solution to each problem, choose the correct answer choice, then darken the appropriate oval on your answer sheet. Do not spend too much time on any one problem. Answer as many problems as you can and then work on the remaining problems within the time limit for this test. Check pages 458–463 for answers and explanations.

Unless the problem indicates otherwise:

- figures are NOT necessarily drawn to scale
- geometric figures are plane figures
- a *line* is a straight line
- an *average* is the arithmetic mean

You may use a calculator.

---

1. 35, 84, and 119 are all multiples of what number?

   A. 4
   B. 5
   C. 6
   D. 7
   E. 8

2. A circle has a radius of 6 inches. What is the length of the diameter, in feet?

   F. $\frac{1}{2}$ foot

   G. 1 foot

   H. $\frac{1}{2}\pi$ feet

   J. π feet

   K. 6π feet

3. Which of the following inequalities creates the graph shown below?

   A. $-2 > x > 2$
   B. $-2 \leq x < 2$
   C. $-2 < x \leq 2$
   D. $2 \leq x < -2$
   E. $2 < x \leq -2$

4. Which of the following is equivalent to $(5x^2y^4)(3xy)^2$?

   F. $45x^2y^4$
   G. $45x^4y^6$
   H. $15x^4y^6$
   J. $15x^2y^4$
   K. $45x^2y^6$

GO ON TO THE NEXT PAGE.

5. In the figure below ∠x and ∠y are congruent. What is the measure of ∠x?

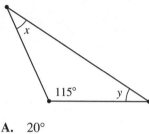

A. 20°
B. 30°
C. 32.5°
D. 40.5°
E. 65°

6. 35 is what percent of 140?

F. 75%
G. 60%
H. 55%
J. 30%
K. 25%

7. Stacey is traveling cross-country and wants to know how far she will have to travel. The map key says that every inch represents 60 miles. If Stacey measures 50 inches on the map, how many miles will the trip be?

A. 2,000 miles
B. 2,500 miles
C. 3,000 miles
D. 3,500 miles
E. 4,000 miles

8. What is the 10th term in the sequence of "gnome" numbers shown below?

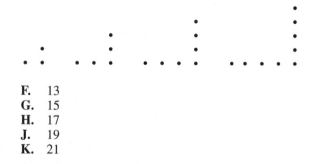

F. 13
G. 15
H. 17
J. 19
K. 21

9. In the figure below the measure of angle x is 20° less than the measure of angle y. What is the measure of angle y?

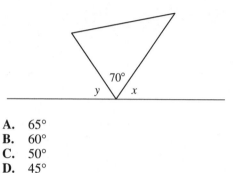

A. 65°
B. 60°
C. 50°
D. 45°
E. 40°

10. Dennis weighed 175 pounds. He lost 12% of this weight in a month. How much did he weigh at the end of the month?

F. 170 pounds
G. 165 pounds
H. 163 pounds
J. 154 pounds
K. 150 pounds

11. If $a = 2$ and $b = 4$, then what is the value of $\frac{(a + b)^2(a^2 - b^2)}{a^2} \times b^2$?

A. 6,912
B. 1,728
C. 864
D. −1,728
E. −6,912

12. There are 4 brown socks, 7 blue socks, and 2 green socks in a drawer. What is the probability that a sock randomly picked from the drawer will be blue?

F. $\frac{2}{13}$
G. $\frac{4}{13}$
H. $\frac{7}{13}$
J. $\frac{13}{7}$
K. $\frac{13}{4}$

GO ON TO THE NEXT PAGE.

13. What is the value of $\frac{3^3 - 4^2}{2^4 - 1^8}$?

   A. $\frac{11}{8}$

   B. $\frac{15}{11}$

   C. $\frac{11}{15}$

   D. $\frac{8}{11}$

   E. $\frac{1}{7}$

14. If $x = 2$, $y = 3$, and $z = -1$, what is the value of $\frac{xy - xyz}{yz - xy}$?

   F. $1\frac{3}{4}$

   G. 0

   H. $-\frac{1}{3}$

   J. $-\frac{3}{4}$

   K. $-1\frac{1}{3}$

15. The triangle below is a 30–60–90 triangle. If $AB = 4\sqrt{3}$, what is the length of BC?

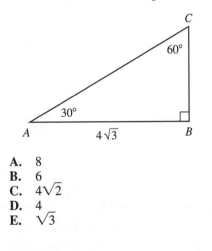

   A. 8
   B. 6
   C. $4\sqrt{2}$
   D. 4
   E. $\sqrt{3}$

16. For all a and b, what is the sum of $3a^3b^2 - 2ab^2$ and $2a^3b^2 + ab^2$?

   F. $a^3b^2 + 3ab^2$
   G. $5a^6b^4 - 3a^4b^4$
   H. $5a^3b^2 - ab^2$
   J. $a^3b^2 - 3ab^2$
   K. $a^4b^2 + 3a^4b^4$

17. The sum of Chad's age (C) and Roberta's age (R) is 51. If Chad is 5 years older than Roberta, which of the following equations can be solved to find Roberta's age?

   A. $R + (R - 5) = 51$
   B. $C(C - 5) = 51$
   C. $2R + 5 = 51$
   D. $R + C = 51$
   E. $C - R = 51$

18. In the figure below $\overleftrightarrow{AB} \parallel \overleftrightarrow{CD}$. What is the measure of $\angle ACB$?

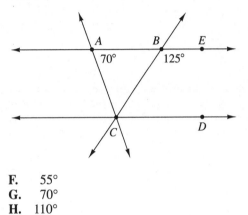

   F. $55°$
   G. $70°$
   H. $110°$
   J. $125°$
   K. $180°$

19. What is the sum of the solutions to $(x - 3)(x + 2) = 0$?

   A. -2
   B. -1
   C. 1
   D. 3
   E. 5

20. What is the probability of being dealt a face card, of which there are 12, from a standard deck of 52 cards?

   F. $\frac{1}{26}$

   G. $\frac{1}{13}$

   H. $\frac{1}{4}$

   J. $\frac{3}{13}$

   K. $\frac{3}{8}$

GO ON TO THE NEXT PAGE.

21. Which of the following is the simplified form of $2x^2 - (x^2 - 3x) + 4x$?

   A. $x^2 + 7x$

   B. $3x^2 - x$

   C. $x^2 - x$

   D. $3x^2 + 7x$

   E. $3x^2 - 7x$

22. If $|x - 3| = 1$, what is the sum of all the solutions for x?

   F. -6

   G. -2

   H. 2

   J. 4

   K. 6

23. The circle graph below shows the number of A's, B's, C's, D's, and F's that were earned on an English test. What is the sum of $\angle\theta$ and $\angle\alpha$?

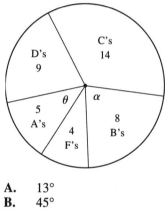

   A. $13°$

   B. $45°$

   C. $72°$

   D. $85°$

   E. $117°$

24. What point on the graph of $y = x^3 - 2x + 1$, has an x-coordinate of 5?

   F. $(5,112)$

   G. $(5,114)$

   H. $(5,116)$

   J. $(5,118)$

   K. $(5,120)$

25. What is the slope of the line with the equation $2x - 4y = 8$?

   A. -2

   B. $\dfrac{1}{4}$

   C. $\dfrac{1}{2}$

   D. 4

   E. 8

26. The center of a line segment on a number line is 6. If the line segment has a length of 8, what is the sum of the endpoints of the line segment?

   F. 2

   G. 6

   H. 10

   J. 12

   K. 22

27. A clothing store is having a sale with 30% off all shirts. If a person buys a shirt at the sale for $14, what was the original price of the shirt?

   A. $16

   B. $17

   C. $18

   D. $19

   E. $20

28. What is the total area of the figure shown below?

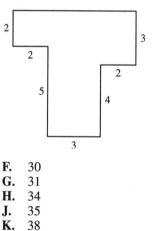

   F. 30

   G. 31

   H. 34

   J. 35

   K. 38

29. What is the value of $\tan\theta$ in the right triangle shown below?

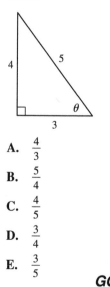

   A. $\dfrac{4}{3}$

   B. $\dfrac{5}{4}$

   C. $\dfrac{4}{5}$

   D. $\dfrac{3}{4}$

   E. $\dfrac{3}{5}$

GO ON TO THE NEXT PAGE.

30. In the right triangle below, what is the length of side b?

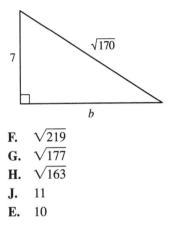

F. $\sqrt{219}$

G. $\sqrt{177}$

H. $\sqrt{163}$

J. 11

E. 10

31. A computer store sells two different types of computers, A and B. Computer A costs \$2,000 and Computer B costs \$2,200. If a total of 27 computers are sold and \$56,800 is collected from these sales, how many type A computers are sold?

A. 11

B. 12

C. 13

D. 14

E. 15

32. In the figure below, what is the length of AD?

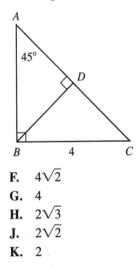

F. $4\sqrt{2}$

G. 4

H. $2\sqrt{3}$

J. $2\sqrt{2}$

K. 2

33. In the figure below $\triangle ABC$ is similar to $\triangle DEF$. What is the length of DE?

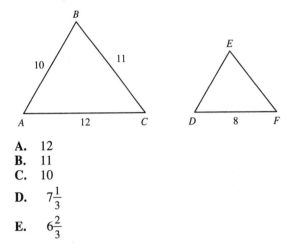

A. 12

B. 11

C. 10

D. $7\frac{1}{3}$

E. $6\frac{2}{3}$

34. What is the area, in square units, of $\triangle ABC$ shown below?

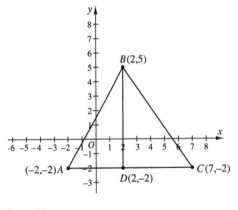

F. 63

G. 59

H. 55

J. 31.5

K. 24.5

35. What is the slope of the line having the equation $y = \frac{3}{2}x - \frac{4}{5}$?

A. $\frac{3}{2}$

B. $\frac{4}{5}$

C. $\frac{2}{3}$

D. $-\frac{4}{5}$

E. $-\frac{5}{4}$

GO ON TO THE NEXT PAGE.

36. Which one of the following number lines contains the solutions for the equation $\dfrac{-4 + x}{3} = \dfrac{2 - x}{x}$?

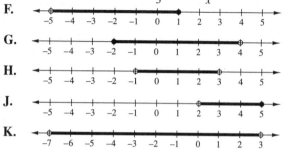

F.

G.

H.

J.

K.

37. If $\triangle ABC$ is similar to $\triangle DEF$ in the diagram below, then $m\angle D = $?

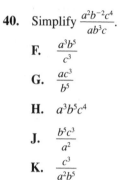

A. 80°
B. 60°
C. 40°
D. 30°
E. 10°

38. Which of the following equations forms the graph below?

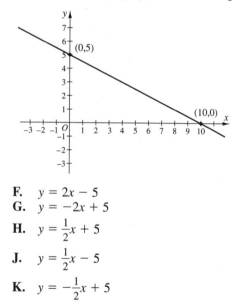

F. $y = 2x - 5$
G. $y = -2x + 5$
H. $y = \dfrac{1}{2}x + 5$
J. $y = \dfrac{1}{2}x - 5$
K. $y = -\dfrac{1}{2}x + 5$

39. What is the smallest number in the solution set to $3x + 5 \le 5x + 1$?

A. $\{-2\}$
B. $\{-1\}$
C. $\{0\}$
D. $\{1\}$
E. $\{2\}$

40. Simplify $\dfrac{a^2b^{-2}c^4}{ab^3c}$.

F. $\dfrac{a^3b^5}{c^3}$

G. $\dfrac{ac^3}{b^5}$

H. $a^3b^5c^4$

J. $\dfrac{b^5c^3}{a^2}$

K. $\dfrac{c^3}{a^2b^5}$

41. In the figure below, each side of the square is tangent to the circle at the midpoint of the side. What is the area of the circle?

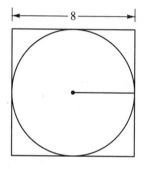

A. 4π
B. 8π
C. 12π
D. 16π
E. 32π

42. A person buys five different types of fruit at the grocery store. Below is a list of the amount and type of fruit and the price.

Fruit	Price
3 peaches	$1.44
7 plums	$2.24
4 apples	$1.48
6 nectarines	$1.74
5 bananas	$2.35

Which type of fruit is the least expensive per piece?

F. peaches
G. plums
H. apples
J. nectarines
K. bananas

GO ON TO THE NEXT PAGE.

43. For which real values of x is the following expression defined?

$$\frac{\sqrt{3y} - 9}{\sqrt{25} + x}$$

A. All real values of x

B. All real values of x except 3

C. All real values of x except -3 and 3

D. All real values of x except -5

E. All real values of x except -5 and 5

44. A triangle has a base that is $\frac{1}{2}$ its height, and an area of 36. What is the sum of the base and the height?

F. 10
G. 12.5
H. 15
J. 18
K. 20

45. Which of the following choices is the complete factored form of $3x^2y^3 + 12x^5y^2 - 6x^3y^3$?

A. $3x^2y^2(y + 4x^3 - 2xy)$
B. $x^3y^3(3 + 4x^2 - 2xy)$
C. $3xy(xy^2 + 4x^4y - 2x^2y^2)$
D. $3x^3y^3(1 + 4x^2 - 2xy)$
E. $x^2y^2(3y + 12x^3 - 6xy)$

46. The volume of a sphere is given by the formula $V = \frac{4}{3}\pi r^3$. Which of the following choices expresses r in terms of V?

F. $r = \left(V \times \frac{4}{3}\right) \div \pi$

G. $r = \sqrt[3]{V \times \frac{4}{3} \times \pi}$

H. $r = \sqrt[3]{\left(V \times \frac{3}{4}\right) \div \pi}$

J. $r = V \times \frac{4}{3} \times \pi$

K. $r = \sqrt{\left(V \times \frac{3}{4}\right) \div \pi}$

47. If $\frac{a}{6} - \frac{a}{9} = \frac{2}{3}$, then $a = ?$

A. 9
B. 10
C. 11
D. 12
E. 13

48. The roof of the house shown below is a right triangle. The front of the house forms a square. Which of the following is the correct relationship among angles and sides in the figure?

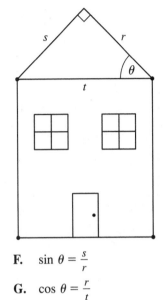

F. $\sin \theta = \frac{s}{r}$

G. $\cos \theta = \frac{r}{t}$

H. $\tan \theta = \frac{t}{r}$

J. $\cot \theta = \frac{r}{t}$

K. $\sec \theta = \frac{t}{s}$

49. What are the center and the radius of the circle having the equation $(x + 2)^2 + (y - 5)^2 = 81$?

A. $c = (2,5)$ and $r = 81$
B. $c = (-2,-5)$ and $r = 9$
C. $c = (-2,5)$ and $r = 9$
D. $c = (-2,5)$ and $r = 81$
E. $c = (2,-5)$ and $r = 9$

50. If y is a real number, what is the smallest possible value of x in the equation $y = \sqrt{4 - x^2}$?

F. -6
G. -5
H. -4
J. -3
K. -2

51. $(2.36 \times 10^2) + (4.31 \times 10^3) = ?$

A. 4.546×10^2
B. 6.67×10^2
C. 4.546×10^3
D. 4.546×10^5
E. 6.67×10^5

GO ON TO THE NEXT PAGE.

52. $\left(\dfrac{1}{\sec 2\theta}\right)\left(\cos^2\theta + \dfrac{1}{\csc^2\theta}\right) = ?$

    F. $1 - \tan\theta$

    G. $2\sin\dfrac{\theta}{2} \times \cos\dfrac{\theta}{2}$

    H. $\csc 2\theta - \cot\theta$

    J. $2\sin\theta \times \cos\theta$

    K. $1 - 2\sin^2\theta$

53. Which of the following choices represents the distance (d) from the point 4 to a point (x) on a number line?

    A. $|x + 4| = d$
    B. $x - 4 = d$
    C. $|x - 4| = d$
    D. $(x - 4)^2 = d$
    E. $(x + 4)^2 = d$

54. What is the slope of the line containing the points (3,6) and (5,2)?

    F. -2
    G. -1
    H. $-\dfrac{1}{2}$
    J. 1
    K. 2

55. What type of figure is formed by connecting the points $A(2,6)$, $B(-2,3)$, $C(7,3)$, and $D(4,6)$ on a coordinate plane?

    A. Rhombus
    B. Parallelogram
    C. Square
    D. Trapezoid
    E. Rectangle

56. What is the sum of the factors of the polynomial $x^2 + 2x - 15$?

    F. $x + 2$
    G. $2x + 2$
    H. $2x + 5$
    J. $x - 3$
    K. $2x + 3$

57. The supplement of angle θ is between 0° and 90°. Which of the following could be the values of the functions $\cos\theta$ and $\sec\theta$?

    A. $+\dfrac{1}{2}, +2$

    B. $+\dfrac{1}{2}, -2$

    C. $-\dfrac{1}{2}, +2$

    D. $-\dfrac{1}{2}, -2$

    E. $+2, -\dfrac{1}{2}$

58. Two cars leave point A and travel to point C. Car 1 travels in a straight line to point C. Car 2 travels in a straight line from point A to point B, a distance of 9 kilometers. From point B, Car 2 turns 90° and goes in a straight line to point C, a distance of 12 kilometers. How far did Car 1 travel?

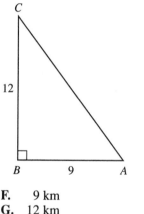

    F. 9 km
    G. 12 km
    H. 15 km
    J. 17 km
    K. 20 km

59. What is the maximum number of zeros that exist for the function $f(x) = 5x^5 - 4x^3 + 2x^2 - x + 11$? (Zeros are the values of x where $f(x) = 0$.)

    A. 6
    B. 5
    C. 3
    D. 2
    E. 1

GO ON TO THE NEXT PAGE.

60. The sales of a new product increased at a constant rate for 6 months. Then the sales leveled off for 4 months. After that the sales decreased for 6 months at the same constant rate as they had increased. The graph of the sales of the new item most resembles which of the following graphs?

F.

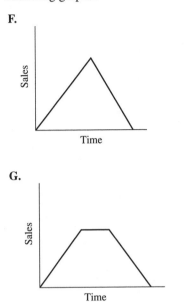

G.

H.

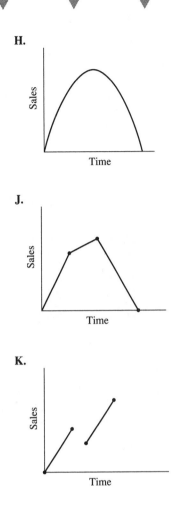

J.

K.

END OF TEST 2

✔ *ANSWERS*

1.	D	16.	H	31.	C	46.	H
2.	G	17.	C	32.	J	47.	D
3.	C	18.	F	33.	E	48.	G
4.	G	19.	C	34.	J	49.	C
5.	C	20.	J	35.	A	50.	K
6.	K	21.	A	36.	G	51.	C
7.	C	22.	K	37.	B	52.	K
8.	K	23.	E	38.	K	53.	C
9.	A	24.	H	39.	E	54.	F
10.	J	25.	C	40.	G	55.	D
11.	D	26.	J	41.	D	56.	G
12.	H	27.	E	42.	J	57.	D
13.	C	28.	G	43.	D	58.	H
14.	K	29.	A	44.	J	59.	B
15.	D	30.	J	45.	A	60.	G

ACT IV MATHEMATICS Answers Explained

1. **D.** $7 \times 5 = 35$ $7 \times 12 = 84$ $7 \times 17 = 119$

2. **G.** Find the length of the diameter in inches.
   Radius (r) = 6 inches
   Diameter (d) = $2r$
   $d = 2 \times 6$ inches
   $d = 12$ inches
   Find the length of the diameter in feet.
   12 inches = 1 foot, therefore d = 1 foot

3. **C.** The graph shows all the numbers greater than -2 and less than or equal to 2. $-2 < x \leq 2$.

4. **G.**

$(5x^2y^4)(3xy)^2$	Write the expression.
$(5x^2y^4)(9x^2y^2)$	Square the second factor.
$45x^4y^6$	Multiply the two factors.

5. **C.** $m\angle x + m\angle y + 115° = 180°$ because the sum of the angle measures of a triangle is 180°.

$m\angle x = m\angle y$ because $\angle x \cong \angle y$	
$m\angle x + m\angle x + 115° = 180°$	Substitute.
$2m\angle x + 115° = 180°$	Add.
$2m\angle x = 65°$	Subtract 115°.
$m\angle x = 32.5°$	Divide by 2.

6. **K.**

$35 = 140x$, where x is a percent in decimal form.	
$35 \div 140 = x$	Divide by 140.
$0.25 = x = 25\%$	

7. **C.** $\frac{60 \text{ miles}}{1 \text{ inch}} \times 50$ inches = 3,000 miles

8. **K.** The gnome numbers are odd numbers starting at 3.

order	1	2	3	4	5	6	7	8	9	10
gnome	3	5	7	9	11	13	15	17	19	21

   The 10th "gnome" number is 21.

9. **A.**

$m\angle x = m\angle y - 20°$	
$m\angle x + m\angle y + 70° = 180°$ because they form a straight angle.	
$m\angle y - 20° + m\angle y + 70° = 180°$	Substitute.
$2m\angle y + 50° = 180°$	
$2m\angle y = 130°$	Subtract 50°.
$m\angle y = 65°$	Divide by 2.

10. **J.** Find out how much weight Dennis lost. Let x represent the number of pounds lost.

$x = 175$ pounds $\times 0.12$

$x = 21$ pounds

Find Dennis's new weight.

175 pounds $-$ 21 pounds $=$ 154 pounds

11. **D.**

$a = 2$ and $b = 4$

$\dfrac{(a + b)^2(a^2 - b^2)}{a^2} \times b^2$

$= \dfrac{(2 + 4)^2(2^2 - 4^2)}{2^2} \times 4^2$ Substitute.

$= \dfrac{(6)^2(4 - 16)}{4} \times 16$

$= \dfrac{36 \times (-12)}{4} \times 16$

$= \dfrac{-6{,}912}{4} = -1{,}728$

12. **H.**

$P(\text{Event}) = \dfrac{\text{the number of times the event could occur}}{\text{the total number of events that could occur}}$

$P(\text{blue sock}) = \dfrac{7}{13}$, because there are 7 blue socks and a total of 13 socks.

13. **C.** $\dfrac{3^3 - 4^2}{2^4 - 1^8} = \dfrac{27 - 16}{16 - 1} = \dfrac{11}{15}$

14. **K.**

$x = 2 \quad y = 3 \quad z = -1$

$\dfrac{xy - xyz}{yz - xy}$

$= \dfrac{(2 \times 3) - [2 \times 3 \times (-1)]}{[3 \times (-1)] - (2 \times 3)}$

$= \dfrac{6 - (-6)}{-3 - 6}$ Substitute.

$= -\dfrac{12}{9}$

$= -1\dfrac{1}{3}$

15. **D.** The sides of a 30–60–90 triangle have the relationships shown in the diagram below.

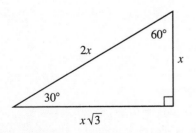

Therefore $BC = 4$.

16. **H.**

$$\begin{array}{r} 3a^3b^2 - 2ab^2 \\ + \ 2a^3b^2 + \ ab^2 \\ \hline 5a^3b^2 - \ ab^2 \end{array}$$

17. **C.** Write an equation for each condition in the problem.

The sum of Chad's age and Roberta's age is 51: $C + R = 51$

Chad is 5 years older than Roberta: $C = 5 + R$

Substitute $5 + R$ for C in the first equation. Then combine like terms.

$C + R = 51$

$(5 + R) + R = 51$

$5 + 2R = 51$, or $2R + 5 = 51$

18. **F.** ABC forms a triangle. First find the measure of $\angle ABC$. This angle is supplementary to $\angle EBC$, so its measure is $180° - m\angle EBC$. $180° - 125° = 55°$.

Now find the measure of $\angle ACB$. The sum of the measures of the angles in a triangle is $180°$.

$70° + 55° + m\angle ACB = 180°$

$125° + m\angle ACB = 180°$

$m\angle ACB = 55°$

19. **C.**

$(x - 3)(x + 2) = 0$

$x - 3 = 0 \quad$ or $\quad x + 2 = 0$

$x = 3 \qquad\qquad x = -2$

Add the solutions. $3 + (-2) = 1$

20. **J.**

$P(\text{Event}) = \dfrac{\text{the number of times the event could occur}}{\text{the total number of events that could occur}}$

$P(\text{Face Card}) = \dfrac{\text{number of face cards}}{\text{total number of cards}} = \dfrac{12}{52} = \dfrac{3}{13}$

21. **A.** $2x^2 - (x^2 - 3x) + 4x = 2x^2 - x^2 + 3x + 4x$

$= x^2 + 7x$

22. **K.** $|x - 3|$ can be rewritten as two separate equations.

$x - 3 = 1 \qquad -(x - 3) = 1$

$x = 4 \qquad\quad x - 3 = -1$

$x = 2$

Add the solutions. $4 + 2 = 6$

23. **E.** Find the measure of $\angle \theta$.

There were a total of 40 scores, 5 of which were A's. Therefore, $\dfrac{5}{40} = 0.125$ or 12.5% of the scores were A's.

$m\angle \theta = 0.125 \times 360° = 45°$

Find the measure of $\angle \alpha$.

There were a total of 40 scores, 8 of which were B's. Therefore, $\dfrac{8}{40} = 0.2$ or 20% of the scores were B's.

$m\angle \alpha = 0.2 \times 360° = 72°$

Find $m\angle \theta + m\angle \alpha$.

$45° + 72° = 117°$

24. H.

$x = 5$

$y = x^3 - 2x + 1$

$y = 5^3 - 2(5) + 1$ Substitute.

$y = 125 - 10 + 1 = 116$

The coordinates are (5,116).

25. C. Write the equation in slope-intercept form, $y = mx + b$, where m = slope.

$2x - 4y = 8$

$\quad -4y = -2x + 8$ Subtract $2x$.

$\qquad y = \frac{1}{2}x - 2$ Divide by -4.

26. J. Since the length of the line segment is 8, the length from the midpoint to each endpoint is 4.

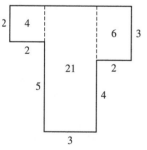

$6 - 4 = 2$ $6 + 4 = 10$

The endpoints are 2 and 10.

Add the endpoints. $2 + 10 = 12$

27. E. Since the shirt was 30% off the original price, the sale price is 70% of the original price.

x = original price of shirt

$x \times 0.7 = \$14$

$x = \dfrac{\$14}{0.7} = \$20$

28. G. Split the figure up into rectangles and find the area of each.

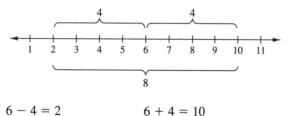

Add the areas. $4 + 21 + 6 = 31$

29. A. $\tan \theta = \dfrac{\text{opposite}}{\text{adjacent}} = \dfrac{4}{3}$

30. J. Use the Pythagorean Theorem.

$a = 7 \quad b = ? \quad c = \sqrt{170}$

$a^2 + b^2 = c^2$

$7^2 + b^2 = (\sqrt{170})^2$ Substitute.

$49 + b^2 = 170$

$\quad b^2 = 121$

$\quad b = \sqrt{121}$

$\quad b = 11$

31. C. Write two equations.

$$A + B = 27$$
$$2{,}000A + 2{,}200B = 56{,}800$$
$$B = 27 - A \quad \text{Solve for B in the first equation.}$$

$2{,}000A + 2{,}200(27 - A) = 56{,}800$ Substitute in the second equation.

$2{,}000A + 59{,}400 - 2{,}200A = 56{,}800$ Solve for A.

$\qquad -200A = -2{,}600$

$\qquad\qquad A = 13$

Alternative solution:

Multiply the equation $A + B = 27$ by $-\$2{,}200$.

$$-\$2{,}200A - \$2{,}200B = -\$59{,}400$$

Write as simultaneous equations and solve.

$-\$2{,}200A - \$2{,}200B = -\$59{,}400$
$+ \ \$2{,}000A + \$2{,}200B = \ \ \$56{,}800$

Add. $-\$200A = -\$2{,}600$

Divide by $-\$200$. $A = 13$

32. J.

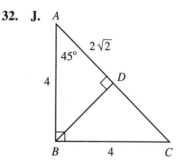

Find the length of AB.

$\triangle ABC$ is an isosceles right triangle, therefore the legs have equal measures.

$AB = BC = 4$

Find the length of AD.

$\triangle ADB$ is an isosceles right triangle, where $\overline{AB}$ is the hypotenuse and $\overline{AD}$ and $\overline{BD}$ are the legs. As mentioned above, the legs of an isosceles right triangle are equal, so $AD = BD$.

Use the Pythagorean Theorem.

$(AD)^2 + (BD)^2 = (AB)^2$

$(AD)^2 + (AD)^2 = (AB)^2$ Substitute AD for BD.

$\qquad 2(AD)^2 = 16$

$\qquad\quad (AD)^2 = 8$

$\qquad\qquad AD = \sqrt{8} = 2\sqrt{2}$

Alternative solution:

As stated above, $AB = BC = 4$ and $\angle A = 45°$.

$\triangle ABD$ is an isosceles right triangle. Therefore, each leg is equal to one-half of the hypotenuse times $\sqrt{2}$.

$AD = 2\sqrt{2}$

33. **E.**

$AB = 10 \qquad AC = 12 \qquad DF = 8$

The ratios of corresponding sides of similar triangles are proportional.

$\dfrac{DF}{AC} = \dfrac{DE}{AB}$ \qquad Write a proportion.

$\dfrac{8}{12} = \dfrac{DE}{10}$ \qquad Substitute known values.

$12 \times DE = 10 \times 8$ \qquad Cross multiply.

$DE = \dfrac{80}{12} = 6\dfrac{2}{3}$

34. **J.** Find the height (h) of the triangle.

$\overline{DB}$ is the height of the triangle. Since the x-values are the same at the two points, the height is the difference in y-values: $h = |5 - (-2)| = |5 + 2| = |7| = 7$

Find the base (b) of the triangle.

$\overline{AC}$ is the base of the triangle. Since the y-values are the same at the two points, the length of the base is the difference in x-values: $b = |-2 - 7| = |-9| = 9$

Find the area of the triangle.

$A = \dfrac{1}{2}bh$

$= \dfrac{1}{2} \times 9 \times 7$

$= \dfrac{1}{2} \times 63$

$= 31.5$

35. **A.** The slope-intercept form is $y = mx + b$, where m is the slope and b is the y-intercept.

In the equation $y = \dfrac{3}{2}x - \dfrac{4}{5}$, $m = \dfrac{3}{2}$.

36. **G.** $\dfrac{-4 + x}{3} = \dfrac{2 - x}{x}$

$-4x + x^2 = 6 - 3x$ \qquad Cross multiply.

$x^2 - x - 6 = 0$ \qquad Simplify.

$(x - 3)(x + 2) = 0$ \qquad Factor.

$x - 3 = 0 \quad$ or $\quad x + 2 = 0$

$x = 3 \qquad\qquad x = -2$

$x = \{-2, 3\}$

The graph shown in choice G contains both points -2 and 3.

37. **B.** Find the measure of $\angle A$.

$m\angle B = 80° \qquad\qquad m\angle C = 40°$

$m\angle A + m\angle B + m\angle C = 180°$ because the sum of the angles of a triangle is 180°.

$m\angle A + 80° + 40° = 180°$ \qquad Substitute.

$m\angle A + 120° = 180°$

$m\angle A = 60°$

Find the measure of $\angle D$.

$\triangle ABC$ and $\triangle DEF$ are similar, so corresponding angles have the same measure.

$\angle A$ and $\angle D$ are corresponding angles, therefore, $m\angle D = m\angle A = 60°$

$m\angle D = 60°$

38. **K.** Find the y-intercept.

The graph goes through the point (0,5), therefore, the y-intercept (b) is equal to 5.

Find the slope. Use two points on the graph: $(x_1, y_1) = (0,5)$ and $(x_2, y_2) = (10,0)$.

$m = \dfrac{y_2 - y_1}{x_2 - x_1}$

$= \dfrac{0 - 5}{10 - 0}$

$= \dfrac{-5}{10}$

$= -\dfrac{1}{2}$

Write the equation in slope-intercept form:

$y = mx + b$

$m = -\dfrac{1}{2} \quad$ and $\quad b = 5$

$y = -\dfrac{1}{2}x + 5$

39. **E.**

$3x + 5 \le 5x + 1$

$-2x \le -4$

$-x \le -2$

$x \ge 2$

The smallest number in the solution set is 2.

40. **G.** Simplifying means eliminating negative exponents and radicals.

$\dfrac{a^2b^{-2}c^4}{ab^3c} = \dfrac{a^2c^4}{ab^3b^2c} = \dfrac{ac^3}{b^5}$

41. **D.** Find the radius of the circle.

Since the circle touches the square at the midpoints, the radius of the circle is equal to half the length of a side ($s = 8$).

$r = \dfrac{1}{2} \times s$

$= \dfrac{1}{2} \times 8$

$= 4$

Find the area of the circle.

$A = \pi r^2$

$A = \pi(4)^2 = 16\pi$

42. **J.** $p =$ price per one piece of fruit

peaches: \quad $p = \$1.44 \div 3 = \$0.48$

plums: \quad $p = \$2.24 \div 7 = \$0.32$

apples: \quad $p = \$1.48 \div 4 = \$0.37$

nectarines: $p = \$1.74 \div 6 = \mathbf{\$0.29}$

bananas: \quad $p = \$2.35 \div 5 = \$0.47$

The nectarines are the cheapest per piece.

43. D.

$\dfrac{\sqrt{3y}-9}{\sqrt{25}+x}$ is defined except when the denominator,

$\sqrt{25}+x$, equals zero.

$\sqrt{25}+x=0$

$5+x=0$

$x=-5$

The expression is defined except when $x=-5$.

44. J. Find the height.

$A=36$

$b=\dfrac{1}{2}h$

$A=\dfrac{1}{2}bh$

$36=\dfrac{1}{2}\times\dfrac{1}{2}\times h\times h$ Substitute.

$36=\dfrac{1}{4}h^2$

$144=h^2$ Multiply by 4.

$\sqrt{144}=h$ Take the square root of each side.

$12=h$

Find the base.

$b=\dfrac{1}{2}h$

$=\dfrac{1}{2}(12)$

$=6$

Add b and h.

$b+h=6+12=18$

45. A. Factor out the largest common factor of all the coefficients and the smallest exponent for each variable.

$3x^2y^3+12x^5y^2-6x^3y^3=3x^2y^2(y+4x^3-2xy)$

46. H.

$V=\dfrac{4}{3}\pi r^3$

$V\times\dfrac{3}{4}=\pi r^3$ Multiply by $\dfrac{3}{4}$.

$\left(V\times\dfrac{3}{4}\right)\div\pi=r^3$ Divide by π.

$\sqrt[3]{\left(V\times\dfrac{3}{4}\right)\div\pi}=\sqrt[3]{r^3}$ Take the cube root.

$\sqrt[3]{\left(V\times\dfrac{3}{4}\right)\div\pi}=r$

47. D.

$\dfrac{a}{6}-\dfrac{a}{9}=\dfrac{2}{3}$

$3a-2a=12$ Multiply by 18.

$a=12$

48. G.

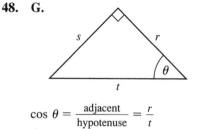

$\cos\theta=\dfrac{\text{adjacent}}{\text{hypotenuse}}=\dfrac{r}{t}$

In the diagram, r is adjacent to angle θ and t is the hypotenuse.

49. C. The general form of the equation of a circle is $(x-h)^2+(y-k)^2=r^2$, where (h,k) is the center and r is the radius.

The equation of the circle in this problem is $(x+2)^2+(y-5)^2=81$.

Rewrite it to match the general form.
$[x-(-2)]^2+[y-(+5)]^2=9^2$

Therefore the center (c) is $(-2,5)$ and the radius $(r)=9$.

50. K. The square root of a negative number is not a real number. In this case, any number smaller than -2 will produce a negative number under the radical.

Let $x=-2$.

$y=\sqrt{4-x^2}$

$=\sqrt{4-(-2)^2}$

$=\sqrt{4-4}$

$=0$

Let $x=-3$.

$y=\sqrt{4-x^2}$

$=\sqrt{4-(-3)^2}$

$=\sqrt{4-9}$

$=\sqrt{-5}$

This is not a real number.

Therefore, if y is a real number, -2 is the smallest possible value for x.

51. C. The best way to solve this problem is to rewrite the values in standard form and use your calculator to add them. Then change the sum to scientific notation.

$(2.36\times10^2)+(4.31\times10^3)=$

$236+4,310=4,546$

$4,546=4.546\times10^3$

52. K.

$$\left(\frac{1}{\sec 2\theta}\right)\left(\cos^2 \theta + \frac{1}{\csc^2 \theta}\right) = ?$$

$= (\cos 2\theta)(\cos^2 \theta + \sin^2 \theta)$	Use trigonometric identities.
$= (\cos 2\theta) \times 1$	Use trigonometric identity.
$= (1 - 2\sin^2 \theta) \times 1$	Use trigonometric identity.
$= 1 - 2\sin^2 \theta$	

53. C.

$$|x - 4| = d$$

If, for example, $x = 1$, the distance to 4 is $|1 - 4| = |-3| = 3$. If $x = 7$, the distance to 4 is $|7 - 4| = |3| = 3$.

54. F.

Slope $(m) = \dfrac{y_2 - y_1}{x_2 - x_1}$

$(x_1, y_1) = (3,6)$ and $(x_2, y_2) = (5,2)$

Substitute known values. $m = \dfrac{2 - 6}{5 - 3} = \dfrac{-4}{2} = -2$

55. D.

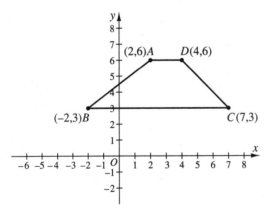

The figure graphed above is a quadrilateral with $\overline{AD} \parallel \overline{BC}$. This is the only special feature of this quadrilateral. These points form a trapezoid when connected.

56. G. Factor. $x^2 + 2x - 15 = (x - 3)(x + 5)$
Add the factors. $(x - 3) + (x + 5) = x - 3 + x + 5$
$\qquad\qquad\qquad\qquad\qquad\qquad = 2x + 2$

57. D. Say that the supplement of θ is $60°$. Then the angle is $120°$, which is in the second quadrant as shown below.

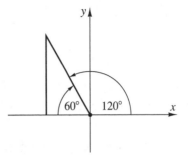

For angles in the second quadrant, the cosine and secant are negative. Choice D shows the correct combination of signs.

58. H. Use the Pythagorean Theorem to find the length of AC.

$$(AB)^2 + (BC)^2 = (AC)^2$$
$$9^2 + 12^2 = (AC)^2$$
$$81 + 144 = (AC)^2$$
$$225 = (AC)^2$$
$$\sqrt{225} = AC$$
$$15 = AC$$

59. B. The maximum number of zeros is equal to the largest power in the polynomial.

60. G. The graph in choice G shows a steady increase, followed by a level period, followed by a steady decrease.

Chapter 17 ▪

Scoring the Mathematics Tests

This chapter shows you how to find the ACT scale scores and subscores.

The **raw score** on a test is the number correct. Use the charts provided to convert the raw score for each test to a scale score. Charts on following pages show you how to find the Mathematics subscores.

Scale scores are the scores reported to colleges. Because different ACTs have different difficulty levels, the same raw score does not always convert to the same scale score. The scale scores here are approximations and are given only to familiarize you with the process of converting raw scores to scale scores. The scale scores for these tests will almost certainly be different from the scale scores on the ACT you take.

Scoring the Mathematics Tests

Use the chart below to convert the raw score for each Mathematics test to a scale score. Charts on following pages show you how to find the Mathematics subscores.

The highest possible raw score is 60; the lowest is 0. The highest possible scale score is 36; the lowest is 1. In the chart below, a raw score of 60 yields a scale score of 36. A raw score of 0 yields a scale score of 1.

Mathematics Scale Scores

Raw Score	Scale Score	Raw Score	Scale Score	Raw Score	Scale Score	Raw Score	Scale Score
60	36	44	26	28	18	12	13
59	33	43	25	27	18	11	13
58	32	42	25	26	18	10	13
57	31	41	24	25	17	9	12
56	31	40	24	24	17	8	12
55	30	39	24	23	17	7	11
54	30	38	23	22	16	6	11
53	29	37	23	21	16	5	10
52	29	36	22	20	16	4	9
51	28	35	22	19	16	3	7
50	28	34	21	18	15	2	5
49	27	33	20	17	15	1	3
48	27	32	20	16	15	0	1
47	27	31	20	15	14		
46	26	30	19	14	14		
45	26	29	19	13	14		

Convert your raw scores to scale scores:

Mathematics Diagnostic ACT	Mathematics Model ACT II	Mathematics Model ACT III	Mathematics Model ACT IV
Raw score _____	Raw score _____	Raw score _____	Raw score _____
Scale score _____	Scale score _____	Scale score _____	Scale score _____

What Does Your Score Mean?

The chart below shows the approximate percent of students who receive a particular scale score or below. The percent of students at or below a score shows the percentile rank of that score.

Use the chart this way. Find your Mathematics scale score in the chart. Then find the percentage of students who scored at or below that score. Say your Mathematics scale score was 18. In the chart, 18 matches 42%. This means that about 42% of the students will get a score of 18 or lower, and that 58% of the students will get a score above 18.

Mathematics Test—Percent of Students At or Below a Scale Score

Scale Score	At or below	Scale Score	At or below	Scale Score	At or below	Scale Score	At or below
36	99%	27	89%	18	42%	9	1%
35	99%	26	85%	17	33%	8	1%
34	99%	25	81%	16	23%	7	1%
33	99%	24	77%	15	14%	6	1%
32	99%	23	72%	14	8%	5	1%
31	98%	22	67%	13	4%	4	1%
30	96%	21	61%	12	1%	3	1%
29	95%	20	56%	11	1%	2	1%
28	92%	19	49%	10	1%	1	1%

DIAGNOSTIC ACT
Mathematics Scoring Key
Check the box of each correct answer.

Item	Answer	Pre-Algebra/ Elementary Algebra	Intermediate Algebra/ Coordinate Geometry	Plane Geometry/ Trigonometry
1	C			☐
2	H	☐		
3	E	☐		
4	G	☐		
5	C	☐		
6	F		☐	
7	D			☐
8	K	☐		
9	D	☐		
10	K	☐		
11	E	☐		
12	G		☐	
13	C			☐
14	G	☐		
15	B			☐
16	J	☐		
17	D			☐
18	F			☐
19	D	☐		
20	F	☐		
21	D		☐	
22	H	☐		
23	B			☐
24	J	☐		
25	A		☐	
26	H			☐
27	E		☐	
28	H		☐	
29	C	☐		
30	F			☐
31	C	☐		
32	G			☐
33	A	☐		
34	K			☐
35	C	☐		
36	F		☐	
37	D	☐		
38	G	☐		
39	C			☐

Item	Answer	Pre-Algebra/ Elementary Algebra	Intermediate Algebra/ Coordinate Geometry	Plane Geometry/ Trigonometry
40	K		☐	
41	D	☐		
42	H		☐	
43	A		☐	
44	G			☐
45	D			☐
46	F	☐		
47	C		☐	
48	G		☐	
49	B		☐	
50	F			☐
51	A	☐		
52	G		☐	
53	B	☐		
54	H		☐	
55	E			☐
56	H			☐
57	B		☐	
58	G		☐	
59	B			☐
60	F		☐	

Number Correct:

Pre-Algebra/Elementary Algebra   _____
Intermediate Algebra/Coordinate Geometry   _____
Plane Geometry/Trigonometry   _____
   Total   _____

MODEL ACT II
Mathematics Scoring Key
Check the box of each correct answer.

Item	Answer	Pre-Algebra/ Elementary Algebra	Intermediate Algebra/ Coordinate Geometry	Plane Geometry/ Trigonometry
1	A			☐
2	H	☐		
3	D	☐		
4	H	☐		
5	D			☐
6	K	☐		
7	C	☐		
8	G	☐		
9	C	☐		
10	J	☐		
11	B	☐		
12	H	☐		
13	C			☐
14	J	☐		
15	A	☐		
16	H			☐
17	D	☐		
18	J			☐
19	B		☐	
20	J	☐		
21	E	☐		
22	H		☐	
23	D	☐		
24	H			☐
25	B	☐		
26	G		☐	
27	B			☐
28	K		☐	
29	E			☐
30	K	☐		
31	C	☐		
32	J			☐
33	C		☐	
34	F		☐	
35	B	☐		
36	K			☐
37	D		☐	
38	H		☐	
39	A		☐	

Item	Answer	Pre-Algebra/ Elementary Algebra	Intermediate Algebra/ Coordinate Geometry	Plane Geometry/ Trigonometry
40	J	☐		
41	A		☐	
42	K		☐	
43	A	☐		
44	G	☐		
45	A			☐
46	H	☐		
47	A		☐	
48	G			☐
49	E			☐
50	H		☐	
51	D			☐
52	F		☐	
53	C			☐
54	K		☐	
55	B			☐
56	G		☐	
57	A			☐
58	H		☐	
59	A			☐
60	H		☐	

Number Correct:

Pre-Algebra/Elementary Algebra   _____

Intermediate Algebra/Coordinate Geometry   _____

Plane Geometry/Trigonometry   _____

    Total   _____

MODEL ACT III
Mathematics Scoring Key
Check the box of each correct answer.

Item	Answer	Pre-Algebra/ Elementary Algebra	Intermediate Algebra/ Coordinate Geometry	Plane Geometry/ Trigonometry
1	B	☐		
2	J	☐		
3	C	☐		
4	H			☐
5	C	☐		
6	F	☐		
7	D			☐
8	J	☐		
9	A	☐		
10	G			☐
11	A	☐		
12	H	☐		
13	D	☐		
14	H	☐		
15	B			☐
16	F			☐
17	D	☐		
18	F	☐		
19	B	☐		
20	H		☐	
21	D			☐
22	F	☐		
23	B	☐		
24	G		☐	
25	C	☐		
26	J		☐	
27	E			☐
28	H		☐	
29	B			☐
30	K	☐		
31	D		☐	
32	G	☐		
33	E	☐		
34	G			☐
35	D			☐
36	G		☐	
37	A			☐
38	H		☐	
39	C	☐		

Item	Answer	Pre-Algebra/ Elementary Algebra	Intermediate Algebra/ Coordinate Geometry	Plane Geometry/ Trigonometry
40	F		☐	
41	A		☐	
42	J		☐	
43	C		☐	
44	J			☐
45	A	☐		
46	K	☐		
47	D			☐
48	G	☐		
49	A			☐
50	H		☐	
51	B		☐	
52	H			☐
53	D			☐
54	H			☐
55	C			☐
56	K		☐	
57	E		☐	
58	H		☐	
59	D		☐	
60	J		☐	

Number Correct:

Pre-Algebra/Elementary Algebra ⎯⎯⎯⎯⎯

Intermediate Algebra/Coordinate Geometry ⎯⎯⎯⎯⎯

Plane Geometry/Trigonometry ⎯⎯⎯⎯⎯

  Total ⎯⎯⎯⎯⎯

MODEL ACT IV
Mathematics Scoring Key
Check the box of each correct answer.

Item	Answer	Pre-Algebra/ Elementary Algebra	Intermediate Algebra/ Coordinate Geometry	Plane Geometry/ Trigonometry
1	D	☐		
2	G			☐
3	C		☐	
4	G	☐		
5	C			☐
6	K	☐		
7	C	☐		
8	K		☐	
9	A			☐
10	J	☐		
11	D	☐		
12	H	☐		
13	C	☐		
14	K	☐		
15	D			☐
16	H	☐		
17	C	☐		
18	F			☐
19	C	☐		
20	J	☐		
21	A	☐		
22	K		☐	
23	E			☐
24	H		☐	
25	C		☐	
26	J	☐		
27	E	☐		
28	G			☐
29	A			☐
30	J			☐
31	C		☐	
32	J			☐
33	E			☐
34	J	☐		
35	A		☐	
36	G		☐	
37	B			☐
38	K		☐	
39	E		☐	

Item	Answer	Pre-Algebra/ Elementary Algebra	Intermediate Algebra/ Coordinate Geometry	Plane Geometry/ Trigonometry
40	G		☐	
41	D			☐
42	J	☐		
43	D		☐	
44	J			☐
45	A	☐		
46	H	☐		
47	D	☐		
48	G			☐
49	C		☐	
50	K		☐	
51	C	☐		
52	K			☐
53	C		☐	
54	F		☐	
55	D		☐	
56	G	☐		
57	D			☐
58	H			☐
59	B		☐	
60	G	☐		

Number Correct:

Pre-Algebra/Elementary Algebra   _____

Intermediate Algebra/Coordinate Geometry   _____

Plane Geometry/Trigonometry   _____

    Total   _____

The Composite Score

To find your composite score, add the scale scores of the four tests (English, Mathematics, Reading, and Science Reasoning) and divide by 4. You can divide each test's scale score by 4 to find out how many points it will contribute to the composite score.

Section IV ▪

Science Reasoning

Chapter 18

The Four-Step Approach to Taking the **ACT** Science Reasoning Test

The ACT Science Reasoning Test is not a test about science facts. It is a test of your ability to read and understand science materials. The more experience you have reading science, the better. However, you do not need advanced science courses to be successful. This chapter shows you how to do your absolute best on the ACT Science Reasoning Test.

In the first part of this chapter, you'll read about the passage types and question types on the ACT. Then you'll learn about a four-step approach to tackling the passages on the Science Reasoning Test and actually see the steps applied to a practice passage. Using the four steps will help you answer the questions more systematically, faster, and with greater accuracy. In Chapters 19, and 20, and 21, you'll take a closer look at each of the three types of passages. These chapters will give you specific helpful tips—and give you a chance to practice using the four steps.

■ Passage and Question Types

There are three types of passages and three levels of questions on the Science Reasoning Test.

Passage Types.   There are seven passages on the test. These passages may be about biology, chemistry, physics, astronomy, geology, or meteorology. However, there are just three types of passages. The following capsule descriptions of the three passage types will give you an idea of what to expect.

Data Representation.   Data representation passages include graphs, tables, diagrams, charts, figures, and illustrations. You might see a traditional bar, line, or circle graph, or you might see an illustration showing the labeled skeleton of the human hand. You might see a graph that shows the relationship between two variables, or you might see a tree diagram that describes how certain traits were inherited. In each case, you will be able to answer the question from the information given.

Questions for data representation passages typically ask you to read data, interpret data, or explain the science that underlies the represented data.

Research Summary.   Research summary passages may supply the designs of experiments or give a summary of experimental results. These passages may also include graphs, tables, diagrams, charts, figures, and illustrations that show experimental results.

Questions for research summary passages typically ask you about the following:

1. Appropriateness of the experimental design
2. The impact of modifications in the design
3. The scientific concepts reflected in the experiment
4. The relationship between the experimental data and the concepts
5. The meaning of the results or the implications for future research

Conflicting Viewpoints. Conflicting viewpoints passages present two or more views that are inconsistent with one another. You are never asked to determine which viewpoint is correct. In fact, one or more of the viewpoints presented may be obviously incorrect. These passages may also include graphs, tables, diagrams, charts, figures, and illustrations.

Questions for conflicting viewpoints passages typically ask you about the following:

1. Scientific ideas or assumptions discussed in the passages
2. The similarities or differences among the viewpoints
3. Whether certain results or facts are consistent with one of the viewpoints
4. Which diagram best illustrates one of the viewpoints

Question Types. The questions on the ACT Science Reasoning Test address three different levels of thinking or reasoning.

Understanding. This type of question asks you to find information or use it at a basic level. Often, the answers can be found in the passage. Sometimes, you may have to make an inference, extrapolate data from a graph, or grasp the concepts underlying a passage.

Analysis. This type of question requires you to draw a conclusion based on the information in the passage. You may also be asked to explain experimental results or describe the relationship among data or information.

Generalization. These questions require you to go beyond the information given. To answer a generalization question, you need to apply the information in the passage to new situations or circumstances.

The Four Steps

Use these four steps as you read science reasoning passages and answer questions. Use step 1 just once for each passage. Then use step 2, step 3, and step 4 for each question.

1. Skim the passage and look over any charts and graphs. Identify the passage as data representation, research summary, or conflicting viewpoints.
2. Read the questions and all of the answers.
3. Eliminate obviously incorrect answers.
4. Choose the correct answer from the remaining choices.

Step 1. Skim and identify the passage. Do not read the passage carefully at first! You don't want to waste time reading material you will not need to answer the question. You should be able to complete this step in less than a minute.

Skim the passage to find what each paragraph in the passage is all about. Only read the first and last sentences. You don't want to know the details yet.

Look over any charts and diagrams just enough to get an idea of what they are about. Don't try to read or interpret the data just yet. You just want to know what sort of data is available.

Identify the passage as data representation, research summary, or conflicting viewpoints.

Step 2. Read the question and all the answers. Read each question and all the answer choices. Be sure you are clear about what the question is asking. You want to answer the question on the test, not some other question.

Step 3. Eliminate obviously incorrect answers. Cross off any answers you're sure are incorrect. Eliminating incorrect answers is a big help in determining the correct answer.

Step 4. Choose the correct answer from the remaining choices. Choose the answer that is most correct. Go back to the passage to confirm your choice. If you don't know the correct answer, guess. There is no penalty for guessing. Never leave an answer choice blank.

Let's apply these four steps to the following sample passage and its questions. This passage and questions are as difficult as the most challenging material you'll encounter on the actual ACT.

ACT Sample Passage With Answers Explained

Scientists are interested in evaluating water quality in areas used for farming (cropped areas) and in other regions of the research area. Cropped areas are heavily fertilized with nitrogen fertilizers. Water was sampled at four different types of wells: field wells, upgradient wells, wetland wells, and off-site wells. Water flows horizontally past field wells. Water flows down from upgradient wells. Wetland wells are in a marsh area. Off-site wells are outside the research area. Water was collected each month for a year.

Figure 1 shows the research area and part of the surrounding area.

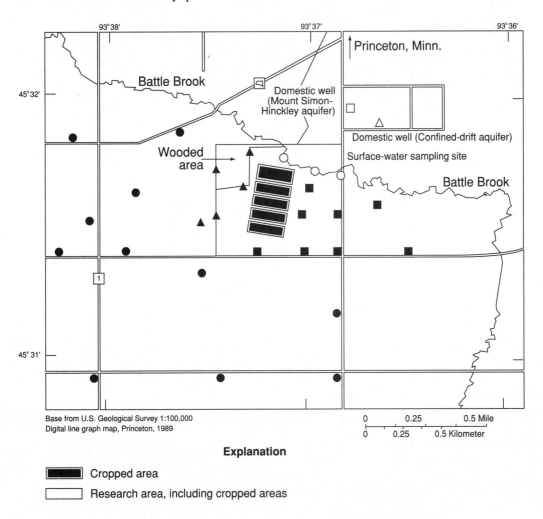

Figure 1

Base from U.S. Geological Survey 1:100,000 Digital line graph map, Princeton, 1989

Explanation

| | Cropped area |
| | Research area, including cropped areas |

Observation wells completed in the surficial aquifer at or near the water table–

■ Field well and identifier   ○ Wetland well and identifier

▲ Upgradient well and identifier   ● Off-site well and identifier

Figure 2 shows the results of the research. Cations are positively charged ions. Anions are negatively charged ions.

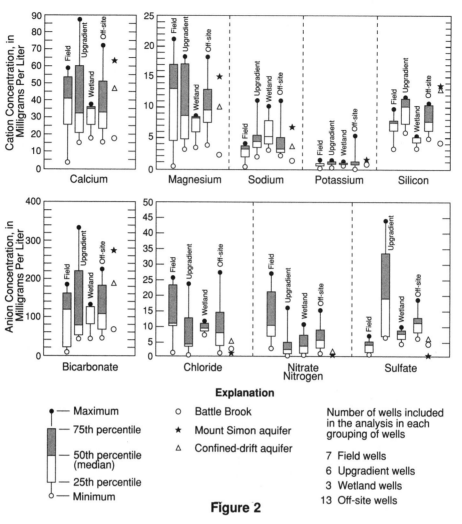

Figure 2

1. Figure 1 shows that the area of the experimental region is about:

   A. 10 kilometers.
   B. 0.25 square mile.
   C. 0.14 square kilometer.
   D. 1 square mile.

2. Which of the following best explains the reason why nitrate concentrations are highest in field wells?

   F. The water flows down from the upgradient wells, taking most of the nitrate with it.
   G. Field wells are located in cropped areas or in the runoff from cropped areas.
   H. Water flows horizontally past field wells, so more nitrate adheres to the soil and is left behind.
   J. Field wells have the highest concentration of chloride.

3. Which of the following statements best represents the results of this investigation?

   A. The maximum amount of bicarbonate in upgradient wells is about the same as the maximum amount of sulfate in upgradient wells.
   B. The median amount of calcium in off-site wells is about twice the maximum amount of nitrate in off-site wells.
   C. The median amount of calcium in upgradient wells is about twice the minimum amount of magnesium in upgradient wells.
   D. The maximum amount of chloride in field wells is about twice the maximum amount of chloride in off-site wells.

Use the Four Steps

Let's use the four steps to examine the passage and answer the questions. The first reaction to a passage is often, "Wow!—that's a lot." It is actually a lot *less* than it seems. Use the steps. Don't be overwhelmed by all the information.

Step 1. Skim the passage and any charts, graphs and diagrams. Identify the passage as data representation, research summary, or conflicting viewpoints.

Use this step only once. Use each of the other steps for each question.

There is one paragraph, one drawn map, and one chart with box-and-whisker plots. The paragraph says that this passage is about water quality. The map shows a research area, a cropped area, well locations, and some other details. The box-and-whisker plots show results in milligrams per liter.

The passage is a research summary.

That's all the information you need from this step. You'll come back to the paragraph, the map, and the chart when you answer the questions.

Step 2. Read the question and all answers.

1. Figure 1 shows that the area of the experimental region is about:
   A. 10 kilometers.
   B. 0.25 square mile.
   C. 0.14 square kilometer.
   D. 1 square mile.

This question asks for the area of the experimental region shown in Figure 1. Glance at the answers. A quick look back at the figure shows that the area of the experimental region is not shown on the map. You're going to have to figure it out.

Step 3. Eliminate obviously incorrect answers.

You can eliminate answer choice A right away. Kilometers are units of distance, not area.

Step 4. Choose the correct answer from the remaining choices.

The area of the experimental region looks like a square. The quickest way to find the area of a square is to use the length of a side. The scale just below the map to the right shows distances on the map in miles and kilometers. Use your fingers or the edge of a piece of paper to determine the length of a side of the experimental region on the map. "Measure" this length against the scale. The length of the side of the experimental region is about 0.5 mile.

The formula for the area of a square is $A = s^2$, where s is the length of a side. $(0.5)^2 = 0.25$.
That's choice B.

The correct answer is B, 0.25 square mile.

Step 2. Read the question and all answers.

2. Which of the following best explains the reason why nitrate concentrations are highest in field wells?
   F. The water flows down from the upgradient wells, taking most of the nitrate with it.
   G. Field wells are located in cropped areas or in the runoff from cropped areas.
   H. Water flows horizontally past field wells, so more nitrate adheres to the soil and is left behind.
   J. Field wells have the highest concentration of chloride.

You have to explain why the nitrate concentrations are highest in field wells. The reason is not directly explained in the paragraph. This is an analysis question.

Step 3. Eliminate obviously incorrect answers.

Eliminate answer J. There is nothing in this passage that shows a cause-and-effect relationship between these two substances.

Step 4. Choose the correct answer from the remaining choices.

Consider the remaining answers one at a time.

F. This doesn't seem to make sense. If the water takes the nitrate with it, it would take other substances too. This answer does not explain the presence of high concentrations of nitrate.

G. A well located in a cropped area will have very high concentrations of nitrate because these fields are fertilized heavily with nitrate fertilizer. This may be the best answer. Let's check answer H.

H. This answer doesn't make sense either. If nitrate is left behind, other substances would be left behind too.

The correct answer is G. It is the best answer of the four choices given.

Step 2. Read the question and all answers.

3. Which of the following statements best represents the results of this investigation?

A. The maximum amount of bicarbonate in up-gradient wells is about the same as the maximum amount of sulfate in upgradient wells.

B. The median amount of calcium in off-site wells is about twice the maximum amount of nitrate in off-site wells.

C. The median amount of calcium in upgradient wells is about twice the minimum amount of magnesium in upgradient wells.

D. The maximum amount of chloride in field wells is about twice the maximum amount of chloride in off-site wells.

This question asks you to compare the graphs in Figure 2. A glance at Figure 2 reveals a problem. Four different scales are used. All are milligrams per liter, but one scale goes from 0–90, one from 0–25, one from 0–400, and one from 0–50. Any comparison will have to be made very carefully.

Step 3. Eliminate obviously incorrect answers.

You can eliminate two answers.

A. The two quantities shown in the answer appear to be at the same level in the plot. However, the scales are different and these quantities are not close at all.

D. The quantities are on the same scale and are about the same. One quantity is obviously not twice the other, so this answer can be eliminated.

Step 4. Choose the correct answer from the remaining choices.

Let's work through answers B and C to determine which one is correct. We'll use the actual quantities to be sure.

B. The median amount of calcium in off-site wells is about 31 milligrams per liter. The maximum amount of nitrate in off-site wells is about 15 milligrams per liter. The first quantity is about twice the second. This is probably the correct answer, but let's check the other to make sure.

C. The median amount of calcium in upgradient wells is about 30 milligrams per liter. The minimum amount of magnesium in upgradient wells is about 3 milligrams per liter. The first quantity is about 10 times the second.

The correct answer is B.

Important strategies for passing the Science Reasoning ACT will be reviewed on pages 511–512, before the Diagnostic Science Reasoning ACT, in Chapter 22.

Chapter 19 ∙

Data Representation

Data representation passages are about graphs, tables, and figures. However, these passages are not the only places where you will find graphic representations of data. You are likely to encounter graphs, tables, and figures in research summary and conflicting viewpoint passages as well. This chapter shows you the types of data representations you will find and explains how to interpret them. Then you will see how to answer actual data representation questions.

▪ Graphs

Most ACT graphs are line graphs plotted against two labeled axes. These graphs usually do not include detailed information. Here are some examples.

Line Graphs

Here's how to read graph A. Look at the entire graph. It is a line graph. There are two axes. The vertical axis shows the volume of gas. The arrow shows that volume increases as you move up the vertical axis. The horizontal axis shows the gas temperature. The arrow shows that gas temperature increases as you move to the right along the horizontal axis.

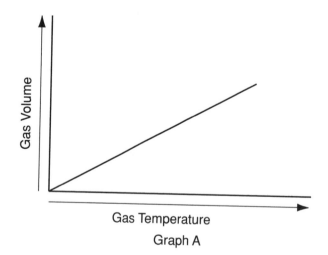

Graph A

Graph A shows the relationship between the volume of a gas and the temperature of a gas. Specifically, the graph shows that there is a positive relationship between the volume of a gas and its temperature. As the temperature increases, volume increases. As temperature decreases, volume decreases.

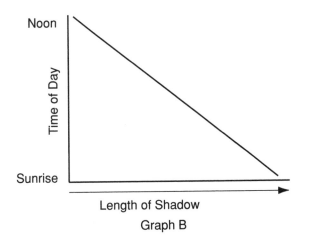

Length of Shadow

Graph B

Here's how to read graph B. Look at the entire graph. It is a line graph. There are two axes. The vertical axis shows the time of day from sunrise at the bottom of the axis to noon at the top of the axis. The horizontal axis shows the length of a shadow. The arrow indicates that the length of the shadow increases as you move out the horizontal axis.

The graph shows the relationship between the time of day and the length of a shadow. Specifically, the graph shows that there is an inverse relationship between the time of day and the length of the shadow. As the day progresses from sunrise to noon, the length of the shadow grows smaller. When the sun shines from the side at sunrise, the shadow is longest. As the sun gets closer to being directly overhead, the shadow gets shorter. When the sun is overhead at noon, the shadow is shortest.

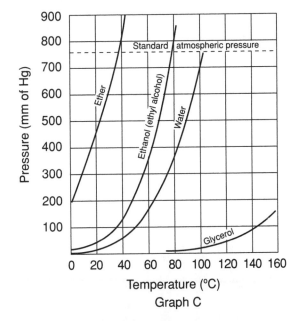

Temperature (°C)

Graph C

Here's how to read graph C. Look at the entire graph. It is a line graph with several curved lines. There are two axes. The vertical axis shows surface pressure. The horizontal axis shows temperature. The four curved lines represent the boiling points for four different liquids.

The graph shows that the boiling point for each liquid depends on the liquid's surface pressure. For example, the boiling point of ether is about 0°C when the pressure is 200 mm of mercury (Hg). The graph shows that the boiling points of ether, ethanol, and water increase slowly with an increase in surface pressure. It also shows that the boiling point of glycerol increases markedly with an increase in surface pressure.

The graph shows there is a positive relationship between pressure and boiling point for each liquid. The higher the pressure, the higher the boiling point.

Let's look at some ways you will be asked to interpret line graphs on the ACT.

Extrapolation and Interpolation.   Extrapolation means looking beyond the information given in a graph to arrive at a value. Interpolation means looking between the information given to arrive at a value. This is simpler than it sounds. Consider this sample extrapolation question about graph C.

Question: What is the boiling point of water at 900 millimeters of mercury?

Answer: The line for water does not extend to 900 millimeters of mercury. You have to extrapolate. Following the curve of the existing line, use your pencil to extend the graph for water to the 900 mm pressure line. Now draw a straight vertical line down from the point where the extended graph touches the 900 mm pressure line to the temperature scale on the horizontal axis. This vertical line touches the horizontal axis at about 110°C. Thus the boiling point for water at 900 mm of mercury is about 110°C.

Now let's try an interpolation question.

Question: At what pressure is the boiling point of ether 15°C?

Answer: On the graph, temperature is marked off in increments of 20°C. 15°C is not directly shown. You have to interpolate. First, determine that 15° is on the horizontal axis, three-fourths of the way between 0° and 20°. Next, draw a straight vertical line up from this point until it touches the graph for ether. The point at which the vertical line touches the ether graph is the boiling point of ether at 15°C. Finally, draw a straight horizontal line from the 15°C boiling point to the vertical axis to determine the pressure at the 15°C boiling point. The horizontal line touches the vertical axis at about 375 mm of mercury. Thus the pressure at the boiling point of ether at 15°C is about 375 mm of mercury.

Bar Graphs

The bar graph represents information by the length or height of a bar.

This bar graph shows the air temperature and water temperature in the same location for seven different times of the day.

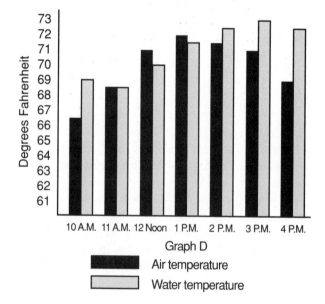

Graph D

■ Air temperature
□ Water temperature

Here's how to read graph D. Look at the entire graph. It is a bar graph. There are two axes. The horizontal axis shows the time of the day by hours from 10 A.M. to 4 P.M. The vertical axis shows the temperature in degrees Fahrenheit. The black bar shows air temperature and the gray bar shows water temperature.

The graph shows air and water temperatures at different times of the day. Specifically, the graph shows that from 10 A.M. to 11 A.M., air temperature rises and water temperature drops. After 11 A.M., air and water temperatures both move up until 1 P.M. Then the water temperature moves up as the air temperature moves down until 3 P.M. Between 3 P.M. and 4 P.M., air and water temperatures both move down.

Here are some sample questions and answers about this bar graph.

Question: At which time is the difference between air and water temperature the greatest?

Answer: A quick examination shows that the biggest difference between the two bars occurs at 4 P.M. The answer is 4 P.M.

Question: According to the graph, what is the water temperature when the air temperature is 69°?

Answer: First, find when the air temperature is 69°. The air temperature is represented by the black bar. Put the edge of a piece of paper right through 69 on the vertical axis and straight across the graph. The paper will touch the top of the black bar for 4 P.M. Now read the water temperature for 4 P.M. It's about 72°. The answer is 72°.

Question: Which sequence of times shows the four highest air temperatures in order from greatest to least?

Answer. Find the greatest air temperature (black bar). Write a 1 above it on the graph. Then find and label the next three highest air temperatures.

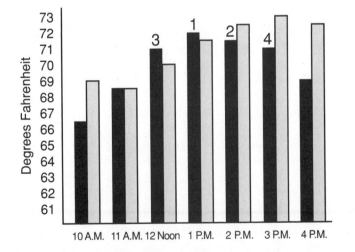

Write the time associated with the highest air temperature first; then write the time associated with the second highest temperature next, and so on. The correct sequence of time is: 1 P.M., 2 P.M., 12 noon, 3 P.M.

Question: Based on the graph, what was the likely air temperature at 11:30 A.M.?

Answer: This is an interpolation question. Remember, in interpolation and extrapolation questions, we use the available data to make our best estimate of the answer.

Draw a diagonal line from the top of the 11 A.M. air temperature bar to the top of the 12 noon air temperature bar. Mark 11:30 halfway between 11 and 12 on the horizontal axis. Draw a line from the 11:30 point straight up to the diagonal line. Then draw a line straight over to the vertical axis. The temperature at 11:30 was most likely 70°F.

■ Tables and Figures

You may see some fairly involved tables and figures on the ACT. The secret—do not look at the table or figure too carefully before you know what the question is. Scan the table or figure quickly so that you have an idea what it contains. Then read the question. Go back to the table or figure and just look for the information you need to answer the question.

Woodport Lake Showing Seven Locations

A—Inlet	B3—Boat dock	D—Bathing beach
B1—Village	C—Center of lake	E—Dam
B2—Farm		

Table F. Water Depth in Meters in Seven Locations on Woodport Lake on the First Day of Four Months

Location	*January*	*April*	*July*	*November*
A	2.9	4.2	4.1	3.6
B1	1.8	1.9	2.2	1.7
B2	4.8	6.2	7.0	5.3
B3	0.9	3.2	3.2	4.8
C	9.5	10.2	12.3	11.6
D	0.5	0.8	1.6	1.2
E	8.6	8.7	8.8	8.6

If you were actually taking the ACT at this moment, you would now be reading the questions, having looked over the diagram and table just long enough to see that the diagram shows locations at a lake and that the table shows the depth of the water at these locations at different times of the year. Remember: Do not waste test time analyzing tables and figures before you know what the questions are. Scan the tables and figures quickly, then read the questions. Go back to the tables and figures and examine them more carefully after you have figured out what information you need.

Since you are now studying for a test, and not taking it, you should make sure you know how to read the diagram and the table. The diagram shows that water enters Woodport Lake through a stream (location A) and leaves through a dam (location E). The center of the lake is called location C. Other locations on the lake, named for features beside the lake, are Village (location B1), Farm (location B2), Boat dock (location B3), and Bathing beach (location D).

The table shows the water depth in meters at each of the seven labeled locations on the lake on the first day of January, April, July, and November. To find the depth at a particular location and time—say at Farm (location B2) in July—you must first locate the row with the appropriate label. In this case, B2 is the fourth row in the table, between B1 and B3. Follow this row across the table from left to right until you encounter the box in the column labeled July. In this case, July is the fourth column in the table, between April and November. The number in the box, 7.0, is the depth of the water in meters. So the depth of the water at Farm (location B2) in July is 7.0 meters.

EXAMPLES

Question: Summarize the change in water depth from all sampling sites from January to April.

Answer: All the sites had at least a small increase in water depth. The increases noted in locations A and B2 are somewhat larger than the others, and the increase noted at location B3 is much larger than the others.

Question: What is the difference between the January and November water depths at the dam?

Answer: In January, the depth of the water at the dam (location E) is 8.6 meters. In November, it is 8.6 meters. To find the difference, if any, subtract 8.6 from 8.6.

$$8.6 - 8.6 = 0$$

The difference between the January and November water depths at the dam is 0 meters.

Question: What is the best explanation for the water depths recorded at location E?

Answer: Location E is the dam. Water depths at this location show none of the fluctuations found at other locations. The best explanation is that the top of the dam is a little more than 8 meters from the bottom of the lake. As water exceeds 8 meters, it flows over the dam. We can't be absolutely certain, but this seems to be the best explanation.

This data representation passage and questions might appear on an actual ACT. This section shows you how to use the four-step approach to answer each question.

PASSAGE I

Over a period of one year, scientists measured the amounts of oxidants in the air in parts per hundred million (pphm) at five different testing stations. They took their measurements at different times during the year. At the same time they asked people living near these stations to report the amount of eye irritation they experienced on a scale from 0 to 100. The results of this study are summarized in the graph below.

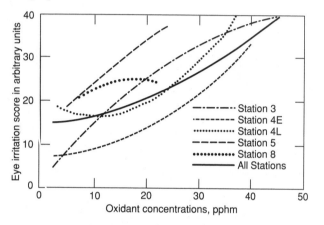

1. Which of the following statements best represents the changes in eye irritation noted at Station 4L as the oxidant concentration increases?

   A. The amount of eye irritation increases and then decreases.
   B. The amount of eye irritation decreases and then increases.
   C. The amount of eye irritation is directly related to an increase in oxidant.
   D. The amount of eye irritation is inversely related to an increase in oxidant.

2. From the data in the graph, it can be concluded that the amount of eye irritation:

   F. is directly related to the concentration of oxidants.
   G. varies from station to station.
   H. is proportional to the concentration of oxidants.
   J. varies according to the time of the year.

3. What conclusion might one draw from the data in the graph about the impact of oxidant concentrations on eye irritation?

   A. Higher concentrations of oxidants will usually result in a person's eyes being more irritated.
   B. Higher concentrations of oxidants will always result in a person's eyes being more irritated.
   C. You can't tell much about how much a person's eye will be irritated from the concentration of oxidants.
   D. Lower concentrations of oxidants may result in a person's eyes being more irritated.

4. Which of the stations most nearly shows a direct relationship between oxidant levels and eye irritation?

   F. Station 3
   G. Station 4E
   H. Station 4L
   J. Station 5

5. If the oxidant level at Station 5 rose to 40, what is the level of eye irritation most likely to be?

   A. 33
   B. 43
   C. 53
   D. 63

Use the Four Steps to Answer the Questions.

Let's use the four steps to answer the questions based on the graph.

Step 1. Skim and identify the passage.

The passage is about the relationship between oxidants and eye irritation. The results are shown on a graph. The graph shows the results from five stations. The graph lines generally go from lower left to upper right. This is a data representation passage.

That's all you need to know for now. Let's apply steps 2, 3, and 4 to each of the questions.

Step 2. Read the question and all answers.

1. Which of the following statements best represents the changes in eye irritation noted at Station 4L as the oxidant concentration increases?

Question 1 is an understanding question. It asks you to describe the data in the graph. Each answer gives a very clear statement about the graph.

The term "directly related" in answer C means the graph would be a straight line from lower left to upper right. The term "inversely related" in answer D means the graph would be a straight line from upper left to lower right.

Step 3. Eliminate obviously incorrect answers.

Answers C and D must be incorrect. All the stations would have to have straight-line graphs for these answers to be correct.

Cross off these answers right on your test like this:

1. Which of the following statements best represents the changes in eye irritation noted at Station 4L as the oxidant concentration increases?

- **A.** The amount of eye irritation increases and then decreases.
- **B.** The amount of eye irritation decreases and then increases.
- **C.** The amount of eye irritation is directly related to an increase in oxidant.
- **D.** The amount of eye irritation is inversely related to an increase in oxidant.

Step 4. Choose the correct answer from the remaining choices.

The graph for station 4L is represented by the smaller dots. The line goes down a little and then starts up. Choice B best describes the graph for station 4L.

The correct answer is B. Write a big letter B next to the question in your test booklet.

1. Which of the following statements best represents the changes in eye irritation noted at Station 4L as the oxidant concentration increases?

- **A.** The amount of eye irritation increases and then decreases.
- **B.** The amount of eye irritation decreases and then increases.
- **C.** The amount of eye irritation is directly related to an increase in oxidant.
- **D.** The amount of eye irritation is inversely related to an increase in oxidant.

Step 2. Read the question and all answers.

  2. From the data in the graph, it can be concluded
     that the amount of eye irritation:

   F. is directly related to the concentration of
        oxidants.
   G. varies from station to station.
   H. is proportional to the concentration of
        oxidants.
   J. varies according to the time of the year.

Question 2 is an analysis question. It asks you to draw a conclusion from the data shown in the graph. Each answer gives a different conclusion about the data. The term "directly related" in answer F means that the graphs will be straight lines. The term "proportional" in answer H also means that the graphs will be straight lines.

Step 3. Eliminate obviously incorrect answers.

Answers F and H must be incorrect. For these answers to be correct, all the graphs would have to be straight lines—but none of them are.
Cross off answer choices F and H in your test booklet.

  2. From the data in the graph it can be concluded
     that the amount of eye irritation:

   ~~**F.**~~ is directly related to the concentration of
        oxidants.
   G. varies from station to station.
   ~~**H.**~~ is proportional to the concentration of
        oxidants.
   J. varies according to the time of the year.

Step 4. Choose the correct answer from the remaining choices.

Choice G seems most correct. The amount of eye irritation does vary widely from station to station. But before you decide, let's take a look at choice J. This answer is incorrect. You might mistakenly choose it if you assumed that the oxidant concentrations went up as the year progressed. However, there is no basis for believing this to be true. The graph does not provide any information about time.

The correct answer is G. Write a big letter G next to the question in your test booklet.

  2. From the data in the graph it can be concluded that the
     amount of eye irritation:
G   ~~**F.**~~ is directly related to the concentration of oxidants.
       G. varies from station to station.
       ~~**H.**~~ is proportional to the concentration of oxidants.
       J. varies according to the time of the year.

From now on, you won't see examples of crossing out incorrect answers or writing the correct answers next to the question in your test booklet. You have the idea.

Step 2. Read the question and all answers.

  3. What conclusion might one draw from the data in
     the graph about the impact of oxidant concentra-
     tions on eye irritation?

Question 3 is a generalization question. It asks for a conclusion about the relationship between eye irritation and oxidant levels that goes beyond the data in the graph. Each answer gives a very clear generalization.

Step 3. Eliminate obviously incorrect answers.

Answer C must be wrong. Oxidant concentration clearly has an impact on eye irritation. Answer D is also wrong. It contradicts the information in the graph.
Cross off answers C and D in the test booklet.

Step 4. Choose the correct answer from the remaining choices.

Look at answers A and B carefully. They are the same except for one word. Choice A says "usually" and choice B says "always."
On the graph, higher oxidant levels usually, but not always, mean higher levels of eye irritation. So one could reasonably conclude that higher oxidant levels will usually, but not always, mean that a person's eyes will be more irritated.

The correct answer is A. Write a big letter A in your test booklet.

Step 2. Read the question and all answers.

4. Which of the stations most nearly shows a direct relationship between oxidant levels and eye irritation?

Question 4 is an understanding question. It asks for a correct reading of information from the graph. The question asks which station most nearly shows a direct relationship between oxidant level and eye irritation. The graph for that station will look most like a straight line.

Step 3. Eliminate incorrect answers.

We're going to answer this question right here. The only graph that looks like a straight line is the graph for Station 5. This is answer J. The rest of the answers are incorrect.

Step 4. Choose the correct answer from the remaining choices.

Answer J—Station 5—is correct.

The correct answer is J. Write a big letter J in your test booklet.

Step 2. Read the question and all answers.

5. If the oxidant level at Station 5 rose to 40, what is the level of eye irritation most likely to be?

Question 5 is an understanding question. However, the diagram you were given does not extend as far as you need to answer the question. You need to extrapolate. The most accurate way of doing this is to mark up the graph in your test booklet. Refer to the marked-up graph that follows as you read about how to extrapolate from the diagram you were given.

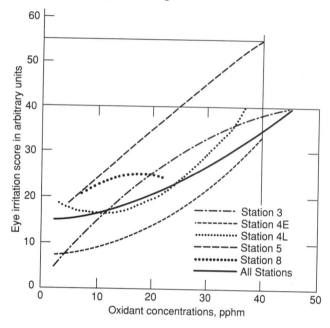

1. Mark the oxidant level of 40 pphm. Draw a straight vertical line from 40 on the horizontal axis up well past the upper edge of the graph. To help you draw the line, use the straight edge of a piece of paper. When you take the ACT, you will not be allowed to use a ruler—but there is no rule against using the edge of your scoring sheet or a pencil to create a makeshift straight edge.

2. Extend the line for Station 5 so that it crosses over the vertical line that marks the oxidant level of 40. You may find a straight edge helpful here, too.

3. Extend the vertical axis and mark eye irritation levels of 50 and 60. Be sure that these levels have exactly the same spacing as the marks given on the original graph. If you put the new marks too far apart or too close together, it will be difficult to obtain the correct answer.

4. Draw a straight horizontal line from the point where the Station 5 line crosses the 40 line to the new part of the vertical axis. This will give you level of eye irritation when the oxidant level at Station 5 is 40.

Step 3. Eliminate obviously incorrect answers.

Answers A is incorrect. The correct level will be off the graph and so has to be greater than 40.

Cross off answer choice A in your test booklet.

Step 4. Choose the correct answer from the remaining choices.

The marked-up graph shows that the irritation level is between 50 and 60, which is answer choice C.

The correct answer is C. Write a big letter C in your test booklet.

Chapter 20 ■

Research Summaries

Research summary questions are based on descriptions of experiments. The descriptions may include graphs, tables, and figures.

You'll use the same steps to answer these questions as you did for data representation questions. Answering research summary questions is a combination of reading comprehension; reading graphs, tables, and figures; and analyzing research design. Analyzing research design is the only new skill and it is discussed below.

■ Analyzing Research Design

All research has a design. Following are descriptions of the most common research designs.

Experiment

In an experiment, the experimenter is usually trying to find or test the effect of one thing on something else. The thing the experimenter wants to test the effect *of* is called the independent variable. The thing the experimenter wants to test the effect *on* is the dependent variable.

For example, the experimenter may test the effect of a particular dog food supplement. In that experiment, the independent variable is the amount of food supplement. The dependent variable is the size of the dog.

An experimental design should include both experimental and control groups. The experimental group consists of dogs who get doses of the food supplement. The control group consists of dogs who do not get the food supplement but are otherwise treated exactly the same as the experimental group. This means the two groups of dogs live under the same conditions and receive the same amounts of food. The only difference is whether or not they receive the food supplement.

At the beginning of an experiment, the experimental and control groups must be as similar to one another as possible. It would not be a fair test of the independent variable if conditions that already existed were likely to affect the dependent variable more in one of the groups than in the other groups. For example, suppose that the dogs getting the food supplement were younger, healthier, and more likely to grow than the dogs in the control group. If, during the course of the experiment, the dogs in the experimental group grew more than the dogs in the control group, it would be impossible to draw any conclusions about the effect of the food supplement. The food supplement might have helped the dogs grow. It might have had no effect at all—the experimental dogs might grow larger simply because they were more likely to do so from the start. Perhaps the food supplement even harmed the dogs in some way, and the difference between the experimental and control groups is not so great as it might have been. It is impossible to tell.

There are two ways to make sure that the experimental and control groups are as similar to one another as possible at the start of the experiment. The first is to make sure that all the subjects in the experiment are basically the same. The second is to randomly assign subjects to the experimental and control groups. For example, the experimenter may test the dog food supplement using two groups of six-month old beagles. Alternatively, the experimenter may assign each of the dogs for the experiment to the experimental or the control group at random. Some dogs may grow more quickly than others. However, random assignment makes it likely that the groups will be evenly matched overall. One group does not start out with an "advantage" over the other.

Observation

In this form of research, the experimenter makes observations and takes measurements of existing situations. For example, the experimenter may want to determine the effect of rainfall on plant growth outside the laboratory. The experimenter cannot actually replicate rainfall on plant growth because he or she cannot control the amount of rainfall.

So the experimenter needs to find different places with the appropriate rainfall totals and measure plant growth in these places. The experimenter also needs to choose sites that have as much in common as possible. If the experimenter can control all the independent variables except rainfall, then some reasonable conclusions may be drawn.

Correlation

Correlation is a way of showing how strongly two variables are related. A correlation of 1 means they are completely related. A correlation of 0 means they are completely unrelated.

Correlation does not mean cause and effect. A correlation of 1 between two variables does not mean that one necessarily causes the other. A correlation of 0 does not mean that one prevents the other. For example, height and weight in people have a positive correlation. However, the height does not cause the weight or vice versa. It's just that they are most likely to be strongly related.

■ Reading Research Summaries

Use the same four steps, but include these questions in step 1:

> What is the research design—experiment, observation, or correlation?
> What are the variables?
> What are the controls?
> What are the results?

The answers to these questions will help you answer the multiple-choice ACT test questions.

Here is a research summary passage that might appear on an actual ACT. This section shows how to use the four steps to answer the questions.

This research summary passage and questions might appear on an actual ACT. This section shows you how to use the four-step approach to answer each question.

PASSAGE I

First Experiment

The amount of energy stored in a food is measured in calories. A calorie is the amount of heat it takes to raise one gram of water 1°C. The number of calories present in food can be measured in the bomb calorimeter pictured below.

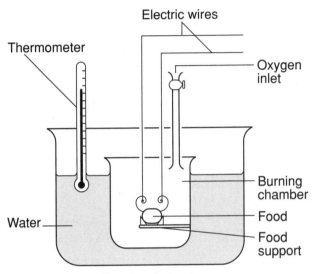

Bomb calorimeter

The calories in a 0.5-kilogram (0.5-kg) sample of a particular dog food were measured when mixed with different amounts of the same dog food supplement. The number of calories for different mixtures of dog food and dog food supplement are given in the table below.

Table 1

Amount of supplement in grams (g) added to 0.5 kg of dog food	Calories
0	281
20	303
40	323
60	343
80	362
100	382
120	404

Second Experiment

Next, the experimenters tested the food supplement to determine if it would affect growth. Fifty essentially identical dogs participated in the experiment. Ten of the dogs were fed 1 kg of food per day consisting of 100% dog food. Ten other dogs were fed 1 kg of food per day consisting of 95% dog food and 5% food supplement. Groups of ten dogs received 90%/10%, 85%/15%, and 80%/20% mixtures. The average weight gain of dogs in each group after two months is shown in Table 2.

Table 2

Percent of dog food/ Percent of supplement	Average weight gain in kg
100/0	2.8
95/5	3.7
90/10	4.8
85/15	4.7
80/20	4.6

1. When comparing calories to the amount of dog food supplement in the first experiment, it would be reasonable to observe that:
   A. as the number of calories increased, the amount of food supplement decreased.
   B. when the total amount of dog food and supplement weighed more than 1 kilogram, the number of calories went over 400.
   C. as the percent of supplement approaches 0, the number of calories is lowest.
   D. as the amount of food supplement increases, the amount of food decreases.

2. What is there about the report of the results for the second experiment that might make the results difficult to interpret?
   F. Average weight gains are reported.
   G. Weight gain is shown in kilograms and tenths of a kilogram instead of grams.
   H. The results do not show weight gain for 70% dog food and 30% supplement.
   J. The results are shown in a table instead of on a line graph.

3. According to the results of the first experiment, the number of calories in the dog food/supplement mix increases about:

A. 10 calories per each 20-gram increase in supplement.
B. 20 calories per each 10-gram increase in supplement.
C. 5 calories per each 10-gram increase in supplement.
D. 1 calorie per each 1-gram increase in supplement.

4. Based on the results of the second experiment, which of the following average weight gains might have been expected if one of the groups of ten dogs had received a dog food/supplement mix of 97.5%/2.5%?

F. 0.9 kg
G. 3.25 kg
H. 3.6 kg
J. 6.5 kg

5. Which of the following graphs best represents the results of the second experiment?

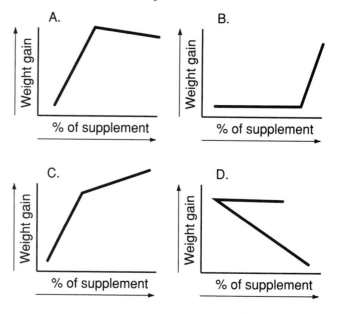

Use the Four Steps to answer the questions.

Step 1. Skim and identify the passage. Answer the research design questions.

This passage is a research summary with two experiments. The first experiment is about the change in calories when food supplement is added to dog food. The second experiment is about the effect of food supplement on weight gain. The results are shown in tables.

What is the research design?
Experiment.

What are the variables?
In the first experiment, the independent variable is the amount of food supplement and the dependent variable is the number of calories.

In the second experiment, the independent variable is the percent of food supplement and the dependent variable is the amount of weight gain.

What are the controls?
In the first experiment, the type of dog food and type of food supplement are the same. In the second experiment, the type of dog food and food supplement are the same for each group of dogs. The dogs in the experiment are all "essentially identical."

What are the results?
The results for the first experiment show that the number of calories increases at a fairly constant rate as more food supplement is added. The results for the second experiment show that the average weight of the dogs increases and then holds steady as the percent of food supplement increases.

That's what you need to know for now. Let's apply steps 2, 3, and 4 to each of the questions about the sample passage.

Step 2. Read the question and all answers.

     1. When comparing calories to the amount of dog food supplement in the first experiment, it would be reasonable to observe that:

Each of the answer choices is a clear statement about the table.

Step 3. Eliminate obviously incorrect answers.

Most of the work for this question will be done at this step. Let's consider the answer choices one at a time.

Choice A is incorrect because as the number of calories increases, so does the amount of food supplement. Choice B is also incorrect because the total weight of the dog food and food supplement never goes over 1 kilogram and the number of calories is still greater than 400.

Step 4. Choose the correct answer from the remaining choices.

You are left with choices C and D. Choice D is tricky because it is correct for the second experiment, but not correct for the first experiment. Since this question is about the first experiment, you are left with Choice C.

Choice C is correct. The calories are lowest as the amount of food supplement approaches 0.

The correct answer is C.

Step 2. Read the question and all answers.

     2. What is there about the report of the results for the second experiment that might make the results difficult to interpret?

The question asks you to identify something about the report of the results from the second experiment that might make it difficult to interpret the results. Note that the question asks about the report of the results, not the experiment or the results themselves.

Step 3. Eliminate obviously incorrect answers.

Choice G is not correct because reporting the weight gains in grams would add only a tiny amount of information and the actual measurements may not have been taken in grams.

Choice H is not correct because the description of the experiment shows that there was no group for 70% food/30% supplement. Therefore, these results could not have been reported.

Choice J is not correct because the data would not change if they were plotted on a line graph.

Step 4. Choose the correct answer from the remaining choices.

The work is done. Choice F is the only choice remaining, but let's double check to be sure. Choice F is correct because showing only the average can hide a lot of information. For example, the actual weight gains in each group could be bunched together or spread far apart. A single very high or very low weight gain could also have a significant impact on the average.

The correct answer is F.

Step 2. Read the question and all answers.

3. According to the results of the first experiment, the number of calories in the dog food/supplement mix increases about:

The question asks about the increase in the number of calories. The question uses the word "about," so the answer will probably not be exact. The question can be answered from the table.

Let's do some calculations now. There are two things to look at, and you'll have to write on the table itself.

The amount of supplement increases by 20 grams from sample to sample. Figure out how the number of calories increases from sample to sample. Write the increases on the table.

Table 1

Amount of supplement in grams (g) added to 0.5 kg of dog food	Calories	
0	281	
		22
20	303	
		20
40	323	
		20
60	343	
		19
80	362	
		20
100	382	
		22
120	404	

There is some variation, but the number of calories goes up about 20 calories for every 20-gram increase in the food supplement. Since 20 calories for 20 grams is not included in the answer, we'll have to do some more calculations.

Step 3. Eliminate obviously incorrect answers.

Choice A is incorrect because it says 10 calories for 20 grams, which you know does not match the 20 calories for 20 grams increase.

Choice B is incorrect because it says 20 calories for 10 grams, which you know does not match the 20 calories for 20 grams increase.

Step 4. Choose the correct answer from the remaining choices.

Write 20/20 as a fraction. Does that fraction mean the same as 5/10 (choice C) or 1/1 (choice D)? It means the same as 1/1.

The correct answer is D.

Step 2. Read the question and all answers.

    4. Based on the results of the second experiment, which of the following average weight gains might be expected if one of the groups of ten dogs received a dog food/supplement mix of 97.5%/2.5%?

The question asks for the expected results from an experiment that was not conducted. This question is an interpolation question. You have to find a value that falls between two other values on the chart.

You have to calculate to answer this question.

The dog food/supplement mix of 97.5%/2.5% falls halfway between 100%/0% and 95%/5%. The average weight gain for 100%/0% was 2.8 kg and the average weight gain for 95%/5% was 3.7 kg.

Find the average of 2.8 and 3.7.

$$2.8 + 3.7 = 6.5. \div 2 = 3.25$$

The expected weight gain for 97.5%/2.5% is 3.25 kg. This must be the correct answer and it appears as an answer choice.

Step 3. Eliminate obviously incorrect answers.

Since you determined the correct answer in step 2, this step is not necessary.

Choices F, H, and J are obviously incorrect.

Step 4. Choose the correct answer from the remaining choices.

The correct answer is 3.25 kg, choice G.

The correct answer is G.

Step 2. Read the question and all answers.

    5. Which of the following graphs best represents the results of the second experiment?

The question asks you to choose a graph that reflects the results from the second experiment. The graphs in the answers suggest direction only and do not contain actual data. The vertical axis shows weight gain and the horizontal axis shows the percent of supplement.

In the second experiment, the percent of food supplement increases at a steady rate. The average weight gain goes up and then comes down a little.

Once you identify the pattern, you may see that this pattern is shown in choice A.

Step 3. Eliminate obviously incorrect answers.

Choice B can't be correct because it shows weight gain staying constant and then going up. Choice C can't be correct because it shows weight gain going up quickly and then going up slowly. Choice D can't be correct because it shows weight gain increasing as the amount of supplement decreases.

Step 4. Choose the correct answer from the remaining choices.

Choice A shows the graph that matches the pattern of weight increase.

The correct answer is A.

Chapter 21

Conflicting Viewpoints

Conflicting viewpoint questions are based on passages that present hypotheses, viewpoints, or ideas that are mutually exclusive. Mutually exclusive means that the hypotheses or viewpoints cannot both be true, although both of them may be false. Note that questions for these passages do not ask you to choose which hypothesis is correct or which viewpoint you agree with.

Conflicting viewpoints passages usually consist of text, but these passages may also include figures and graphs. Answering conflicting viewpoints questions requires you to use the reading comprehension skills. It also helps to have a working science vocabulary.

The Science Vocabulary List

The Science Reasoning Test does not assume you know much science. However, the definitions in the list of words below may help you to answer conflicting viewpoints and other questions.

Don't try to memorize the words and meanings. Just scan them several times to familiarize yourself with the words and their meanings. Even if you don't come across these words on the test, you'll be familiar with science terms.

The Science List

absorption   process by which end products of digestion move from the small intestine into the blood

acid   a compound that dissociates in water to form hydrogen ions

active transport   energy-requiring process that moves materials through a cell membrane

adaptation   a characteristic of an organism that enables it to survive

aerobic respiration   an energy-releasing process of cells that requires oxygen

allergen   a foreign substance that causes an allergic reaction

allergy   overreaction of the immune system to a foreign substance

amino acid   the organic building unit of polypeptides and proteins

anaerobic respiration   an energy-releasing process that does not require oxygen

antibiotic   a substance produced by certain organisms that prevents the growth and multiplication of microorganisms

antibody   a substance produced by the body that counteracts an antigen

antigen   a foreign protein that stimulates the body to form antibodies

aorta   the largest artery of the body; carries oxygenated blood from the left ventricle of the heart to most body organs

atom   an atom has a nucleus that contains neutrons (neutral charge) and protons (positive charge); electrons (negative charge) move around the nucleus

atomic number   the number of protons in the nucleus of an atom of an element

autonomic nervous system   the branch of the nervous system that regulates certain internal responses

base   a compound that dissociates in water to form hydroxide ions

biome   a large climate region composed of a group of ecosystems

calorie   the amount of heat needed to raise the temperature of one gram of water one degree Celsius

carbohydrate   an organic compound that contains carbon, hydrogen, and oxygen; includes sugars, starches, glycogen, and cellulose

catalyst   a substance that changes (usually speeds up) the rate of a chemical reaction without itself being permanently changed

cell membrane   the living outer layer of a cell through which substances pass into and out of the cell; also called the plasma membrane

cellular respiration   the reactions within a cell that release and store energy

cerebellum   the part of the human brain, located behind and below the cerebrum, that controls muscular coordination

cerebrum   the largest part of the human brain; it is involved in sensation, memory, voluntary action, and intelligence

chemical bond   the force of attraction that holds atoms together and stores chemical energy

chemical reaction   any process that results in the production of different substances with new properties

chlorophyll   a complex green pigment that captures light energy, for use in photosynthesis

chromosome   a structure, composed primarily of DNA, that contains the genes

commensalism   a symbiotic relationship in which one organism benefits and the other is not harmed

compound   two or more elements combined chemically in definite proportions by weight

cytoplasm   most of the cellular material located between the nucleus and the cell membrane

dendrite   an extension of the cell body of a nerve cell that forms a synapse with an adjoining nerve cell

density   mass per unit of volume

differentiation   the transformation of embryonic cells into the specialized cells of different tissues

diffusion   the movement of molecules from a region of greater concentration to a region of lesser concentration

digestion   a chemical process that changes complex food molecules into simple food molecules

diploid number   the normal, or species, number of chromosomes characteristic of the body cells of an organism; it is usually designated as $2n$

DNA (deoxyribonucleic acid)   the hereditary material in cells

dominant trait   a hereditary trait that shows itself when its form of its gene is present

ecology   the study of the relationships between organisms and their environment

element   a substance that cannot be chemically changed into a simpler substance; all atoms of an element have the same number of protons

embryo   an organism in the early stages of growth and differentiation

enzyme   an organic catalyst that lowers the activation energy of a reaction, thus speeding up the reaction

eukaryote   any cell or organism that has a membrane enclosing its genetic material

evolution   change over time

excretion   the removal of metabolic wastes from cells and from body fluids

fission   division of a parent cell into two or more daughter cells

fungus (*plural,* **fungi**)   an organism, sometimes parasitic, often saprophytic—living off the nutrients in dead organisms; examples are molds, mildews, and mushrooms

gas   a substance that takes the shape and fills the volume of its container

gene   the portion of a chromosome that carries the genetic information for a specific trait

half-life   the time required for half the atoms in a radioactive specimen to change to stable end products

histamine   a chemical compound formed by cells in response to certain antigens; produces allergy symptoms

homeostasis   the tendency of a living system (organism) to maintain the stability of its internal environment

hydrolysis   a reaction in which a large complex molecule reacts with water to form two simpler molecules

immunity   the ability of the body to combat disease-causing organisms

imprinting   a behavior pattern exhibited by certain animals in response to a stimulus received early in life

inorganic   relating to substances that were never alive; or, relating to compounds that lack carbon and hydrogen

invertebrate   an animal without a backbone

kinetic energy   the energy of motion

lipid   any fat or oil; a fat-soluble organic compound composed of fatty acid molecules and glycerol molecules

liquid   a substance that takes the shape of the vessel that contains it but does not necessarily fill its volume

lymph nodes (lymph glands)   small structures located along the lymph vessels that help protect the body by producing some white blood cells and filtering out bacteria

malleable   the ability of a metal to have its shape permanently changed by applying a force, e.g., hammering

marsupial   a pouched mammal

mass   the measure of the amount of matter in an object; the mass of an object remains the same regardless of the force of gravity

medulla   the part of the brain stem, connecting the brain to the spinal cord, that controls involuntary activities such as breathing and heartbeat

meiosis   a cell division process that reduces the diploid number ($2n$) of chromosomes to the haploid, or monoploid, or number (n)

metabolism   the sum of the building-up and tearing-down reactions that occur in cells

mitosis   the cell division process that duplicates nuclear material (chromosomes) and distributes the material equally between daughter cells

molecule   the smallest unit of an element or a compound, two or more atoms covalently bonded

mutualism   a symbiotic relationship in which both species benefit

Newton's Laws of Motion:

   First   A body will stay in its present state of rest or motion until acted upon by an outside force.

   Second   The change in acceleration of a body is proportional to the force applied.

   Third   For every action there is an equal and opposite reaction.

nucleic acids   DNA and RNA (composed of nucleotides); they both control heredity and protein synthesis

nucleus   the cell organelle that controls the cell's activities and contains DNA

nutrients   molecules that provide energy and/or raw materials for growth, such as proteins, carbohydrates, fats, water, and vitamins

organ   several tissues that work together to perform a function

organic   relating to compounds that contain carbon and hydrogen

osmosis   the movement of water molecules through a semipermeable membrane from a region of greater concentration to a region of lesser concentration

ovary   the egg-producing female reproductive organ of plants or animals

ovule   a reproductive structure in seed plants; after fertilization, the ovule develops into a seed

oxidation   the chemical union of oxygen with a substance; a loss of electrons

parasitism   a symbiotic relationship in which one organism (the parasite) benefits and the other (the host) is harmed

pepsin   a protein-splitting enzyme in gastric juice

periodic event   an event that occurs at regular time intervals

permeability   the extent to which a membrane allows different molecules to pass through it

pH   a measure of the acidity of a solution; a pH of 7 is neutral, less than 7 is acidic, and greater than 7 is basic

phagocyte   a white blood cell that engulfs and ingests foreign matter

phloem   tissue in plants that conducts food

photosynthesis   the process in which energy is used to form carbohydrate and oxygen from carbon dioxide and water

pistil   the female part of a flower; contains the ovary, style, and stigma

plasma   the liquid portion of blood; contains water and dissolved materials

progesterone   a hormone that builds up the lining of the uterus and stimulates the growth of blood vessels in the uterus

prokaryote   any cell that lacks a membrane enclosing its genetic material

protein   a complex organic molecule composed of a chain of amino acids

pulmonary circulation   the circulation of the blood through the lungs and back to the heart

radioactive   refers to elements that emit particles and radiation during the spontaneous disintegration of their nuclei

recessive trait   a hereditary trait that does not appear in an individual if the dominant form of the gene is present

reflection   occurs when light bounces off a surface; when light is reflected from a flat mirror, the angle of *incidence* (light striking the mirror) equals the angle of *reflection* (light leaving the mirror)

reproduction   the life activity by which organisms produce offspring

respiration (cellular)   the process in which carbohydrates react with oxygen, which releases energy and produces water and carbon dioxide

RNA (ribonucleic acid)   a single chain of nucleotides patterned from a DNA template

saprophyte   an organism that obtains food by absorbing organic matter from dead or decaying organisms

seed   a ripened ovule; contains an embryo plant and stored food

semicircular canals   structures located in each inner ear that detect changes in body movement and help to maintain balance

solid   matter that has a definite shape and volume

solution   a homogeneous mixture formed when one substance dissolves in another

solvent   a substance in which a solute dissolves to form a solution

spore   an asexual cell that can withstand unfavorable conditions and is capable of producing a new organism

stamen   the male reproductive organ of a flower; consists of the filament and the anther, which produces pollen grains

stimulus   any change in an organism's internal or external environment

symbiosis   a permanent relationship between two different organisms living together; one organism lives either on, in, or near the other

synapse   the space between the end brush of one neuron and the cell body of another neuron across which a nerve impulse (signal) passes

systemic circulation   the circulation of the blood through all parts of the body except the lungs

taiga   the land biome south of the tundra, which is characterized by coniferous forests

thorax   in insects, the region of the body between the head and abdomen; in humans, the chest region

tissue   a group of similar cells that carry out a specialized activity

tropism   an automatic response of a plant or part of a plant to an environmental stimulus such as sunlight or gravity

tundra   the land biome located north of the taiga biome, which is characterized by permanently frozen subsoil

umbilical cord   a structure in mammals that connects a fetus with a placenta

uterus   a muscular organ in female mammals in which an embryo develops

vertebra (*plural,* **vertebrae**)   one of the bones of the spinal column

vertebrate   an animal with a backbone

virus   an infectious particle showing some characteristics of life that survives only as a parasite in a host cell

viscous   describes a material that flows slowly

vitamin   an organic nutrient that usually cannot be manufactured by the body; most function to assist enzymes

water cycle   the movement of water from the atmosphere to the ground, through organisms, and back to the atmosphere

weight   the measure of the gravitational force that attracts an object; the weight of an object changes as the force of gravity changes

xylem   water-conducting tissue in plants

yolk   stored food material in an egg cell

zygote   a fertilized egg cell

■ Reading Conflicting Viewpoints Passages

Use the same four steps you have used for other types of passages. Include these four questions in Step 1.

> What is the fundamental issue?
> What is each scientist's position?
> What points does each scientist use to support his or her position?
> What flaws are there in each scientist's position?

   Here is a conflicting viewpoints passage. This section shows how to use the four steps to answer the questions.

This conflicting viewpoints passage and questions might appear on an actual ACT. This section shows you how to use the four-step approach to answer each question.

PASSAGE I

Scientist 1

I am writing this brief passage to explain why Earth has a primary atmosphere. By this I mean that Earth's atmosphere of 79% nitrogen, 20% oxygen, and trace amounts of water vapor and other gases existed essentially in that form from the time of Earth's formation.

Earth was formed some 4 billion years ago and forms of life began to appear about 3.5 billion years ago. Even this very early life required an atmosphere with a great deal of nitrogen and oxygen. The prokaryotic cells that formed first were anaerobic (could respire without oxygen), but this probably means that these cells were formed in seawater, which contains very little oxygen.

Earth's gravity held the oxygen-rich atmosphere close to the surface. This process explains why humans and other life-forms rely on the oxygen in Earth's atmosphere for survival.

Scientist 2

Earth has a secondary atmosphere, which means that Earth's atmosphere was formed some time after the planet was formed and after life first appeared. There was little if any oxygen in Earth's atmosphere 4 billion years ago.

In fact, the first life-forms to appear in the atmosphere were anaerobic prokaryotes, bacteria that do not need oxygen to live. Research conducted in the 1950s showed that amino acids and life could form in the oxygen-poor atmosphere found on Earth 3 or 4 billion years ago.

What happened was that volcanic eruptions released great quantities of nitrogen and oxygen into the atmosphere. Later, green plants released even more oxygen into the atmosphere. These quantities of nitrogen and oxygen were held in the atmosphere by Earth's gravitational field, resulting in the 79% nitrogen/20% oxygen atmosphere found today near Earth's surface.

1. Which of the following calls into question part of Scientist 1's argument that Earth has a primary atmosphere?
   A. Krypton is one of the trace gases in Earth's atmosphere.
   B. The first life on Earth was formed in freshwater.
   C. Earth's atmosphere contains about 79% nitrogen and 20% oxygen.
   D. Life on Earth probably began less than 3.5 billion years ago.

2. One point which both scientists agree on is:
   F. the universe is more than 7 billion years old.
   G. life-forms became healthier through aerobic exercise.
   H. anaerobic respiration preceded aerobic respiration in life on Earth.
   J. Earth's atmosphere has not changed significantly since life-forms first appeared on Earth.

3. Regardless of which scientist is correct:
   A. the atmosphere grows less dense as you get farther from Earth.
   B. Earth's atmosphere has always contained about 79% nitrogen.
   C. green plants existed about 3.5 billion years ago.
   D. aerobic respiration preceded anaerobic respiration in life on Earth.

4. If a scientist finds that all cells living in the atmosphere at a particular time in Earth's history were anaerobic prokaryotes, the scientist could reasonably conclude that:
   F. Earth has a primary atmosphere.
   G. there was very little oxygen in the atmosphere at that time.
   H. Earth was more than 4 billion years old at that time.
   J. these cells were the predecessors of reptiles or mammals.

5. The scientist who is correct is definitively indicated by:
   A. actual samples of prokaryotes.
   B. actual samples of Earth's current atmosphere from high altitudes.
   C. actual samples of seawater from very shallow and very deep parts of the ocean.
   D. actual samples of the atmosphere from 3.5 billion years ago.

Use the Four Steps to Answer the Questions.

Step 1. Skim and identify the passage. Answer the conflicting viewpoints questions.

This passage is a conflicting viewpoints question presented entirely in text form.

What is the fundamental issue?

Does the Earth have a primary or secondary atmosphere?

Primary atmosphere means it was the same at the beginning as it is today.

Secondary atmosphere means that Earth's atmosphere changed to become the atmosphere of today.

What is each scientist's position?

Scientist 1 says primary atmosphere.
Scientist 2 says secondary atmosphere.

What points does each scientist use to support his or her position?

Scientist 1: The first cells—anaerobic prokaryotes—must have formed in the ocean. Gravity holds the atmosphere close to Earth's surface.

Scientist 2: The first cells did not use oxygen. Research shows amino acids, and probably life, can form without oxygen.

What flaws are in each scientist's position?

Scientist 1: No absolute proof. No explanation for why first cells are anaerobic despite a plentiful supply of oxygen. The link between gravity holding the atmosphere and organisms needing oxygen is not explained, and seems questionable.

Scientist 2: No absolute proof.

Step 2. Read the question and all answers.

1. Which of the following calls into question part of Scientist 1's argument that Earth has a primary atmosphere?

To call an argument into question, you need to find something wrong with the argument. Sometimes, you do this by pointing out a flaw in reasoning. Other times, you do this by pointing out contradictions—either in the argument itself, or between the argument and other statements. In this case, you need to find the statement that contradicts something in Scientist 1's argument.

Step 3. Eliminate obviously incorrect answers.

The passage has nothing to do with krypton, so you can eliminate choice A. Scientist 1 agrees that Earth's atmosphere contains about 79% nitrogen and 20% oxygen, so you can eliminate choice C.

Step 4. Choose the correct answer from the remaining choices.

Choice B, which states that life was first formed in freshwater, directly contradicts Scientist 1's statement that life first formed in seawater. Choice D, which states that life began less than 3.5 billion years ago, is not that different from Scientist 1's statement that forms of life began to appear 3.5 billion years ago. In addition, the timing of life's appearance has little to do with whether Earth has a primary or secondary atmosphere.

The correct answer is B.

Step 2. Read the question and all answers.

    2. One point which both scientists agree on is:

The questions asks for a point on which both scientists agree. The answers are not directly from the passage.

Step 3. Eliminate incorrect answers.

    Choice G is incorrect. There's nothing in either passage about aerobic exercise. Be careful not to be drawn to this choice just because it includes the word *aerobic*.

    Choice J is also incorrect. This statement is exactly what the scientists *do not* agree on.

    That leaves choices F and H.

Step 4. Choose the correct answer from the remaining choices.

    Both scientists appear to agree that Earth is about 4 billion years old. But is there anything in these passages to indicate that they agree that the universe is more than 7 billion years old? No. Choice F must be incorrect. This leaves choice H. Just to be safe, let's make sure it makes sense. Both scientists say that prokaryotes were the first life-forms on Earth. Both also say that these prokaryotes used a form of anaerobic respiration, which means they agree that anaerobic respiration came first. So they must also agree that anaerobic respiration preceded aerobic respiration.

    The correct answer is H.

Step 2. Read the question and all answers.

    3. Regardless of which scientist is correct:

The question asks for a statement that is true regardless of which scientist is correct. The answer choices seem straightforward, but remember the answer must be based on the passage. **Do not use your own opinion.**

Step 3. Eliminate obviously incorrect answers.

    Choice B is incorrect. This is what the scientists disagree about.

    Choice D is incorrect. As you determined in the previous question, both scientists agree that this statement is false.

    That leaves choices A and C.

Step 4. Choose the correct answer from the remaining choices.

    Choice A states that the atmosphere grows less dense as one gets farther from Earth. Both scientists agree that gravity holds the atmosphere close to Earth. The force of gravity grows weaker as you get farther from Earth. This answer makes sense and seems to be the correct answer.

    Choice C states that green plants existed 3.5 billion years ago. There is nothing in this passage to indicate that green plants existed 3.5 billion years ago, around the time life first appeared. In fact, Scientist 2 says the green plants came "later." This answer is not correct.

    The correct answer is A.

Step 2. Read the question and all answers.

4. If a scientist finds that all cells living in the atmosphere at a particular time in Earth's history were anaerobic prokaryotes, the scientist could reasonably conclude that:

This question asks you to draw an inference. What if all the cells living at a particular time were prokaryotes?

You know both scientists agree that prokaryotes were the first life-forms.

Step 3. Eliminate incorrect answers.

Choice H is incorrect. Prokaryotes have nothing to do with Earth being more than 4 billion years old.

Choice J is incorrect. Nothing in the passage says anything about these cells being the forerunners of plants or animals.

That leaves choices F and G.

Step 4. Choose the correct answer from the remaining choices.

If Earth has a primary atmosphere, this would mean that the anaerobic prokaryotes were living in a high-oxygen environment. You might recall from a biology class that some anaerobes cannot survive if oxygen is present, and that aerobic respiration is much more effective than anaerobic respiration. It doesn't make sense that anaerobes would survive in an atmosphere poisonous to them, or that they would not have been quickly supplanted by aerobic organisms, which are better suited for a high-oxygen environment. Choice F is not correct.

Choice G is essentially the opposite of choice F. Since it makes sense for anaerobic organisms to be in a low-oxygen environment, choice G is a reasonable answer.

The correct answer is G.

Step 2. Read the question and all answers.

5. The scientist who is correct is definitively indicated by:

The question asks you to identify the statement that would settle this disagreement. The answers are general statements.

Step 3. Eliminate obviously incorrect answers.

Answer A is incorrect. Actual samples of prokaryotes would not tell you which scientist was correct. The scientists agree about the makeup of these cells.

Answer B is incorrect. There is nothing about Earth's current atmosphere that would resolve this dispute. The scientists agree about Earth's current atmosphere.

Answer C is incorrect. The makeup of seawater is not a central issue in this dispute.

Only choice D remains.

Step 4. Choose the correct answer from the remaining choices.

Choice D is obviously the correct choice. Any actual sample of the atmosphere from 3.5 billion years ago would settle this disagreement once and for all. The scientists would know the actual composition of the atmosphere from that time and know whether Earth has a primary or secondary atmosphere.

The correct answer is D.

Chapter 22 ▪

Diagnostic Science Reasoning ACT: Model Science Reasoning **ACT I**

▪ Strategies for Passing the Science Reasoning ACT

Use these strategies as you take the Science Reasoning ACT. Remember to follow the test preparation strategies on pages 477–479.

Use the four steps.

Use these four steps described on pages 477–483 and reviewed below.

Step 1. Skim the passage and look over any charts and graphs.

> Don't try to read the entire passage. Just skim over the written information and familiarize yourself with any figures.

Identify the passage as data representation, research summary, or conflicting viewpoints. The questions are easier to answer once you have identified the type of passage you are reading.

Step 2. Read the questions and possible answers.

The questions and possible answers point to the parts of the passage you should concentrate on.

Step 3. Eliminate those answers that are obviously incorrect.

Every answer you can eliminate increases your chance of selecting the correct answer.

Step 4. Choose the correct answer from the choices remaining.

Choose the correct answer, if you know it. If not, guess from among the remaining choices. **Never leave an answer blank!** Always select one of the answers. Unlike other similar tests, there is no penalty for selecting an incorrect answer.

Follow the strategies for taking multiple-choice tests.

Pages 478–483 provide strategies for taking any multiple-choice test. Use these strategies as you take the Science Reasoning Test.

Remember, the machine that grades the ACT senses whether you have marked the correct space on the answer sheet. Mark your answer sheet carefully.

Pace yourself.

You have 35 minutes to answer 40 questions. Use 45 seconds to answer each question and you will have about five minutes left to check your answers. For most questions, 45 seconds is ample time to answer a question.

This is a science reasoning test.

The science reasoning test measures your ability to read and understand the information in the passage, including graphs, diagrams, and tables. While experience reading science information is a big help, test questions do not assume that you *know* the science discussed in the passages.

Difficult concepts are explained.

Any difficult terms or concepts you need to know are usually explained in the passage. If you see a term or concept explained, read the explanation carefully. You will probable have to understand the term or concept to answer one of the questions.

Understand enough to answer the questions.

Passages may contain more scientific information than you need to answer the questions. You just have to understand enough to answer the questions. No one expects you to know additional information. Don't even try to learn the material presented.

The ACT measures only your ability to answer the test questions. The ACT does not measure your understanding of the science in the passage.

Draw, sketch, and write on graphs, tables, charts, and diagrams.

Most passages contain graphs, tables, charts, and diagrams. Often, it helps to draw a line, fill in missing information in a table, or sketch a figure. Jot down any notes that will help you answer a question. You can also circle or underline important information and cross out information you do not need. Remember, the test booklet is yours and you can make any marks on it you like.

Diagnostic Science Reasoning ACT

You should take this Diagnostic Science Reasoning ACT after you complete the science review. This test is just like a real ACT. It is the first of four practice ACTs in this book.

Take the Diagnostic ACT under simulated test conditions. Allow 35 minutes to answer the 40 test questions. Tear out the answer sheet on the following page and answer the questions in the section labeled Test 4. Use a pencil to mark the answer sheet.

Use the answer key on pages 528–530 to correct the answer sheet and to review the answer explanations. Then turn to the chart on page 527 to find out which science review sections to reread. The test scoring chart on page 589 shows you how to convert the number correct into ACT scale scores.

Each question is numbered and the answers are lettered. Choose the best answer for each question. Then mark the letter for your answer in the correct space on the answer sheet. If you want to change an answer, completely erase the pencil mark from the first circle and mark another one.

DO NOT leave any answers blank. There is no penalty for guessing incorrectly on the ACT. Remember that the test booklet is yours. You may mark up, write on, or draw on the test booklet.

When you are ready to begin, note the time. Stop in exactly 35 minutes.

ANSWER SHEET

The ACT answer sheet looks something like this one. Use a No. 2 pencil
to completely fill the circle corresponding to the correct answer.
If you erase, erase completely; incomplete erasures may be read as answers.

TEST 1—English

1 Ⓐ Ⓑ Ⓒ Ⓓ	11 Ⓐ Ⓑ Ⓒ Ⓓ	21 Ⓐ Ⓑ Ⓒ Ⓓ	31 Ⓐ Ⓑ Ⓒ Ⓓ	41 Ⓐ Ⓑ Ⓒ Ⓓ	51 Ⓐ Ⓑ Ⓒ Ⓓ	61 Ⓐ Ⓑ Ⓒ Ⓓ	71 Ⓐ Ⓑ Ⓒ Ⓓ
2 Ⓕ Ⓖ Ⓗ Ⓙ	12 Ⓕ Ⓖ Ⓗ Ⓙ	22 Ⓕ Ⓖ Ⓗ Ⓙ	32 Ⓕ Ⓖ Ⓗ Ⓙ	42 Ⓕ Ⓖ Ⓗ Ⓙ	52 Ⓕ Ⓖ Ⓗ Ⓙ	62 Ⓕ Ⓖ Ⓗ Ⓙ	72 Ⓕ Ⓖ Ⓗ Ⓙ
3 Ⓐ Ⓑ Ⓒ Ⓓ	13 Ⓐ Ⓑ Ⓒ Ⓓ	23 Ⓐ Ⓑ Ⓒ Ⓓ	33 Ⓐ Ⓑ Ⓒ Ⓓ	43 Ⓐ Ⓑ Ⓒ Ⓓ	53 Ⓐ Ⓑ Ⓒ Ⓓ	63 Ⓐ Ⓑ Ⓒ Ⓓ	73 Ⓐ Ⓑ Ⓒ Ⓓ
4 Ⓕ Ⓖ Ⓗ Ⓙ	14 Ⓕ Ⓖ Ⓗ Ⓙ	24 Ⓕ Ⓖ Ⓗ Ⓙ	34 Ⓕ Ⓖ Ⓗ Ⓙ	44 Ⓕ Ⓖ Ⓗ Ⓙ	54 Ⓕ Ⓖ Ⓗ Ⓙ	64 Ⓕ Ⓖ Ⓗ Ⓙ	74 Ⓕ Ⓖ Ⓗ Ⓙ
5 Ⓐ Ⓑ Ⓒ Ⓓ	15 Ⓐ Ⓑ Ⓒ Ⓓ	25 Ⓐ Ⓑ Ⓒ Ⓓ	35 Ⓐ Ⓑ Ⓒ Ⓓ	45 Ⓐ Ⓑ Ⓒ Ⓓ	55 Ⓐ Ⓑ Ⓒ Ⓓ	65 Ⓐ Ⓑ Ⓒ Ⓓ	75 Ⓐ Ⓑ Ⓒ Ⓓ
6 Ⓕ Ⓖ Ⓗ Ⓙ	16 Ⓕ Ⓖ Ⓗ Ⓙ	26 Ⓕ Ⓖ Ⓗ Ⓙ	36 Ⓕ Ⓖ Ⓗ Ⓙ	46 Ⓕ Ⓖ Ⓗ Ⓙ	56 Ⓕ Ⓖ Ⓗ Ⓙ	66 Ⓕ Ⓖ Ⓗ Ⓙ	
7 Ⓐ Ⓑ Ⓒ Ⓓ	17 Ⓐ Ⓑ Ⓒ Ⓓ	27 Ⓐ Ⓑ Ⓒ Ⓓ	37 Ⓐ Ⓑ Ⓒ Ⓓ	47 Ⓐ Ⓑ Ⓒ Ⓓ	57 Ⓐ Ⓑ Ⓒ Ⓓ	67 Ⓐ Ⓑ Ⓒ Ⓓ	
8 Ⓕ Ⓖ Ⓗ Ⓙ	18 Ⓕ Ⓖ Ⓗ Ⓙ	28 Ⓕ Ⓖ Ⓗ Ⓙ	38 Ⓕ Ⓖ Ⓗ Ⓙ	48 Ⓕ Ⓖ Ⓗ Ⓙ	58 Ⓕ Ⓖ Ⓗ Ⓙ	68 Ⓕ Ⓖ Ⓗ Ⓙ	
9 Ⓐ Ⓑ Ⓒ Ⓓ	19 Ⓐ Ⓑ Ⓒ Ⓓ	29 Ⓐ Ⓑ Ⓒ Ⓓ	39 Ⓐ Ⓑ Ⓒ Ⓓ	49 Ⓐ Ⓑ Ⓒ Ⓓ	59 Ⓐ Ⓑ Ⓒ Ⓓ	69 Ⓐ Ⓑ Ⓒ Ⓓ	
10 Ⓕ Ⓖ Ⓗ Ⓙ	20 Ⓕ Ⓖ Ⓗ Ⓙ	30 Ⓕ Ⓖ Ⓗ Ⓙ	40 Ⓕ Ⓖ Ⓗ Ⓙ	50 Ⓕ Ⓖ Ⓗ Ⓙ	60 Ⓕ Ⓖ Ⓗ Ⓙ	70 Ⓕ Ⓖ Ⓗ Ⓙ	

TEST 2—Mathematics

1 Ⓐ Ⓑ Ⓒ Ⓓ Ⓔ	9 Ⓐ Ⓑ Ⓒ Ⓓ Ⓔ	17 Ⓐ Ⓑ Ⓒ Ⓓ Ⓔ	25 Ⓐ Ⓑ Ⓒ Ⓓ Ⓔ	33 Ⓐ Ⓑ Ⓒ Ⓓ Ⓔ	41 Ⓐ Ⓑ Ⓒ Ⓓ Ⓔ	49 Ⓐ Ⓑ Ⓒ Ⓓ Ⓔ	57 Ⓐ Ⓑ Ⓒ Ⓓ Ⓔ
2 Ⓕ Ⓖ Ⓗ Ⓙ Ⓚ	10 Ⓕ Ⓖ Ⓗ Ⓙ Ⓚ	18 Ⓕ Ⓖ Ⓗ Ⓙ Ⓚ	26 Ⓕ Ⓖ Ⓗ Ⓙ Ⓚ	34 Ⓕ Ⓖ Ⓗ Ⓙ Ⓚ	42 Ⓕ Ⓖ Ⓗ Ⓙ Ⓚ	50 Ⓕ Ⓖ Ⓗ Ⓙ Ⓚ	58 Ⓕ Ⓖ Ⓗ Ⓙ Ⓚ
3 Ⓐ Ⓑ Ⓒ Ⓓ Ⓔ	11 Ⓐ Ⓑ Ⓒ Ⓓ Ⓔ	19 Ⓐ Ⓑ Ⓒ Ⓓ Ⓔ	27 Ⓐ Ⓑ Ⓒ Ⓓ Ⓔ	35 Ⓐ Ⓑ Ⓒ Ⓓ Ⓔ	43 Ⓐ Ⓑ Ⓒ Ⓓ Ⓔ	51 Ⓐ Ⓑ Ⓒ Ⓓ Ⓔ	59 Ⓐ Ⓑ Ⓒ Ⓓ Ⓔ
4 Ⓕ Ⓖ Ⓗ Ⓙ Ⓚ	12 Ⓕ Ⓖ Ⓗ Ⓙ Ⓚ	20 Ⓕ Ⓖ Ⓗ Ⓙ Ⓚ	28 Ⓕ Ⓖ Ⓗ Ⓙ Ⓚ	36 Ⓕ Ⓖ Ⓗ Ⓙ Ⓚ	44 Ⓕ Ⓖ Ⓗ Ⓙ Ⓚ	52 Ⓕ Ⓖ Ⓗ Ⓙ Ⓚ	60 Ⓕ Ⓖ Ⓗ Ⓙ Ⓚ
5 Ⓐ Ⓑ Ⓒ Ⓓ Ⓔ	13 Ⓐ Ⓑ Ⓒ Ⓓ Ⓔ	21 Ⓐ Ⓑ Ⓒ Ⓓ Ⓔ	29 Ⓐ Ⓑ Ⓒ Ⓓ Ⓔ	37 Ⓐ Ⓑ Ⓒ Ⓓ Ⓔ	45 Ⓐ Ⓑ Ⓒ Ⓓ Ⓔ	53 Ⓐ Ⓑ Ⓒ Ⓓ Ⓔ	
6 Ⓕ Ⓖ Ⓗ Ⓙ Ⓚ	14 Ⓕ Ⓖ Ⓗ Ⓙ Ⓚ	22 Ⓕ Ⓖ Ⓗ Ⓙ Ⓚ	30 Ⓕ Ⓖ Ⓗ Ⓙ Ⓚ	38 Ⓕ Ⓖ Ⓗ Ⓙ Ⓚ	46 Ⓕ Ⓖ Ⓗ Ⓙ Ⓚ	54 Ⓕ Ⓖ Ⓗ Ⓙ Ⓚ	
7 Ⓐ Ⓑ Ⓒ Ⓓ Ⓔ	15 Ⓐ Ⓑ Ⓒ Ⓓ Ⓔ	23 Ⓐ Ⓑ Ⓒ Ⓓ Ⓔ	31 Ⓐ Ⓑ Ⓒ Ⓓ Ⓔ	39 Ⓐ Ⓑ Ⓒ Ⓓ Ⓔ	47 Ⓐ Ⓑ Ⓒ Ⓓ Ⓔ	55 Ⓐ Ⓑ Ⓒ Ⓓ Ⓔ	
8 Ⓕ Ⓖ Ⓗ Ⓙ Ⓚ	16 Ⓕ Ⓖ Ⓗ Ⓙ Ⓚ	24 Ⓕ Ⓖ Ⓗ Ⓙ Ⓚ	32 Ⓕ Ⓖ Ⓗ Ⓙ Ⓚ	40 Ⓕ Ⓖ Ⓗ Ⓙ Ⓚ	48 Ⓕ Ⓖ Ⓗ Ⓙ Ⓚ	56 Ⓕ Ⓖ Ⓗ Ⓙ Ⓚ	

TEST 3—Reading

1 Ⓐ Ⓑ Ⓒ Ⓓ	6 Ⓕ Ⓖ Ⓗ Ⓙ	11 Ⓐ Ⓑ Ⓒ Ⓓ	16 Ⓕ Ⓖ Ⓗ Ⓙ	21 Ⓐ Ⓑ Ⓒ Ⓓ	26 Ⓕ Ⓖ Ⓗ Ⓙ	31 Ⓐ Ⓑ Ⓒ Ⓓ	36 Ⓕ Ⓖ Ⓗ Ⓙ
2 Ⓕ Ⓖ Ⓗ Ⓙ	7 Ⓐ Ⓑ Ⓒ Ⓓ	12 Ⓕ Ⓖ Ⓗ Ⓙ	17 Ⓐ Ⓑ Ⓒ Ⓓ	22 Ⓕ Ⓖ Ⓗ Ⓙ	27 Ⓐ Ⓑ Ⓒ Ⓓ	32 Ⓕ Ⓖ Ⓗ Ⓙ	37 Ⓐ Ⓑ Ⓒ Ⓓ
3 Ⓐ Ⓑ Ⓒ Ⓓ	8 Ⓕ Ⓖ Ⓗ Ⓙ	13 Ⓐ Ⓑ Ⓒ Ⓓ	18 Ⓕ Ⓖ Ⓗ Ⓙ	23 Ⓐ Ⓑ Ⓒ Ⓓ	28 Ⓕ Ⓖ Ⓗ Ⓙ	33 Ⓐ Ⓑ Ⓒ Ⓓ	38 Ⓕ Ⓖ Ⓗ Ⓙ
4 Ⓕ Ⓖ Ⓗ Ⓙ	9 Ⓐ Ⓑ Ⓒ Ⓓ	14 Ⓕ Ⓖ Ⓗ Ⓙ	19 Ⓐ Ⓑ Ⓒ Ⓓ	24 Ⓕ Ⓖ Ⓗ Ⓙ	29 Ⓐ Ⓑ Ⓒ Ⓓ	34 Ⓕ Ⓖ Ⓗ Ⓙ	39 Ⓐ Ⓑ Ⓒ Ⓓ
5 Ⓐ Ⓑ Ⓒ Ⓓ	10 Ⓕ Ⓖ Ⓗ Ⓙ	15 Ⓐ Ⓑ Ⓒ Ⓓ	20 Ⓕ Ⓖ Ⓗ Ⓙ	25 Ⓐ Ⓑ Ⓒ Ⓓ	30 Ⓕ Ⓖ Ⓗ Ⓙ	35 Ⓐ Ⓑ Ⓒ Ⓓ	40 Ⓕ Ⓖ Ⓗ Ⓙ

TEST 4—Science Reasoning

1 Ⓐ Ⓑ Ⓒ Ⓓ	6 Ⓕ Ⓖ Ⓗ Ⓙ	11 Ⓐ Ⓑ Ⓒ Ⓓ	16 Ⓕ Ⓖ Ⓗ Ⓙ	21 Ⓐ Ⓑ Ⓒ Ⓓ	26 Ⓕ Ⓖ Ⓗ Ⓙ	31 Ⓐ Ⓑ Ⓒ Ⓓ	36 Ⓕ Ⓖ Ⓗ Ⓙ
2 Ⓕ Ⓖ Ⓗ Ⓙ	7 Ⓐ Ⓑ Ⓒ Ⓓ	12 Ⓕ Ⓖ Ⓗ Ⓙ	17 Ⓐ Ⓑ Ⓒ Ⓓ	22 Ⓕ Ⓖ Ⓗ Ⓙ	27 Ⓐ Ⓑ Ⓒ Ⓓ	32 Ⓕ Ⓖ Ⓗ Ⓙ	37 Ⓐ Ⓑ Ⓒ Ⓓ
3 Ⓐ Ⓑ Ⓒ Ⓓ	8 Ⓕ Ⓖ Ⓗ Ⓙ	13 Ⓐ Ⓑ Ⓒ Ⓓ	18 Ⓕ Ⓖ Ⓗ Ⓙ	23 Ⓐ Ⓑ Ⓒ Ⓓ	28 Ⓕ Ⓖ Ⓗ Ⓙ	33 Ⓐ Ⓑ Ⓒ Ⓓ	38 Ⓕ Ⓖ Ⓗ Ⓙ
4 Ⓕ Ⓖ Ⓗ Ⓙ	9 Ⓐ Ⓑ Ⓒ Ⓓ	14 Ⓕ Ⓖ Ⓗ Ⓙ	19 Ⓐ Ⓑ Ⓒ Ⓓ	24 Ⓕ Ⓖ Ⓗ Ⓙ	29 Ⓐ Ⓑ Ⓒ Ⓓ	34 Ⓕ Ⓖ Ⓗ Ⓙ	39 Ⓐ Ⓑ Ⓒ Ⓓ
5 Ⓐ Ⓑ Ⓒ Ⓓ	10 Ⓕ Ⓖ Ⓗ Ⓙ	15 Ⓐ Ⓑ Ⓒ Ⓓ	20 Ⓕ Ⓖ Ⓗ Ⓙ	25 Ⓐ Ⓑ Ⓒ Ⓓ	30 Ⓕ Ⓖ Ⓗ Ⓙ	35 Ⓐ Ⓑ Ⓒ Ⓓ	40 Ⓕ Ⓖ Ⓗ Ⓙ

Diagnostic Science Reasoning ACT

Follow these steps to evaluate your Diagnostic Science Reasoning ACT.

1. Turn to pages 528–530 to mark your answer sheet and review the answer explanations.

2. Circle the number of each incorrect answer on the answer sheet.

3. Write the number of correct answers in the space provided at the bottom of this sheet.

4. Use the chart on page 527 to find out which science review sections to reread.

5. Use the chart on page 589 to convert the number of correct answers to an ACT scale score.

Number Correct _____   Science Reasoning Scale Score _____

Diagnostic Science Reasoning ACT

Diagnostic Science Reasoning ACT

40 Questions—35 Minutes

INSTRUCTIONS: Each of the seven science passages on this test is followed by questions. Choose the best answer for each question based on the passage. Then fill in the appropriate circle on the answer sheet. You can look back at the passages as often as you want.

You CANNOT use a calculator on this test.

Check pages 528–530 for answers and explanations.

PASSAGE I Graphing

Oxidant concentration in the air and people's physical symptoms were monitored for twelve months. The following graphs show the data for five symptoms: eye irritation, excessive tearing, sore throat, dyspnea (shortness of breath), and headache. The graphs show by month and year the percentage of people complaining of a particular symptom when the oxidant concentration was more than 0.10 parts per million (ppm) and when it was less than 0.10 ppm.

The months of the year are classified as follows:

Winter: December, January, February

Spring: March, April, May

Summer: June, July, August

Fall: September, October, November

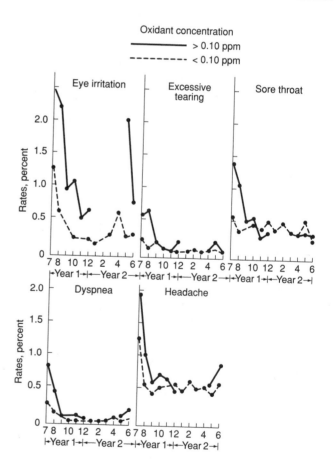

Figure 1

GO ON TO THE NEXT PAGE.

1. According to Figure 1, which symptom shows the smallest difference in percentage of complaints for the two oxidant concentrations during the month of June?

    A. Excessive tearing
    B. Sore throat
    C. Dyspnea
    D. Headache

2. According to Figure 1, the pattern of sore throat complaints when the oxidant level is above 0.10 ppm is most like that for:

    F. headache complaints when the oxidant level is above 0.10 ppm.
    G. excessive tearing complaints when the oxidant level is below 0.10 ppm.
    H. eye irritation complaints when the oxidant level is above 0.10 ppm.
    J. dyspnea complaints when the oxidant level is above 0.10 ppm.

3. According to the graph in Figure 1, the percentage of people complaining about eye irritation when the oxidant concentration is above 0.10 ppm during the coming fall would be about:

    A. 0.8%.
    B. 1%.
    C. 8%.
    D. 10%.

4. Which of the following best explains why the lowest rate of eye irritation for oxidant concentration under 0.10 ppm occurred during year 2 of this study?

    F. Instruments were not correctly calibrated at the beginning of the study.
    G. People stopped reporting symptoms as the study progressed.
    H. Air was more polluted during the first year than during the second year.
    J. Data were collected for January in year 2 of the study.

5. Data for eye irritation when the oxidant concentration is greater than 0.10 ppm are missing for January through April. Which of the following is the most reasonable explanation for the missing data?

    A. During these months, there were no complaints about eye irritation when oxidant concentration was above 0.10 ppm.
    B. During these months, oxidant concentration was always below 0.10 ppm.
    C. During these months, oxidant concentration was always above 0.10 ppm.
    D. During these months, eye irritation could not be measured because there were too many other causes of eye irritation.

GO ON TO THE NEXT PAGE.

PASSAGE II   Graphing

Scientists studied the climate during the most recent glacial period by taking readings from North Atlantic sediment cores. Figure 1 shows the age of the sediments, the thickness of the sediments, and the distance covered by the glacier from west to east across western Scandinavia. Scientists are interested in the relationship between the amount of sediment and the area covered by the glacier.

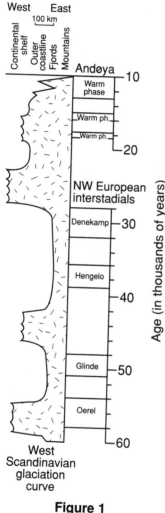

Figure 1

6. According to Figure 1, what is the age in years of the thinnest layer of sediment in the core sample?
   F. 10,000
   G. 20,000
   H. 45,000
   J. 55,000

7. When was glacial coverage less than 100 km?
   A. 18,000–30,000 years ago and 42,000–50,000 years ago
   B. 30,000–36,000 years ago and 58,000–60,000 years ago
   C. 30,000–42,000 years ago and 48,000–54,000 years ago
   D. 10,000–28,000 years ago and 54,000–60,000 years ago

8. For which of the named periods on the right of the graph were there no increases in the amount of sediment at the North Atlantic site from the beginning of the period?
   F. Denekamp
   G. Hengelo
   H. Glinde
   J. Oerel

9. During the most recent glacial period, glacial coverage was greatest around:
   A. 17,000 B.C.
   B. 19,000 B.C.
   C. 23,000 B.C.
   D. 30,000 B.C.

10. During the time between the Glinde period and the Hengelo period, approximately when would have the fjords first been covered by the glacier?
    F. 28,000 years ago
    G. 38,000 years ago
    H. 47,000 years ago
    J. 54,000 years ago

GO ON TO THE NEXT PAGE.

PASSAGE III

A study compared several physical characteristics of elderly recreational runners with those of elderly and younger controls. The elderly runners (ER) were men over 60 and were all nonsmokers. The elderly controls (EC) were men over 60 who did not engage in recreational running and included smokers and nonsmokers. The young controls (YC) were men under 20 who did not engage in recreational running and included smokers and nonsmokers.

Study 1

Scientists measured several skinfolds in millimeters and calculated the sum of those measurements. The scientists then determined the oxygen intake of each group, shown below in Table 1.

Table 1

	Sum of skinfolds	*Oxygen intake*
Young controls	157 mm	40 ml/kg
Elderly controls	124 mm	29 ml/kg
Elderly runners	86 mm	39 ml/kg

Study 2

Scientists measured the white blood cells (WBC) in blood samples from the smokers and nonsmokers. They obtained specific counts for two types of white blood cells: lymphocytes, which can produce antibodies and change into other types of blood cells; and eosinophils, which increase in number when the body is exposed to foreign proteins. The results are shown in Table 2.

Table 2

	Total WBC	*Lymphocytes*	*Eosinophils*
YC nonsmokers	6,100 ± 400	2,480 ± 160	190 ± 40
YC smokers	7,730 ± 470	3,270 ± 430	360 ± 60
EC nonsmokers	5,930 ± 430	1,970 ± 230	190 ± 30
EC smokers	7,490 ± 640	3,320 ± 410	220 ± 30
ER nonsmokers	5,430 ± 210	1,970 ± 100	150 ± 10

11. A nonsmoker with high lymphocyte production is most likely:

    A. young.
    B. elderly.
    C. a runner.
    D. a nonrunner.

12. Which of the following groups is most likely to have the lowest level of eosinophils?

    F. Smokers
    G. Elderly
    H. Runners
    J. Young

13. Which of the following conclusions can be drawn from these studies?

    A. Oxygen intake is lower in smokers than in nonsmokers.
    B. Skinfold measurements are lower in smokers than in nonsmokers.
    C. Total WBC is higher in smokers than in nonsmokers.
    D. Lymphocyte production is lower in smokers than in nonsmokers.

14. Which of the following graphs best represents the eosinophil counts for Study 2?

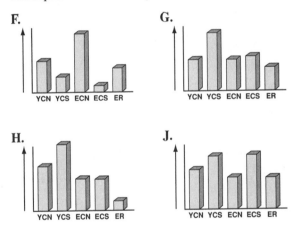

15. According to this study, which of the following groups could realize the most benefit from weight loss?

    A. Young runners
    B. Young nonrunners
    C. Elderly nonrunners
    D. Elderly runners

16. Based on the study, which of the following offers the best advice for lowering the lymphocyte levels of men?

    F. Men should engage in a program of weight reduction, exercise, and a low-fat diet.
    G. Men should engage in an exercise program and stop smoking.
    H. Men should stop smoking.
    J. Older men should exercise, but younger men do not need to exercise because they do not benefit from it.

GO ON TO THE NEXT PAGE.

PASSAGE IV Graphing

Four test landfill sites were prepared to determine the effect of industrial wastes on water absorption, alkalinity, and metal concentrations. All four sites contained the same amount and type of municipal waste and were located within a mile of one another. Site 1 was used as a control and no additional material was added. Sewage sludge was added to Site 2, battery production waste was added to Site 3, and inorganic pigment waste was added to Site 4.

Over the course of a year, measurements were taken to determine what percent of the maximum water capacity the site had maintained. Scientists measured each site's alkalinity, and its concentration of copper and nickel.

Table 1 Concentrations in mg per kg

	Alkalinity	Nickel	Copper
Site 1	2080	182	17
Site 2	5820	236	36
Site 3	3008	287	53
Site 4	4420	938	134

17. According to the study, the best model for studying the effects of increased copper concentration in landfills would be a site that:

A. contains only municipal waste.
B. receives wastes from an inorganic pigment factory.
C. receives wastes from a battery factory.
D. is very alkaline.

18. Which of the following best describes the pattern of water saturation seen in Figure 1?

F. Site 1 was saturated with water more quickly than the other sites.
G. Site 4 was saturated with water more quickly than the other sites.
H. Site 3 was saturated with water at about the same rate as Site 2.
J. Site 2 was saturated with water at about the same rate as Site 4.

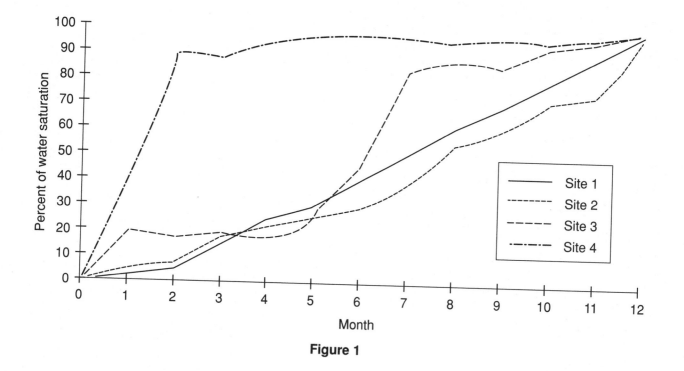

Figure 1

GO ON TO THE NEXT PAGE.

19. What conclusion can be drawn from the results in Figure 1?

    A. More rain fell on Sites 3 and 4 than on Sites 1 and 2.
    B. The municipal waste was more porous in Site 3 than in the other sites.
    C. Adding battery waste decreases the amount of time for saturation to occur.
    D. Site 1 absorbed water at about the same rate for all twelve months.

20. On the basis of these experimental results, which of the following is least likely to delay the saturation of a landfill site?

    F. Increasing the amount of copper
    G. Adding sewage sludge
    H. Increasing alkalinity
    J. Adding inorganic pigment

21. Suppose that scientists created a fifth test site and added about half the amount of inorganic pigment that was added to Site 4. When compared to Site 1, saturation in the new site will probably occur:

    A. more quickly, because adding inorganic pigment speeds up saturation.
    B. more slowly, because adding inorganic pigment increases the concentration of nickel.
    C. more quickly, because adding inorganic pigment greatly increases alkalinity.
    D. more slowly, because adding inorganic pigment slows down saturation.

22. The alkalinity of a site was most affected by:

    F. municipal refuse.
    G. sewage sludge.
    H. battery production waste.
    J. inorganic pigment waste.

GO ON TO THE NEXT PAGE.

PASSAGE V Theory

The following positions present two views about the development of Earth's landmasses.

Position 1

Earth's continents were once a single landmass. Over the millennia, this single landmass separated into several large regions and drifted to form the continents we are familiar with today. For confirmation of this, one need only look at the places where the edges of continents, now thousands of miles apart, fit together nicely. Other evidence of moving continents is seen in similarities between rock formations, such as certain ones in western Africa and eastern South America.

Earth's crust is composed of plates that move across a subsurface of molten rock. The slow movement of these enormous plates moves Earth's landmasses with them, so the surface will look vastly different in another 100 million years.

Position 2

Earth's continents have always looked pretty much as they do today. There may have been some changes caused by the lowering and raising of ocean levels and by the movement of glaciers. However, the movement of water and ice are the only forces that create significant continental changes. Some people have pointed out continental coastlines that seem to "fit into" coastlines of other continents, but a person can identify many more places where coastlines do not fit together.

It is true that huge plates are in motion on Earth's surface. The only noticeable reactions to this plate movement are zones of earthquakes and volcanoes in certain parts of the world, such as the Pacific's "ring of fire." As a result, Earth's surface will be different in 100 million years only if the oceans are at different depths or if glaciers cover more or less of Earth's surface.

23. Which of the following pieces of evidence for Position 1 does Position 2 fail to address?

   A. Plate movement
   B. Matching coastlines
   C. Similar rock formations
   D. Volcanoes

24. Which of the following diagrams best illustrates Position 1's point about the development of Earth's landmasses?

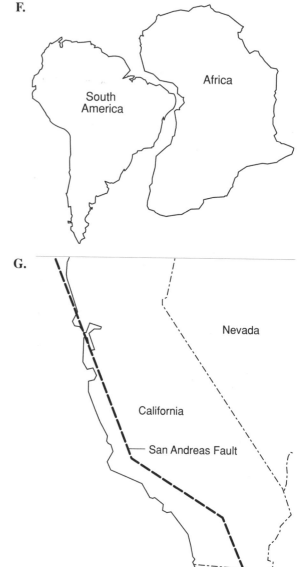

GO ON TO THE NEXT PAGE.

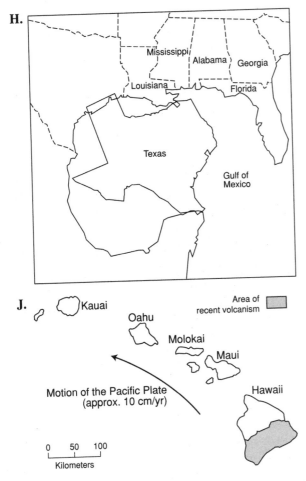

H.

J. Kauai

Oahu

Molokai

Maui

Hawaii

Area of recent volcanism

Motion of the Pacific Plate
(approx. 10 cm/yr)

0   50   100
Kilometers

25. In each of the positions, the discussion about plate movement refers to a process called plate

A. melting.
B. tectonics.
C. gravity.
D. mineralization.

26. Which of the following statements might a person use to support Position 2?

F. Plate movement is random, so it is impossible to predict the future positions of the continents.
G. Plate movements consist of small oscillations, so there is no net change in position.
H. Plates move across a subsurface of molten rock, but the temperatures of this molten rock cannot be exactly determined.
J. Plate movement often leaves the separated pieces of a former continent with vastly different rock formations because small landmasses break off between the continents as they move.

27. A person who believed Position 1 would say that the movement of glaciers on Earth's surface was:

A. a factor in shaping landmasses, not moving them.
B. another example of plate movement.
C. the reason why the plates moved, not where they moved.
D. the reason why parts of continents appear to "fit" together.

28. Which of the following best supports Position 2?

F. Evidence that water levels were lower millions of years ago
G. Evidence of extinct underwater volcanoes caused by plate movement
H. Evidence of active underwater volcanoes caused by plate movement
J. Evidence of a glacier that pushed apart a continent

29. The author of Position 2 points out that many continental features do not "fit" nicely into one another. A supporter of Position 1 is most likely to say which of the following to explain why these continental mismatches do not disprove Position 1?

A. The icebergs that caused the continents to move ground away much of the evidence of continental matching.
B. Pieces of continents broke off to form islands; erosion and the ocean's action erased all but the largest continental matches.
C. Volcanoes under the ocean covered the ocean floor with lava beds.
D. The huge plates that move across Earth's surface created many smaller continents out of what was originally one large continent.

GO ON TO THE NEXT PAGE.

PASSAGE VI Graphing

Chemical reactions can involve the rearrangement or breaking of chemical bonds. In order for a reaction to occur, a minimum amount of kinetic energy must be present. That minimum amount of kinetic energy is called the point of activation energy. Faster-moving particles have higher kinetic energy. Figure 1 shows a graph for the kinetic energy of a substance. The graph shows that a relatively small number of particles have a high kinetic energy. Figure 2 shows that the point of energy activation is at point *A*. The total amount of particles that have this activation energy is represented by the area under the curve.

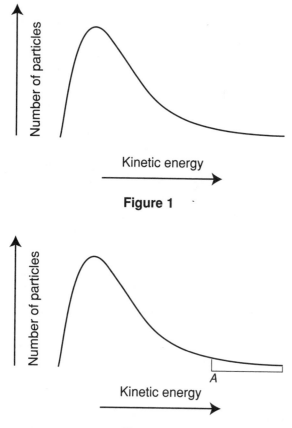

Figure 1

Figure 2

30. According to Figure 2, which of the following is true?
   F. Most of the particles in the sample have relatively high levels of kinetic energy.
   G. The particles with kinetic energies to the right of point *A* may react with other particles.
   H. Kinetic energy increases as one moves from right to left.
   J. None of the particles in the sample possesses the minimum energy for reaction.

31. Figure 3 shows the kinetic energy for the same substance as Figures 1 and 2, except that the temperature is higher. Based on Figure 3, which of the following statements could be made about this reaction?

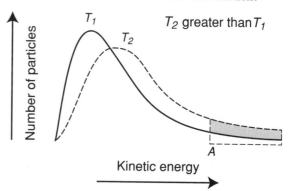

Figure 3

   A. When the temperature is higher, the point of energy of activation is higher.
   B. When the temperature is higher, more particles have higher activation energy.
   C. When the temperature is higher, fewer particles have higher activation energy.
   D. When the temperature is higher, the point of energy activation is lower.

32. According to Figure 3, raising the temperature of a substance increases the number of particles that possess the minimum energy for reaction by:
   F. increasing the overall number of particles in the substance.
   G. speeding up the particles.
   H. enlarging the particles.
   J. reducing the minimum energy for reaction.

GO ON TO THE NEXT PAGE.

33. Based on Figure 4, which of the following statements could be made about this reaction?

Points of activation energy

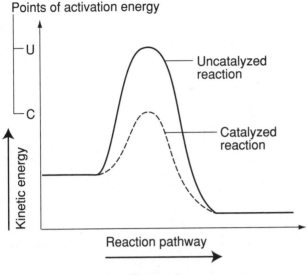

Figure 4

A. When a catalyst is added, the point of energy of activation is higher.
B. When a catalyst is added, more particles have higher activation energy.
C. When a catalyst is added, fewer particles have higher activation energy.
D. When a catalyst is added, the point of energy activation is lower.

34. For the reaction shown in Figure 3, how would a 10°C increase in temperature affect the energy profile for the catalyzed reaction?

F. The activation energy will be lower because faster-moving particles are activated sooner.
G. The activation energy will be the same because the catalyst will balance out the effect of the temperature increase.
H. The activation energy will be the same because temperature does not change activation energy.
J. The activation energy will be higher because high temperatures cause catalysts to break down.

GO ON TO THE NEXT PAGE.

PASSAGE VII Experiment

The law of conservation of matter states that matter may be neither created nor destroyed. One way to interpret this law is to say that atoms are conserved in a chemical reaction.

The French chemist Antoine Lavoisier was the first to devise a procedure to make sure that none of the products of burning were lost. In his experiments, he demonstrated that when matter is burned, the mass of the products (the ashes and gases formed by burning) is equal to the mass of the reactants (the matter that is burned plus the oxygen consumed during burning).

Scientists used the following experiment to confirm the law of conservation of matter. First, they heated 4.0 grams (g) of mercury in a sealed chamber containing air until all of the mercury had reacted. This reaction, which used 0.3 g of oxygen, produced 4.3 g of mercuric oxide. The number of atoms in 4.0 g of mercury plus the number of atoms in 0.3 g of oxygen is the same as the number of atoms in 4.3 g of mercuric oxide.

35. Astronauts in a spacecraft perform the experiment described in the passage in a free-fall environment. Can the results of their experiment be used to confirm the law of conservation of matter?

   A. No, conservation of matter applies only to matter at Earth's surface.
   B. No, measurements in space are affected by low gravity.
   C. Yes, an object's weight is the same in free-fall as it is on Earth.
   D. Yes, grams are a measure of mass.

36. A total of 3.5 g of air was used in this experiment. How much air was left at the end of the experiment?

   F. 3.2 g
   G. 3.5 g
   H. 3.8 g
   J. 4.3 g

37. Other scientists performed the same experiment, but this time, they added 5 g of oxygen to the air in the chamber. Will this change lead to a different experimental result?

   A. Yes, because adding oxygen might cool the air, which would necessitate a higher temperature setting to cause the mercury to react with oxygen.
   B. Yes, because larger amounts of oxygen would combine with mercury.
   C. No, because the same amount of oxygen would combine with the mercury to form mercuric oxide.
   D. No, because adding extra oxygen to the chamber would create a separate chemical reaction between the oxygen and other elements in the air.

38. Which of the following would indicate that matter was not conserved?

   F. The amounts of reactants are not equal.
   G. The amounts of reactants are equal.
   H. The mass of reactants minus the mass of products equals 1.
   J. The mass of reactants minus the mass of products equals 0.

39. A scientist conducts the experiment described above in an open container, but the experiment yields only 3.8 g of mercuric oxide in the container. The most likely reason the results were different is that

   A. oxygen from the air mixed with oxygen in the container.
   B. the temperature needed to consume the mercury was higher.
   C. matter escaped from the open container.
   D. oxygen escaped from the container.

40. A scientist conducts the experiment exactly as described in the passage, but uses 12 g of mercury. Upon measuring the amount of mercuric oxide produced by the experiment, the scientist found that there were 12.87 g instead of the expected 12.9 g. The best explanation for this result is

   F. the scientist made an error in measurement.
   G. the scientist made an error in experimental procedure.
   H. there was too much oxygen.
   J. the chamber is the wrong size.

END OF TEST 4

Diagnostic Science Reasoning Act I Checklist

Answer	Check if missed	Review this section	Pages
1. A	☐	Line Graphs	484–486
2. H	☐	Line Graphs	484–486
3. A	☐	Line Graphs	484–486
4. J	☐	Line Graphs; Tables	484–486; 488–489
5. B	☐	Line Graphs	484–486
6. F	☐	Line Graphs; Data Representation	484–494
7. C	☐	Line Graphs; Data Representation	484–494
8. G	☐	Line Graphs; Data Representation	484–494
9. A	☐	Line Graphs; Data Representation	484–494
10. H	☐	Line Graphs; Data Representation	484–494
11. A	☐	Tables and Figures	488–494
12. H	☐	Reading Research Summaries; Tables and Figures	495–502; 488–494
13. C	☐	Reading Research Summaries; Tables and Figures	495–502; 488–494
14. G	☐	Bar Graphs	486–488
15. B	☐	Research Design	495–496
16. H	☐	Research Design	495–496
17. B	☐	Tables and Figures	488–494
18. G	☐	Line Graphs	484–486
19. D	☐	Line Graphs	484–486
20. J	☐	Line Graphs	484–486
21. A	☐	Reading Research Summaries; Tables and Figures	495–502; 488–494
22. G	☐	Reading Research Summaries; Tables and Figures	495–502; 488–494
23. C	☐	Reading Conflicting Viewpoints Questions	506–510
24. F	☐	Data Representation	484–494
25. B	☐	Reading Conflicting Viewpoints Questions	506–510
26. G	☐	Reading Conflicting Viewpoints Questions	506–510
27. A	☐	Reading Conflicting Viewpoints Questions	506–510
28. J	☐	Reading Conflicting Viewpoints Questions	506–510
29. B	☐	Reading Conflicting Viewpoints Questions	506–510
30. G	☐	Line Graphs	484–486
31. B	☐	Line Graphs	484–486
32. G	☐	Line Graphs; Research Design	484–486; 495–496
33. D	☐	Line Graphs; Research Design	484–486; 495–496
34. H	☐	Line Graphs; Research Design	484–486; 495–496
35. D	☐	Research Design	495–496
36. F	☐	Data Representation	484–494
37. C	☐	Reading Research Summaries	495–502
38. H	☐	Research Design	495–496
39. C	☐	Research Design	495–496
40. F	☐	Research Design	495–496

PASSAGE I

1. **A.** A review of the graphs shows that the percentage of complaints of excessive tearing was the same for both oxidant levels during July.

2. **H.** The pattern for eye irritation complaints is most similar to the pattern for sore throat complaints when the oxidant level is above 0.10 ppm. Both show a sharp decline from July to September, a slight rise from September to October, a moderate decline from October to November, and a slight rise from November to December.

3. **A.** The eye irritation rates for the three fall months are approximately 0.9%, 1.1%, and 0.5%, so the average must be less than 1% and greater than 0.5%. The only answer that fits these criteria is 0.8%. (The average for these three months can also be obtained mathematically. (0.9% + 1.1% + 0.5%) ÷ 3 = 0.83%; the closest answer is 0.8%.)

4. **J.** The lowest rate of eye irritation for oxidant levels under 0.1 ppm occurred during January, and the data for January were collected during the second year of the study. The other explanations either contradict, or cannot be proven by, the data in Figure 1.

5. **B.** If there were no complaints about eye irritation when oxidant concentration was above 0.10 ppm, 0 would be plotted on the graph. Thus choice A is not correct. Choice C is not correct, because if the oxidant concentration was always above 0.10 ppm, data would be plotted on the graph. If data existed, but went off the top of the chart, the researchers would have adjusted the chart so that the scale on the vertical axis could accommodate all the data. Choice D is not correct, because data exist for oxidant concentration below 0.10 ppm. Because all the other answers can be eliminated, choice B must be correct.

PASSAGE II

6. **F.** The thinnest layer of sediment appears near the top of the graph, close to the 10,000-year mark.

7. **C.** The other time ranges include periods when the glacier extended more that 100 km from the mountains to as far as the continental shelf.

8. **G.** The amount of sediment during the Hengelo period is always less than at the beginning of the period. All of the other periods show an increase in sediment depth from the beginning of the period.

9. **A.** The most recent glacial period reached its height about 19,000 years ago, or about 17,000 B.C.

10. **H.** Note that time "starts" at the bottom of the graph. Starting at the end of the Glinde period and moving up toward the Hengelo period, the glacier would have first covered the fjords about 47,000 years ago.

PASSAGE III

11. **A.** The lymphocyte levels for nonsmokers are YC = 2,480, EC = 1,970, and ER = 1,970. The highest readings come from the young controls.

12. **H.** To make it easier for you to compare numbers, write them next to the answer choices:

    A. Smokers: 360, 220
    B. Elderly: 190, 220, 150
    C. Runners: 150
    D. Young: 190, 360

    Sorting the data to match the answer choices makes it readily apparent that runners are most likely to have the lowest eosinophil levels.

13. **C.** Check the accuracy of each answer in the list.

A. NOT TRUE.	The study does not compare the oxygen intake of smokers vs. nonsmokers so no conclusion can be drawn.
B. NOT TRUE.	Same reasoning as choice A.
C. TRUE.	The total WBC in smokers: 7,730, 7,490
	The total WBC in nonsmokers: 6,100, 5,930, 5,430
D. NOT TRUE.	Lymphocytes in smokers: 3,270, 3,320
	Lymphocytes in nonsmokers: 2,480, 1,970, 1,970

14. **G.** To help you determine which chart is the best match, jot down the eosinophil counts for each group.

190	360	190	220	150
YCN	YCS	ECN	ECS	ER

The first (YCN) and the third (ECN) counts are the same. The second (YCS) count is the highest. The fourth (ECS) count is a little higher than the first and third counts, and the fifth (ER) count is the lowest.

15. **B.** Skinfold measurements are highest in the young control group.

16. **H.** Lymphocyte levels are lowest among nonsmokers. Choice F is not correct because there is no evidence in this study that weight reduction or a low-fat diet affects lymphocyte production. Choices G and J are not correct because lymphocyte levels are the same for elderly runners and elderly nonrunners, which indicates that exercise has little to do with lymphocyte production.

PASSAGE IV

17. **B.** The site with the highest copper concentration is Site 4, which contains inorganic pigment wastes.

18. **G.** Site 4 was saturated more quickly than the other sites, which supports the statement in choice G and contradicts the statement in choice F. Site 2 was saturated at a slower rate than the other sites, so choices H and J are incorrect.

19. **D.** Since the sites are close together, choice A is unlikely to be true. Choice B can be eliminated because the description of the study states that all four sites contained the same amount and type of municipal waste. Because the graph shows that Site 3 jumped from about 50% saturation to 80% saturation from June to July, choice C is incorrect.

20. **J.** Least likely to delay saturation means the same as most likely to speed saturation. Site 4 (inorganic pigment waste added) was saturated first, so inorganic pigment waste is the least likely to delay saturation.

21. **A.** Adding inorganic pigment waste speeded saturation in the study, so saturation will occur more quickly in the new site than it did in Site 1. Thus choices B and D can be eliminated. The results for Site 2 indicate that increased alkalinity does not appear to speed up saturation, so choice C can also be eliminated. Only choice A remains.

22. **G.** Alkalinity is highest in Site 2, where sewage sludge was added to the municipal waste in the landfill.

PASSAGE V

23. **C.** The writer of Position 2 does comment on the plate movement (choice A) and coastline matching (choice B) arguments presented by the writer of Position 1. The writer of Position 1 never mentions volcanoes (choice D). Choice C is the only choice remaining.

24. **F.** The other diagrams do not show continental "matches."

25. **B.** You may remember the term *tectonics* from the science classes you have taken. Even if you don't recall this term, you can readily eliminate choices A and C. Choice D, mineralization, refers to the formation of minerals, and not to plate movement.

26. **G.** Consider each choice to determine if it explains why plate movement does not cause the movement of continents. Remember, the explanation need not be correct.
    F.   NO. Random movement does not explain why continental shift would not take place.
    G.   YES. *Oscillate* means to move back and forth. If plates moved back and forth, continents would not move far as a result of oscillation.
    H.   NO. Inability to determine the temperature of the lava does not explain why plate movement would not take place.
    J.   NO. Creation of small landmasses does not explain why continents do not move.

27. **A.** According to Position 1, glaciers are neither an example of nor a cause of continental movement. That eliminates choices B, C, and D.

28. **J.** Position 2 holds that any continental changes were caused by the movement of water or by glacial action. Only choice J supports that view.

29. **B.** Choice B is the only answer that explains why continents formed by plate movement do not all "fit."

PASSAGE VI

30. **G.** Choices F, H, and J directly contradict the information in the graph.

31. **B.** The point of activation energy is the same regardless of temperature. However, the white area in this diagram is larger than the white area in the other diagram, showing that more particles have this higher activation energy when the temperature is higher.

32. **G.** Recall from the passage that faster-moving particles have higher kinetic energy.

33. **D.** The graph shows that the activation energy is lower when a catalyst is used. So using a catalyst lowers the point of activation energy.

34. **H.** Temperature does not change the activation energy for a reaction. However, temperature increases the number of particles that have the minimum energy required for reaction.

PASSAGE VII

35. **D.** The laws of physics, including the law of conservation of mass, apply everywhere. The astronauts' results will be the same as those of their colleagues on Earth. Although the weight of objects changes with changes in gravity, their mass—the amount of matter that makes them up—is constant.

36. **F.** Part of the air—0.3 g of oxygen, to be exact—was used in the reaction. To determine the amount of air at the end of the experiment, subtract the amount of oxygen that reacted with the mercury from the amount of air at the beginning of the experiment. $(3.5 - 0.3 = 3.2)$

37. **C.** Mercury, not oxygen, is the limiting factor in this experiment. No matter how much extra oxygen is available, the 4 g of mercury will react with exactly 0.3g of oxygen to form 4.3 g of mercuric oxide.

38. **H.** Conservation of matter means that the difference between the mass of the reactants and the mass of the products is zero (0). Choice H is the only answer that is an exception to this rule. An experiment that produced this result would not confirm the law of conservation of matter. Choice G is the only answer that confirms this law. Choices F and J do not indicate whether or not the experiment confirms the law.

39. **C.** The clue to this answer are the words "open container." A successful conservation of matter experiment assumes that you can capture and measure all the new products that are formed. In this experiment, some of the products escaped.

40. **F.** The other choices can be eliminated because the question states that the experiment was conducted exactly as described in the passage. In addition, it is reasonable to believe that such a small difference from the expected result is due to a slight error in measurement.

Section V ■

Model Science Reasoning Tests

Chapter 23 ▪

Model Science Reasoning **ACT II**
With Answers Explained

This Science Reasoning Model ACT II is just like a real ACT. Take this test after you take the Diagnostic Science Reasoning ACT.

Take this model test under simulated test conditions. Allow 35 minutes to answer the 40 Science Reasoning items. Use a pencil to mark the answer sheet on the next page, and answer the questions in the Test 4 (Science Reasoning) section.

Use the answer key on page 548 to mark the answer sheet. Review the answer explanations on pages 548–551. You may decide to retake this test later. There are additional answer sheets for this purpose following page 605.

The test scoring chart on page 589 shows you how to convert the number correct to ACT scale scores.

DO NOT leave any answers blank. There is no penalty for guessing on the ACT. Remember that the test is yours. You may mark up, write on, or draw on the test.

When you are ready, note the time and begin.

ANSWER SHEET

The ACT answer sheet looks something like this one. Use a No. 2 pencil
to completely fill the circle corresponding to the correct answer.
If you erase, erase completely; incomplete erasures may be read as answers.

TEST 1—English

1 Ⓐ Ⓑ Ⓒ Ⓓ	11 Ⓐ Ⓑ Ⓒ Ⓓ	21 Ⓐ Ⓑ Ⓒ Ⓓ	31 Ⓐ Ⓑ Ⓒ Ⓓ	41 Ⓐ Ⓑ Ⓒ Ⓓ	51 Ⓐ Ⓑ Ⓒ Ⓓ	61 Ⓐ Ⓑ Ⓒ Ⓓ	71 Ⓐ Ⓑ Ⓒ Ⓓ
2 Ⓕ Ⓖ Ⓗ Ⓙ	12 Ⓕ Ⓖ Ⓗ Ⓙ	22 Ⓕ Ⓖ Ⓗ Ⓙ	32 Ⓕ Ⓖ Ⓗ Ⓙ	42 Ⓕ Ⓖ Ⓗ Ⓙ	52 Ⓕ Ⓖ Ⓗ Ⓙ	62 Ⓕ Ⓖ Ⓗ Ⓙ	72 Ⓕ Ⓖ Ⓗ Ⓙ
3 Ⓐ Ⓑ Ⓒ Ⓓ	13 Ⓐ Ⓑ Ⓒ Ⓓ	23 Ⓐ Ⓑ Ⓒ Ⓓ	33 Ⓐ Ⓑ Ⓒ Ⓓ	43 Ⓐ Ⓑ Ⓒ Ⓓ	53 Ⓐ Ⓑ Ⓒ Ⓓ	63 Ⓐ Ⓑ Ⓒ Ⓓ	73 Ⓐ Ⓑ Ⓒ Ⓓ
4 Ⓕ Ⓖ Ⓗ Ⓙ	14 Ⓕ Ⓖ Ⓗ Ⓙ	24 Ⓕ Ⓖ Ⓗ Ⓙ	34 Ⓕ Ⓖ Ⓗ Ⓙ	44 Ⓕ Ⓖ Ⓗ Ⓙ	54 Ⓕ Ⓖ Ⓗ Ⓙ	64 Ⓕ Ⓖ Ⓗ Ⓙ	74 Ⓕ Ⓖ Ⓗ Ⓙ
5 Ⓐ Ⓑ Ⓒ Ⓓ	15 Ⓐ Ⓑ Ⓒ Ⓓ	25 Ⓐ Ⓑ Ⓒ Ⓓ	35 Ⓐ Ⓑ Ⓒ Ⓓ	45 Ⓐ Ⓑ Ⓒ Ⓓ	55 Ⓐ Ⓑ Ⓒ Ⓓ	65 Ⓐ Ⓑ Ⓒ Ⓓ	75 Ⓐ Ⓑ Ⓒ Ⓓ
6 Ⓕ Ⓖ Ⓗ Ⓙ	16 Ⓕ Ⓖ Ⓗ Ⓙ	26 Ⓕ Ⓖ Ⓗ Ⓙ	36 Ⓕ Ⓖ Ⓗ Ⓙ	46 Ⓕ Ⓖ Ⓗ Ⓙ	56 Ⓕ Ⓖ Ⓗ Ⓙ	66 Ⓕ Ⓖ Ⓗ Ⓙ	
7 Ⓐ Ⓑ Ⓒ Ⓓ	17 Ⓐ Ⓑ Ⓒ Ⓓ	27 Ⓐ Ⓑ Ⓒ Ⓓ	37 Ⓐ Ⓑ Ⓒ Ⓓ	47 Ⓐ Ⓑ Ⓒ Ⓓ	57 Ⓐ Ⓑ Ⓒ Ⓓ	67 Ⓐ Ⓑ Ⓒ Ⓓ	
8 Ⓕ Ⓖ Ⓗ Ⓙ	18 Ⓕ Ⓖ Ⓗ Ⓙ	28 Ⓕ Ⓖ Ⓗ Ⓙ	38 Ⓕ Ⓖ Ⓗ Ⓙ	48 Ⓕ Ⓖ Ⓗ Ⓙ	58 Ⓕ Ⓖ Ⓗ Ⓙ	68 Ⓕ Ⓖ Ⓗ Ⓙ	
9 Ⓐ Ⓑ Ⓒ Ⓓ	19 Ⓐ Ⓑ Ⓒ Ⓓ	29 Ⓐ Ⓑ Ⓒ Ⓓ	39 Ⓐ Ⓑ Ⓒ Ⓓ	49 Ⓐ Ⓑ Ⓒ Ⓓ	59 Ⓐ Ⓑ Ⓒ Ⓓ	69 Ⓐ Ⓑ Ⓒ Ⓓ	
10 Ⓕ Ⓖ Ⓗ Ⓙ	20 Ⓕ Ⓖ Ⓗ Ⓙ	30 Ⓕ Ⓖ Ⓗ Ⓙ	40 Ⓕ Ⓖ Ⓗ Ⓙ	50 Ⓕ Ⓖ Ⓗ Ⓙ	60 Ⓕ Ⓖ Ⓗ Ⓙ	70 Ⓕ Ⓖ Ⓗ Ⓙ	

TEST 2—Mathematics

1 Ⓐ Ⓑ Ⓒ Ⓓ Ⓔ	9 Ⓐ Ⓑ Ⓒ Ⓓ Ⓔ	17 Ⓐ Ⓑ Ⓒ Ⓓ Ⓔ	25 Ⓐ Ⓑ Ⓒ Ⓓ Ⓔ	33 Ⓐ Ⓑ Ⓒ Ⓓ Ⓔ	41 Ⓐ Ⓑ Ⓒ Ⓓ Ⓔ	49 Ⓐ Ⓑ Ⓒ Ⓓ Ⓔ	57 Ⓐ Ⓑ Ⓒ Ⓓ Ⓔ
2 Ⓕ Ⓖ Ⓗ Ⓙ Ⓚ	10 Ⓕ Ⓖ Ⓗ Ⓙ Ⓚ	18 Ⓕ Ⓖ Ⓗ Ⓙ Ⓚ	26 Ⓕ Ⓖ Ⓗ Ⓙ Ⓚ	34 Ⓕ Ⓖ Ⓗ Ⓙ Ⓚ	42 Ⓕ Ⓖ Ⓗ Ⓙ Ⓚ	50 Ⓕ Ⓖ Ⓗ Ⓙ Ⓚ	58 Ⓕ Ⓖ Ⓗ Ⓙ Ⓚ
3 Ⓐ Ⓑ Ⓒ Ⓓ Ⓔ	11 Ⓐ Ⓑ Ⓒ Ⓓ Ⓔ	19 Ⓐ Ⓑ Ⓒ Ⓓ Ⓔ	27 Ⓐ Ⓑ Ⓒ Ⓓ Ⓔ	35 Ⓐ Ⓑ Ⓒ Ⓓ Ⓔ	43 Ⓐ Ⓑ Ⓒ Ⓓ Ⓔ	51 Ⓐ Ⓑ Ⓒ Ⓓ Ⓔ	59 Ⓐ Ⓑ Ⓒ Ⓓ Ⓔ
4 Ⓕ Ⓖ Ⓗ Ⓙ Ⓚ	12 Ⓕ Ⓖ Ⓗ Ⓙ Ⓚ	20 Ⓕ Ⓖ Ⓗ Ⓙ Ⓚ	28 Ⓕ Ⓖ Ⓗ Ⓙ Ⓚ	36 Ⓕ Ⓖ Ⓗ Ⓙ Ⓚ	44 Ⓕ Ⓖ Ⓗ Ⓙ Ⓚ	52 Ⓕ Ⓖ Ⓗ Ⓙ Ⓚ	60 Ⓕ Ⓖ Ⓗ Ⓙ Ⓚ
5 Ⓐ Ⓑ Ⓒ Ⓓ Ⓔ	13 Ⓐ Ⓑ Ⓒ Ⓓ Ⓔ	21 Ⓐ Ⓑ Ⓒ Ⓓ Ⓔ	29 Ⓐ Ⓑ Ⓒ Ⓓ Ⓔ	37 Ⓐ Ⓑ Ⓒ Ⓓ Ⓔ	45 Ⓐ Ⓑ Ⓒ Ⓓ Ⓔ	53 Ⓐ Ⓑ Ⓒ Ⓓ Ⓔ	
6 Ⓕ Ⓖ Ⓗ Ⓙ Ⓚ	14 Ⓕ Ⓖ Ⓗ Ⓙ Ⓚ	22 Ⓕ Ⓖ Ⓗ Ⓙ Ⓚ	30 Ⓕ Ⓖ Ⓗ Ⓙ Ⓚ	38 Ⓕ Ⓖ Ⓗ Ⓙ Ⓚ	46 Ⓕ Ⓖ Ⓗ Ⓙ Ⓚ	54 Ⓕ Ⓖ Ⓗ Ⓙ Ⓚ	
7 Ⓐ Ⓑ Ⓒ Ⓓ Ⓔ	15 Ⓐ Ⓑ Ⓒ Ⓓ Ⓔ	23 Ⓐ Ⓑ Ⓒ Ⓓ Ⓔ	31 Ⓐ Ⓑ Ⓒ Ⓓ Ⓔ	39 Ⓐ Ⓑ Ⓒ Ⓓ Ⓔ	47 Ⓐ Ⓑ Ⓒ Ⓓ Ⓔ	55 Ⓐ Ⓑ Ⓒ Ⓓ Ⓔ	
8 Ⓕ Ⓖ Ⓗ Ⓙ Ⓚ	16 Ⓕ Ⓖ Ⓗ Ⓙ Ⓚ	24 Ⓕ Ⓖ Ⓗ Ⓙ Ⓚ	32 Ⓕ Ⓖ Ⓗ Ⓙ Ⓚ	40 Ⓕ Ⓖ Ⓗ Ⓙ Ⓚ	48 Ⓕ Ⓖ Ⓗ Ⓙ Ⓚ	56 Ⓕ Ⓖ Ⓗ Ⓙ Ⓚ	

TEST 3—Reading

1 Ⓐ Ⓑ Ⓒ Ⓓ	6 Ⓕ Ⓖ Ⓗ Ⓙ	11 Ⓐ Ⓑ Ⓒ Ⓓ	16 Ⓕ Ⓖ Ⓗ Ⓙ	21 Ⓐ Ⓑ Ⓒ Ⓓ	26 Ⓕ Ⓖ Ⓗ Ⓙ	31 Ⓐ Ⓑ Ⓒ Ⓓ	36 Ⓕ Ⓖ Ⓗ Ⓙ
2 Ⓕ Ⓖ Ⓗ Ⓙ	7 Ⓐ Ⓑ Ⓒ Ⓓ	12 Ⓕ Ⓖ Ⓗ Ⓙ	17 Ⓐ Ⓑ Ⓒ Ⓓ	22 Ⓕ Ⓖ Ⓗ Ⓙ	27 Ⓐ Ⓑ Ⓒ Ⓓ	32 Ⓕ Ⓖ Ⓗ Ⓙ	37 Ⓐ Ⓑ Ⓒ Ⓓ
3 Ⓐ Ⓑ Ⓒ Ⓓ	8 Ⓕ Ⓖ Ⓗ Ⓙ	13 Ⓐ Ⓑ Ⓒ Ⓓ	18 Ⓕ Ⓖ Ⓗ Ⓙ	23 Ⓐ Ⓑ Ⓒ Ⓓ	28 Ⓕ Ⓖ Ⓗ Ⓙ	33 Ⓐ Ⓑ Ⓒ Ⓓ	38 Ⓕ Ⓖ Ⓗ Ⓙ
4 Ⓕ Ⓖ Ⓗ Ⓙ	9 Ⓐ Ⓑ Ⓒ Ⓓ	14 Ⓕ Ⓖ Ⓗ Ⓙ	19 Ⓐ Ⓑ Ⓒ Ⓓ	24 Ⓕ Ⓖ Ⓗ Ⓙ	29 Ⓐ Ⓑ Ⓒ Ⓓ	34 Ⓕ Ⓖ Ⓗ Ⓙ	39 Ⓐ Ⓑ Ⓒ Ⓓ
5 Ⓐ Ⓑ Ⓒ Ⓓ	10 Ⓕ Ⓖ Ⓗ Ⓙ	15 Ⓐ Ⓑ Ⓒ Ⓓ	20 Ⓕ Ⓖ Ⓗ Ⓙ	25 Ⓐ Ⓑ Ⓒ Ⓓ	30 Ⓕ Ⓖ Ⓗ Ⓙ	35 Ⓐ Ⓑ Ⓒ Ⓓ	40 Ⓕ Ⓖ Ⓗ Ⓙ

TEST 4—Science Reasoning

1 Ⓐ Ⓑ Ⓒ Ⓓ	6 Ⓕ Ⓖ Ⓗ Ⓙ	11 Ⓐ Ⓑ Ⓒ Ⓓ	16 Ⓕ Ⓖ Ⓗ Ⓙ	21 Ⓐ Ⓑ Ⓒ Ⓓ	26 Ⓕ Ⓖ Ⓗ Ⓙ	31 Ⓐ Ⓑ Ⓒ Ⓓ	36 Ⓕ Ⓖ Ⓗ Ⓙ
2 Ⓕ Ⓖ Ⓗ Ⓙ	7 Ⓐ Ⓑ Ⓒ Ⓓ	12 Ⓕ Ⓖ Ⓗ Ⓙ	17 Ⓐ Ⓑ Ⓒ Ⓓ	22 Ⓕ Ⓖ Ⓗ Ⓙ	27 Ⓐ Ⓑ Ⓒ Ⓓ	32 Ⓕ Ⓖ Ⓗ Ⓙ	37 Ⓐ Ⓑ Ⓒ Ⓓ
3 Ⓐ Ⓑ Ⓒ Ⓓ	8 Ⓕ Ⓖ Ⓗ Ⓙ	13 Ⓐ Ⓑ Ⓒ Ⓓ	18 Ⓕ Ⓖ Ⓗ Ⓙ	23 Ⓐ Ⓑ Ⓒ Ⓓ	28 Ⓕ Ⓖ Ⓗ Ⓙ	33 Ⓐ Ⓑ Ⓒ Ⓓ	38 Ⓕ Ⓖ Ⓗ Ⓙ
4 Ⓕ Ⓖ Ⓗ Ⓙ	9 Ⓐ Ⓑ Ⓒ Ⓓ	14 Ⓕ Ⓖ Ⓗ Ⓙ	19 Ⓐ Ⓑ Ⓒ Ⓓ	24 Ⓕ Ⓖ Ⓗ Ⓙ	29 Ⓐ Ⓑ Ⓒ Ⓓ	34 Ⓕ Ⓖ Ⓗ Ⓙ	39 Ⓐ Ⓑ Ⓒ Ⓓ
5 Ⓐ Ⓑ Ⓒ Ⓓ	10 Ⓕ Ⓖ Ⓗ Ⓙ	15 Ⓐ Ⓑ Ⓒ Ⓓ	20 Ⓕ Ⓖ Ⓗ Ⓙ	25 Ⓐ Ⓑ Ⓒ Ⓓ	30 Ⓕ Ⓖ Ⓗ Ⓙ	35 Ⓐ Ⓑ Ⓒ Ⓓ	40 Ⓕ Ⓖ Ⓗ Ⓙ

Model Science Reasoning ACT II

40 Questions—35 Minutes

INSTRUCTIONS: Each of the seven science passages on this test is followed by questions. Choose the best answer for each question based on the passage. Then fill in the appropriate circle on the answer sheet. You can look back at the passages as often as you want.

You CANNOT use a calculator on this test.

Check pages 548–551 for answers and explanations.

PASSAGE I

Scientists conducted an experiment to determine how fast PCBs are degraded by microbes. PCB-degrading microbes were isolated from soils and mixed with dry river sediment previously contaminated with PCBs. The percent degradation for each sample was measured after 1 day, 5 days, and 10 days. Scientists classified the PCBs according to the chlorine content of the PCB isomer.

Experiment 1

The degradation of a monochloro-isomer sample was compared with that of a dichloro-isomer sample. The results are reported in the table below.

Table 1

Sample	Number of chlorine atoms	Percent degradation after		
		1 day	5 days	10 days
A	1	100	100	100
B	2	51	93	99

Experiment 2

The degradation of a trichloro-isomer sample was compared with that of a tetrachloro-isomer sample. The results are reported in the table below.

Table 2

Sample	Number of chlorine atoms	Percent degradation after		
		1 day	5 days	10 days
C	3	49	71	71
D	4	11	48	46

Experiment 3

The degradation of a sample containing both dichloro-isomers and trichloro-isomers was measured.

Table 3

Sample	Number of chlorine atoms	Percent degradation after		
		1 day	5 days	10 days
E	2 + 3	55	71	74

1. The tables show the percent degradation for each sample. What is the minimum degradation after 5 days for any of the samples?

   A. 11%
   B. 46%
   C. 48%
   D. 71%

2. A reasonable explanation of why monochloro-isomers and dichloro-isomers might degrade faster than trichloro-isomers and tetrachloro-isomers is that molecules with fewer chlorine atoms:

   F. are less complex than ones with more chlorine atoms.
   G. degraded over a longer period of time than other molecules.
   H. are more complex than molecules with more chlorine atoms.
   J. are least likely to be degraded by the microbes in this experiment.

GO ON TO THE NEXT PAGE.

3. Which of these conclusions is supported by Experiment 3?

   A. When a sample contains isomers with two different chlorine values, the degradation pattern most closely resembles the isomer with the higher value.

   B. When a sample contains isomers with two different chlorine values, the degradation pattern most closely resembles the isomer with the sum of the values.

   C. When a sample degrades between 70% and 80% after 10 days, that sample contains isomers with two different chlorine values.

   D. When a sample degrades more than 70% after 10 days, then the number of chlorines in all the isomers in the sample is 1 or 2.

4. Based on the results reported above, it is most likely that using microbes to reclaim farmland contaminated by PCBs:

   F. holds great promise because over 50% of the PCBs were degraded during this study.

   G. has to be studied further because one of these samples showed that the percent of PCBs degraded started to decline.

   H. has to be studied further because the PCBs with 1 and 2 chlorines showed a lower degradation rate.

   J. holds great promise because the microbe treatments were successful when tried out on 5-acre farm sites.

5. How could the experiment be redesigned to provide more information about the usefulness of microbial treatments for cleansing PCB-contaminated soil?

   A. Use soil contaminated with bacterial waste as well as with PCBs.

   B. Apply chemicals that might also degrade PCBs.

   C. Use samples with these numbers of chlorines: 1 + 2 and 1 + 3.

   D. Measure the percent of PCB degradation over longer periods of time.

6. Suppose water inhibits the microbial degradation of PCBs. Imagine that a second series of tests is performed on samples of wet mud from the bottom of a river that have exactly the same type and amount of PCBs as the samples in the original experiment. What results would you expect from the microbial treatment of the samples of wet mud?

   F. PCBs would degrade at a faster rate than in the original samples.

   G. PCBs would degrade at the same rate as in the original samples.

   H. PCBs would degrade at a slower rate than in the original samples.

   J. Tetra- and tri-isomers would degrade at a slower rate, but mono- and di-isomers would degrade at a faster rate than in the original samples.

GO ON TO THE NEXT PAGE.

PASSAGE II

Scientists agree that natural gas formation has to do with thermal decomposition of oil and organic matter in shales. Gas and oil researchers conducted this experiment to find out whether gas and oil are produced together.

Experiment 1

Researchers collected natural gas and oil samples from shale in eight sites and computed the hydrogen index for each sample. A high hydrogen index indicates that the shale is "immature" and that oil generation has just begun. A low hydrogen index indicates that the shale is "mature" and that oil generation is largely complete. The researchers arranged the sites in order from least mature to most mature and reported their results in Figure 1. The figure also shows the gas/oil ratio in cubic feet of gas/barrel of oil.

Figure 1

Increasing maturity of organic matter →

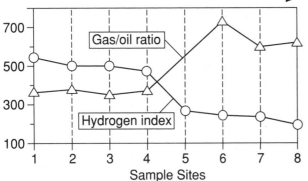

Experiment 2

The researchers measured the percentage of methane, ethane, propane, and butane in each gas sample. The relative percentage of these different gases is presented in Figure 2.

Figure 2

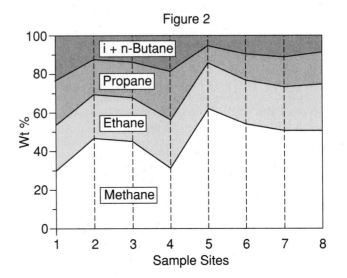

7. Which of the following sites has the smallest percentage of propane gas?

   A. Site 1
   B. Site 4
   C. Site 5
   D. Site 7

8. At Site 1, how much ethane would be contained in 1,000 cubic feet of gas?

   F. 500 cubic feet
   G. 400 cubic feet
   H. 300 cubic feet
   J. 200 cubic feet

GO ON TO THE NEXT PAGE.

9. Which of the following graphs best represents the relationship between the hydrogen index and the gas/oil ratio?

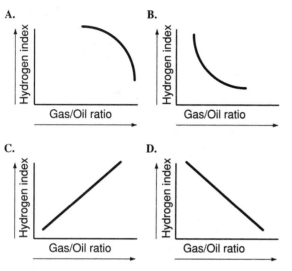

A.

B.

C.

D.

10. A sample collected at Site 6 is likely to have:
    I. about 700 times more barrels of oil than cubic feet of gas.
    II. about 700 times more cubic feet of gas than barrels of oil.
    III. less than 40% of the gas as methane.
    IV. more than 40% of the gas as ethane.

    F. I and III
    G. I
    H. II
    J. II and IV

11. Which of the following results is NOT in keeping with the data presented here about hydrogen index and percentage of gas volume?

    A. The higher the hydrogen index, the lower the methane percentage.
    B. The hydrogen index decreases as the maturity of organic matter increases.
    C. Butane makes up the lowest percentage of gas at every site.
    D. The gas/oil ratio is lower than the hydrogen index at half the sites.

12. Which of the following conclusions could scientists reach from these studies?

    F. Methane gas inhibits the production of oil, because the percentage of methane gas increases as the hydrogen index decreases.
    G. The gas/oil ratio is higher at Sites 6, 7, and 8 because gas production increases faster than oil production as organic matter matures.
    H. Methane gas makes up the largest percentage of gas because the hydrogen index decreases as organic matter matures.
    J. The methane gas percentage is highest at Site 5 because organic matter is not maturing.

GO ON TO THE NEXT PAGE.

PASSAGE III

The amount of oxidant (NO_x) and ozone (O_3) was measured every hour from January in Year 1 through May in Year 2. The oxidant was measured in milligrams per cubic meter of air while ozone was measured in parts per billion. The figure below shows the highest hourly measurement of ozone in each month, the lowest hourly measurement of ozone in each month, and the monthly average of ozone and oxidant levels.

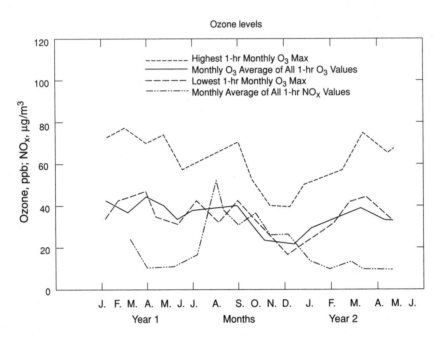

13. What is the relationship between average oxidant level and the highest hourly ozone level?

    A. As oxidant levels increase, the ozone level decreases.
    B. As oxidant levels increase, the ozone level increases.
    C. Oxidant levels and ozone levels are generally unrelated.
    D. There is a direct relationship between ozone levels and oxidant levels.

14. During which month was the average monthly ozone level lowest?

    F. December
    G. February
    H. April
    J. June

15. A finding that particulate (visible) oxidants were at their lowest levels in August would be consistent with a finding that:

    A. ozone levels were highest.
    B. ozone levels were lowest.
    C. gaseous oxidant levels were highest.
    D. gaseous oxidant levels were lowest.

16. The amount of oxygen in the air is reduced by the amount of oxidant. Based on this information and the figure, in which of the months listed would the air be the most oxygen rich?

    F. May
    G. August
    H. November
    J. January

17. Researchers frequently use oxidant/ozone ratios to measure air quality. Using the data in the figure above for monthly average oxidant and ozone levels, in which of the following months would the oxidant/ozone level be highest?

    A. June
    B. April
    C. February
    D. November

GO ON TO THE NEXT PAGE.

PASSAGE IV

During meiosis, chromosomes may exchange portions of their chromatids in a process called crossing-over. Crossing-over produces new combinations of genes. The figure below shows meiosis with crossing-over in a female fruit fly (*left*) and meiosis without crossing-over in a male fruit fly (*right*). Note that this cross involves genes for body color and wing size. Each gene has two forms, or alleles. The alleles for body color are yellow (*Y*) and black (*y*). The alleles for wing size are long (*L*) and short (*l*).

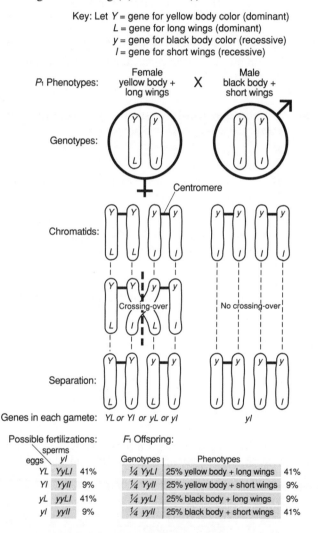

Key: Let *Y* = gene for yellow body color (dominant)
 L = gene for long wings (dominant)
 y = gene for black body color (recessive)
 l = gene for short wings (recessive)

18. The alleles in the male's gametes are:

- **F.** dominant.
- **G.** recessive.
- **H.** dominant and recessive.
- **J.** neither dominant nor recessive.

19. Which of the following best explains the results of crossing-over noted in the figure at left?

- **A.** Chromosomes change genotypes, resulting in genes that are neither dominant nor recessive.
- **B.** Dominant and recessive genes are randomly mixed, resulting in an unpredictable combination of dominant and recessive genes.
- **C.** Chromatids exchange dominant and recessive genes, resulting in gametes that include mixtures of dominant and recessive genes.
- **D.** Recessive genes are absorbed by dominant genes resulting in gametes that include mixtures of dominant and recessive genes.

20. The diagrams at the bottom of the accompanying figure show the types of offspring that might result from a mating between the "crossed-over" female and the "noncrossed-over" male. Which of the following combinations of eggs and sperm could result in a fly with a yellow body and short wings?

- **F.** *YYLL*
- **G.** *Yyll*
- **H.** *yyLL*
- **J.** *YyLL*

21. Which of the following statements best explains why a mating between the two flies shown in this diagram could not result in a fertilization with four dominant genes?

- **A.** The cross-over noted for the female fly created gametes that included gametes with both recessive and dominant genes.
- **B.** The original genes of the male fly included only recessive genes.
- **C.** Fertilization with four dominant genes cannot result when cross-over is noted in the female fly.
- **D.** Four dominant genes would create an aberration to the species and these aberrations never occur in normal fertilization.

22. Scientists find that fertilizations with both dominant traits are about 50% more likely to occur than the number of phenotypes predicted. About what percent of the offspring will have long wings and yellow bodies?

- **F.** 12.5%
- **G.** 25.5%
- **H.** 37.5%
- **J.** 50.5%

GO ON TO THE NEXT PAGE.

PASSAGE V

Two types of gases contribute significantly to global warming: greenhouse gases and chlorofluorocarbons. Greenhouse gases—carbon dioxide, methane, and nitrous oxide, for example—keep Earth from radiating off all the heat it absorbs from the sun. As the levels of greenhouse gases in the air increase, the atmosphere retains more heat, and Earth's temperature increases. The rising levels of greenhouse gases in the atmosphere are attributed primarily to emissions produced by burning fossil fuels. China, South Africa, and India use the highest amounts of fossil fuels.

Chlorofluorocarbons (CFCs) were once widely used as aerosol propellants, but are now used primarily in cooling refrigeration equipment. CFCs react with the ozone layer in the stratosphere, breaking down the ozone and admitting more harmful radiation to the atmosphere. Most CFCs are released into the atmosphere through industrial emissions.

A simulation of Earth's atmosphere before 1850 yielded the following readings:

Table 1
Atmospheric Concentration of Gases
Before 1850

Gas	Concentration
Carbon dioxide	275 ppmv
Methane	0.7 ppmv
Nitrous oxide	0.285 ppmv
CFC-11	0
CFC-12	0

ppmv = parts per million by volume

When scientists measured these substances in the modern atmosphere, they obtained the following results:

Table 2
Atmospheric Concentration of Gases
in 1995

Gas	Concentration
Carbon dioxide	345 ppmv
Methane	1.7 ppmv
Nitrous oxide	0.304 ppmv
CFC-11	0.22 ppbv
CFC-12	0.38 ppbv

ppmv = parts per million by volume
ppbv = parts per billion by volume

23. Do the data in the tables and the information in the passage support the argument that global warming will significantly decrease in the next century if agricultural practices are revised?

    A. Yes, because agricultural practices are listed as one of the contributors of greenhouse gases to the environment.
    B. Yes, because carbon dioxide emissions are a significant source of greenhouse gases.
    C. No, because there is no evidence in the passage that agricultural practices contribute significantly to the concentration of greenhouse gases.
    D. No, because agricultural practices do not contribute to the concentration of greenhouse gases.

24. If China eliminated its use of fossil fuels in the next century, one might reasonably expect:

    F. a reduction in CFC-11.
    G. the sun to heat up.
    H. a reduction in global carbon dioxide emissions.
    J. the sun to cool down.

25. According to the data in the tables, a person who was alive in 1825 and who visited Earth today would find that:

    A. carbon dioxide concentrations have doubled.
    B. carbon dioxide concentrations have increased by about $\frac{3}{4}$.
    C. carbon dioxide concentrations have increased by about $\frac{1}{2}$.
    D. carbon dioxide concentrations have increased by about $\frac{1}{4}$.

26. Which of the following is most likely to result in a decline in the amount of CFCs in the atmosphere?

    F. Reduce the temperature of the atmosphere.
    G. Reduce solar emissions.
    H. Reduce industrial emissions.
    J. Revise agricultural practices.

GO ON TO THE NEXT PAGE.

27. How might scientists design an experiment that would allow them to control the amount of carbon dioxide emissions?

   A. Choose parts of the world with different amounts of carbon dioxide emissions.
   B. Choose parts of the world with the same amount of carbon dioxide emissions.
   C. Construct a laboratory experiment in which they created their own carbon dioxide emissions.
   D. Construct a laboratory experiment in which burning coal was placed in a chamber containing air for varying periods of time.

28. Which of the following assumptions was an important factor when the scientists re-created the atmosphere that existed on Earth prior to 1850?

   F. Earth's population was about 1,000,000,000.
   G. Water covered about 2/3 of Earth's surface.
   H. The distance from the sun to Earth was about 93,000,000 miles.
   J. It is expected that CO_2 emissions will increase to about 10 billion tons in 2030.

GO ON TO THE NEXT PAGE.

PASSAGE VI

Earth's original atmosphere probably consisted mainly of a mixture of the gases ammonia (NH_3), methane (CH_4), hydrogen (H_2), and water vapor (H_2O). Today, Earth's atmosphere consists primarily of the gases nitrogen (N_2), oxygen (O_2), carbon dioxide (CO_2), and water vapor (H_2O). Earth is much cooler than when the original atmosphere was present. Our planet has been bombarded by a steady stream of heat, ultraviolet rays, gamma rays, and cosmic rays. Sometime during Earth's development, organic compounds—the chemicals of life—developed.

Hypothesis 1

Life was formed in Earth's early atmosphere. As a result of pressure and heat, igneous and sedimentary rocks were changed into harder and denser metamorphic rocks. Specks of metamorphic rock were weathered away and swept into the atmosphere by wind. While there, these compounds combined with other bits of rock and mixed with other gases. The result of this weathering and interaction in the atmosphere along with lightning strikes led to the development of the first life-forms.

Hypothesis 2

Life was formed in Earth's ancient seas. Rain fell on the land, dissolved minerals from the rocks, and carried the minerals to the seas. Lightning from storms and energy from the sun caused the gas molecules of the original atmosphere to form compounds such as amino acids. These compounds rained into the seas to form an organic "soup" in which developed the organic compounds that led to the development of the first life-forms.

29. For Hypothesis 1 to be true:

   A. the results of the interactions among the rocks and other bits of matter in the atmosphere must produce inorganic compounds.
   B. the results of the interactions among the rocks and bits of other matter in the atmosphere must produce organic compounds.
   C. the gas molecules in the original atmosphere must be organic.
   D. the gas molecules in the original atmosphere must be inorganic.

30. Scientists conducted an experiment to assess Hypothesis 2. The scientists replicated Earth's original atmosphere in a large sealed glass enclosure. At one end of the tank, a pool of water (A) was surrounded by rock formations. At the other end of the tank, a pool of water (B) stood alone. The entire enclosure was subjected to continuous rain and to electronic strobes replicating lightning. Which of the following results lends support to Hypothesis 2?

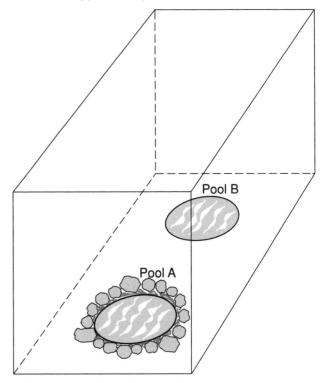

   F. Inorganic compounds form in water pool A.
   G. Inorganic compounds form in water pool B.
   H. An organic "soup" forms in water pool A.
   J. An organic "soup" forms in water pool B.

31. The change in Earth's atmosphere from ammonia–methane–hydrogen to nitrogen–oxygen–carbon dioxide suggests that:

   A. the early photosynthetic organisms had an enormous impact on the chemical makeup of the atmosphere.
   B. ammonia and methane combined to form nitrogen while Earth was being drenched with rain.
   C. respiration of mammals during the time of the early atmosphere added nitrogen, oxygen, and carbon dioxide to the atmosphere.
   D. carbon dioxide from burning fossil fuels during the time of the early atmosphere was added to the atmosphere.

GO ON TO THE NEXT PAGE.

32. Scientists conducted another experiment that replicated Earth's original atmosphere in a large sealed glass enclosure. A large shallow pool at the bottom of the enclosure represented Earth's ancient seas. The "sea" contained amino acids and DNA. At the conclusion of the experiment, scientists found that additional organic material, a first step in the development of life, had been formed in the pool. Which of the following conclusions is NOT supported by this experiment?

I.   Hypothesis 1 is supported because the ingredients for life formed in the sea.
II. Hypothesis 2 is supported because the ingredients for life formed in the sea.
III. Amino acids are required for life to form.

F. I
G. I and III
H. II and III
J. III

33. Which of the following would most clearly SUPPORT Hypothesis 1?

A. Amino acids react to form the ingredients for life in conditions replicating Earth's early atmosphere.
B. Nitrogen content of the early atmosphere declined rapidly.
C. Lightning storms were common in Earth's early atmosphere.
D. Igneous rocks were common during early stages of Earth's development.

34. The two hypotheses share the common assumption that:

F. Earth's atmosphere changed dramatically from its original composition.
G. minerals from rocks were carried to the seas.
H. the atmosphere and the seas were places in which life originally formed on Earth.
J. organic material was not originally present in Earth's early atmosphere.

35. Which of the following would REFUTE Hypothesis 1?

I.   Inorganic material combined with gases cannot form organic material.
II. Earth's early atmosphere had no wind.
III. Lightning was not present in the atmosphere.

A. I, II, and III
B. I and III
C. II and III
D. III

GO ON TO THE NEXT PAGE.

PASSAGE VII

Exploration of the Greenland ice cap reveals 17 warm (interstadial) periods in the past 60,000 years. Glacial caps finally receded about 10,000 years ago. As a benchmark, scientists took oxygen isotope readings from different layers of the ice cap. To determine if these warm periods appeared elsewhere on Earth, a deep hole was dug off the southern coast of California. Scientists examined the stratified sample and used evidence of life-forms to estimate the oxygen concentration. The samples from Greenland and from off the California coast were independently dated. The figure below shows the results of these investigations with the interstadial periods shaded gray.

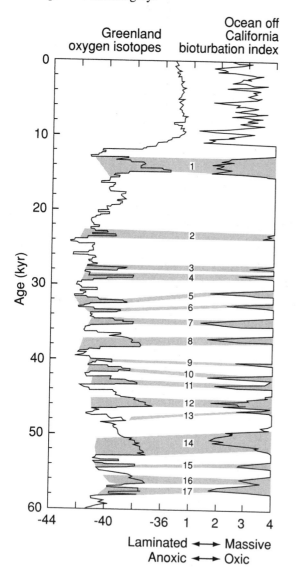

To further investigate the relationship between the Greenland and California data, the age of interstadials from Greenland data was plotted against age of life-form data from California. The figure below shows those data.

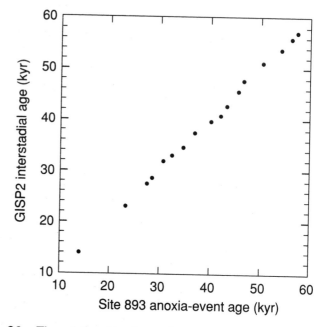

36. The relationship shown between the Greenland and California age data in the figure above is best described as:

  F. proportional.
  G. strongly negative.
  H. strongly positive.
  J. directly proportional.

37. A lower numbered oxygen isotope reading and a lower bioturbation index indicate a warmer atmospheric temperature. For which of the following interstadial periods does the bioturbation index indicate the highest temperature?

  A. 1
  B. 7
  C. 14
  D. 17

GO ON TO THE NEXT PAGE.

38. Bioturbation indices usually vary with the depth of a sediment, as shown in the figure below. According to these data, how would you describe sediment with bioturbation indices 0–2?

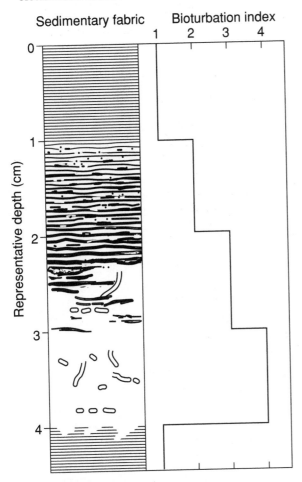

Sedimentary fabric   Bioturbation index

39. For which noninterstadial period does the Greenland data indicate higher temperatures?

   A. About 8,000–10,000 years ago
   B. About 12,000–14,000 years ago
   C. About 18,000–20,000 years ago
   D. About 36,000–38,000 years ago

40. Which of the following is the correct order from least to greatest of the highest bioturbation indices for interstadial periods 1, 2, 3, 6, 12, and 16?

   F. 3, 2, 6, 16, 12, 1
   G. 1, 12, 3, 2, 16, 6
   H. 2, 3, 12, 16, 6, 1
   J. 1, 12, 6, 16, 3, 2

END OF TEST 4

   F. Sediment with indices 0–2 tends to be more diffused and undifferentiated.
   G. Sediment with indices 0–2 tends to be more granular with vertical layers.
   H. Sediment with indices 0–2 tends to vary in appearance depending on the depth of the sediment.
   J. Sediment with indices 0–2 tends to be more uniform and stratified.

Model Science Reasoning ACT II

ANSWERS

1. C	11. A	21. B	31. A
2. F	12. G	22. H	32. G
3. A	13. C	23. C	33. A
4. G	14. F	24. H	34. J
5. D	15. C	25. D	35. A
6. H	16. F	26. H	36. H
7. C	17. D	27. C	37. B
8. J	18. G	28. F	38. J
9. A	19. C	29. B	39. A
10. H	20. G	30. H	40. J

ACT II Science Reasoning Answers Explained

PASSAGE I

1. **C.** The lowest rate of degradation after 5 days is 48% for sample D. Note that 11% is the degradation rate after 1 day for sample C and that 46% is the rate after 10 days for that same sample.

2. **F.** It makes sense that more complex isomers take longer to degrade. Answer F is the only reasonable choice from the four answers available.

3. **A.** Consider the accuracy of each answer.
   A. TRUE.   Sample E contains isomers with two different values and the pattern of degradation most clearly resembles the sample with 3 chlorines.
   B. FALSE.   There is no information about a sample with 5 chlorines so we can't draw this conclusion.
   C. FALSE.   You can't work backward from the data in this manner. What is more, it is reasonable from the data here that a sample with 3, or even 4, chlorines could degrade between 70% and 80% after 10 days.
   D. FALSE.   This answer contradicts the data from the experiment.

4. **G.** The percent of PCBs degraded declined in sample D and remained the same in sample C from 5 days to 10 days. This decline raises the question, "Would the percent degradation remain constant or continue to decline after 10 days, leaving the farmland contaminated?"

5. **D.** Samples C and D seem to indicate that PCB degradation may stop or even start to reverse itself after 5 days. Whenever an ACT question asks how to improve an experiment, the answer is almost always more or longer trials. To provide more information, measure the percent degradation over a longer period of time.

6. **H.** If water inhibits microbial degradation of PCBs, the PCBs in the wet mud will be broken down at a slower rate than those in the original samples.

7. **C.** Visual inspection reveals that the thickness of the propane band is narrowest at Site 5.

8. **J.** The band for ethane gas at Site 1 extends from 30% to 50%. This means the band represents about 20%. 20% of 1,000 cubic feet is 200 cubic feet.

9. **A.** The hydrogen index starts out slightly higher than the gas/oil ratio, but then the hydrogen index gets progressively lower as the gas/oil ratio gets higher. This pattern is best represented by graph A.

10. **H.** The gas/oil ratio is cubic feet of gas/barrels of oil. The gas/oil ratio for Site 6 is about 700 cubic feet of gas for 1 barrel of oil. So there is about 700 times as many cubic feet of gas as there are barrels of oil, as reflected in II. The percent of methane at Site 6 is more than 50% and the band shows the percent of ethane at that site is about 20%. II is the only correct statement and H is the correct answer.

11. **A.** Remember you are looking for the one INCORRECT answer. Consider each answer in turn.
    - A. INCORRECT. The hydrogen percentage goes down and the methane percentage goes up and down.
    - B. CORRECT.   The hydrogen index moves steadily down and the maturity of organic materials moves steadily up from left to right on the graph.
    - C. CORRECT.   Visual inspection reveals that butane is the lowest percentage of gas at each site.
    - D. CORRECT.   The gas/oil ratio is lower than the hydrogen index at 4 of the 8 sites.

12. **G.** Consider each answer in turn.
    - F. NOT TRUE.   There is nothing in these data to support this conclusion and the "because" part is not true.
    - G. TRUE.   The gas/oil ratio increases when the amount of gas increases faster than the amount of oil.
    - H. NOT TRUE.   There is no demonstrated cause-effect relationship between the percent of methane gas and the hydrogen index.
    - J. NOT TRUE.   The methane gas percentage actually drops a little as the organic matter matures.

PASSAGE III

13. **C.** The lines for these two values move up and down independent of each other.

14. **F.** Trace the low point of this line to the scale on the horizontal axis to determine when the ozone low occurs. You will find that it occurs between September and February. December is the only month listed among the choices that falls between September and February.

15. **C.** Something must explain the highest oxidant levels noted in August. If particulate levels are lowest, then other oxidant levels must be highest.

16. **F.** Oxidant levels are lower in May than in any of the other months, so the air is most oxygen rich during that month.

17. **D.** The oxidant/ozone level is greater than 1 when the oxidant level is higher than the ozone level and less than 1 when the ozone level is higher. In the list of choices, November is the only month listed in which the oxidant level exceeds the ozone level. Thus it is the only month with a ratio greater than 1.

18. **G.** The male's gametes contain only *y* and *l* alleles. In standard genetics notation, lowercase letters represent recessive alleles. In addition, the key on the diagram identifies *y* and *l* as recessive.

19. **C.** The diagram shows chromatids exchanging dominant and recessive genes. The gametes that result include some that have a mixture of dominant and recessive genes.

20. **G.** The traits yellow body and short wings represent a fly with dominant and recessive genes. Both recessive genes for wing length must be present for a fly to have this recessive trait. Only choice G has both alleles for this recessive trait.

21. **B.** For all the genes in the fertilization to be dominant, both the male and female flies must have dominant genes.

22. **H.** The phenotype for yellow bodies and long wings for these offspring comes from the alleles for the dominant traits. This combination accounts for 25% of the phenotypes. A 50% increase on 25% is 12.5%. 12.5% + 25% = 37.5%. So about 37.5% of the offspring will have long wings and yellow bodies.

PASSAGE V

23. **C.** The passage and table do not tell you what proportion of Earth's greenhouse gases result from agricultural practices. If agricultural practices contribute only a small fraction of Earth's greenhouse gases, revising agricultural practices will have little effect on global warming.

24. **H.** Elimination of fossil fuels in a huge country such as China might reasonably reduce carbon dioxide emissions. A reduction in the use of fossil fuels on Earth could never have an impact on the heat of the sun.

25. **D.** Carbon dioxide concentrations rose from 275 ppmv before 1850 to 345 ppmv today. The increase is 345 − 275 = 70. 70 divided by 275 is about 0.25, or about $1/4$.

26. **H.** The paragraph at the beginning of the passage specifically mentions industrial emissions of CFCs.

27. **C.** This choice is the only one that allows scientists to control the amount of carbon dioxide emissions. In the other choices, the amount of emissions is not under their control.

28. **F.** Earth's population is five times today what it was around 1850. This population increase is an important factor in determining fossil fuel usage. Choices G and H are the same now as in 1850 and would not be important factors in determining the makeup of the atmosphere. Choice J gives information about the future, which is not useful in determining the makeup of Earth's atmosphere before 1850.

PASSAGE VI

29. **B.** For Hypothesis 1 to be true, the results of the interaction in the atmosphere must produce organic compounds and eventually life.

30. **H.** Hypothesis 2 states that an organic "soup" formed in water following a mineralization run-off from rocks. The circumstances at pool A represent the circumstances in Hypothesis 2.

31. **A.** Choice B is false because ammonia and methane do not form nitrogen. Choices C and D are not correct because mammals and fossil fuels were not around when Earth's early atmosphere was present.

32. **G.** I is not supported. Note, however, that it is not disproved—just because the ingredients for life formed in the sea does not mean they could not form in the atmosphere. III is not supported. Just because amino acids were present when additional organic material was formed does not mean these amino acids are required for life to form. II is supported because the formation of the ingredients for life makes it more believable that life was formed in the sea.

33. **A.** If interactions in the experimental atmosphere create the ingredients for life, the hypothesis that life started in the atmosphere is supported. Hypothesis 2 states that amino acids were formed in the atmosphere and that these amino acids led to the formation of life in the sea. Since amino acids are organic materials, then the writer of Hypothesis 2 actually supports Hypothesis 1.

34. **J.** Consider each answer in turn.

    F. FALSE. This statement is true, but neither hypothesis assumes this statement.

    G. FALSE. This is the assumption from Hypothesis 2 only.

    H. FALSE. These are not assumptions. This compound conclusion is not reached in either hypothesis.

    J. TRUE. However life formed, each hypothesis assumes that it was not present originally.

35. **A.** All three refute Hypothesis 1, which asserts that inorganic material that combined with gases led to the formation of life and that wind and lightning were present in the early atmosphere.

PASSAGE VII

36. **H.** A lower left to upper right line shows a strongly positive relationship. To be proportional, the line would have to be perfectly straight. To be directly proportional, the line would have to be perfectly straight from lower left to upper right. A negative relationship slopes from upper left to lower right.

37. **B.** The lowest biturbation index (highest temperature) is for interstadial period 7. The lower the index, the more the peak extends to the left. Use the edge of a piece of paper to check this answer.

38. **J.** These sediments are more stratified (horizontal) and uniform (spaces between sediments are the same).

39. **A.** The Greenland sites indicates the highest temperature about 8,000–10,000 years ago, a noninterstadial period. Recall from question 37 that the lower the oxygen isotope reading and the lower the biturbation index, the higher the temperature.

40. **J.** Use the edge of a piece of paper to confirm this answer. Remember that the lower the index, the longer the peak on the graph.

Chapter 24 ▪

Model Science Reasoning **ACT III**
With Answers Explained

This Science Reasoning Model ACT III is just like a real ACT. Take this test after you take Model ACT II.

Take this model test under simulated test conditions. Allow 35 minutes to answer the 40 Science Reasoning items. Use a pencil to mark the answer sheet on the next page, and answer the questions in the Test 4 (Science Reasoning) section.

Use the answer key on page 566 to mark the answer sheet. Review the answer explanations on pages 566–568. You may decide to retake this test later. There are additional answer sheets for this purpose following page 605.

The test scoring chart on page 589 shows you how to convert the number correct to ACT scale scores.

DO NOT leave any answers blank. There is no penalty for guessing on the ACT. Remember that the test is yours. You may mark up, write on, or draw on the test.

When you are ready, note the time and begin.

ANSWER SHEET

The ACT answer sheet looks something like this one. Use a No. 2 pencil
to completely fill the circle corresponding to the correct answer.
If you erase, erase completely; incomplete erasures may be read as answers.

TEST 1—English

1 Ⓐ Ⓑ Ⓒ Ⓓ	11 Ⓐ Ⓑ Ⓒ Ⓓ	21 Ⓐ Ⓑ Ⓒ Ⓓ	31 Ⓐ Ⓑ Ⓒ Ⓓ	41 Ⓐ Ⓑ Ⓒ Ⓓ	51 Ⓐ Ⓑ Ⓒ Ⓓ	61 Ⓐ Ⓑ Ⓒ Ⓓ	71 Ⓐ Ⓑ Ⓒ Ⓓ
2 Ⓕ Ⓖ Ⓗ Ⓙ	12 Ⓕ Ⓖ Ⓗ Ⓙ	22 Ⓕ Ⓖ Ⓗ Ⓙ	32 Ⓕ Ⓖ Ⓗ Ⓙ	42 Ⓕ Ⓖ Ⓗ Ⓙ	52 Ⓕ Ⓖ Ⓗ Ⓙ	62 Ⓕ Ⓖ Ⓗ Ⓙ	72 Ⓕ Ⓖ Ⓗ Ⓙ
3 Ⓐ Ⓑ Ⓒ Ⓓ	13 Ⓐ Ⓑ Ⓒ Ⓓ	23 Ⓐ Ⓑ Ⓒ Ⓓ	33 Ⓐ Ⓑ Ⓒ Ⓓ	43 Ⓐ Ⓑ Ⓒ Ⓓ	53 Ⓐ Ⓑ Ⓒ Ⓓ	63 Ⓐ Ⓑ Ⓒ Ⓓ	73 Ⓐ Ⓑ Ⓒ Ⓓ
4 Ⓕ Ⓖ Ⓗ Ⓙ	14 Ⓕ Ⓖ Ⓗ Ⓙ	24 Ⓕ Ⓖ Ⓗ Ⓙ	34 Ⓕ Ⓖ Ⓗ Ⓙ	44 Ⓕ Ⓖ Ⓗ Ⓙ	54 Ⓕ Ⓖ Ⓗ Ⓙ	64 Ⓕ Ⓖ Ⓗ Ⓙ	74 Ⓕ Ⓖ Ⓗ Ⓙ
5 Ⓐ Ⓑ Ⓒ Ⓓ	15 Ⓐ Ⓑ Ⓒ Ⓓ	25 Ⓐ Ⓑ Ⓒ Ⓓ	35 Ⓐ Ⓑ Ⓒ Ⓓ	45 Ⓐ Ⓑ Ⓒ Ⓓ	55 Ⓐ Ⓑ Ⓒ Ⓓ	65 Ⓐ Ⓑ Ⓒ Ⓓ	75 Ⓐ Ⓑ Ⓒ Ⓓ
6 Ⓕ Ⓖ Ⓗ Ⓙ	16 Ⓕ Ⓖ Ⓗ Ⓙ	26 Ⓕ Ⓖ Ⓗ Ⓙ	36 Ⓕ Ⓖ Ⓗ Ⓙ	46 Ⓕ Ⓖ Ⓗ Ⓙ	56 Ⓕ Ⓖ Ⓗ Ⓙ	66 Ⓕ Ⓖ Ⓗ Ⓙ	
7 Ⓐ Ⓑ Ⓒ Ⓓ	17 Ⓐ Ⓑ Ⓒ Ⓓ	27 Ⓐ Ⓑ Ⓒ Ⓓ	37 Ⓐ Ⓑ Ⓒ Ⓓ	47 Ⓐ Ⓑ Ⓒ Ⓓ	57 Ⓐ Ⓑ Ⓒ Ⓓ	67 Ⓐ Ⓑ Ⓒ Ⓓ	
8 Ⓕ Ⓖ Ⓗ Ⓙ	18 Ⓕ Ⓖ Ⓗ Ⓙ	28 Ⓕ Ⓖ Ⓗ Ⓙ	38 Ⓕ Ⓖ Ⓗ Ⓙ	48 Ⓕ Ⓖ Ⓗ Ⓙ	58 Ⓕ Ⓖ Ⓗ Ⓙ	68 Ⓕ Ⓖ Ⓗ Ⓙ	
9 Ⓐ Ⓑ Ⓒ Ⓓ	19 Ⓐ Ⓑ Ⓒ Ⓓ	29 Ⓐ Ⓑ Ⓒ Ⓓ	39 Ⓐ Ⓑ Ⓒ Ⓓ	49 Ⓐ Ⓑ Ⓒ Ⓓ	59 Ⓐ Ⓑ Ⓒ Ⓓ	69 Ⓐ Ⓑ Ⓒ Ⓓ	
10 Ⓕ Ⓖ Ⓗ Ⓙ	20 Ⓕ Ⓖ Ⓗ Ⓙ	30 Ⓕ Ⓖ Ⓗ Ⓙ	40 Ⓕ Ⓖ Ⓗ Ⓙ	50 Ⓕ Ⓖ Ⓗ Ⓙ	60 Ⓕ Ⓖ Ⓗ Ⓙ	70 Ⓕ Ⓖ Ⓗ Ⓙ	

TEST 2—Mathematics

1 Ⓐ Ⓑ Ⓒ Ⓓ Ⓔ	9 Ⓐ Ⓑ Ⓒ Ⓓ Ⓔ	17 Ⓐ Ⓑ Ⓒ Ⓓ Ⓔ	25 Ⓐ Ⓑ Ⓒ Ⓓ Ⓔ	33 Ⓐ Ⓑ Ⓒ Ⓓ Ⓔ	41 Ⓐ Ⓑ Ⓒ Ⓓ Ⓔ	49 Ⓐ Ⓑ Ⓒ Ⓓ Ⓔ	57 Ⓐ Ⓑ Ⓒ Ⓓ Ⓔ
2 Ⓕ Ⓖ Ⓗ Ⓙ Ⓚ	10 Ⓕ Ⓖ Ⓗ Ⓙ Ⓚ	18 Ⓕ Ⓖ Ⓗ Ⓙ Ⓚ	26 Ⓕ Ⓖ Ⓗ Ⓙ Ⓚ	34 Ⓕ Ⓖ Ⓗ Ⓙ Ⓚ	42 Ⓕ Ⓖ Ⓗ Ⓙ Ⓚ	50 Ⓕ Ⓖ Ⓗ Ⓙ Ⓚ	58 Ⓕ Ⓖ Ⓗ Ⓙ Ⓚ
3 Ⓐ Ⓑ Ⓒ Ⓓ Ⓔ	11 Ⓐ Ⓑ Ⓒ Ⓓ Ⓔ	19 Ⓐ Ⓑ Ⓒ Ⓓ Ⓔ	27 Ⓐ Ⓑ Ⓒ Ⓓ Ⓔ	35 Ⓐ Ⓑ Ⓒ Ⓓ Ⓔ	43 Ⓐ Ⓑ Ⓒ Ⓓ Ⓔ	51 Ⓐ Ⓑ Ⓒ Ⓓ Ⓔ	59 Ⓐ Ⓑ Ⓒ Ⓓ Ⓔ
4 Ⓕ Ⓖ Ⓗ Ⓙ Ⓚ	12 Ⓕ Ⓖ Ⓗ Ⓙ Ⓚ	20 Ⓕ Ⓖ Ⓗ Ⓙ Ⓚ	28 Ⓕ Ⓖ Ⓗ Ⓙ Ⓚ	36 Ⓕ Ⓖ Ⓗ Ⓙ Ⓚ	44 Ⓕ Ⓖ Ⓗ Ⓙ Ⓚ	52 Ⓕ Ⓖ Ⓗ Ⓙ Ⓚ	60 Ⓕ Ⓖ Ⓗ Ⓙ Ⓚ
5 Ⓐ Ⓑ Ⓒ Ⓓ Ⓔ	13 Ⓐ Ⓑ Ⓒ Ⓓ Ⓔ	21 Ⓐ Ⓑ Ⓒ Ⓓ Ⓔ	29 Ⓐ Ⓑ Ⓒ Ⓓ Ⓔ	37 Ⓐ Ⓑ Ⓒ Ⓓ Ⓔ	45 Ⓐ Ⓑ Ⓒ Ⓓ Ⓔ	53 Ⓐ Ⓑ Ⓒ Ⓓ Ⓔ	
6 Ⓕ Ⓖ Ⓗ Ⓙ Ⓚ	14 Ⓕ Ⓖ Ⓗ Ⓙ Ⓚ	22 Ⓕ Ⓖ Ⓗ Ⓙ Ⓚ	30 Ⓕ Ⓖ Ⓗ Ⓙ Ⓚ	38 Ⓕ Ⓖ Ⓗ Ⓙ Ⓚ	46 Ⓕ Ⓖ Ⓗ Ⓙ Ⓚ	54 Ⓕ Ⓖ Ⓗ Ⓙ Ⓚ	
7 Ⓐ Ⓑ Ⓒ Ⓓ Ⓔ	15 Ⓐ Ⓑ Ⓒ Ⓓ Ⓔ	23 Ⓐ Ⓑ Ⓒ Ⓓ Ⓔ	31 Ⓐ Ⓑ Ⓒ Ⓓ Ⓔ	39 Ⓐ Ⓑ Ⓒ Ⓓ Ⓔ	47 Ⓐ Ⓑ Ⓒ Ⓓ Ⓔ	55 Ⓐ Ⓑ Ⓒ Ⓓ Ⓔ	
8 Ⓕ Ⓖ Ⓗ Ⓙ Ⓚ	16 Ⓕ Ⓖ Ⓗ Ⓙ Ⓚ	24 Ⓕ Ⓖ Ⓗ Ⓙ Ⓚ	32 Ⓕ Ⓖ Ⓗ Ⓙ Ⓚ	40 Ⓕ Ⓖ Ⓗ Ⓙ Ⓚ	48 Ⓕ Ⓖ Ⓗ Ⓙ Ⓚ	56 Ⓕ Ⓖ Ⓗ Ⓙ Ⓚ	

TEST 3—Reading

1 Ⓐ Ⓑ Ⓒ Ⓓ	6 Ⓕ Ⓖ Ⓗ Ⓙ	11 Ⓐ Ⓑ Ⓒ Ⓓ	16 Ⓕ Ⓖ Ⓗ Ⓙ	21 Ⓐ Ⓑ Ⓒ Ⓓ	26 Ⓕ Ⓖ Ⓗ Ⓙ	31 Ⓐ Ⓑ Ⓒ Ⓓ	36 Ⓕ Ⓖ Ⓗ Ⓙ
2 Ⓕ Ⓖ Ⓗ Ⓙ	7 Ⓐ Ⓑ Ⓒ Ⓓ	12 Ⓕ Ⓖ Ⓗ Ⓙ	17 Ⓐ Ⓑ Ⓒ Ⓓ	22 Ⓕ Ⓖ Ⓗ Ⓙ	27 Ⓐ Ⓑ Ⓒ Ⓓ	32 Ⓕ Ⓖ Ⓗ Ⓙ	37 Ⓐ Ⓑ Ⓒ Ⓓ
3 Ⓐ Ⓑ Ⓒ Ⓓ	8 Ⓕ Ⓖ Ⓗ Ⓙ	13 Ⓐ Ⓑ Ⓒ Ⓓ	18 Ⓕ Ⓖ Ⓗ Ⓙ	23 Ⓐ Ⓑ Ⓒ Ⓓ	28 Ⓕ Ⓖ Ⓗ Ⓙ	33 Ⓐ Ⓑ Ⓒ Ⓓ	38 Ⓕ Ⓖ Ⓗ Ⓙ
4 Ⓕ Ⓖ Ⓗ Ⓙ	9 Ⓐ Ⓑ Ⓒ Ⓓ	14 Ⓕ Ⓖ Ⓗ Ⓙ	19 Ⓐ Ⓑ Ⓒ Ⓓ	24 Ⓕ Ⓖ Ⓗ Ⓙ	29 Ⓐ Ⓑ Ⓒ Ⓓ	34 Ⓕ Ⓖ Ⓗ Ⓙ	39 Ⓐ Ⓑ Ⓒ Ⓓ
5 Ⓐ Ⓑ Ⓒ Ⓓ	10 Ⓕ Ⓖ Ⓗ Ⓙ	15 Ⓐ Ⓑ Ⓒ Ⓓ	20 Ⓕ Ⓖ Ⓗ Ⓙ	25 Ⓐ Ⓑ Ⓒ Ⓓ	30 Ⓕ Ⓖ Ⓗ Ⓙ	35 Ⓐ Ⓑ Ⓒ Ⓓ	40 Ⓕ Ⓖ Ⓗ Ⓙ

TEST 4—Science Reasoning

1 Ⓐ Ⓑ Ⓒ Ⓓ	6 Ⓕ Ⓖ Ⓗ Ⓙ	11 Ⓐ Ⓑ Ⓒ Ⓓ	16 Ⓕ Ⓖ Ⓗ Ⓙ	21 Ⓐ Ⓑ Ⓒ Ⓓ	26 Ⓕ Ⓖ Ⓗ Ⓙ	31 Ⓐ Ⓑ Ⓒ Ⓓ	36 Ⓕ Ⓖ Ⓗ Ⓙ
2 Ⓕ Ⓖ Ⓗ Ⓙ	7 Ⓐ Ⓑ Ⓒ Ⓓ	12 Ⓕ Ⓖ Ⓗ Ⓙ	17 Ⓐ Ⓑ Ⓒ Ⓓ	22 Ⓕ Ⓖ Ⓗ Ⓙ	27 Ⓐ Ⓑ Ⓒ Ⓓ	32 Ⓕ Ⓖ Ⓗ Ⓙ	37 Ⓐ Ⓑ Ⓒ Ⓓ
3 Ⓐ Ⓑ Ⓒ Ⓓ	8 Ⓕ Ⓖ Ⓗ Ⓙ	13 Ⓐ Ⓑ Ⓒ Ⓓ	18 Ⓕ Ⓖ Ⓗ Ⓙ	23 Ⓐ Ⓑ Ⓒ Ⓓ	28 Ⓕ Ⓖ Ⓗ Ⓙ	33 Ⓐ Ⓑ Ⓒ Ⓓ	38 Ⓕ Ⓖ Ⓗ Ⓙ
4 Ⓕ Ⓖ Ⓗ Ⓙ	9 Ⓐ Ⓑ Ⓒ Ⓓ	14 Ⓕ Ⓖ Ⓗ Ⓙ	19 Ⓐ Ⓑ Ⓒ Ⓓ	24 Ⓕ Ⓖ Ⓗ Ⓙ	29 Ⓐ Ⓑ Ⓒ Ⓓ	34 Ⓕ Ⓖ Ⓗ Ⓙ	39 Ⓐ Ⓑ Ⓒ Ⓓ
5 Ⓐ Ⓑ Ⓒ Ⓓ	10 Ⓕ Ⓖ Ⓗ Ⓙ	15 Ⓐ Ⓑ Ⓒ Ⓓ	20 Ⓕ Ⓖ Ⓗ Ⓙ	25 Ⓐ Ⓑ Ⓒ Ⓓ	30 Ⓕ Ⓖ Ⓗ Ⓙ	35 Ⓐ Ⓑ Ⓒ Ⓓ	40 Ⓕ Ⓖ Ⓗ Ⓙ

Model Science Reasoning ACT III

40 Questions—35 Minutes

INSTRUCTIONS: Each of the seven science passages on this test is followed by questions. Choose the best answer for each question based on the passage. Then fill in the appropriate circle on the answer sheet. You can look back at the passages as often as you want.

You CANNOT use a calculator on this test.

Check pages 566–568 for answers and explanations.

PASSAGE I

The Savannah sparrow, which migrates at night, has magnetic and star compasses to help it find its way. It also orients itself by using visual cues when there are clear sunset skies. Scientists interested in understanding how the birds calibrate, or fine-tune, their navigation systems captured 39 Savannah sparrows, 11 adult birds (older than a year) and 28 immature birds. The scientists then exposed the birds to normal and shifted magnetic fields.

Experiment 1

All the birds were placed in cages with unshifted magnetic fields until they had experienced four clear days and four clear nights. This experiment constituted the control. Before the birds were subjected to the control, they exhibited a NNE orientation of about 340°. After exposure to this control treatment, they showed the same orientation.

Experiment 2

Seven adult birds and twelve immature birds were exposed to a 90° counterclockwise magnetic shift. The birds were put in a cage inside a Rubens coil, which shifted the magnetic field from magnetic north (360°) to magnetic west (270°). The birds were left in the cages with this shifted field until they had experienced four clear days and four clear nights. After exposure to this experimental treatment, birds showed an average orientation of about 250°.

Experiment 3

Four adult birds and sixteen immature birds were exposed to a 90° clockwise magnetic shift. The birds were put in a cage inside a Rubens coil, which shifted the magnetic field from magnetic north (360°) to magnetic east (90°). The birds were left in the cages with this shifted field until they had experienced four clear days and four clear nights. After exposure to this experimental treatment, the birds showed an average orientation of about 70°.

1. According to these data, what was the birds' orientation prior to the 90° clockwise magnetic shift?

    A. 70°
    B. 140°
    C. 250°
    D. 340°

2. Unshifted magnetic fields:

    F. retain the birds' initial perception of magnetic north.
    G. always change the birds' perception of magnetic north by 90°.
    H. may change the birds' perception of magnetic north by any number of degrees.
    J. change the birds' perception of magnetic north only on clear days.

GO ON TO THE NEXT PAGE.

3. The results of Experiment 2 could be used to support which of the following hypotheses?

   A. Birds with magnetic north shifted 90° clockwise will shift their orientation to face east.
   B. Birds with magnetic north shifted 90° in any direction will shift their orientation to face in that direction.
   C. Birds with magnetic north shifted 90° in any direction will shift their orientation 90° in that direction.
   D. Birds with magnetic north shifted 90° clockwise will shift their orientation to face in that direction.

4. How might the results have been different if the birds were placed in sealed cages and could not see the sun or the stars?

   F. Most birds would reorient themselves on the basis of magnetic direction alone.
   G. Most birds would not reorient themselves.
   H. Most birds would reorient themselves only at night.
   J. Most birds would reorient themselves only when in the cage.

5. If magnetic north was shifted 60° counterclockwise in Experiment 2, then the birds would shift their orientation to:

   A. 40°.
   B. 300°.
   C. 250°.
   D. 280°.

6. In the reports of the results of these experiments, the scientists imply that:

   F. all birds shifted their orientation exactly the same amount.
   G. there were no significant differences between immature and adult birds.
   H. there were significant differences in the amount of orientation shift from experiment to experiment.
   J. birds exposed to different amounts of magnetic shifts would reorient themselves 90° clockwise or 90° counterclockwise.

PASSAGE II

Scientists and engineers are always trying to find good places to mine for valuable ore. Scientists bored five test holes in a rock deposit. They hoped to use the information from these borings to identify the best place to mine for uranium.

Figure 1

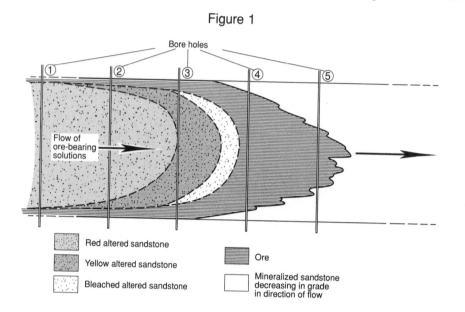

Bore holes

Flow of ore-bearing solutions

Red altered sandstone

Yellow altered sandstone

Bleached altered sandstone

Ore

Mineralized sandstone decreasing in grade in direction of flow

Experiment 1

Scientists took a longitudinal sample from each boring site and measured the percent of pyrite/marcasite, uranium, iron, and carbon. The results are shown in Table 1.

Table 1. Concentration in Percents

	Site				
	1	2	3	4	5
Pyrite/marcasite	0	0.2	0.5	3.0	3.8
Uranium	0.04	0.08	0.4	3.0	3.5
Iron	0.8	0.6	0.5	1.2	2.4
Carbon	0.7	1.2	2.3	1.1	1.4

Experiment 2

Scientists took a longitudinal sample from each boring site and measured the amount of selenium and vanadium in parts per million (ppm). The results are shown in Table 2.

Table 2. Concentrations in Parts per Million

	Site				
	1	2	3	4	5
Selenium	10	40	100	490	10
Vanadium	80	110	130	600	1,400

7. Miners trying to find the highest concentration of uranium ore would drill:

   A. at the place where the ore body was thickest.
   B. near the narrow section of the ore body.
   C. where the ore body was surrounded by unaltered sandstone.
   D. where the stone is entirely mineralized sandstone.

8. Of all the measurements taken, which appears to be LEAST useful in predicting the presence of uranium?

   F. Iron
   G. Carbon
   H. Pyrite/marcasite
   J. Vanadium

GO ON TO THE NEXT PAGE.

9. Scientists drill four more test holes as shown in the diagram below. Based on the data from the previous bore holes, which of these test holes will yield material with the highest concentrations of vanadium?

Figure 2

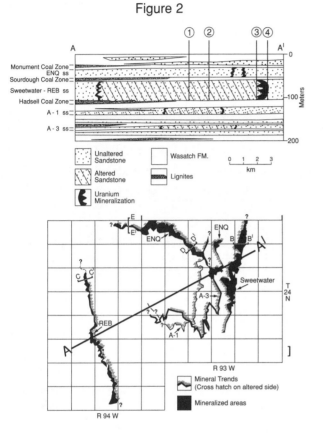

A. Bore hole 1
B. Bore hole 2
C. Bore hole 3
D. Bore hole 4

10. Scientists drilled a bore hole beyond hole 5 in the region shown in the first figure and found no signs of uranium ore. The material most likely found was:

F. red altered sandstone.
G. yellow altered sandstone.
H. bleached altered sandstone.
J. mineralized sandstone.

11. Which of the following best describes the action that results in the depositing of uranium ore in the formation shown in the first figure?

A. The ore is deposited by glaciation.
B. The ore is created through a process of fossilization.
C. The ore precipitates out of solutions as they flow through the ground.
D. The ore is created in the same way as coal and oil.

12. The direction(s) of flow of the ore-bearing solutions in the second figure is most likely:

F. toward A but away from A'.
G. toward A' but away from A.
H. away from both A and A'.
J. toward both A and A'.

GO ON TO THE NEXT PAGE.

PASSAGE III

Scientists conducted experiments in which balls of different masses were dropped in a vacuum in actual Earth gravity and in simulated moon gravity. The results of their experiments are shown below.

Experiment	Mass	Height off surface	Gravity	Period of fall
1	1 kg	9.8 m	Earth	1 second
2	2 kg	9.8 m	Earth	1 second
3	10 kg	44.1 m	Earth	3 seconds
4	0.25 kg	44.1 m	Earth	3 seconds
5	1 kg	0.98 m	Moon	1 second
6	2 kg	8.82 m	Moon	3 seconds
7	10 kg	8.82 m	Moon	3 seconds

13. The results of Experiments 1–4 demonstrate that the period of the fall depends on:

- **A.** the mass of the ball.
- **B.** the height off the surface.
- **C.** the composition of the ball.
- **D.** the density of the ball.

14. Based on the experiments, a person can conclude that when the experiments are conducted on the moon instead of on Earth:

- **F.** the height from the surface increases.
- **G.** the mass of the ball increases.
- **H.** the rate of fall decreases.
- **J.** the time to fall decreases.

15. Which of the following changes in conditions could change the results of Experiments 1–4?

- **A.** Cubes being dropped instead of balls
- **B.** Balls dropped in actual moon gravity
- **C.** Height being measured in feet instead of meters
- **D.** Balls not being dropped in a vacuum

16. A ball with a mass of 20 kilograms was dropped on the moon and took about 1 second to reach the surface. How high was the ball above the surface when it was dropped?

- **F.** 1 meter
- **G.** 2 meters
- **H.** 3 meters
- **J.** 4 meters

17. If it takes 1 second for a ball to fall 9.8 meters, why doesn't a ball fall about 30 meters in 3 seconds?

- **A.** The ball accelerates as it falls.
- **B.** The ball's mass decreases in lower gravity.
- **C.** The gravitational attraction is much greater closer to Earth and applies extra force to the fall.
- **D.** The ball alters its molecular composition as it gets closer to the surface.

GO ON TO THE NEXT PAGE.

PASSAGE IV

Scientists are studying the impact of farms and industries on river pollution. The diagram shows a river, along with the placement of a farm, a factory, and a tributary feeding the river. Five monitoring stations are numbered on the diagram.

On one day, a reading was taken at each monitoring station in the morning, and another reading was taken in the afternoon. Readings were taken for water flow, alkalinity, sewage sludge, and nitrate. The results are reported in the table.

Figure 1

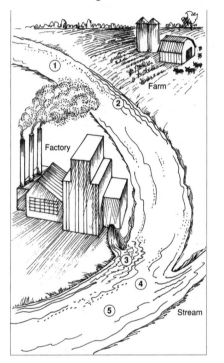

Site	Water flow, ft/sec	Alkalinity, mg/kg	Sewage sludge, mg/kg	Nitrate, mg/kg
A.M.				
1	7	200	700	65
2	5	190	2,600	900
3	13	5,400	2,550	880
4	16	4,900	1,800	780
5	10	4,800	1,750	620
P.M.				
1	8	190	550	60
2	6	180	2,400	700
3	21	6,700	2,300	640
4	27	6,400	1,700	520
5	19	6,200	1,400	380

18. Which of the following statements best explains the change in nitrate levels noted at Site 2?

   F. The sand on the riverbed at Site 2 contains a large amount of nitrate.
   G. Nitrate fertilizer is used on the farm.
   H. Smoke from the factory carries nitrate down to the water.
   J. Nitrate from the tributary backs up the river to Site 2.

19. Based on the data, which of the following conclusions can be drawn about the factory?

   I. The factory discharges warm water into the river.
   II. The factory is more active in the afternoon than in the morning.
   III. The factory gives off sewage sludge.
   IV. The factory increases the alkalinity of the river about 100%.

   A. II
   B. IV
   C. I and III
   D. I and II

20. Of the choices below, it is most likely that the factory:

   F. manufactures alkaline batteries.
   G. manufactures matches.
   H. electroplates car bumpers.
   J. is a sewage treatment plant.

21. Which of the following statements best explains the readings at Site 4?

   I. The sewage sludge reading is high because of sewage sludge runoff at this site.
   II. The alkalinity level drops because of the water from the tributary.
   III. The nitrate level is lowest in the afternoon because of the effects of sunlight.
   IV. Water flow is faster in the afternoon because of a spring near Site 4.

   A. I
   B. II
   C. II and III
   D. III and IV

22. Which of the sites most closely serves the function of a control?

   F. Site 1
   G. Site 2
   H. Site 3
   J. Site 4

PASSAGE V

Crickets use a calling song to communicate. The sound is received through the central membrane in the cricket's thorax. The frequency of the sound received by the cricket's ear can be determined by the speed of tympanal vibrations. The following experiments were conducted to determine the cricket's ability to hear sounds at different frequencies.

Experiment 1

Crickets with intact central membranes were exposed to sounds from 3.5 to 5.25 kHz. The amplitude of the sounds was measured by laser vibrometry and classified on an amplitude scale from 0 to 10. The results of the experiment are shown in the following table.

Frequency, kHz	Amplitude	Frequency, kHz	Amplitude
3.5	0.25	4.5	9.8
3.625	0.25	4.625	7
3.75	3.0	4.75	4.6
3.875	4.1	4.875	3.6
4.0	4.4	5.0	3.2
4.125	4.6	5.125	2.1
4.25	5.8	5.25	1.7
4.375	7.4		

Experiment 2

Crickets with perforated central membranes were exposed to sounds from 3.5 to 5.25 kHz. The amplitude of the sounds was measured by laser vibrometry and classified on an amplitude scale from 0 to 10. The results of the experiment are shown in the table below.

Frequency, kHz	Amplitude	Frequency, kHz	Amplitude
3.5	1.75	4.5	2.0
3.625	2.0	4.625	2.2
3.75	2.2	4.75	2.3
3.875	2.6	4.875	2.4
4.0	2.4	5.0	2.2
4.125	2.0	5.125	0.25
4.25	1.8	5.25	0
4.375	1.9		

GO ON TO THE NEXT PAGE.

23. The relationship between the frequency and amplitude from Experiment 1 is best characterized by the statement:

    A. As the amplitude goes up, the frequency goes down.
    B. As the frequency goes down, the amplitude goes up then down.
    C. As the frequency goes up, the amplitude goes down then up.
    D. As the frequency goes up, the amplitude goes down.

24. Which of the following conclusions is most directly supported by the results from Experiment 1?

    F. Crickets most likely communicate at a frequency of 4.5 kHz.
    G. Crickets can hear sounds from frequencies of 0.25 kHz to 1.7 kHz.
    H. Crickets most likely communicate at 9.8 kHz.
    J. Crickets can hear sounds from 0 kHz to 5.25 kHz.

25. If Experiment 1 were repeated with the central membrane removed, the predicted amplitude for a frequency of 4.0 kHz would be:

    A. 4.4
    B. 2.4
    C. 2
    D. 0

26. Which of the graphs below best represents the results of Experiment 1?

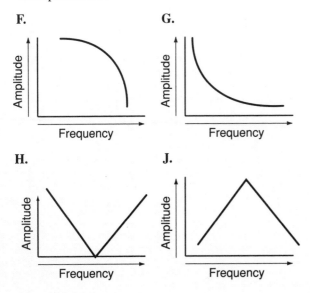

F.
G.
H.
J.

27. Which of the following conclusions is supported by the results of Experiment 2?

    A. Crickets hear sounds from 3.5 kHz to 5.0 kHz at about the same amplitude.
    B. As the frequency of sounds goes up, the amplitude of the sounds goes up and then goes down.
    C. An intact central membrane is most important for crickets to hear sounds from 3.875 kHz to 5.0 kHz.
    D. Crickets can't hear sounds above a frequency of 3.0 kHz.

28. Suppose that scientists conducted another experiment about cricket hearing. The experimenters found that crickets with perforated central membranes heard equally well from every angle around them. Crickets with intact membranes heard best from 0° to 180° around them, with 0° being directly in front of the crickets. Which of the following is the most reasonable conclusion?

    F. Crickets hear other crickets best when the other crickets are on their right.
    G. Crickets hear other crickets best when the temperature is from 0° to 180°.
    H. Crickets hear other crickets best when their central membrane is perforated.
    J. The frequency at which crickets communicate depends on the direction of the communication song.

GO ON TO THE NEXT PAGE.

PASSAGE VI

In the following paragraphs, two archaeologists discuss their theories about the age of the Sphinx.

Archaeologist 1

The Sphinx is about 4,500 years old. From 2700 B.C. to 2150 B.C., the Old Kingdom of Egypt flourished in the Nile River Valley. Pharaohs ruled over a civilization that had writing and a calendar. During this period Egyptians built massive pyramids and the Sphinx in a desert region called the Valley of the Kings. Blocks cut away to form the Sphinx were used to create temples at the front of the Sphinx.

The face of the Sphinx bears a striking similarity to a pharaoh of this time and it is believed that the face of the Sphinx was created in his image. We know from artifacts found in the area that the Sphinx was carved during the time of a civilization much like the one present during the Old Kingdom.

Radiocarbon dating of pieces of organic material found near the hindquarters of the Sphinx indicate that it dates from about 4,000 to 5,000 years ago. Other dating techniques confirm these dates.

Archaeologist 2

The Sphinx is about 10,000 years old. About 10,000 years ago, the head of the sphinx was a yardank (an outcropping of rock). It protruded from what was then a lush, green valley in Egypt. During that time, an advanced civilization carved the yardank into some human or animal form. Over time this carving was covered over and then uncovered during the time of the Old Kingdom when that whole region was a desert. Workers during that time added the paws and a body to the Sphinx.

The face of the Sphinx shows very significant water erosion caused by long periods of persistent rain. We know that the Sphinx has been in a desert climate for the past 4,500 years. It was about 4,000 years prior to that when that part of Egypt would have had significant amounts of rainfall.

Radar readings also indicate that the ground in front of the Sphinx is weathered to a depth twice as great as the back of the Sphinx, suggesting that the front is twice as old as the back.

29. Assume that Archaeologist 2 is correct. What could account for the radiocarbon dating mentioned by Archaeologist 1?

    A. The part dated was constructed before the head was carved.
    B. The face did not actually look like the pharaoh.
    C. The part dated was constructed after the head was carved.
    D. A similar civilization had existed about 5,000 years earlier.

30. Which of the following discoveries would most effectively support the position of Archaeologist 2?

    F. Discovery of an Egyptianlike civilization about 8000 B.C.
    G. Discovery of an Egyptianlike civilization about 3000 B.C.
    H. Historic records showing that yardanks existed about 10,000 years ago
    J. Historic records showing that it rained heavily on the Sphinx from 3000 B.C. to 1000 B.C.

31. When one encounters such different interpretations of the age of a famous structure, it is safe to assume that:

    A. one of the theories is based completely on false information.
    B. each theory contains some facts and some inaccuracies.
    C. one of the theories is deliberately misleading.
    D. each theory is probably incorrect.

32. One fact that both archaeologists implicitly agree on is that:

    F. Egyptian civilization first reached its peak about 3500 B.C.
    G. the Sphinx bears a striking resemblance to an Egyptian pharaoh.
    H. the Sphinx is at least 4,000 years old.
    J. the Old Kingdom of Egypt dates from about 2,500 years ago.

GO ON TO THE NEXT PAGE.

33. Assume that the Sphinx's head was a yardank. Archaeologist 1 could incorporate this information in his or her theory by saying:

    A. the yardank was carved several thousand years before the rest of the Sphinx was constructed.

    B. the underground rivers eroded the yardank during the years that it was buried under the grassy valley.

    C. the yardank was carved at the same time as the rest of the Sphinx and then it was all buried.

    D. the yardank was connected to a large, underground stone formation that became the Sphinx.

34. Which of the following statements is least consistent with the theory of Archaeologist 2?

    F. Yardanks have been found in other parts of Egypt.

    G. The New Kingdom (about 1500–1100 B.C.) occupied most of the coast of the Mediterranean.

    H. The civilization in Egypt during the Old Kingdom reached heights far beyond any known previously on Earth.

    J. Reports state that troops in the army of Alexander the Great broke off the nose of the Sphinx.

35. Egyptologists studying the Sphinx discover a crevice that goes deep into the body of the Sphinx. They put a microtelevision camera that goes deep into the Sphinx into a small chamber. In the chamber, the scientists see the carving of a pharaoh that they know conclusively to be from the Old Kingdom. How would this discovery affect the views of the two archaeologists?

    A. It would refute the position of Archaeologist 2 because it would show that the Sphinx originally dates from the Old Kingdom.

    B. It would refute the position of Archaeologist 1 because it would show that the Sphinx originally dates from more than 5,000 years ago.

    C. It would not refute the position of Archaeologist 2 because it would show that the body of the Sphinx could originally date from 5,000 years ago.

    D. It would not refute the position of Archaeologist 1 because it would show that the entire Sphinx originally dates from about 5,000 years ago.

GO ON TO THE NEXT PAGE.

PASSAGE VII

Sometime in the past, a meteorite crashed to Earth on the southern tip of the Yucatan peninsula in Mexico, forming the Chicxulub crater. This force of the meteorite's impact "shocked" the quartz in the Cretaceous (K-T) boundary and ejected shocked quartz grains at least 10,000 kilometers (km) from the crater. The drawings below show the four phases of the Chicxulub cratering event.

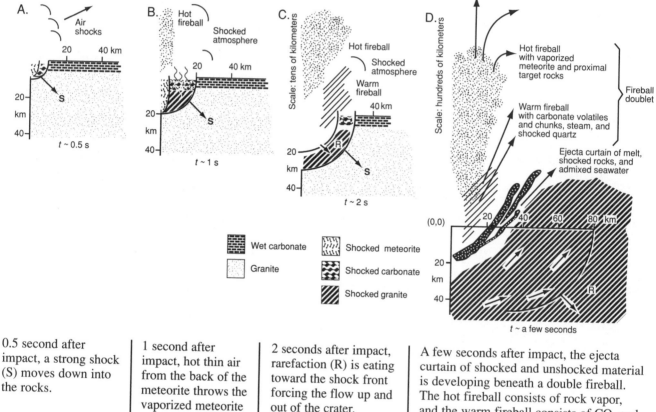

0.5 second after impact, a strong shock (S) moves down into the rocks.

1 second after impact, hot thin air from the back of the meteorite throws the vaporized meteorite and earth back out of the cavity.

2 seconds after impact, rarefaction (R) is eating toward the shock front forcing the flow up and out of the crater.

A few seconds after impact, the ejecta curtain of shocked and unshocked material is developing beneath a double fireball. The hot fireball consists of rock vapor, and the warm fireball consists of CO_2 and H_2O vapor with trapped rock fragments.

GO ON TO THE NEXT PAGE.

Following the impact, the ejecta launched from the crater was propelled as far away as Clear Creek in Colorado. The schematic below shows the angle at which the ejecta was launched from the crater, the time it took to reach Clear Creek, and the layers of material the ejecta deposited at Clear Creek.

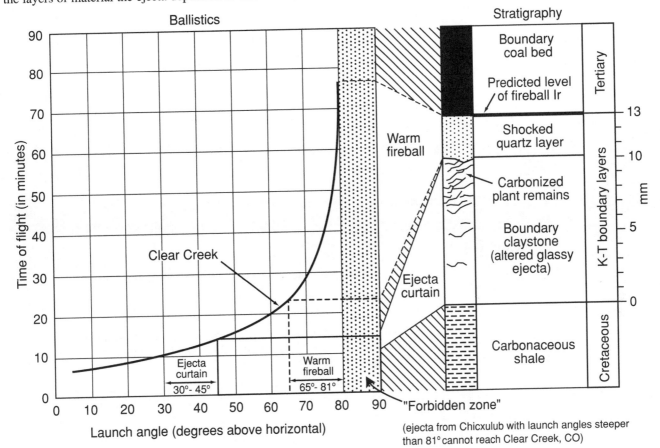

36. According to the first figure, how deep is the crater 2 seconds after impact?

- **F.** 10 km
- **G.** 20 km
- **H.** 40 km
- **J.** 55 km

37. According to the data in the second figure, material from the ejecta curtain would be:

- **A.** airborne about 50 minutes.
- **B.** airborne about 20 minutes.
- **C.** deposited about 4–7 mm above the surface.
- **D.** deposited about 0–10 mm above the surface.

38. The vaporized meteorite:

- **F.** would never reach Clear Creek.
- **G.** would reach Clear Creek in about 12 minutes.
- **H.** would reach Clear Creek in about 22 minutes.
- **J.** would reach Clear Creek in about 52 minutes.

39. A few seconds after impact, how much material had the blast removed from the area of impact?

- **A.** 600 square kilometers
- **B.** 1,200 square kilometers
- **C.** 1,800 square kilometers
- **D.** 2,400 square kilometers

40. When comparing the warm fireball to the ejecta curtain, one can conclude that:

- **F.** the warm fireball contributed more to the deposits than the ejecta curtain.
- **G.** the warm fireball was airborne longer than the ejecta curtain.
- **H.** the ejecta curtain was launched at a steeper angle than the fireball.
- **J.** the ejecta curtain arrived later at Clear Creek than the fireball.

END OF TEST 4

Model Science Reasoning ACT III

ANSWERS

1. D	11. C	21. B	31. B
2. F	12. J	22. F	32. H
3. C	13. B	23. B	33. D
4. G	14. H	24. F	34. H
5. D	15. D	25. D	35. C
6. G	16. F	26. J	36. G
7. B	17. A	27. C	37. D
8. G	18. G	28. F	38. F
9. D	19. A	29. C	39. A
10. J	20. F	30. F	40. G

ACT III Science Reasoning Answers Explained

PASSAGE I

1. **D.** The birds' initial orientation is stated in the description of the control treatment.

2. **F.** The experiment shows that unshifted magnetic fields do not change the birds' perception of magnetic north.

3. **C.** Choices A, B, and D are false because the birds with magnetic fields shifted 90° in any direction never shift their orientation to face in that direction. C is correct because it says that birds will shift their orientation in that direction.

4. **G.** Most birds would not reorient themselves because they need to see the stars and other visual cues to point to the new "north." Without these cues, they would not know that the direction of "north" had changed.

5. **D.** The birds start out at 340°. A shift of 60° counterclockwise reorients them to 340° − 60° = 280°.

6. **G.** Consider each answer in turn.
   F. FALSE. The scientists report averages, implying there was variation.
   G. TRUE. The scientists do not report differences between immature and adult birds, implying there were no significant differences.
   H. FALSE. The results show that the amount of orientation shift was about 90° for both experiments.
   J. FALSE. The results indicate that birds reorient themselves about the same amount and direction as the magnetic shift.

PASSAGE II

7. **B.** The concentration of uranium is highest at Site 5, which is near the narrow tip of the ore.

8. **G.** Carbon measurements move up and down from site to site and appear not to be related to the presence of uranium.

9. **D.** Vanadium appears in concentrations at the end of the uranium ore nearest the unaltered sandstone.

10. **J.** Figure 1 clearly shows mineralized sandstone as the only material that occurs when there is no uranium ore.

11. **C.** The diagram shows the direction of flow of the ore-bearing solution. The diagram of the ore deposits shows the bulging from the force of this flow.

12. **J.** Figure 2 shows uranium deposits at either end of each altered sandstone formation. This indicates that ore-bearing solutions flow in both directions shown on the chart.

PASSAGE III

13. **B.** Balls dropped from the same height fall in the same time regardless of the weight, composition, or density of the ball.

14. **H.** Objects falling on the moon fall at a slower rate than objects falling on Earth.

15. **D.** Dropping balls in a vacuum eliminates drag from the atmosphere that might slow larger balls.

16. **F.** A 20-kilogram ball that fell to the surface in 1 second would have been dropped from the same height as a 1-kilogram ball that fell to the surface in 1 second.

17. **A.** The ball accelerates (moves faster) as it nears Earth's surface.

PASSAGE IV

18. **G.** Most fertilizers are nitrate. None of the other explanations is plausible.

19. **A.** The water flow and the alkalinity are highest around the factory in the afternoon. This indicates that the factory is most active in the afternoon. None of the other choices are supported by the data.

20. **F.** Consider each answer in turn.
    F. TRUE. The factory could make alkaline batteries because there is a heavy alkaline runoff.
    G. FALSE. Match manufacture would result in a sulfur runoff.
    H. FALSE. Electroplating would create a chrome or other metallic runoff.
    J. FALSE. There is no sewage runoff from the factory.

21. **B.** Water from the tributary causes the alkalinity level to drop and disperses the alkaline factory runoff. The data does not support any of the other answers.

22. **F.** Site 1 shows the water conditions before being subjected to the effects of farmland, factory runoff, and a tributary.

PASSAGE V

23. **B.** This statement correctly describes the pattern in the table from Experiment 1. Note that the amplitude goes up, then down as the frequency goes up.

24. **F.** Crickets hear sounds at 4.5 kHz at the highest amplitude, making it likely that they communicate at that frequency.

25. **D.** Crickets need their central membranes to hear. If the membrane is removed, crickets can't hear.

26. **J.** The graph correctly shows that the amplitude goes up and then down as the frequency goes up.

27. **C.** Perforating the central membrane reduces the amplitude of sounds in the 3.875 kHz– 5.0 kHz frequency range.

28. **F.** The direction from 0° to 180° around them refers to the cricket's right side.

PASSAGE VI

29. **C.** Radiocarbon dating was not reported for the face of the Sphinx.

30. **F.** An Egyptian-like civilization about 8000 B.C. would show that there was a civilization in place about 10,000 years ago to build a figure like the Sphinx.

31. **B.** Conflicting theories based on information 5,000 to 10,000 years old usually contain both facts and inaccuracies.

32. H. One thinks the Sphinx is 4,500 years old and the other thinks it is 10,000 years old. So they both agree that the Sphinx is *at least* 4,000 years old.

33. D. The other three statements support Archaeologist 2.

34. H. If the Old Kingdom reached heights far beyond those of any other prior civilization, then the Sphinx could not have been built before the Old Kingdom.

35. C. Consider each answer in turn.

    A. FALSE.   Archaeologist 2 says the body was constructed during the Old Kingdom.

    B. FALSE.   The discovery does not suggest that the body of the Sphinx is more than 5,000 years old.

    C. TRUE.   Archaeologist 2 says that the body of the Sphinx was built during the Old Kingdom, and the discovery does not refute this view.

    D. FALSE.   This discovery does not show that the entire Sphinx originally dates from about 5,000 years ago.

PASSAGE VII

36. G. The vertical scale shows that the bottom of the crater (the top of the shocked granite) is at about 25 km.

37. D. The ejecta curtain, which would be deposited about 0–10 mm above the surface.

38. F. The vaporized meteorite was a part of the hot fireball, and this material did not reach Clear Creek.

39. A. The blast removed about 40 km from the surface to a depth of about 30 km. The area of that rectangle is about 1,200 square kilometers. The triangle representing the material removed is about half the size of that rectangle, or 600 square kilometers.

40. G. Table 2 shows that the ejecta curtain was airborne for about 10 or 15 minutes, while the warm fireball was airborne for more than 45 minutes.

Chapter 25 ■

Model Science Reasoning **ACT IV**
With Answers Explained

This Science Reasoning Model ACT IV is just like a real ACT. Take this test after you take Model ACT III.

Take this model test under simulated test conditions. Allow 35 minutes to answer the 40 Science Reasoning items. Use a pencil to mark the answer sheet on the next page, and answer the questions in the Test 4 (Science Reasoning) section.

Use the answer key on page 586 to mark the answer sheet. Review the answer explanations on pages 586–588. You may decide to retake this test later. There are additional answer sheets for this purpose following page 605.

The test scoring chart on page 589 shows you how to convert the number correct to ACT scale scores.

DO NOT leave any answers blank. There is no penalty for guessing on the ACT. Remember that the test is yours. You may mark up, write on, or draw on the test.

When you are ready, note the time and begin.

ANSWER SHEET

The ACT answer sheet looks something like this one. Use a No. 2 pencil
to completely fill the circle corresponding to the correct answer.
If you erase, erase completely; incomplete erasures may be read as answers.

TEST 1—English

1 Ⓐ Ⓑ Ⓒ Ⓓ	11 Ⓐ Ⓑ Ⓒ Ⓓ	21 Ⓐ Ⓑ Ⓒ Ⓓ	31 Ⓐ Ⓑ Ⓒ Ⓓ	41 Ⓐ Ⓑ Ⓒ Ⓓ	51 Ⓐ Ⓑ Ⓒ Ⓓ	61 Ⓐ Ⓑ Ⓒ Ⓓ	71 Ⓐ Ⓑ Ⓒ Ⓓ
2 Ⓕ Ⓖ Ⓗ Ⓙ	12 Ⓕ Ⓖ Ⓗ Ⓙ	22 Ⓕ Ⓖ Ⓗ Ⓙ	32 Ⓕ Ⓖ Ⓗ Ⓙ	42 Ⓕ Ⓖ Ⓗ Ⓙ	52 Ⓕ Ⓖ Ⓗ Ⓙ	62 Ⓕ Ⓖ Ⓗ Ⓙ	72 Ⓕ Ⓖ Ⓗ Ⓙ
3 Ⓐ Ⓑ Ⓒ Ⓓ	13 Ⓐ Ⓑ Ⓒ Ⓓ	23 Ⓐ Ⓑ Ⓒ Ⓓ	33 Ⓐ Ⓑ Ⓒ Ⓓ	43 Ⓐ Ⓑ Ⓒ Ⓓ	53 Ⓐ Ⓑ Ⓒ Ⓓ	63 Ⓐ Ⓑ Ⓒ Ⓓ	73 Ⓐ Ⓑ Ⓒ Ⓓ
4 Ⓕ Ⓖ Ⓗ Ⓙ	14 Ⓕ Ⓖ Ⓗ Ⓙ	24 Ⓕ Ⓖ Ⓗ Ⓙ	34 Ⓕ Ⓖ Ⓗ Ⓙ	44 Ⓕ Ⓖ Ⓗ Ⓙ	54 Ⓕ Ⓖ Ⓗ Ⓙ	64 Ⓕ Ⓖ Ⓗ Ⓙ	74 Ⓕ Ⓖ Ⓗ Ⓙ
5 Ⓐ Ⓑ Ⓒ Ⓓ	15 Ⓐ Ⓑ Ⓒ Ⓓ	25 Ⓐ Ⓑ Ⓒ Ⓓ	35 Ⓐ Ⓑ Ⓒ Ⓓ	45 Ⓐ Ⓑ Ⓒ Ⓓ	55 Ⓐ Ⓑ Ⓒ Ⓓ	65 Ⓐ Ⓑ Ⓒ Ⓓ	75 Ⓐ Ⓑ Ⓒ Ⓓ
6 Ⓕ Ⓖ Ⓗ Ⓙ	16 Ⓕ Ⓖ Ⓗ Ⓙ	26 Ⓕ Ⓖ Ⓗ Ⓙ	36 Ⓕ Ⓖ Ⓗ Ⓙ	46 Ⓕ Ⓖ Ⓗ Ⓙ	56 Ⓕ Ⓖ Ⓗ Ⓙ	66 Ⓕ Ⓖ Ⓗ Ⓙ	
7 Ⓐ Ⓑ Ⓒ Ⓓ	17 Ⓐ Ⓑ Ⓒ Ⓓ	27 Ⓐ Ⓑ Ⓒ Ⓓ	37 Ⓐ Ⓑ Ⓒ Ⓓ	47 Ⓐ Ⓑ Ⓒ Ⓓ	57 Ⓐ Ⓑ Ⓒ Ⓓ	67 Ⓐ Ⓑ Ⓒ Ⓓ	
8 Ⓕ Ⓖ Ⓗ Ⓙ	18 Ⓕ Ⓖ Ⓗ Ⓙ	28 Ⓕ Ⓖ Ⓗ Ⓙ	38 Ⓕ Ⓖ Ⓗ Ⓙ	48 Ⓕ Ⓖ Ⓗ Ⓙ	58 Ⓕ Ⓖ Ⓗ Ⓙ	68 Ⓕ Ⓖ Ⓗ Ⓙ	
9 Ⓐ Ⓑ Ⓒ Ⓓ	19 Ⓐ Ⓑ Ⓒ Ⓓ	29 Ⓐ Ⓑ Ⓒ Ⓓ	39 Ⓐ Ⓑ Ⓒ Ⓓ	49 Ⓐ Ⓑ Ⓒ Ⓓ	59 Ⓐ Ⓑ Ⓒ Ⓓ	69 Ⓐ Ⓑ Ⓒ Ⓓ	
10 Ⓕ Ⓖ Ⓗ Ⓙ	20 Ⓕ Ⓖ Ⓗ Ⓙ	30 Ⓕ Ⓖ Ⓗ Ⓙ	40 Ⓕ Ⓖ Ⓗ Ⓙ	50 Ⓕ Ⓖ Ⓗ Ⓙ	60 Ⓕ Ⓖ Ⓗ Ⓙ	70 Ⓕ Ⓖ Ⓗ Ⓙ	

TEST 2—Mathematics

1 Ⓐ Ⓑ Ⓒ Ⓓ Ⓔ	9 Ⓐ Ⓑ Ⓒ Ⓓ Ⓔ	17 Ⓐ Ⓑ Ⓒ Ⓓ Ⓔ	25 Ⓐ Ⓑ Ⓒ Ⓓ Ⓔ	33 Ⓐ Ⓑ Ⓒ Ⓓ Ⓔ	41 Ⓐ Ⓑ Ⓒ Ⓓ Ⓔ	49 Ⓐ Ⓑ Ⓒ Ⓓ Ⓔ	57 Ⓐ Ⓑ Ⓒ Ⓓ Ⓔ
2 Ⓕ Ⓖ Ⓗ Ⓙ Ⓚ	10 Ⓕ Ⓖ Ⓗ Ⓙ Ⓚ	18 Ⓕ Ⓖ Ⓗ Ⓙ Ⓚ	26 Ⓕ Ⓖ Ⓗ Ⓙ Ⓚ	34 Ⓕ Ⓖ Ⓗ Ⓙ Ⓚ	42 Ⓕ Ⓖ Ⓗ Ⓙ Ⓚ	50 Ⓕ Ⓖ Ⓗ Ⓙ Ⓚ	58 Ⓕ Ⓖ Ⓗ Ⓙ Ⓚ
3 Ⓐ Ⓑ Ⓒ Ⓓ Ⓔ	11 Ⓐ Ⓑ Ⓒ Ⓓ Ⓔ	19 Ⓐ Ⓑ Ⓒ Ⓓ Ⓔ	27 Ⓐ Ⓑ Ⓒ Ⓓ Ⓔ	35 Ⓐ Ⓑ Ⓒ Ⓓ Ⓔ	43 Ⓐ Ⓑ Ⓒ Ⓓ Ⓔ	51 Ⓐ Ⓑ Ⓒ Ⓓ Ⓔ	59 Ⓐ Ⓑ Ⓒ Ⓓ Ⓔ
4 Ⓕ Ⓖ Ⓗ Ⓙ Ⓚ	12 Ⓕ Ⓖ Ⓗ Ⓙ Ⓚ	20 Ⓕ Ⓖ Ⓗ Ⓙ Ⓚ	28 Ⓕ Ⓖ Ⓗ Ⓙ Ⓚ	36 Ⓕ Ⓖ Ⓗ Ⓙ Ⓚ	44 Ⓕ Ⓖ Ⓗ Ⓙ Ⓚ	52 Ⓕ Ⓖ Ⓗ Ⓙ Ⓚ	60 Ⓕ Ⓖ Ⓗ Ⓙ Ⓚ
5 Ⓐ Ⓑ Ⓒ Ⓓ Ⓔ	13 Ⓐ Ⓑ Ⓒ Ⓓ Ⓔ	21 Ⓐ Ⓑ Ⓒ Ⓓ Ⓔ	29 Ⓐ Ⓑ Ⓒ Ⓓ Ⓔ	37 Ⓐ Ⓑ Ⓒ Ⓓ Ⓔ	45 Ⓐ Ⓑ Ⓒ Ⓓ Ⓔ	53 Ⓐ Ⓑ Ⓒ Ⓓ Ⓔ	
6 Ⓕ Ⓖ Ⓗ Ⓙ Ⓚ	14 Ⓕ Ⓖ Ⓗ Ⓙ Ⓚ	22 Ⓕ Ⓖ Ⓗ Ⓙ Ⓚ	30 Ⓕ Ⓖ Ⓗ Ⓙ Ⓚ	38 Ⓕ Ⓖ Ⓗ Ⓙ Ⓚ	46 Ⓕ Ⓖ Ⓗ Ⓙ Ⓚ	54 Ⓕ Ⓖ Ⓗ Ⓙ Ⓚ	
7 Ⓐ Ⓑ Ⓒ Ⓓ Ⓔ	15 Ⓐ Ⓑ Ⓒ Ⓓ Ⓔ	23 Ⓐ Ⓑ Ⓒ Ⓓ Ⓔ	31 Ⓐ Ⓑ Ⓒ Ⓓ Ⓔ	39 Ⓐ Ⓑ Ⓒ Ⓓ Ⓔ	47 Ⓐ Ⓑ Ⓒ Ⓓ Ⓔ	55 Ⓐ Ⓑ Ⓒ Ⓓ Ⓔ	
8 Ⓕ Ⓖ Ⓗ Ⓙ Ⓚ	16 Ⓕ Ⓖ Ⓗ Ⓙ Ⓚ	24 Ⓕ Ⓖ Ⓗ Ⓙ Ⓚ	32 Ⓕ Ⓖ Ⓗ Ⓙ Ⓚ	40 Ⓕ Ⓖ Ⓗ Ⓙ Ⓚ	48 Ⓕ Ⓖ Ⓗ Ⓙ Ⓚ	56 Ⓕ Ⓖ Ⓗ Ⓙ Ⓚ	

TEST 3—Reading

1 Ⓐ Ⓑ Ⓒ Ⓓ	6 Ⓕ Ⓖ Ⓗ Ⓙ	11 Ⓐ Ⓑ Ⓒ Ⓓ	16 Ⓕ Ⓖ Ⓗ Ⓙ	21 Ⓐ Ⓑ Ⓒ Ⓓ	26 Ⓕ Ⓖ Ⓗ Ⓙ	31 Ⓐ Ⓑ Ⓒ Ⓓ	36 Ⓕ Ⓖ Ⓗ Ⓙ
2 Ⓕ Ⓖ Ⓗ Ⓙ	7 Ⓐ Ⓑ Ⓒ Ⓓ	12 Ⓕ Ⓖ Ⓗ Ⓙ	17 Ⓐ Ⓑ Ⓒ Ⓓ	22 Ⓕ Ⓖ Ⓗ Ⓙ	27 Ⓐ Ⓑ Ⓒ Ⓓ	32 Ⓕ Ⓖ Ⓗ Ⓙ	37 Ⓐ Ⓑ Ⓒ Ⓓ
3 Ⓐ Ⓑ Ⓒ Ⓓ	8 Ⓕ Ⓖ Ⓗ Ⓙ	13 Ⓐ Ⓑ Ⓒ Ⓓ	18 Ⓕ Ⓖ Ⓗ Ⓙ	23 Ⓐ Ⓑ Ⓒ Ⓓ	28 Ⓕ Ⓖ Ⓗ Ⓙ	33 Ⓐ Ⓑ Ⓒ Ⓓ	38 Ⓕ Ⓖ Ⓗ Ⓙ
4 Ⓕ Ⓖ Ⓗ Ⓙ	9 Ⓐ Ⓑ Ⓒ Ⓓ	14 Ⓕ Ⓖ Ⓗ Ⓙ	19 Ⓐ Ⓑ Ⓒ Ⓓ	24 Ⓕ Ⓖ Ⓗ Ⓙ	29 Ⓐ Ⓑ Ⓒ Ⓓ	34 Ⓕ Ⓖ Ⓗ Ⓙ	39 Ⓐ Ⓑ Ⓒ Ⓓ
5 Ⓐ Ⓑ Ⓒ Ⓓ	10 Ⓕ Ⓖ Ⓗ Ⓙ	15 Ⓐ Ⓑ Ⓒ Ⓓ	20 Ⓕ Ⓖ Ⓗ Ⓙ	25 Ⓐ Ⓑ Ⓒ Ⓓ	30 Ⓕ Ⓖ Ⓗ Ⓙ	35 Ⓐ Ⓑ Ⓒ Ⓓ	40 Ⓕ Ⓖ Ⓗ Ⓙ

TEST 4—Science Reasoning

1 Ⓐ Ⓑ Ⓒ Ⓓ	6 Ⓕ Ⓖ Ⓗ Ⓙ	11 Ⓐ Ⓑ Ⓒ Ⓓ	16 Ⓕ Ⓖ Ⓗ Ⓙ	21 Ⓐ Ⓑ Ⓒ Ⓓ	26 Ⓕ Ⓖ Ⓗ Ⓙ	31 Ⓐ Ⓑ Ⓒ Ⓓ	36 Ⓕ Ⓖ Ⓗ Ⓙ
2 Ⓕ Ⓖ Ⓗ Ⓙ	7 Ⓐ Ⓑ Ⓒ Ⓓ	12 Ⓕ Ⓖ Ⓗ Ⓙ	17 Ⓐ Ⓑ Ⓒ Ⓓ	22 Ⓕ Ⓖ Ⓗ Ⓙ	27 Ⓐ Ⓑ Ⓒ Ⓓ	32 Ⓕ Ⓖ Ⓗ Ⓙ	37 Ⓐ Ⓑ Ⓒ Ⓓ
3 Ⓐ Ⓑ Ⓒ Ⓓ	8 Ⓕ Ⓖ Ⓗ Ⓙ	13 Ⓐ Ⓑ Ⓒ Ⓓ	18 Ⓕ Ⓖ Ⓗ Ⓙ	23 Ⓐ Ⓑ Ⓒ Ⓓ	28 Ⓕ Ⓖ Ⓗ Ⓙ	33 Ⓐ Ⓑ Ⓒ Ⓓ	38 Ⓕ Ⓖ Ⓗ Ⓙ
4 Ⓕ Ⓖ Ⓗ Ⓙ	9 Ⓐ Ⓑ Ⓒ Ⓓ	14 Ⓕ Ⓖ Ⓗ Ⓙ	19 Ⓐ Ⓑ Ⓒ Ⓓ	24 Ⓕ Ⓖ Ⓗ Ⓙ	29 Ⓐ Ⓑ Ⓒ Ⓓ	34 Ⓕ Ⓖ Ⓗ Ⓙ	39 Ⓐ Ⓑ Ⓒ Ⓓ
5 Ⓐ Ⓑ Ⓒ Ⓓ	10 Ⓕ Ⓖ Ⓗ Ⓙ	15 Ⓐ Ⓑ Ⓒ Ⓓ	20 Ⓕ Ⓖ Ⓗ Ⓙ	25 Ⓐ Ⓑ Ⓒ Ⓓ	30 Ⓕ Ⓖ Ⓗ Ⓙ	35 Ⓐ Ⓑ Ⓒ Ⓓ	40 Ⓕ Ⓖ Ⓗ Ⓙ

Model Science Reasoning ACT IV

40 Questions—35 Minutes

INSTRUCTIONS: This test has seven passages. Each passage is followed by five to seven questions. After reading a passage, choose the best answer to each question and fill in the corresponding circle on your answer sheet. You may look back at the passages as often as you like.

You CANNOT use a calculator on this test.

Check pages 586–588 for answers and explanations.

PASSAGE I

Scientists planning an experimental flood release from the Glen Canyon Dam conducted studies to show the impact of the release on the Colorado River. The accompanying graph and table show the results of these studies.

Study 1

A steady flow of 8,000 cubic feet per second (cfs) through the dam will be maintained for 4 days. Beginning on March 26, the flow will be increased over 10 hours to a steady flow of 45,000 cfs. The steady flow of 45,000 cfs will be maintained for 167 hours and then be reduced over 42 hours to 8,000 cfs. The scientists made the accompanying graph to estimate hydrograph readings in cubic feet per second at the dam and at Colorado River stream flow gauging stations.

Figure 1

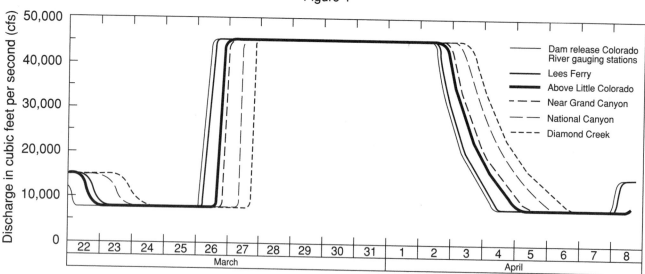

Study 2

The scientists also made a graph to show the duration of the flows along the river.

Figure 2

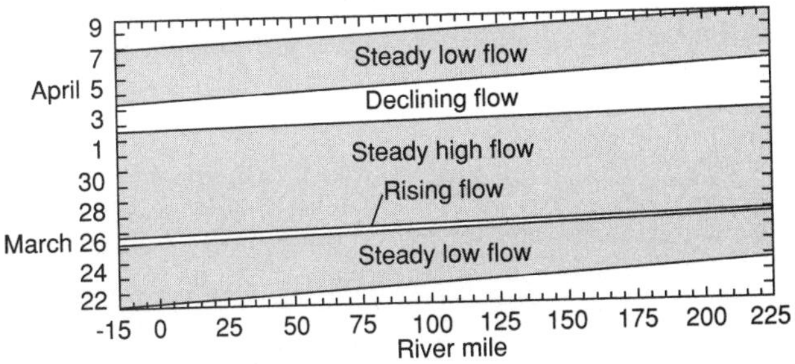

The location of different stream flow gauging stations is given in the table.

Station	River mile
Glen Canyon Dam	−15
Lees Ferry	0
Above Little Colorado	62
Near Grand Canyon	88
National Canyon	166
Diamond Creek	225

1. If the dam release occurred on midnight March 24 instead of midnight March 26, which of the following choices best represents the hydrograph reading at the dam on April 2?

   A. 45,000 cfs
   B. 25,000 cfs
   C. 19,000 cfs
   D. 8,000 cfs

2. The hydrograph readings at each station go up abruptly as the release reaches the station, but go down gradually as the water release subsides. Which of the following choices gives the best explanation for this outcome?

   F. Gravity slows the water as it moves down the river, creating a differential flow.
   G. Water molecules bond together, making the water denser as it flows down the river.
   H. The water release was stopped more slowly than it was started.
   J. The stream flow gauging stations are different distances apart.

3. Which of the following conclusions can be drawn from the flow periods shown in Figure 2?

   A. The water moved down the river at a constant rate.
   B. Water at the end of the release reached the last gauging station more slowly than water at the beginning of the release.
   C. The declining flow will last a shorter time at Diamond Creek than at the dam.
   D. The declining flow will last a longer time at Diamond Creek than at the dam.

4. Will the data from this simulated release be close to the actual release?

   F. Yes, because the scientists have accurate models to plot the simulated release.
   G. Yes, because computers permit simulations to be very accurate.
   H. No, because no simulation can come close to matching the real thing.
   J. No, because the hydrographs may be broken at some of the gauging stations during the actual release.

5. Which assumption in the simulation is critical to successful modeling of the release?

   A. The duration of the flows at river mile 100
   B. The time of day of the initial release
   C. The distance downstream to the gauging stations
   D. The duration of the releases at the dam

6. The scientists conducting the simulation conclude that the experimental data about stream flow heights are accurate. Which of the following would help confirm that conclusion?

   F. Measuring the current water heights at the gauging stations
   G. Measuring water heights at gauging stations during thunderstorms
   J. Replicating the simulation using different release amounts and durations
   H. Conducting a simulation on a working model of the dam and river

GO ON TO THE NEXT PAGE.

PASSAGE II

Scientists use the geologic time scale shown below to help date events from more than one-half billion years ago to the present.

Geologic Period	Millions of Years Ago
Quaternary	Present
	1.6
Tertiary	66
Cretaceous	138
Jurassic	205
Triassic	240
Permian	290
Pennsylvanian	330
Mississippian	360
Devonian	410
Silurian	435
Ordovician	500
Cambrian	570

Study 1

In order to help date events and fossils, and place them on the geologic time scale, scientists studied isotope decay. Scientists found that some atoms decay at a steady rate, and determined the time it takes for half of the parent atoms to decay to daughter atoms. This time is called the half-life. The half-life of some useful atoms is shown in the isotope dating table.

Study 2

Scientists used isotopic dating and other techniques to place animal and plant species on the geologic time scale, as shown in the fossil succession chart.

Isotope Dating

Isotope		Half-life of parent (years)	Useful range (years)
Parent	Daughter		
Carbon-14	Nitrogen-14	5,730	100–30,000
Potassium-40	Argon-40	1.3 billion	100,000–4.6 billion
Rubidium-87	Strontium-87	47 billion	10 million–4.6 billion
Uranium-238	Lead-206	4.5 billion	10 million–4.6 billion
Uranium-235	Lead-207	710 million	10 million–4.6 billion

Fossil Succession

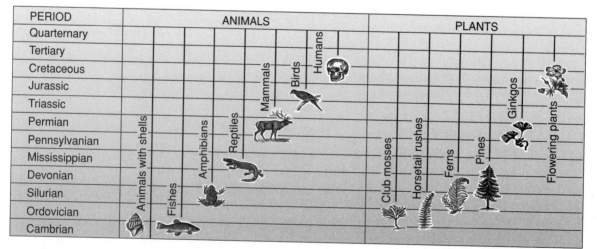

GO ON TO THE NEXT PAGE.

7. A researcher believes that a particular type of plant appeared about 50 million years before reptiles first appeared on Earth. If the researcher is correct, during which period did the plant first appear?

   A. Permian
   B. Pennsylvanian
   C. Mississippian
   D. Devonian

8. Based on the information in the studies, one can generalize that if one knows the half-life of the parent isotope, the substance can be correctly dated:

   F. to a particular geologic period.
   G. if the numbers of remaining parent atoms and daughter atoms are known.
   H. within 100,000 years.
   J. if one knows the geologic period in which the half-life was reached.

9. Which isotopes would be needed to date substances from the earliest part of the geologic scale until about 250,000 years ago?

   A. Carbon-14
   B. Potassium-40
   C. Rubidium-87
   D. Uranium-238

10. The fossil succession chart shows that reptiles and mammals were both present on Earth from about 210,000,000 years ago. Would Study 2 have to be modified to estimate the length of time mammals and reptiles coexisted on Earth?

   F. Yes, because Study 2 does not show when specific reptiles, such as dinosaurs, and mammals both inhabited Earth.
   G. Yes, because Study 2 does not show when more advanced mammals appeared on Earth and the presence of these advanced mammals is of great interest to researchers.
   H. No, because Study 2 gives enough information to estimate the length of time mammals and reptiles coexisted.
   J. No, because humans are mammals and human fossils don't appear earlier than the Tertiary Period.

11. New data reveal that fossils for shelled animals are entirely missing in one part of the United States for the Cretaceous Period. Does a conclusion of faulty isotope dating explain these new data?

   A. Yes, because the fossil succession chart clearly shows that these fossils have been found for these dates.
   B. Yes, because isotope dating can frequently have dating errors of millions of years.
   C. No, because shelled animals were not present for most of the Cretaceous Period.
   D. No, because these fossils may just not be present in this area.

12. Scientists determine that all of the carbon-14 atoms in a substance have turned to nitrogen-14 atoms. Based on this information, the scientists can reasonably conclude:

   F. the substance is about 5,730 years old.
   G. the substance is between 100 and 30,000 years old.
   H. the age of the substance is unknown, but it is more than 57,300 years old.
   J. the age of the substance can be determined by uranium dating.

GO ON TO THE NEXT PAGE.

PASSAGE III

Gravity is a force that attracts bodies of matter to one another. Large bodies of matter, such as Earth, the Earth's moon, and the planets, exert substantial amounts of gravity. The mass of a body of matter is constant. The weight of an object is the mass of the object times the gravity. The weight of a body of matter varies with the amount of gravity.

Experiment 1

Scientists simulated the gravity of Earth, Earth's moon, Mars, Neptune, and Pluto. They used solid balls that weighed 5 and 7 pounds on Earth and found the weight of these objects on the moon, Mars, Neptune, and Pluto. The results of this experiment are shown in Table 1.

Table 1. Weight of Two Objects on the Surface of Various Celestial Bodies, Pounds

	Object A	Object B
Earth	5.0	7.0
Earth's moon	0.85	1.19
Mars	1.9	2.66
Neptune	5.65	7.91
Pluto	0.35	0.49

Experiment 2

Using the same experimental conditions, the scientists dropped the objects from a height of 80 meters. The results of this experiment are shown in Table 2.

Table 2. Time for Two Objects to Free Fall 80 Meters on the Surface of Various Celestial Bodies, Seconds

	Object A	Object B
Earth	4	4
Earth's moon	23.5	23.5
Mars	10.5	10.5
Neptune	3.5	3.5
Pluto	57	57

13. To confirm the force of gravity, an object weighing 80 pounds on Earth was weighed in Pluto's gravity field. The weight of this object on Pluto is most like the weight of a 5-pound object on which celestial body shown in Table 1?

    A. Earth's moon
    B. Mars
    C. Neptune
    D. Pluto

14. An object weighing 10 pounds is dropped from 80 meters, in gravity twice as strong as the gravity on Earth's moon. To the nearest second, about how long will it take the object to reach the ground?

    F. 6 seconds
    G. 9 seconds
    H. 12 seconds
    J. 15 seconds

15. A scientist knows the weight of an object on a planet not listed here. The scientist also knows how long the object takes to free fall to the surface of that planet and to free fall to the surface of Earth. Can the scientist closely estimate the weight of the object on Earth?

    A. Yes, because free-fall time is proportional to a planet's gravity.
    B. Yes, because the weight of an object determines the free-fall time.
    C. No, because the weight of an object does not determine the free-fall time.
    D. No, because free-fall times are not related to the planet's gravity.

GO ON TO THE NEXT PAGE.

16. Four objects are dropped at the same time from a height of 80 meters. According to the data in Table 1 and Table 2, which of the following objects would free fall to the surface most quickly?

  F. A ball weighing 25 pounds on Earth's moon
  G. A cube weighing 100 pounds on Pluto
  H. A ball weighing 2 pounds on Neptune
  J. A cube weighing 1 pound on Mars

17. Which of the following graphs best represents the relationship between height and free-fall time?

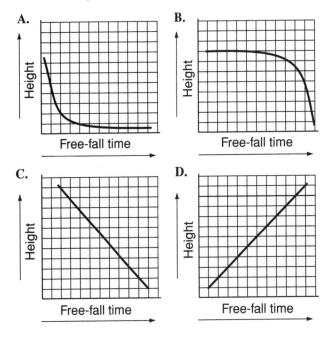

A.

B.

C.

D.

GO ON TO THE NEXT PAGE.

PASSAGE IV

Chemists study acid-base reactions.

Acids dissociate in water to produce hydrogen ions (H^+).

Bases dissociate in water to produce hydroxide ions (OH^-).

pH measures the degree of acidity of a solution.

pOH measures the degree of basicity of a solution.

Solutions with a pH less than 7 are acidic.
Solutions with a pH more than 7 are basic.
The sum of pH and pOH is 14 (pH + pOH = 14).

Scientists add indicators to determine the pH of a solution. The table below shows how some indicators react in solutions. The reaction takes place only in the pH range of color change.

Reactions to Indicators

Indicator	Color in acid	pH range of color change	Color in base
Congo red	Blue	3–5	Red
Methyl orange	Red	3.2–4.5	Yellow
Methyl red	Red	4.3–6	Yellow
Litmus	Red	4.5–8.2	Blue
Bromthymol blue	Yellow	6.0–7.6	Blue
Phenolphthalein	Colorless	8.3–10	Pink
Alizarin yellow	Yellow	10.1–12.1	Red

18. Which of the following indicators would be best to identify whether a substance with a pH near 7 was an acid or a base?

    F. Congo red
    G. Bromthymol blue
    H. Phenolphthalein
    J. Alizarin yellow

19. A solution turns red when mixed with the indicator alizarin yellow. According to the information in the table, what is the minimum pOH of the solution?

    A. 1.9
    B. 2.9
    C. 3.9
    D. 4.9

20. The indicator methyl red is mixed with a solution and the solution turns red. The indicator litmus is mixed with a solution and the solution turns red. What conclusion can a researcher draw about the pH of the solution?

    F. The pH is 6.
    G. The pH is 4.3 to 4.5.
    H. The pH is 4.5 to 6.
    J. The pH is 4.3 to 6.

21. Based on the information in the table, which of the following reactions indicates that a solution is a base?

    A. A reaction to congo red
    B. A reaction to methyl red
    C. A reaction to litmus
    D. A reaction to alizarin yellow

GO ON TO THE NEXT PAGE.

22. A beaker contains 50 milliliters (mL) of a strong acid solution. A researcher adds 100 mL of a strong base, 10 mL at a time, and measures the pH of the solution after each addition of the base. The graph at right shows the results of this experiment. The equivalence point on the graph is where there are exactly 50 mL of base and 50 mL of acid. Which of the following conclusions can the researcher draw from the graph?

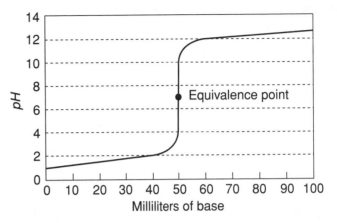

F. The pH of a solution is 7 when the solution contains 50 mL of a strong acid.

G. The solution is a base below the equivalence point on the graph.

H. The pOH of the solution is 14 after 100 mL of the base is added.

J. Most of the pH change is accounted for near the equivalence point.

GO ON TO THE NEXT PAGE.

PASSAGE V

Caltech has a number of seismic stations in southern California, including one at Pasadena. The Pasadena site monitors seismic activity all over the world. Each day, the Pasadena station reports the most significant seismic activity in the world. This event is reported as the seismic record of the day.

General information and the seismograph records for three days in a recent year are shown in the following tables and figures.

The seismograph record shows the duration of the event in seconds. The three traces show ground displacement. The top two traces show the horizontal (back-and-forth) ground movement. The bottom trace shows the vertical (up-and-down) ground movement. The scales vary from one seismograph record to the other.

Study 1

Region:	Northern Peru
Date:	October 28
Time (gmt):	6:15:18
Moment magnitude:	7.2
Latitude:	4.3 S
Longitude:	76.6 E
Depth (km):	125

Study 2

Region:	California–Nevada border region
Date:	November 2
Time (gmt):	8:51:54
Moment magnitude:	5.2
Latitude:	37.8 N
Longitude:	118.1 E
Depth (km):	55

GO ON TO THE NEXT PAGE.

Study 3

Region:	Xizang
Date:	November 8
Time (gmt):	10:2:48
Magnitude:	7.5
Latitude:	35.0 N
Longitude:	87.3 W
Depth (km):	150

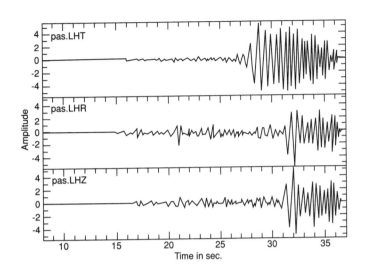

23. Relative to the seismograph traces for Study 1 and Study 3, the maximum, positive vertical displacement amplitude in Study 3 is about:

  A. twice the maximum positive vertical displacement amplitude in Study 1.
  B. three times the maximum positive vertical displacement amplitude in Study 1.
  C. four times the maximum positive vertical displacement amplitude in Study 1.
  D. five times the maximum positive vertical displacement amplitude in Study 1.

24. Each seismograph uses a different amplitude scale. These different scales make it difficult to directly compare seismograph readings, but they are used because:

  F. the time of the event in seconds varies from event to event and different scales permit the entire length of the event to be shown most accurately.
  G. the positive traces on the scale are often greater than the negative traces on the scale and different scales permit these differences to be shown most accurately.
  H. the traces have to be shown on the same size strip for each event and different scales permit every event to be shown most accurately on this size strip.
  J. the scale has to be the same for the entire length of the trace and different scales permit the scale to be the same length for each trace.

25. Suppose scientists deliberately caused the seismic activity in Study 2. Which of the following describes an experimental result?

  A. The time
  B. The moment magnitude
  C. The latitude
  D. The depth

26. Asked to describe the seismic event in Study 2, a student correctly says:

  F. it is the strongest of the three seismic events described in these studies.
  G. there was stronger side-to-side than up-and-down movement.
  H. the event lasted about 4 minutes.
  J. the event occurred along the San Andreas fault.

27. All three studies combined show that the magnitude of an event is related to which of the following factors?

  A. The depth of the event
  B. The length of the event
  C. The hemisphere of the event
  D. The time between peak traces

28. A scientist studies the traces of other seismic events and identifies a trace in which the distance between the highest and lowest vertical amplitudes is about 7. This is most like the distance between the highest and lowest vertical amplitudes in which of the three studies described above?

  F. Study 1
  G. Study 2
  H. Study 3
  J. Study 2 and Study 3

GO ON TO THE NEXT PAGE.

PASSAGE VI

Cover crops are planted after a harvest to protect the soil against erosion. These crops are usually plowed under the following spring. The following table gives information about cover crops and recommended planting practices in Ohio.

Ohio Vegetable Farm Cover Crops

Crop	Pounds/Bushel	Pounds/Acre	Comments
Non-legumes			
Rye	60	90	Most widely used cover crop. It germinates easily in the fall and survives severe winters.
Perennial ryegrass	34	15–20	Ryegrass can be seeded at the last cultivation of sweet corn, peppers, or eggplant. Plow in nitrogen with the ryegrass sod.
Field corn	56	50–60	Field corn can be drilled solid with a corn drill. Field corn can be used as a summer cover crop following early harvested spring vegetable crops.
Winter barley	48	80–100	Use in southern Ohio where winter killing is less severe. Root growth is not as extensive as rye or ryegrass.
Legumes			
Sweet clover	60	16–20	Use sweet clover for summer seeding. Use lime on soil to produce a pH of 6.5 –7.0 to ensure successful growth.
Red clover	60	10–15	Red clover can be established in soil with lower pH than required for sweet clover.
Soybeans	60	90–100	Use as a summer cover crop. Soybeans have rapid growth but a limited root system in comparison to other legumes.
Alfalfa	60	15–20	Use this crop in rotation when it can stand for more than one year. Alfalfa needs lime and other minerals for good growth.

29. According to the information in the table, cover crops:

   A. need more pounds of seed per acre when there are fewer pounds of seed per bushel.
   B. are always planted in the fall following harvest.
   C. are sometimes used in the summer.
   D. are better if there are more pounds per bushel.

30. Scientists at the state extension center say that additives help make some cover crops grow successfully. The information for which of these cover crops supports that proposition?

   F. Rye
   G. Perennial ryegrass
   H. Red clover
   J. Alfalfa

GO ON TO THE NEXT PAGE.

31. Researchers are working on a cover crop similar to rye. The new cover crop requires as many pounds of seed per acre as rye, but has half the number of pounds per bushel as rye. About how many bushels of seed for the new cover crop are needed for one acre?

    A. 0.5
    B. 2
    C. 3
    D. 4

32. Which of the following generalizations agrees with the data in the table?

    F. Cover crops grow better when lime is added to the soil.
    G. Perennial ryegrass seeds are smaller than rye seeds.
    H. pH level is most important for growing clover cover crops.
    J. The summer cover crops are legumes.

33. A scientist looking over the information in this table concludes that cover crops that can be harvested are summer cover crops. Information about which of the following cover crops does NOT support this conclusion?

    A. Sweet clover
    B. Field corn
    C. Barley
    D. Soybeans

GO ON TO THE NEXT PAGE.

PASSAGE VII

The oldest known dinosaur fossils are from rocks about 230,000,000 years old, but it is likely that dinosaurs existed on Earth before that time. Mammals were also present when dinosaurs were the dominant species on Earth. Then suddenly, about 65,000,000 years ago, dinosaurs died out, although birds may be the descendants of dinosaurs. There are many theories to explain the cause of dinosaur extinction. Two of these theories are the impact theory and the cooling theory.

Comet or asteroid impact

The dinosaurs became extinct in the aftermath of comet or meteor impact on Earth. About 65,000,000 years ago, a huge meteor or comet struck Earth. NASA scientists have located a huge crater in the Yucatan peninsula in Mexico. These scientists have evidence the impact took place around 65,000,000 years ago. The devastating impact of this type of event was seen recently when fragments of a comet struck Jupiter. The impact created a number of huge explosions, some of which were larger than Earth. Such an explosion could certainly have led to dinosaur extinction.

Global cooling

The dinosaurs became extinct when the temperature of Earth's atmosphere cooled. Dinosaurs were cold-blooded animals. Like modern reptiles, dinosaurs relied on the air temperature to keep their blood warm enough for them to live. About 65,000,000 years ago there was a worldwide climatic cooling. As Earth cooled, dinosaurs started to die. When Earth reached its coolest temperature, most dinosaur species became extinct. The warm-blooded mammals survived and became the dominant large animal life-form.

34. Suppose the dinosaurs' extinction began 75 million years ago. A scientist has a theory that the number of dinosaurs at the end of a 2-million-year period was 40% less than the number of dinosaurs at the beginning of that 2-million-year period. Which of the following graphs correctly represents that theory?

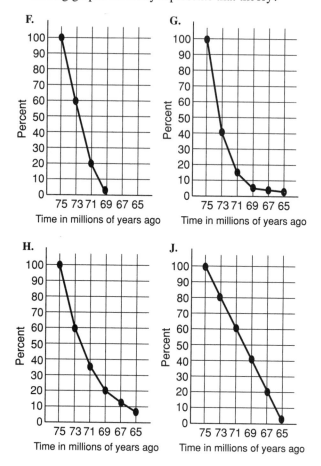

35. The last ice age lasted about 100,000 years and ended about 10,000 years ago. During this time, glaciers covered the northern part of the United States. If the following graph is an accurate representation of the size of Earth's reptile population, which theory of dinosaur extinction is supported?

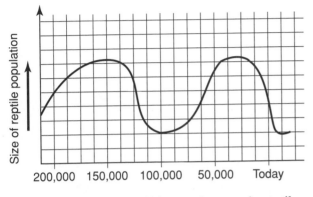

A. Comet or asteroid impact, because the reptile population rose to a high 50,000 years ago
B. Comet or asteroid impact, because the reptile population rose to a high 100,000 years ago
C. Global cooling, because the reptile population dropped to a low 100,000 years ago
D. Global cooling, because the reptile population was lower 100,000 years ago than it was 150,000 years ago

36. If an event described in the impact theory happened in the future, the most likely result would be:

F. destruction of the entire planet Earth.
G. a reduction in the number of green plants.
H. extinction of the mammals.
J. extinction of the reptiles.

37. Which of the following statements points out a significant weakness of one of the theories?

A. The impact theory is weak because it does not describe the size of the meteor or comet.
B. The impact theory is weak because it does not explain why mammals did not become extinct.
C. The global cooling theory is weak because it does not explain why mammals did not become extinct.
D. The global cooling theory is weak because it does not say how cool Earth became.

38. Other scientists believe that the extinction of dinosaurs was caused by a disease. Considering this theory along with the impact and cooling theories, which of the following is a true statement about the consistency or inconsistency of these theories?

F. The impact theory and the disease theory are inconsistent because a disease would not have come after the impact.
G. The global cooling theory and the disease theory are inconsistent because a disease could not have caused global cooling.
H. The global cooling theory and the impact theory are consistent because the impact could have caused cooling.
J. The disease theory and the cooling theory are consistent because the low temperature can cause people to catch a cold.

39. Dinosaurs are only one group of large reptiles that became extinct 65 million years ago. Other large reptiles, including the plesiosaurs and the pterosaurs, became extinct at the same time. Yet crocodiles and snakes survived and flourished. This variable survival rate is not consistent with the impact theory because:

A. when the larger dinosaurs died, owing to inadequate vegetation, the meat-eating dinosaurs also would have died.
B. dinosaurs laid eggs and all the eggs may have broken on impact.
C. all cold-blooded creatures should have been affected in about the same way.
D. humans could not have survived if dinosaurs became extinct.

40. What fact about dinosaurs is required for the global cooling theory, but not required for the impact theory?

F. Dinosaurs did not live in the ocean.
G. Dinosaurs had small brains.
H. Dinosaurs were cold-blooded.
J. Dinosaurs' legs were tucked under their bodies.

END OF TEST 4

Model Science Reasoning ACT IV

ANSWERS

1. D	11. D	21. D	31. C
2. H	12. H	22. J	32. H
3. D	13. C	23. D	33. C
4. F	14. H	24. H	34. H
5. D	15. A	25. B	35. A
6. H	16. H	26. G	36. G
7. D	17. D	27. A	37. B
8. G	18. G	28. G	38. H
9. B	19. A	29. C	39. C
10. H	20. H	30. J	40. H

ACT IV Science Reasoning Answers Explained

PASSAGE I

1. **D.** If the dam release is moved back two days, then the April 2 reading will be the same as the original April 4 reading. For most of April 4, the reading is 8,000 cfs. Therefore, this will be the reading on April 2 for the new release date.

2. **H.** The obvious answer is the correct answer. The water readings decline more gradually because the water flow stops more gradually than it starts.

3. **D.** The graph shows that the declining flow at Diamond Creek (mile 225) lasts a longer period of time than the declining flow at the dam (mile −15).

4. **F.** Simulations based on accurate models accurately predict the real event. Simply using a computer is not enough to ensure an accurate prediction.

5. **D.** Consider the choices in turn:

   A. The duration of the flows at river mile 100 is a consequence of the assumptions.
   B. The time of day of the initial release will not affect the ultimate outcomes.
   C. The distance downstream is known and fixed. It is not an assumption.
   D. The duration of the releases is an assumption in the simulation that affects the outcomes.

6. **H.** Results of a model simulation would help confirm the results of the experiment.

PASSAGE II

7. **D.** According to the chart, reptiles first appeared at the end of the Mississippian Period, about 330 million years ago. This means the plants would first have appeared about 380 million years ago, in the Devonian Period.

8. **G.** With each successive half-life, half of the remaining parent atoms decay to form daughter atoms. To determine the age of a substance, one must know the ratio of parent isotope atoms to daughter isotope atoms.

9. **B.** Potassium-40 is the only isotope that can date elements to about 4.6 billion years and also date them to 250,000 years ago.

10. **H.** The fossil chart gives enough information to estimate the time that reptiles and mammals coexisted. Information about specific organisms in choices F and G and is not needed.

11. **D.** The fossil chart does not mean that every fossil is found everywhere. The "missing" fossils most likely mean that the organism itself was not present in that area.

12. **H.** Once all of the carbon-14 atoms have degraded to nitrogen-14 atoms, carbon-14 dating is not a useful dating technique. The substance is likely to be at least 30,000 years old, but the age of the substance cannot be determined from the information. The substance may be too recent to be dated by uranium dating.

PASSAGE III

13. **C.** Multiply the weight of an object on Earth by 0.07 to find the weight of that object on Pluto: $80 \times 0.07 = 5.6$. This is most like the weight of an object on Neptune that weighs 5 pounds on Earth.

14. **H.** As Table 2 shows, the weight of the object does not affect the time it takes the object to fall to the surface. If the gravity is twice the moon's gravity, then the free-fall time will be half the moon's free-fall time. Thus, $23.5 \div 2 = 11.75$ seconds, or 12 seconds to the nearest second.

15. **A.** Free-fall time is proportional to a planet's gravity. You can confirm this by using information in Table 2. For example, the free-fall time for Object A on Mars is 10.5 seconds and the free-fall time for it on Earth is 4 seconds. Thus, $10.5 \div 4 = 2.625$; and 2.625×1.9 (the weight of the object on Mars) $= 4.9875 \approx 5$. Indeed, Object A weights 5 pounds on Earth. Given free-fall times on Earth and on another planet, and the weight of the object on another planet, you can determine the weight of the object on Earth.

16. **H.** Objects always free fall from 80 meters to the surface at the times shown in Table 2, regardless of weight. Of the choices given, an object will always fall most quickly to the surface of Neptune.

17. **D.** As the height increases, the free-fall time increases at the same rate.

PASSAGE IV

18. **G.** Bromthymol blue reacts in the pH range of 6.0–7.6. This indicator would help identify the substance as an acid or a base.

19. **A.** Use the formula pH + pOH = 14. The table shows the maximum pH is 12.1. The minimum pOH is $14 - 12.1 = 1.9$.

20. **H.** The pH must be within the reaction range of both indicators. The "highest" low number is 4.5 and the "lowest" high number is 6. The pH is in the range of 4.5 to 6.

21. **D.** Alizarin yellow is the only indicator listed that shows a reaction in the pH range of a base.

22. **J.** The pH ranges from 3 to 11, very near the equivalence point. This accounts for most of the pH range.

PASSAGE V

23. **D.** The bottom seismograph scale shows vertical displacement. The maximum positive vertical displacement for Study 1 is about 1. The maximum positive vertical displacement for Study 3 is about 5.

24. **H.** Each trace is shown on a strip the same width. Different scales are used to most accurately show each trace within the strip.

25. **B.** The time, the latitude, and the depth were all set by the experimenters. The moment magnitude is a result of the experiment.

26. **G.** The amplitudes of each of the top two records are stronger than the amplitude of the bottom record. This shows there is more side-to-side (horizontal) movement than up-and-down (vertical) movement.

27. **A.** Deeper events have stronger magnitudes. The magnitudes of these events do not show a relationship with any of the other factors.

28. G. The question asks for the distance between the highest and lowest amplitudes and so this number will always be positive. The bottom trace shows the vertical amplitude.

The vertical amplitude for Study 2 ranges from about −3 to about 4. The distance between them is about 7.

The distance between the highest and lowest vertical amplitudes for Study 1 is about 3.

The distance between the highest and lowest vertical amplitudes for Study 3 is about 10.

PASSAGE VI

29. C. Some cover crops, such as field corn and soybeans, are used in the summer.

Choices A and B are incorrect because these statements are false for many cover crops. The table does not provide information about a relationship between pounds per bushel and quality of a cover crop.

30. J. The comments for alfalfa specifically mention that "Alfalfa needs lime and other minerals for good growth." Choice G, perennial ryegrass, is incorrect because the comments say to plow in nitrogen with the sod, which occurs after the grass has grown.

31. C. If there are half as many pounds per bushel, there are 30 pounds per bushel. It would take 3 bushels to make 90 pounds.

32. H. pH level is mentioned only in connection with clover cover crops.

33. C. The table specifically mentions that barley is for planting in winter, while the table indicates that the other listed crops are appropriate for summer.

PASSAGE VII

34. H. The scientist theorizes that the number of dinosaurs decreases by 40% every two million years. So, the number of dinosaurs 75 million years ago is 100%. The number of dinosaurs 73 million years ago is 40% less, or 60%; and the number of dinosaurs 71 million years ago is 40% less than 60%, or about 36%.

35. A. The ice age lasted from about 110,000 years ago to 10,000 years ago. A reduction in the size of the reptile population for the entire ice age would support the global cooling theory. However, since the reptile population rose to a high during the ice age, the impact theory is supported. Note that just because data support a theory, does NOT mean that the theory is correct.

36. G. A cataclysmic impact would throw up a huge cloud of dust, blocking out sunlight, denying green plants the energy they need to produce carbohydrates, and huge numbers of plants would die.

37. B. Experts have observed that it is easy to come up with scenarios that wipe out most or all of life on Earth; it is diffficult to come up with plausible explanations for why some types of organisms are wiped out while others survive. A weakness of the impact theory is that it seems improbable that a global disaster of that magnitude would have little effect on the mammals.

38. H. The impact could have caused cooling when the dust cloud resulting from the explosion blocked out the sun. These two theories are consistent for the reason stated.

39. C. If the impact caused the dinosaurs to become extinct, it should have caused the other animals to become extinct as well.

40. H. The global cooling theory is based on cold-blooded dinosaurs. If dinosaurs were not cold-blooded, this theory cannot be true.

Chapter 26 ■

SCORING THE SCIENCE REASONING TESTS

This chapter shows you how to find the ACT scale scores and subscores.

The **raw score** on a test is the number correct. Use the charts provided to convert the raw score for each test to a scale score.

Scale scores are the scores reported to colleges. Because different ACTs have different difficulty levels, the same raw score does not always convert to the same scale score. The scale scores here are approximations and are given only to familiarize you with the process of converting raw scores to scale scores. The scale scores for these tests will almost certainly be different from the scale scores on the ACT you take.

Scoring the Science Reasoning Tests

The raw score on the Science Reasoning test is the number of correct answers. There are no Science Reasoning subscores. Use the chart below to convert the raw score for each Science Reasoning test to a scale score.

The highest possible raw score is 40; the lowest is 0. The highest possible scale score is 36; the lowest is 1. In the chart below, a raw score of 40 yields a scale score of 36; a raw score of 0 yields a scale score of 1. A raw score of 26 yields a scale score of 23.

Science Reasoning Scale Scores

Raw Score	Scale Score	Raw Score	Scale Score	Raw Score	Scale Score	Raw Score	Scale Score
40	36	33	27	17–18	18	4	9
39	35	31–32	26	15–16	17	3	8
—	34	29–30	25	14	16	2	7
38	33	28	24	12–13	15	—	6
—	32	26–27	23	10–11	14	—	5
37	31	24–25	22	9	13	1	4
36	30	22–23	21	7–8	12	—	3
35	29	20–21	20	6	11	—	2
34	28	19	19	5	10	0	1

Convert your raw scores to scale scores:

Science Reasoning Diagnostic ACT I	Science Reasoning Model ACT II	Science Reasoning Model ACT III	Science Reasoning Model ACT IV
Raw score _____	Raw score _____	Raw score _____	Raw score _____
Scale score _____	Scale score _____	Scale score _____	Scale score _____

What Does Your Score Mean?

The chart below shows the approximate percent of students who receive a particular scale score or below. The percent of students at or below a score shows the percentile rank of that score.

Use the chart this way. Find your Science Reasoning scale score in the chart. Then find the percentage of students who scored at or below that score. Say your Science Reasoning scale score was 18. In the chart, 18 matches 32%. This means that about 32% of the students will get a score of 18 or lower, and that 68% of the students will get a score above 18.

Science Reasoning Test—Percent of Students At or Below a Scale Score

Scale score	At or below	Scale score	At or below	Scale score	At or below	Scale score	At or below
36	99%	27	92%	18	32%	9	1%
35	99%	26	89%	17	24%	8	1%
34	99%	25	85%	16	15%	7	1%
33	99%	24	80%	15	10%	6	1%
32	99%	23	74%	14	7%	5	1%
31	99%	22	65%	13	4%	4	1%
30	98%	21	59%	12	2%	3	1%
29	97%	20	50%	11	1%	2	1%
28	94%	19	41%	10	1%	1	1%

The Composite Score

To find your composite score, add the scale scores of the four tests (English, Mathematics, Reading, and Science Reasoning) and divide by 4. You can divide each test's scale score by 4 to find out how many points it will contribute to the composite score.

Section VI ■

College Admission

Chapter 27 ∎

Getting Into College

∎ So—You're Thinking of Going to College

The only reason to take the ACT is to get into college. If you're reading this book, you are probably among the 50 to 60 percent of American high school graduates who will go on to higher education.

ACT Benchmark Scores

The ACT uses Benchmark Scores to help you determine the likelihood that you will be successful in college. Benchmark Scores are based on extensive research by ACT on about 100 colleges and 100,000 students. The ACT Benchmark Scores show a 50% chance of getting a B (or higher) and a 75% chance of getting a C (or higher) in these freshman college courses: English Composition, Social Sciences, College Algebra, and Biology. Go to the link below for the most recent ACT Benchmark Scores and additional explanations:

http://www.act.org/research/policymakers/pdf/benchmarks.pdf

You can also visit www.act.org and search for "benchmark scores."

However, the Benchmark Scores are just indicators. They are not a guarantee of good grades. You can score below the benchmarks and do well, or score above the benchmarks and do poorly. It depends on the effort you put into your college work. Colleges have started paying attention to these scores when making admissions decisions. Please note that the ACT requirements at many colleges will vary from the Benchmark scores.

The link below shows the percent of high school graduates in a recent year who achieved at or above each of the Benchmark Scores:

http://www.act.org/news/data/09/benchmarks.html

As of publication, it is somewhat unusual to score at or above all four Benchmarks.

Remember, you decide whether to send your ACT scores to colleges or to use the scores as the basis for further study. Talk with your counselor about how your ACT scores compare to the benchmarks and the expected ACT scores at colleges you would like to attend. Consider taking the ACT again if there is room for improvement. Many students take the ACT more than once. Some students take the ACT several times.

Research shows that additional high school academic coursework can help raise your ACT score. Additional review in this book is also an effective way of raising your Mathematics and Science Reasoning scores. Remember, do not skip practice questions or practice tests, and do not look ahead at the answers. Take the practice tests under simulated test conditions. Use the basic ACT test strategy—when in doubt, eliminate and guess. Never leave any answer blank.

Choosing a College

Attending college is usually a good idea. You will be an educated person, practically a requirement for success in the 21st century. As a college graduate, you will find more and better job opportunities than if you had no college degree.

The average college graduate earns a lot more a year than a person without a college degree, and this earnings gap will widen in the years ahead. An increasing percentage of the jobs available will require a college education as well as technical training and expertise. An undergraduate degree is required if you want to attend professional or graduate school.

This chapter summarizes the steps to follow for college admission. The chapter does not offer suggestions about which high school courses or extracurricular activities will help you get accepted by a college. It goes without saying, though, that the strength of your high school program, your grades, your class rank, teachers' recommendations, and your ACT scores along with your extracurricular activities are important factors in college admission.

According to recent reports, there are over 30,000 high schools in the United States. That means that there will be over 30,000 valedictorians and over 30,000 salutatorians seeking admission to college in the fall. You do not have to go to a competitive college to have a successful career.

Besides the factors listed here, there are lots of social reasons for choosing to apply to a particular college. Perhaps one of your parents went to the college. Or perhaps some of your friends are going there. Choosing a college for social reasons is not always the wrong approach. However, choosing a college solely on the basis of social considerations frequently leads to problems.

Follow these steps even if you think you already know the college or colleges you want to apply to. You may be surprised at the exciting and interesting choices that emerge.

Step I. Determine the type of degree you should pursue.

(Complete by October of your junior year.)

As a high school graduate, you may pursue a two-year associate's degree or a four-year bachelor's degree. Graduate and professional degrees require a bachelor's degree, and many colleges offer programs that combine undergraduate and graduate degrees.

What are your career aspirations?

The type of college degree you will pursue will depend on your interests and the career you plan for yourself. You don't have to know exactly what you want to be, but you should have some general idea before you attend college. The United States Department of Education provides the table below in their publication "Preparing Your Child for College" to help you decide which type of degree to pursue.

COLLEGE DEGREES AND CAREERS

ASSOCIATE'S DEGREE (Two years)	BACHELOR'S DEGREE (Four years)	GRADUATE AND PROFESSIONAL DEGREE (Bachelor's Degree Required)
Surveyor	Teacher	Lawyer
Registered nurse	Accountant	Physician
Dental hygienist	FBI agent	Architect
Medical laboratory technician	Engineer	University professor
Computer technician	Visual artist	Economist
Commercial artist	Journalist	Psychologist
Hotel/restaurant manager	Diplomat	Sociologist
Funeral director	Computer systems analyst	Dentist
Drafter	Insurance agent	Veterinarian
Engineering technician	Pharmacist	Public policy analyst
Automotive mechanic	Dietitian	Geologist
Administrative assistant	Writer	Paleontologist
Cardiovascular technician	Editor	Zoologist
Medical records technician	Graphic designer	Management consultant
Surgical technologist	Social worker	Rabbi
Water and wastewater treatment plant operator	Recreational therapist	Priest
Heating, air-conditioning, and refrigeration technician	Public relations specialist	Minister
	Research assistant	Chiropractor
	Investment banker	Biologist
	Medical illustrator	

List below the careers you might be interested in and the degree you will need for each career type.

Career	Type of Degree
1._____	_____
2._____	_____
3._____	_____

Step II. Determine whether you should apply to a two-year or a four-year college.

(Complete by October of your junior year.)

Regardless of your career choice, you can begin your college career at either a two-year or a four-year college. While most students pursuing a bachelor's degree start at a four-year college, there are good reasons for first attending a two-year college. You may decide to apply to both two-year and four-year colleges.

The two-year college may offer a low-cost alternative for your freshman and sophomore years. It may be easier to gain admission to a four-year college after attending a two-year college. In fact, many public colleges have special admission arrangements for students who have attended public two-year colleges in their state.

The descriptions of two-year and four-year colleges given below may help you choose the type or types of colleges you will apply to.

Two-Year Colleges. These colleges may offer associate's degrees (A.A., A.S., and A.A.S) based on two years of college work. Some two-year colleges offer programs of two years or shorter that lead directly to a career or job. These colleges fall into three categories:

Community colleges are usually public two-year institutions with commuter students from nearby communities and offer a full range of academic and technical courses. Most programs at these colleges lead to an associate's degree.

Junior colleges are usually private two-year institutions with dormitories and usually offer a full range of academic and technical courses. Most programs at these colleges lead to an associate's degree.

Technical colleges may be either public or private and emphasize training in technical fields and careers. Many programs at technical colleges do not lead to an associate's degree.

Four-Year Colleges. These colleges offer bachelor's degrees based on four years of college work. While the emphasis at most four-year colleges is academic, four-year schools also offer programs that lead to careers in technical and other fields.

A university is a four-year school that also grants graduate and professional degrees. Universities may offer programs that lead to both an undergraduate and a graduate degree.

Colleges and universities may be best thought of by the competitive nature of the admissions process. Some colleges and universities are extremely competitive, admitting less than 20 percent of the students who apply. Other colleges and universities have very liberal admissions policies and reject almost no one who has applied for admission. Most schools fall between these two extremes, with varying degrees of admission selectivity.

List below the types of college you are interested in attending:

1. _____

2. _____

STEP III. Determine the characteristics of the college you would like to attend.

(Complete by November of your junior year.)

You may want to attend college in a particular state or a part of the country. Or you may want to attend a college of a particular size.

Write your preliminary choices for the characteristics of the college you would like to attend next to each category below. You can always change your mind.

Field(s) of study:

Include the major fields of study or the careers you are interested in or write *undecided.*

_____

_____

_____

State or states in which you would attend college:

Write the name of every state in which you would attend college or write "all" if you would attend college anywhere in the United States.

_____	_____	_____	_____
_____	_____	_____	_____
_____	_____	_____	

Approximate size of the student body:

Circle one of the choices below.

| Under 500 | 500–2,500 | 2,500–7,500 |
| 7,500–15,000 | Above 15,000 | No preference |

Type of school:

Circle two-year, four-year, or both. Circle private, public, or no preference. If you want to attend a college with a religious affiliation, then write the name of the affiliation in the space provided.

| Two-year | Four-year | Both |
| Public | Private | No preference |

Religious affiliation? _____

Disabilities:

Circle below whether you will need to have services for a physical or learning disability at the college of your choice.

Physical disability   Learning disability

STEP IV. List the colleges that meet your criteria.

(Complete by November of your junior year.)

The best way to get a list of colleges that meet your requirements is to use one of the free search services on the Internet.

Surf the Net for a List of Colleges. At the publication deadline for this book, the college search service through the College Board was the best. It gives you access to a sophisticated college search as well as information about financial aid. (Collegeboard.com. Click on *College Search.*)

Internet resources are constantly being added and updated. You should search to find new sites that will give you information about getting into college.

STEP V. Gather information about colleges on your list.

(Complete by March of your junior year.)

Look over the list of colleges that meet your criteria. Sit down with your parents, teachers, and high school counselor to see if they can think of any other colleges to add to the list. Remove the names of schools you absolutely will not attend. Remember, you're just gathering information.

Write or call each college on your list and request complete information about the school, including its catalog. All the information you get from the college is advertising. This information points out strong points and does not mention weaknesses. But college advertising will provide you with useful general information about a school.

Read about colleges in guidebooks that give students' or others' opinions about the school. In general, avoid books that merely reprint the information provided by the college. You will find a wide array of college guidebooks in the guidance counselor's office and in bookstores. Make sure you are looking at the most recent versions of the books.

Attend college fairs. Colleges send representatives to college fairs throughout the country. You can talk to a representative and pick up literature and applications at these fairs. Your counselor will have information about college fairs in your area.

Talk to people who are attending the college. Your guidance counselor may be able to provide you with names of students who are currently attending a college. Don't hesitate to call the college admissions office and ask for the names of students to talk to. They probably have a list of people they refer to potential students. But don't hesitate to call the students and ask them to just tell you the truth. And if it's a dorm number, you can just talk to the first person who answers the phone.

You can also call directly into an academic department and ask to talk to a faculty member. Do not hesitate to ask the faculty member questions about her or his department.

Some colleges also distribute videotapes about their schools. The videotapes are available directly through the school and some are available free in video rental stores.

STEP VI. Create a final list of five to fifteen schools. Visit each school.

(Complete by October of your senior year.)

Use all the information you have gathered to narrow your list of schools. There is no rule for the number of schools that should be on the list. Your counselor and parents can help you with this list.

You have to visit the schools you're thinking of attending! It is better to visit a college in session. This will give you a chance to attend classes, talk with students, and stay in the dorm. But it is OK to make your first visit during the summer between your junior and senior years. If you eventually decide you may apply to that school, you should visit again in the fall of your senior year.

Your visit to the school will probably include your first formal visit with an admissions counselor. If you are visiting a competitive school or a school that is somewhat of a reach for you, the interview can be an important factor. Your guidance counselor and teachers can help you prepare for this type of interview.

Remember that the admissions office at every college is a sales office. Different offices use different approaches, but their primary job is to attract as many applicants as possible. So it is important for you to get away from the office for a portion of your visit.

So take the official tour. But you also want to talk with individual students and faculty. Make arrangements to meet personally with the chair or one of the faculty of the department you're most interested in. Be sure that the courses you need to graduate are offered frequently enough to finish in four years.

Meet individually with students who attend the school. Find out what it is really like to be at the school. If possible, spend the night in a dorm room and sit in on some classes. If your parents are with you, they can tour the school themselves while you're getting the inside scoop.

STEP VII. Apply to three to eight schools.

(Complete before the admission deadline.)

Narrow your list of schools again to from three to eight choices. Talk to your parents and guidance counselor as you make up this final list. Get several copies of each application. Request letters of recommendation from teachers and others. Do not hesitate to get advice from your guidance counselor and teachers as you carefully complete each application. If possible, type your essays and other information on a word processor so that you can use them over again.

If there is one school you want to go to above all others, apply to that school for early admission. If you are accepted by a college for early admission, the understanding is that you will attend that college. You are supposed to apply to only one school for early admission; if the colleges find out that you have applied to more than one, they may void your applications. If you are not accepted for early admission, your application is placed into the regular applicant pool. Check with the colleges for early admission deadlines.

STEP VIII. Accept an admissions offer.

(By the college's deadline.)

In the spring of your senior year you'll receive your acceptance notices from the colleges you applied to. Accept one of the offers of admission and send in a deposit.

You're on your way!

■ Paying for College

The cost of college has increased at a faster rate than most other costs. Estimated average costs of attending two- and four-year public and private colleges are shown below. About 75 percent of all college students attend public institutions.

Actual tuition and fees can range from under $1,000 to over $30,000 annually and may depend on whether or not you are a resident of a particular state or county. Average room, board, and expenses for two-year public college students are lower because many public two-year college students live at home and commute to school.

Average Estimated College Costs
(Information gathered from several sources)

COLLEGE TYPE	COST*
Public two-year	$ 8,000
Public four-year	14,000
Private two-year	18,000
Private four-year	33,000

*The average extra charge for an out-of-state student at a public college is over $10,000.

Costs

These are averages. The amount you pay for college depends on the tuition, fees, and other expenses at the school you attend. You can control your college costs by attending a less expensive school. In recent years, many of the less expensive state colleges have emerged as highly ranked schools. Many students begin their college careers at two-year schools to control costs.

But every college costs something and someone has to pay for it. There are four ways that college costs can be paid for:

- You or your parents pay.
- You get a scholarship or grant.
- You get a loan.
- You work while attending college (including work-study at the college).

Financial Aid

About two-thirds of all undergraduates receive some aid. Scholarships and grants are the preferred form of financial aid because they do not have to be repaid. Loans do have to be repaid, although portions of some loans can be forgiven if you enter particular forms of national service or certain professions. Work-study may be a good option if there are not better-paid jobs available off campus, and work-study may permit you to work with college faculty in your chosen field.

About half the students who attend college receive some form of financial aid. A total of over $40 billion in financial aid will be available each year you attend college. Some financial aid is need-based, which means you qualify for financial aid if your financial circumstances make it difficult for you to pay for college. Other financial aid is based on merit, which means that a college wants to support you because of your academic or athletic performance or ability.

There are many different sources of financial aid, and it is important to apply for each one you might qualify for. Sources of financial aid change constantly; you should meet with your counselor in your junior year and discuss the types of aid you might qualify for. Look over as many books and on-line sources of information as you can get your hands on to find all the sources of financial aid you qualify for. Contact the colleges you are interested in to find out about any special scholarships they have. There may be local sources of college financial aid.

See below for an on-line source of information about financial aid. You will also find a glossary of admissions and financial aid terms on pages 604–605.

Apply

You have to apply to receive financial aid. It will not happen automatically. File the Free Application for Federal Student Aid (FAFSA) by January of your senior year. Some aid is limited and is given on a first come, first served basis, so apply early.

You can apply on-line at www.fafsa.ed.gov or call 1-800-4FED-AID to obtain an application form. The forms should also be available in your high school or library.

Remember also to file the other financial aid forms required by a college or by private sources of financial aid.

Satisfactory Educational Progress

To be eligible for federal financial aid, and some state and other financial aid programs, you must make satisfactory educational progress. Each school has its own standards that usually include the number of credits you have completed and your grade point average (GPA). Be sure you know your college's requirements so that you do not accidentally lose financial aid eligibility.

On-line Financial Aid Resources

There are many resources on-line for information about financial aid. Be aware that many of these sources may charge a fee for their services.

At press time, an excellent source for financial aid information is the ACT student website. Go to www.actstudent.org and click on Financial Aid at the top of the page for comprehensive information.

Internet resources are constantly being added and updated. You should search to find new sites that will give you information about paying for college.

College and Financial Aid Glossary

This glossary of terms adapted from the Department of Education publication "Preparing Your Child for College" may come in handy as you get ready to go to college.

A.A.: associate of arts; a degree that can be earned at most two-year colleges.

A.A.S.: associate of applied science; a degree that can be earned at some two-year colleges.

B.A., B.S.: bachelor of arts, bachelor of science; degrees earned at four-year colleges, depending on the kinds of courses offered at the particular college.

Default rate: the percentage of students who took out federal student loans to help pay their college expenses but did not repay them properly.

Expected family contribution (EFC): an amount determined by a formula specified by law that indicates how much of a family's financial resources should be available to pay for school. Factors such as taxable and nontaxable income, assets (such as savings and checking accounts), and benefits (for example, unemployment or Social Security) are all considered in this calculation. The EFC is used in determining eligibility for federal need-based aid.

Fees: charges that cover costs not associated with the student's course load, such as costs of some athletic activities, clubs, and special events.

Financial aid: money available from various sources to help students pay for college.

Financial aid package: the total amount of financial aid a student receives. Federal and nonfederal aid such as grants, loans, or work-study are combined to help meet the student's need. Using available resources to give each student the best possible package of aid is one of the major responsibilities of a school's financial aid administrator.

Financial need: In the context of student financial aid, financial need is equal to the cost of education (estimated costs for college attendance and basic living expenses) minus the expected family contribution (the amount a student's family is expected to pay, which varies according to the family's financial resources).

General Educational Development (GED) diploma: the certificate students receive if they have passed a high school equivalency test. Students who do not have a high school diploma but who have a GED will qualify for federal student aid.

Grant: a sum of money given to a student for the purposes of paying at least part of the cost of college. A grant does not have to be repaid.

Loan: a type of financial aid that is available to students and the parents of students. An education loan must be repaid. In many cases, however, payments do not begin until the student finishes school.

Merit-based financial aid: aid given to students who meet requirements not related to financial needs. Most merit-based aid is awarded on the basis of academic performance or potential and is given in the form of scholarships or grants.

Need-based financial aid: aid given to students in need of assistance based on their income and assets and their families' income and assets, as well as some other factors.

Open admissions: most or all students who apply to a school are admitted. At some colleges, anyone who has a high school diploma or a GED can enroll. At other schools, it means that anyone over eighteen can enroll. Open admissions, therefore, can mean slightly different things at different schools.

Pell grants: federal need-based grants, which have been given to just under 4 million students. The maximum Pell grant has been $2,340 annually.

Perkins loans: a federal financial aid program that consists of low-interest loans for undergraduate and graduate students with exceptional financial need. Loans are awarded by the school.

PLUS loans: federal loans that allow parents to borrow money for their children's college education.

Postsecondary: after high school; refers to all programs for high school graduates, including programs at two- and four-year colleges and vocational and technical schools.

Proprietary: describes postsecondary schools that are private and are legally permitted to make a profit. Most proprietary schools offer technical and vocational courses.

PSAT/NMSQT: Preliminary Scholastic Aptitude Test/National Merit Scholarship Qualifying Test; a practice test that helps students prepare for the SAT I Reasoning Test. The PSAT is usually administered to tenth- or eleventh-grade students. Although colleges do not see a student's PSAT/NMSQT score, a student who does very well on this test and who meets many other academic performance criteria may qualify for the National Merit Scholarship program.

ROTC: Reserve Officers Training Corps; a scholarship program wherein the military covers part of the cost of tuition, fees, and textbooks, and also provides a monthly allowance. Scholarship recipients participate in summer training while in college and fulfill a service commitment after college.

SAT I: This test measures a student's mathematical and verbal reasoning abilities. Colleges accept this test or the ACT. Most students take the SAT I or the ACT during their junior or senior year of high school.

SAT II Subject Test: offered in many areas of study, including English, mathematics, many sciences, history, and foreign languages. Many colleges use ACT subtest scores in place of SAT II test scores.

Scholarship: a sum of money given to a student for the purpose of paying at least part of the cost of college. Scholarships may be awarded to students based on their academic achievements or on many other factors.

SEOG (Supplemental Educational Opportunity Grant): a federal award that helps undergraduates with exceptional financial need, awarded by the school. A SEOG does not have to be paid back.

Stafford loans: student loans offered by the federal government. There are two types of Stafford loans—need-based and nonneed-based. Under the Stafford loan programs, students can borrow money to attend school and the federal government will guarantee the loan in case of default. The combined loan limits have been $2,625 for the first year, $3,500 for the second year, and $5,500 for the third or subsequent years.

Transcript: a list of all courses a student has taken with the grades earned in each course. A college will often require a high school transcript when the student applies for admission.

Tuition: money that colleges charge for classroom and other instruction and the use of some facilities such as libraries. Tuition can range from a few hundred dollars per year to more than $20,000. A few colleges do not charge any tuition.

William D. Ford federal direct loans: Under this new program, students may obtain loans directly from their college or university with funds provided by the U.S. Department of Education instead of a bank or other lender.

Work-study programs: Offered by many colleges, they allow students to work part-time during the school year as part of their financial aid package. The jobs are usually on campus and the money earned is used to pay for tuition or other college charges.

Additional Answer Sheets

Several additional answer sheets follow this page. They are provided for your convenience if you wish to retake the tests contained in this review book.

ANSWER SHEET

The ACT answer sheet looks something like this one. Use a No. 2 pencil
to completely fill the circle corresponding to the correct answer.
If you erase, erase completely; incomplete erasures may be read as answers.

TEST 1—English

1 Ⓐ Ⓑ Ⓒ Ⓓ	11 Ⓐ Ⓑ Ⓒ Ⓓ	21 Ⓐ Ⓑ Ⓒ Ⓓ	31 Ⓐ Ⓑ Ⓒ Ⓓ	41 Ⓐ Ⓑ Ⓒ Ⓓ	51 Ⓐ Ⓑ Ⓒ Ⓓ	61 Ⓐ Ⓑ Ⓒ Ⓓ	71 Ⓐ Ⓑ Ⓒ Ⓓ
2 Ⓕ Ⓖ Ⓗ Ⓙ	12 Ⓕ Ⓖ Ⓗ Ⓙ	22 Ⓕ Ⓖ Ⓗ Ⓙ	32 Ⓕ Ⓖ Ⓗ Ⓙ	42 Ⓕ Ⓖ Ⓗ Ⓙ	52 Ⓕ Ⓖ Ⓗ Ⓙ	62 Ⓕ Ⓖ Ⓗ Ⓙ	72 Ⓕ Ⓖ Ⓗ Ⓙ
3 Ⓐ Ⓑ Ⓒ Ⓓ	13 Ⓐ Ⓑ Ⓒ Ⓓ	23 Ⓐ Ⓑ Ⓒ Ⓓ	33 Ⓐ Ⓑ Ⓒ Ⓓ	43 Ⓐ Ⓑ Ⓒ Ⓓ	53 Ⓐ Ⓑ Ⓒ Ⓓ	63 Ⓐ Ⓑ Ⓒ Ⓓ	73 Ⓐ Ⓑ Ⓒ Ⓓ
4 Ⓕ Ⓖ Ⓗ Ⓙ	14 Ⓕ Ⓖ Ⓗ Ⓙ	24 Ⓕ Ⓖ Ⓗ Ⓙ	34 Ⓕ Ⓖ Ⓗ Ⓙ	44 Ⓕ Ⓖ Ⓗ Ⓙ	54 Ⓕ Ⓖ Ⓗ Ⓙ	64 Ⓕ Ⓖ Ⓗ Ⓙ	74 Ⓕ Ⓖ Ⓗ Ⓙ
5 Ⓐ Ⓑ Ⓒ Ⓓ	15 Ⓐ Ⓑ Ⓒ Ⓓ	25 Ⓐ Ⓑ Ⓒ Ⓓ	35 Ⓐ Ⓑ Ⓒ Ⓓ	45 Ⓐ Ⓑ Ⓒ Ⓓ	55 Ⓐ Ⓑ Ⓒ Ⓓ	65 Ⓐ Ⓑ Ⓒ Ⓓ	75 Ⓐ Ⓑ Ⓒ Ⓓ
6 Ⓕ Ⓖ Ⓗ Ⓙ	16 Ⓕ Ⓖ Ⓗ Ⓙ	26 Ⓕ Ⓖ Ⓗ Ⓙ	36 Ⓕ Ⓖ Ⓗ Ⓙ	46 Ⓕ Ⓖ Ⓗ Ⓙ	56 Ⓕ Ⓖ Ⓗ Ⓙ	66 Ⓕ Ⓖ Ⓗ Ⓙ	
7 Ⓐ Ⓑ Ⓒ Ⓓ	17 Ⓐ Ⓑ Ⓒ Ⓓ	27 Ⓐ Ⓑ Ⓒ Ⓓ	37 Ⓐ Ⓑ Ⓒ Ⓓ	47 Ⓐ Ⓑ Ⓒ Ⓓ	57 Ⓐ Ⓑ Ⓒ Ⓓ	67 Ⓐ Ⓑ Ⓒ Ⓓ	
8 Ⓕ Ⓖ Ⓗ Ⓙ	18 Ⓕ Ⓖ Ⓗ Ⓙ	28 Ⓕ Ⓖ Ⓗ Ⓙ	38 Ⓕ Ⓖ Ⓗ Ⓙ	48 Ⓕ Ⓖ Ⓗ Ⓙ	58 Ⓕ Ⓖ Ⓗ Ⓙ	68 Ⓕ Ⓖ Ⓗ Ⓙ	
9 Ⓐ Ⓑ Ⓒ Ⓓ	19 Ⓐ Ⓑ Ⓒ Ⓓ	29 Ⓐ Ⓑ Ⓒ Ⓓ	39 Ⓐ Ⓑ Ⓒ Ⓓ	49 Ⓐ Ⓑ Ⓒ Ⓓ	59 Ⓐ Ⓑ Ⓒ Ⓓ	69 Ⓐ Ⓑ Ⓒ Ⓓ	
10 Ⓕ Ⓖ Ⓗ Ⓙ	20 Ⓕ Ⓖ Ⓗ Ⓙ	30 Ⓕ Ⓖ Ⓗ Ⓙ	40 Ⓕ Ⓖ Ⓗ Ⓙ	50 Ⓕ Ⓖ Ⓗ Ⓙ	60 Ⓕ Ⓖ Ⓗ Ⓙ	70 Ⓕ Ⓖ Ⓗ Ⓙ	

TEST 2—Mathematics

1 Ⓐ Ⓑ Ⓒ Ⓓ Ⓔ	9 Ⓐ Ⓑ Ⓒ Ⓓ Ⓔ	17 Ⓐ Ⓑ Ⓒ Ⓓ Ⓔ	25 Ⓐ Ⓑ Ⓒ Ⓓ Ⓔ	33 Ⓐ Ⓑ Ⓒ Ⓓ Ⓔ	41 Ⓐ Ⓑ Ⓒ Ⓓ Ⓔ	49 Ⓐ Ⓑ Ⓒ Ⓓ Ⓔ	57 Ⓐ Ⓑ Ⓒ Ⓓ Ⓔ
2 Ⓕ Ⓖ Ⓗ Ⓙ Ⓚ	10 Ⓕ Ⓖ Ⓗ Ⓙ Ⓚ	18 Ⓕ Ⓖ Ⓗ Ⓙ Ⓚ	26 Ⓕ Ⓖ Ⓗ Ⓙ Ⓚ	34 Ⓕ Ⓖ Ⓗ Ⓙ Ⓚ	42 Ⓕ Ⓖ Ⓗ Ⓙ Ⓚ	50 Ⓕ Ⓖ Ⓗ Ⓙ Ⓚ	58 Ⓕ Ⓖ Ⓗ Ⓙ Ⓚ
3 Ⓐ Ⓑ Ⓒ Ⓓ Ⓔ	11 Ⓐ Ⓑ Ⓒ Ⓓ Ⓔ	19 Ⓐ Ⓑ Ⓒ Ⓓ Ⓔ	27 Ⓐ Ⓑ Ⓒ Ⓓ Ⓔ	35 Ⓐ Ⓑ Ⓒ Ⓓ Ⓔ	43 Ⓐ Ⓑ Ⓒ Ⓓ Ⓔ	51 Ⓐ Ⓑ Ⓒ Ⓓ Ⓔ	59 Ⓐ Ⓑ Ⓒ Ⓓ Ⓔ
4 Ⓕ Ⓖ Ⓗ Ⓙ Ⓚ	12 Ⓕ Ⓖ Ⓗ Ⓙ Ⓚ	20 Ⓕ Ⓖ Ⓗ Ⓙ Ⓚ	28 Ⓕ Ⓖ Ⓗ Ⓙ Ⓚ	36 Ⓕ Ⓖ Ⓗ Ⓙ Ⓚ	44 Ⓕ Ⓖ Ⓗ Ⓙ Ⓚ	52 Ⓕ Ⓖ Ⓗ Ⓙ Ⓚ	60 Ⓕ Ⓖ Ⓗ Ⓙ Ⓚ
5 Ⓐ Ⓑ Ⓒ Ⓓ Ⓔ	13 Ⓐ Ⓑ Ⓒ Ⓓ Ⓔ	21 Ⓐ Ⓑ Ⓒ Ⓓ Ⓔ	29 Ⓐ Ⓑ Ⓒ Ⓓ Ⓔ	37 Ⓐ Ⓑ Ⓒ Ⓓ Ⓔ	45 Ⓐ Ⓑ Ⓒ Ⓓ Ⓔ	53 Ⓐ Ⓑ Ⓒ Ⓓ Ⓔ	
6 Ⓕ Ⓖ Ⓗ Ⓙ Ⓚ	14 Ⓕ Ⓖ Ⓗ Ⓙ Ⓚ	22 Ⓕ Ⓖ Ⓗ Ⓙ Ⓚ	30 Ⓕ Ⓖ Ⓗ Ⓙ Ⓚ	38 Ⓕ Ⓖ Ⓗ Ⓙ Ⓚ	46 Ⓕ Ⓖ Ⓗ Ⓙ Ⓚ	54 Ⓕ Ⓖ Ⓗ Ⓙ Ⓚ	
7 Ⓐ Ⓑ Ⓒ Ⓓ Ⓔ	15 Ⓐ Ⓑ Ⓒ Ⓓ Ⓔ	23 Ⓐ Ⓑ Ⓒ Ⓓ Ⓔ	31 Ⓐ Ⓑ Ⓒ Ⓓ Ⓔ	39 Ⓐ Ⓑ Ⓒ Ⓓ Ⓔ	47 Ⓐ Ⓑ Ⓒ Ⓓ Ⓔ	55 Ⓐ Ⓑ Ⓒ Ⓓ Ⓔ	
8 Ⓕ Ⓖ Ⓗ Ⓙ Ⓚ	16 Ⓕ Ⓖ Ⓗ Ⓙ Ⓚ	24 Ⓕ Ⓖ Ⓗ Ⓙ Ⓚ	32 Ⓕ Ⓖ Ⓗ Ⓙ Ⓚ	40 Ⓕ Ⓖ Ⓗ Ⓙ Ⓚ	48 Ⓕ Ⓖ Ⓗ Ⓙ Ⓚ	56 Ⓕ Ⓖ Ⓗ Ⓙ Ⓚ	

TEST 3—Reading

1 Ⓐ Ⓑ Ⓒ Ⓓ	6 Ⓕ Ⓖ Ⓗ Ⓙ	11 Ⓐ Ⓑ Ⓒ Ⓓ	16 Ⓕ Ⓖ Ⓗ Ⓙ	21 Ⓐ Ⓑ Ⓒ Ⓓ	26 Ⓕ Ⓖ Ⓗ Ⓙ	31 Ⓐ Ⓑ Ⓒ Ⓓ	36 Ⓕ Ⓖ Ⓗ Ⓙ
2 Ⓕ Ⓖ Ⓗ Ⓙ	7 Ⓐ Ⓑ Ⓒ Ⓓ	12 Ⓕ Ⓖ Ⓗ Ⓙ	17 Ⓐ Ⓑ Ⓒ Ⓓ	22 Ⓕ Ⓖ Ⓗ Ⓙ	27 Ⓐ Ⓑ Ⓒ Ⓓ	32 Ⓕ Ⓖ Ⓗ Ⓙ	37 Ⓐ Ⓑ Ⓒ Ⓓ
3 Ⓐ Ⓑ Ⓒ Ⓓ	8 Ⓕ Ⓖ Ⓗ Ⓙ	13 Ⓐ Ⓑ Ⓒ Ⓓ	18 Ⓕ Ⓖ Ⓗ Ⓙ	23 Ⓐ Ⓑ Ⓒ Ⓓ	28 Ⓕ Ⓖ Ⓗ Ⓙ	33 Ⓐ Ⓑ Ⓒ Ⓓ	38 Ⓕ Ⓖ Ⓗ Ⓙ
4 Ⓕ Ⓖ Ⓗ Ⓙ	9 Ⓐ Ⓑ Ⓒ Ⓓ	14 Ⓕ Ⓖ Ⓗ Ⓙ	19 Ⓐ Ⓑ Ⓒ Ⓓ	24 Ⓕ Ⓖ Ⓗ Ⓙ	29 Ⓐ Ⓑ Ⓒ Ⓓ	34 Ⓕ Ⓖ Ⓗ Ⓙ	39 Ⓐ Ⓑ Ⓒ Ⓓ
5 Ⓐ Ⓑ Ⓒ Ⓓ	10 Ⓕ Ⓖ Ⓗ Ⓙ	15 Ⓐ Ⓑ Ⓒ Ⓓ	20 Ⓕ Ⓖ Ⓗ Ⓙ	25 Ⓐ Ⓑ Ⓒ Ⓓ	30 Ⓕ Ⓖ Ⓗ Ⓙ	35 Ⓐ Ⓑ Ⓒ Ⓓ	40 Ⓕ Ⓖ Ⓗ Ⓙ

TEST 4—Science Reasoning

1 Ⓐ Ⓑ Ⓒ Ⓓ	6 Ⓕ Ⓖ Ⓗ Ⓙ	11 Ⓐ Ⓑ Ⓒ Ⓓ	16 Ⓕ Ⓖ Ⓗ Ⓙ	21 Ⓐ Ⓑ Ⓒ Ⓓ	26 Ⓕ Ⓖ Ⓗ Ⓙ	31 Ⓐ Ⓑ Ⓒ Ⓓ	36 Ⓕ Ⓖ Ⓗ Ⓙ
2 Ⓕ Ⓖ Ⓗ Ⓙ	7 Ⓐ Ⓑ Ⓒ Ⓓ	12 Ⓕ Ⓖ Ⓗ Ⓙ	17 Ⓐ Ⓑ Ⓒ Ⓓ	22 Ⓕ Ⓖ Ⓗ Ⓙ	27 Ⓐ Ⓑ Ⓒ Ⓓ	32 Ⓕ Ⓖ Ⓗ Ⓙ	37 Ⓐ Ⓑ Ⓒ Ⓓ
3 Ⓐ Ⓑ Ⓒ Ⓓ	8 Ⓕ Ⓖ Ⓗ Ⓙ	13 Ⓐ Ⓑ Ⓒ Ⓓ	18 Ⓕ Ⓖ Ⓗ Ⓙ	23 Ⓐ Ⓑ Ⓒ Ⓓ	28 Ⓕ Ⓖ Ⓗ Ⓙ	33 Ⓐ Ⓑ Ⓒ Ⓓ	38 Ⓕ Ⓖ Ⓗ Ⓙ
4 Ⓕ Ⓖ Ⓗ Ⓙ	9 Ⓐ Ⓑ Ⓒ Ⓓ	14 Ⓕ Ⓖ Ⓗ Ⓙ	19 Ⓐ Ⓑ Ⓒ Ⓓ	24 Ⓕ Ⓖ Ⓗ Ⓙ	29 Ⓐ Ⓑ Ⓒ Ⓓ	34 Ⓕ Ⓖ Ⓗ Ⓙ	39 Ⓐ Ⓑ Ⓒ Ⓓ
5 Ⓐ Ⓑ Ⓒ Ⓓ	10 Ⓕ Ⓖ Ⓗ Ⓙ	15 Ⓐ Ⓑ Ⓒ Ⓓ	20 Ⓕ Ⓖ Ⓗ Ⓙ	25 Ⓐ Ⓑ Ⓒ Ⓓ	30 Ⓕ Ⓖ Ⓗ Ⓙ	35 Ⓐ Ⓑ Ⓒ Ⓓ	40 Ⓕ Ⓖ Ⓗ Ⓙ

ANSWER SHEET

The ACT answer sheet looks something like this one. Use a No. 2 pencil to completely fill the circle corresponding to the correct answer.
If you erase, erase completely; incomplete erasures may be read as answers.

TEST 1—English

1 Ⓐ Ⓑ Ⓒ Ⓓ	11 Ⓐ Ⓑ Ⓒ Ⓓ	21 Ⓐ Ⓑ Ⓒ Ⓓ	31 Ⓐ Ⓑ Ⓒ Ⓓ	41 Ⓐ Ⓑ Ⓒ Ⓓ	51 Ⓐ Ⓑ Ⓒ Ⓓ	61 Ⓐ Ⓑ Ⓒ Ⓓ	71 Ⓐ Ⓑ Ⓒ Ⓓ
2 Ⓕ Ⓖ Ⓗ Ⓙ	12 Ⓕ Ⓖ Ⓗ Ⓙ	22 Ⓕ Ⓖ Ⓗ Ⓙ	32 Ⓕ Ⓖ Ⓗ Ⓙ	42 Ⓕ Ⓖ Ⓗ Ⓙ	52 Ⓕ Ⓖ Ⓗ Ⓙ	62 Ⓕ Ⓖ Ⓗ Ⓙ	72 Ⓕ Ⓖ Ⓗ Ⓙ
3 Ⓐ Ⓑ Ⓒ Ⓓ	13 Ⓐ Ⓑ Ⓒ Ⓓ	23 Ⓐ Ⓑ Ⓒ Ⓓ	33 Ⓐ Ⓑ Ⓒ Ⓓ	43 Ⓐ Ⓑ Ⓒ Ⓓ	53 Ⓐ Ⓑ Ⓒ Ⓓ	63 Ⓐ Ⓑ Ⓒ Ⓓ	73 Ⓐ Ⓑ Ⓒ Ⓓ
4 Ⓕ Ⓖ Ⓗ Ⓙ	14 Ⓕ Ⓖ Ⓗ Ⓙ	24 Ⓕ Ⓖ Ⓗ Ⓙ	34 Ⓕ Ⓖ Ⓗ Ⓙ	44 Ⓕ Ⓖ Ⓗ Ⓙ	54 Ⓕ Ⓖ Ⓗ Ⓙ	64 Ⓕ Ⓖ Ⓗ Ⓙ	74 Ⓕ Ⓖ Ⓗ Ⓙ
5 Ⓐ Ⓑ Ⓒ Ⓓ	15 Ⓐ Ⓑ Ⓒ Ⓓ	25 Ⓐ Ⓑ Ⓒ Ⓓ	35 Ⓐ Ⓑ Ⓒ Ⓓ	45 Ⓐ Ⓑ Ⓒ Ⓓ	55 Ⓐ Ⓑ Ⓒ Ⓓ	65 Ⓐ Ⓑ Ⓒ Ⓓ	75 Ⓐ Ⓑ Ⓒ Ⓓ
6 Ⓕ Ⓖ Ⓗ Ⓙ	16 Ⓕ Ⓖ Ⓗ Ⓙ	26 Ⓕ Ⓖ Ⓗ Ⓙ	36 Ⓕ Ⓖ Ⓗ Ⓙ	46 Ⓕ Ⓖ Ⓗ Ⓙ	56 Ⓕ Ⓖ Ⓗ Ⓙ	66 Ⓕ Ⓖ Ⓗ Ⓙ	
7 Ⓐ Ⓑ Ⓒ Ⓓ	17 Ⓐ Ⓑ Ⓒ Ⓓ	27 Ⓐ Ⓑ Ⓒ Ⓓ	37 Ⓐ Ⓑ Ⓒ Ⓓ	47 Ⓐ Ⓑ Ⓒ Ⓓ	57 Ⓐ Ⓑ Ⓒ Ⓓ	67 Ⓐ Ⓑ Ⓒ Ⓓ	
8 Ⓕ Ⓖ Ⓗ Ⓙ	18 Ⓕ Ⓖ Ⓗ Ⓙ	28 Ⓕ Ⓖ Ⓗ Ⓙ	38 Ⓕ Ⓖ Ⓗ Ⓙ	48 Ⓕ Ⓖ Ⓗ Ⓙ	58 Ⓕ Ⓖ Ⓗ Ⓙ	68 Ⓕ Ⓖ Ⓗ Ⓙ	
9 Ⓐ Ⓑ Ⓒ Ⓓ	19 Ⓐ Ⓑ Ⓒ Ⓓ	29 Ⓐ Ⓑ Ⓒ Ⓓ	39 Ⓐ Ⓑ Ⓒ Ⓓ	49 Ⓐ Ⓑ Ⓒ Ⓓ	59 Ⓐ Ⓑ Ⓒ Ⓓ	69 Ⓐ Ⓑ Ⓒ Ⓓ	
10 Ⓕ Ⓖ Ⓗ Ⓙ	20 Ⓕ Ⓖ Ⓗ Ⓙ	30 Ⓕ Ⓖ Ⓗ Ⓙ	40 Ⓕ Ⓖ Ⓗ Ⓙ	50 Ⓕ Ⓖ Ⓗ Ⓙ	60 Ⓕ Ⓖ Ⓗ Ⓙ	70 Ⓕ Ⓖ Ⓗ Ⓙ	

TEST 2—Mathematics

1 Ⓐ Ⓑ Ⓒ Ⓓ Ⓔ	9 Ⓐ Ⓑ Ⓒ Ⓓ Ⓔ	17 Ⓐ Ⓑ Ⓒ Ⓓ Ⓔ	25 Ⓐ Ⓑ Ⓒ Ⓓ Ⓔ	33 Ⓐ Ⓑ Ⓒ Ⓓ Ⓔ	41 Ⓐ Ⓑ Ⓒ Ⓓ Ⓔ	49 Ⓐ Ⓑ Ⓒ Ⓓ Ⓔ	57 Ⓐ Ⓑ Ⓒ Ⓓ Ⓔ
2 Ⓕ Ⓖ Ⓗ Ⓙ Ⓚ	10 Ⓕ Ⓖ Ⓗ Ⓙ Ⓚ	18 Ⓕ Ⓖ Ⓗ Ⓙ Ⓚ	26 Ⓕ Ⓖ Ⓗ Ⓙ Ⓚ	34 Ⓕ Ⓖ Ⓗ Ⓙ Ⓚ	42 Ⓕ Ⓖ Ⓗ Ⓙ Ⓚ	50 Ⓕ Ⓖ Ⓗ Ⓙ Ⓚ	58 Ⓕ Ⓖ Ⓗ Ⓙ Ⓚ
3 Ⓐ Ⓑ Ⓒ Ⓓ Ⓔ	11 Ⓐ Ⓑ Ⓒ Ⓓ Ⓔ	19 Ⓐ Ⓑ Ⓒ Ⓓ Ⓔ	27 Ⓐ Ⓑ Ⓒ Ⓓ Ⓔ	35 Ⓐ Ⓑ Ⓒ Ⓓ Ⓔ	43 Ⓐ Ⓑ Ⓒ Ⓓ Ⓔ	51 Ⓐ Ⓑ Ⓒ Ⓓ Ⓔ	59 Ⓐ Ⓑ Ⓒ Ⓓ Ⓔ
4 Ⓕ Ⓖ Ⓗ Ⓙ Ⓚ	12 Ⓕ Ⓖ Ⓗ Ⓙ Ⓚ	20 Ⓕ Ⓖ Ⓗ Ⓙ Ⓚ	28 Ⓕ Ⓖ Ⓗ Ⓙ Ⓚ	36 Ⓕ Ⓖ Ⓗ Ⓙ Ⓚ	44 Ⓕ Ⓖ Ⓗ Ⓙ Ⓚ	52 Ⓕ Ⓖ Ⓗ Ⓙ Ⓚ	60 Ⓕ Ⓖ Ⓗ Ⓙ Ⓚ
5 Ⓐ Ⓑ Ⓒ Ⓓ Ⓔ	13 Ⓐ Ⓑ Ⓒ Ⓓ Ⓔ	21 Ⓐ Ⓑ Ⓒ Ⓓ Ⓔ	29 Ⓐ Ⓑ Ⓒ Ⓓ Ⓔ	37 Ⓐ Ⓑ Ⓒ Ⓓ Ⓔ	45 Ⓐ Ⓑ Ⓒ Ⓓ Ⓔ	53 Ⓐ Ⓑ Ⓒ Ⓓ Ⓔ	
6 Ⓕ Ⓖ Ⓗ Ⓙ Ⓚ	14 Ⓕ Ⓖ Ⓗ Ⓙ Ⓚ	22 Ⓕ Ⓖ Ⓗ Ⓙ Ⓚ	30 Ⓕ Ⓖ Ⓗ Ⓙ Ⓚ	38 Ⓕ Ⓖ Ⓗ Ⓙ Ⓚ	46 Ⓕ Ⓖ Ⓗ Ⓙ Ⓚ	54 Ⓕ Ⓖ Ⓗ Ⓙ Ⓚ	
7 Ⓐ Ⓑ Ⓒ Ⓓ Ⓔ	15 Ⓐ Ⓑ Ⓒ Ⓓ Ⓔ	23 Ⓐ Ⓑ Ⓒ Ⓓ Ⓔ	31 Ⓐ Ⓑ Ⓒ Ⓓ Ⓔ	39 Ⓐ Ⓑ Ⓒ Ⓓ Ⓔ	47 Ⓐ Ⓑ Ⓒ Ⓓ Ⓔ	55 Ⓐ Ⓑ Ⓒ Ⓓ Ⓔ	
8 Ⓕ Ⓖ Ⓗ Ⓙ Ⓚ	16 Ⓕ Ⓖ Ⓗ Ⓙ Ⓚ	24 Ⓕ Ⓖ Ⓗ Ⓙ Ⓚ	32 Ⓕ Ⓖ Ⓗ Ⓙ Ⓚ	40 Ⓕ Ⓖ Ⓗ Ⓙ Ⓚ	48 Ⓕ Ⓖ Ⓗ Ⓙ Ⓚ	56 Ⓕ Ⓖ Ⓗ Ⓙ Ⓚ	

TEST 3—Reading

1 Ⓐ Ⓑ Ⓒ Ⓓ	6 Ⓕ Ⓖ Ⓗ Ⓙ	11 Ⓐ Ⓑ Ⓒ Ⓓ	16 Ⓕ Ⓖ Ⓗ Ⓙ	21 Ⓐ Ⓑ Ⓒ Ⓓ	26 Ⓕ Ⓖ Ⓗ Ⓙ	31 Ⓐ Ⓑ Ⓒ Ⓓ	36 Ⓕ Ⓖ Ⓗ Ⓙ
2 Ⓕ Ⓖ Ⓗ Ⓙ	7 Ⓐ Ⓑ Ⓒ Ⓓ	12 Ⓕ Ⓖ Ⓗ Ⓙ	17 Ⓐ Ⓑ Ⓒ Ⓓ	22 Ⓕ Ⓖ Ⓗ Ⓙ	27 Ⓐ Ⓑ Ⓒ Ⓓ	32 Ⓕ Ⓖ Ⓗ Ⓙ	37 Ⓐ Ⓑ Ⓒ Ⓓ
3 Ⓐ Ⓑ Ⓒ Ⓓ	8 Ⓕ Ⓖ Ⓗ Ⓙ	13 Ⓐ Ⓑ Ⓒ Ⓓ	18 Ⓕ Ⓖ Ⓗ Ⓙ	23 Ⓐ Ⓑ Ⓒ Ⓓ	28 Ⓕ Ⓖ Ⓗ Ⓙ	33 Ⓐ Ⓑ Ⓒ Ⓓ	38 Ⓕ Ⓖ Ⓗ Ⓙ
4 Ⓕ Ⓖ Ⓗ Ⓙ	9 Ⓐ Ⓑ Ⓒ Ⓓ	14 Ⓕ Ⓖ Ⓗ Ⓙ	19 Ⓐ Ⓑ Ⓒ Ⓓ	24 Ⓕ Ⓖ Ⓗ Ⓙ	29 Ⓐ Ⓑ Ⓒ Ⓓ	34 Ⓕ Ⓖ Ⓗ Ⓙ	39 Ⓐ Ⓑ Ⓒ Ⓓ
5 Ⓐ Ⓑ Ⓒ Ⓓ	10 Ⓕ Ⓖ Ⓗ Ⓙ	15 Ⓐ Ⓑ Ⓒ Ⓓ	20 Ⓕ Ⓖ Ⓗ Ⓙ	25 Ⓐ Ⓑ Ⓒ Ⓓ	30 Ⓕ Ⓖ Ⓗ Ⓙ	35 Ⓐ Ⓑ Ⓒ Ⓓ	40 Ⓕ Ⓖ Ⓗ Ⓙ

TEST 4—Science Reasoning

1 Ⓐ Ⓑ Ⓒ Ⓓ	6 Ⓕ Ⓖ Ⓗ Ⓙ	11 Ⓐ Ⓑ Ⓒ Ⓓ	16 Ⓕ Ⓖ Ⓗ Ⓙ	21 Ⓐ Ⓑ Ⓒ Ⓓ	26 Ⓕ Ⓖ Ⓗ Ⓙ	31 Ⓐ Ⓑ Ⓒ Ⓓ	36 Ⓕ Ⓖ Ⓗ Ⓙ
2 Ⓕ Ⓖ Ⓗ Ⓙ	7 Ⓐ Ⓑ Ⓒ Ⓓ	12 Ⓕ Ⓖ Ⓗ Ⓙ	17 Ⓐ Ⓑ Ⓒ Ⓓ	22 Ⓕ Ⓖ Ⓗ Ⓙ	27 Ⓐ Ⓑ Ⓒ Ⓓ	32 Ⓕ Ⓖ Ⓗ Ⓙ	37 Ⓐ Ⓑ Ⓒ Ⓓ
3 Ⓐ Ⓑ Ⓒ Ⓓ	8 Ⓕ Ⓖ Ⓗ Ⓙ	13 Ⓐ Ⓑ Ⓒ Ⓓ	18 Ⓕ Ⓖ Ⓗ Ⓙ	23 Ⓐ Ⓑ Ⓒ Ⓓ	28 Ⓕ Ⓖ Ⓗ Ⓙ	33 Ⓐ Ⓑ Ⓒ Ⓓ	38 Ⓕ Ⓖ Ⓗ Ⓙ
4 Ⓕ Ⓖ Ⓗ Ⓙ	9 Ⓐ Ⓑ Ⓒ Ⓓ	14 Ⓕ Ⓖ Ⓗ Ⓙ	19 Ⓐ Ⓑ Ⓒ Ⓓ	24 Ⓕ Ⓖ Ⓗ Ⓙ	29 Ⓐ Ⓑ Ⓒ Ⓓ	34 Ⓕ Ⓖ Ⓗ Ⓙ	39 Ⓐ Ⓑ Ⓒ Ⓓ
5 Ⓐ Ⓑ Ⓒ Ⓓ	10 Ⓕ Ⓖ Ⓗ Ⓙ	15 Ⓐ Ⓑ Ⓒ Ⓓ	20 Ⓕ Ⓖ Ⓗ Ⓙ	25 Ⓐ Ⓑ Ⓒ Ⓓ	30 Ⓕ Ⓖ Ⓗ Ⓙ	35 Ⓐ Ⓑ Ⓒ Ⓓ	40 Ⓕ Ⓖ Ⓗ Ⓙ

INDEX

basic elements of, 224–225
calculator tips, 240, 244, 261, 268
subtest, 280–282, 308–310
Please Excuse My Dear Aunt Sally (order of operations), 81
Plus sign (+), 75
Points, 224
collinear, 224
on the coordinate plane, 191
Polygons, 232
regular, 232
Polynomial equations, 170
Polynomials, 135. *See also* Expressions
addition of, 135–136
division of, 137–139
factoring, 141–143
quadratics, 144–145
multiplication of, 136–137
subtraction of, 136
Positive numbers, 62, 75
comparing, 76
computation with, 77–80
convert to scientific notation, 62
Positive slope, 193
Power of 10, 62
Pre-algebra, 44–91
calculator tips, 44, 45, 48, 50, 55, 61, 62, 65, 66, 68, 77, 81, 87, 91, 95, 98, 117
subtest, 123–124, 154–156
Prime numbers, 59–60
Prism, rectangular, 272
Probability, 107
independent and dependent events, 108
Problem solving
with matrices, 182–183
with proportions, 95–96
Products, 111
Proof, 249
concept and techniques of, 249–250
Proportions, 94
solving, 95–96
writing, 95
Pyramid, square, 275
Pythagorean identities, 288
Pythagorean Theorem, 239
Pythagorean triples, 239
multiples of, 239

Q

Quadratic equations, 144–147, 170–171
Quadratic formula, 170
Quadratic inequalities, 173–174
Quadrilaterals, 232–233
concave, 232
convex, 232
relationships among, 233
types of
parallelogram, 233
rectangle, 233
rhombus, 233
square, 233

trapezoid, 232
Quotient identities, 288

R

Radicals
operations with
addition of, 132
division of, 133
multiplication of, 132–133
subtraction of, 132
simplifying, 167
undefined, 167–168
Radius of circle, 255
Ratio, 94
golden, 178
trigonometric, 292, 294
Rational expressions, 167–169
Rationalizing the denominator, 129
Ray, 225
Reading Test, 3, 8, 15
Reciprocal identities, 288
Rectangle, 233
area of, 267
Rectangular prism, 272
Reflection, 261
Reflex angles, 227
Registration, 4
Regular polygon, 232
Research summaries, 12, 477, 495–502
correlation, 496
experiments, 495–496
observation, 496
Rhombus, 233
Right angles, 227, 235
Right triangles, 235, 238–241, 292
hypotenuse of, 198, 238, 239
isosceles, 239
legs of, 238, 239
Right triangle trigonometry, 283
Roots, 167, 170
Rotations, 261
Rounding
of decimals, 53
of whole numbers, 46

S

Sales tax, 90
Same-side exterior angles, 229
Same-side interior angles, 229
SAT
comparison with ACT, 3–4
Writing Test, 3
Scalar multiplication, 181
Scale scores, 464, 589
Scalene triangle, 235
Science Reasoning Test, 3, 8, 12–13
diagnostic, 516–530
model tests, 516–588
overview, 12–13
scoring, 589–590
Science Vocabulary List, 503–506
Scientific notation, 62–63
Scoring, 6

benchmark, 593–594
Mathematics Tests, 464–473
reporting, 6
Science Reasoning Tests, 589–590
special dates for, 5
Secant, 255, 284
Sequences, 178–179
arithmetic, 178–179
Fibonacci, 178
geometric, 179
Side Angle Side (SAS), 249
Side Side Side (SSS), 249
Sides
congruent, 235
corresponding, 245
Similar terms, 135
Similar triangles, 244–245
Simple interest, 126
Simplest form, 65
Simplifying
radicals, 167
rational expressions, 167
Sine, 283
Skew lines, 225
Slope-intercept form, 193–194
Slope
formula, 192
of lines, 192, 193
of parallel lines, 193
of perpendicular lines, 193
SOH-CAH-TOA, 283
Special factors,
difference of cubes, 142
difference of squares, 142
perfect squares, 142
sum of cubes, 142
Special testing arrangements, 5–6
Sphere, 274
Square, 233
area of, 267
perfect, 61
Square numbers, 178
Square pyramid, 275
Square roots, 61–62
Square units, 271
Standard form
of complex numbers, 175
of equation for
circle, 209
ellipse, 210
hyperbola, 212
parabola, 211
Statistics, 98–99
Stem-and-leaf diagram, 102–103
Straight angles, 227
Student ACT website, 3
Subtraction
of linear equations, 163–164
of numbers
decimals, 56
fractions, 69–70
mixed, 69–70
positive and negative, 78–79